# DIRECT AND LARGE-EDDY SIMULATION II

# *ERCOFTAC* SERIES

VOLUME 5

*Series Editors*



*Aims and Scope of the Series*

ERCOFTAC (European Research Community on Flow, Turbulence and Combustion) was founded as an international association with scientific objectives in 1988. ERCOFTAC strongly promotes joint efforts of European research institutes and industries that are active in the field of flow, turbulence and combustion, in order to enhance the exchange of technical and scientific information on fundamental and applied research and design. Each year, ERCOFTAC organizes several meetings in the form of workshops, conferences and summerschools, where ERCOFTAC members and other researchers meet and exchange information.

The ERCOFTAC Series will publish the proceedings of ERCOFTAC meetings, which cover all aspects of fluid mechanics. The series will comprise proceedings of conferences and workshops, and of textbooks presenting the material taught at summerschools.

The series covers the entire domain of fluid mechanics, which includes physical modelling, computational fluid dynamics including grid generation and turbulence modelling, measuring-techniques, flow visualization as applied to industrial flows, aerodynamics, combustion, geophysical and environmental flows, hydraulics, multi-phase flows, non-Newtonian flows, astrophysical flows, laminar, turbulent and transitional flows.

*The titles published in this series are listed at the end of this volume.*

# Direct and Large-Eddy Simulation II

*Proceedings of the ERCOFTAC Workshop held in Grenoble, France, 16–19 September 1996*

*Edited by*


JEAN-PIERRE CHOLLET

*Laboratoire des Encoulements Géophysiques et Industriels,*
*Université Joseph Fourier,*
*Grenoble, France*

PETER R. VOKE

*School of Mechanical and Materials Engineering,*
*The University of Surrey,*
*Guildford, U.K.*

and

LEONHARD KLEISER

*Institute of Fluid Dynamics,*
*ETH Zürich,*
*Switzerland*



SPRINGER SCIENCE+BUSINESS MEDIA, B.V.

A C.I.P. Catalogue record for this book is available from the Library of Congress.

ISBN 978-94-010-6370-8 ISBN 978-94-011-5624-0 (eBook)
DOI 10.1007/978-94-011-5624-0

---

*Printed on acid-free paper*

# Contents

## Separated Flows and Transition



## Thermal and Compressible Flows



## Wall Flows and Channel Flows

## Complex Flows



## Reacting Flows, Combustion

## Flow past a Square Cylinder : Test Case LES2

# Preface

Most of flows of real interest in industrial devices as well as in nature are turbulent and so very difficult to predict or even merely to understand. Improvements of numerical methods and increase in computing power have made numerical simulations, either DNS (direct numerical simulations) or LES (large eddy simulations), efficient tools to provide answers of practical interest while still nurturing more fundamental studies since relying on few hypotheses. Less than three years after the first workshop organized in Guildford, this second workshop was devoted to direct and large eddy simulations. Both DNS and LES are based on the numerical solution of equations for conservation of mass, momentum and energy. Contrary to other kinds of model, they lead to a complete description of the flow, in both space and time, and satisfy well the requirements for studying non-stationary processes while making possible the detailed analysis of flow structures in space. Nevertheless, let us point out basic differences between LES and DNS.

Direct numerical simulation (DNS) relies on numerical solutions of the Navier-Stokes equations without further model. All the scales of motion are calculated with the only assumption of the fluid to be Newtonian, which is never a limitation in all the cases considered in the workshop. Calculations are explicit for scales ranging from those which are smaller than the Kolmogorov scale where diffusive processes prevail to larger scales where non-linearity may develop turbulence. Unfortunately, DNS is feasible only for low values of the Reynolds numbers which means slow velocity or small domains for the development of the flow. A true DNS must also be direct as far as boundary conditions are concerned. For instance, no-slip conditions should be prescribed along walls. These limitations mean DNS is seldom used for flows of industrial interest but the fact that it does not depend on an arbitrary model makes DNS to be preferred when studying basic features of flow dynamics. In this volume, DNS is used this way in order to study the growth of instabilities. Results can be compared to analytical theories and DNS has the superiority of remaining valid long after non-linearities have developed. DNS are also used to study small-scale turbulence, most often with periodic boundaries to model a homogeneous pattern at large scales. DNS makes it possible to detail the behaviour of small scales, e.g. the intermittency of dissipative structures. DNS is also considered when coupling turbulent flow with changes in fluid properties, so chemical processes, which are strongly represented in this volume, are typical of this need of an accurate description of motions at small scales where molecular diffusivity plays an important part in mixing and reacting processes. DNS is expected to keep on making progress, mainly with the growth in computing power. Selecting high order numerical schemes may help to keep this growth reasonable. DNS are commonly used as numerical experiments (not necessarily run by numericists) to provide sets of data. As in laboratory experiments, and for phenomena which are not so sensitive to Reynolds number, DNS may be preferred to LES since it does not depend on a variety of models added to the Navier-Stokes equations.

The simulation of the only large scales is often presented as the compromise to be used as long as computers are not big enough to run DNS. DNS of high Reynolds number flows is indeed beyond the capabilities of present computers and will remain so for a long time, if not for ever. Anyway, whatever the computer power is, it does not often make sense to solve fluid motions at very small scales. Moreover as illustrated here, e.g. for particles and reacting species, numerical simulations do not always aim at computing flow fields. They are often combined with models of other phenomena whose improvement may be preferred to any refinement of space resolution. Subgrid scales of the flow field are not calculated but, due to the non-linearity of the Navier-Stokes equations, they affect the evolution of the large scales which are the only ones to be calculated explicitly. Subgrid scale (SGS) models are usually written using eddy viscosity and diffusivity formulations. In this volume, the dynamical model appears as being the most widely used, either in the original form or under various extensions. Eddy viscosity has not only to be calculated at each time and space step but must be made dependent on the distribution of the energy among the scales around the cut-off. This latter feature makes possible the growth of instabilities in flows with transition to turbulence. It should be noted that there is some contradiction between averaging which is required to stabilize the calculation and the definition of eddy viscosity varying in space. Negative values of the eddy viscosity are allowed by some of the authors and produce an apparent variability and even some back transfer : nevertheless, such a model lacks physical support.

The variety of SGS models may be regarded as less wide than in Guildford, for instance no SGS model explicitly includes backscatter transfers. For turbulent scalar fields, variable values of turbulent Prandtl numbers are usually considered. Extension to turbulent Schmidt numbers dependent on each species is still a subject of discussion. High Reynolds numbers make the treatment of wall boundaries more difficult. Wall laws are often used in applications but are too rough for basic studies which are considered here. No-slip conditions are usually used but the authors themselves admit to under-resolution of the region close to the wall. Moreover grid refinement induces large aspect ratios of the mesh which could help enhance anisotropy of calculated flow fields.

A wide variety of flows has been considered in the workshop. Homogeneous flows are still of special interest for basic studies especially while developing SGS models. Jets and mixing layers are standard shear flows which develop most turbulent features, from instability to fully developed turbulence with a lot of eddy structures. Simulation of wall flows is still a challenge because of the prevalence of molecular viscosity in the close vicinity of the wall and the long distance required for transition to take place. LES and DNS have also been extended to flows whose complexity may be due to wall geometry, thermal effects, fluid compressibility, reacting species, and so forth.

A variety of numerical methods is used. Finite elements are not yet popular (one communication only). Finite differences are widely used. Finite volumes are well adapted to formulations which are supported by conservation properties and are more open to unstructured meshes which could develop in the future. Spectral codes are restricted to simple geometries. Good accuracy is required but high order schemes may not apply to boundaries and give rise to oscillations across strong gradients. Up to now, no SGS model may be viewed as independent of the numerical method used in the LES it is embedded in. Most of numerical schemes exhibit more or less numerical diffusivity which adds to or even takes the place of the subgrid eddy viscosity. The dynamical procedure explicitly uses the numerical scheme to filter at two different scales, but filtering properties of the numerics are not well known and therefore cannot be considered when deriving the formulation of the subgrid eddy viscosity. The problem may become even worse in the future as complex flows require non-homogeneous meshes whose filtering properties may vary significantly from one place to another. LES rely heavily on computing performance. Most LES codes are highly vectorized provided that the grids are regular enough. Parallelisation is slowly making progress for solving the dynamics, and seems more promising when addressing processes which are local in space, like complex chemistry kinetics in reacting flows.

DNS or LES produce huge quantities of data. 2D and 3D pictures are commonly used to observe the organisation of eddy structures. However more quantitative information is needed in order to draw conclusions for applications or merely compare the results to experimental data. Statistics have therefore to be produced but they are observed to converge rather slowly. A compromise has to be reached between high resolution and the multiplicity of realisations (or the time of integration). If statistics in space are performed easily as a post process of recorded files, statistics in time are more expensive, requiring a lot of output which has to be predetermined when starting the run. When comparing results from LES to experiments, second order moments of the fluctuating variables like the Reynolds stress are considered although some discrepancy may be expected between a quantity computed from large scales only in LES to a quantity derived from the full field in experiments. This discrepancy should be even worse for higher order moments, like skewness.

The book is organized in eight chapters corresponding to the sessions of the workshop. Of course the distribution of the contributions among the sessions is somewhat arbitrary since most of contributions are dealing with several subjects of interest for the workshop.

A first chapter is devoted to jets as a standard shear flow with comparisons from experiments. The different authors observe the same eddy structures induced by the varicose and helical instability modes. Strouhal numbers seem to be well predicted: these features seem quite robust and strongly large-scale dependent since they do not seem to be significantly affected by the subgrid scale model in use. On the other hand, the quantitative information is not so good, including bulk quantities like the spreading rate of the jet. The subgrid-scale model is observed to matter both for such quantities and of course for the location of the transition. A test case (LES3) was proposed and was run by two groups. G.Urbin (CEA, Grenoble) ran the Re=19000 case with a LES and the structure function model for SGS and the TRIO VF code (finite volume, third order in space). The resolution was $21 \times 32 \times 96$. The upstream condition was either a no slip condition in the nozzle or a tanh velocity profile. Vortex structures are observed to be axisymmetric tori which tend to evolve through an alternate pairing process. The agreement with the experimental vortex shedding Strouhal number is quite good. On the other hand, disparities of the mean and fluctuating axial velocities were observed, especially close to the nozzle. A.Abba, R. Bucci, C. Cercignani and L.Valdettaro (polit. di Milano) studied the same test case using a cylindrical coordinate system, 2nd order centered finite differences, and a dynamic model stabilized by imposing a positive total viscosity (molecular + turbulent) averaged in the angular direction. The resolution was $54 \times 54 \times 90$ with no slip condition at the inlet. Strouhal numbers were well reproduced but discrepancy is observed in the jet spreading rate. Both groups observed that the subgrid scale model matters but that most of the uncertainty is associated with the specification of the inlet condition. No slip conditions in the nozzle require too many nodes while not being really typical of the development of boundary layers in the nozzle. The documentation of the inlet experimental conditions appears as the most critical point. Moderate values of the Reynolds number seem also to be preferred to higher values in order to better control the transition.

Subgrid models are confronted with basic problems such as the commutation between the filter and the derivative, or improvement in the description of small-scale turbulence through a higher order of the numerical approximation. Extension of subgrid models to anisotropic turbulence and subgrid models of scalar mixing are also considered. DNS, and with some reservation also LES, are used as tools appropriate for studying transition since they fully describe the evolution with time. The SGS is observed to be important and most dynamical models are suitable for switching off the drain of energy towards the small scales before the transition. Transition studies were presented in various cases : in free flows, especially in the jets; along a plate; and in a wake. Structures associated with transition processes seem to be adequately reproduced. More arguable is the quantitative information related to the exact location where the transition occurs. Such a location of course depends on the inlet and outer flow conditions. More surprising are the discrepancies in the values of bulk quantities, such as the rate of expansion of free shear layers.

Although LES and DNS are widely used for thermal and compressible flows, few contributions were presented at this workshop. Thermal convection was simulated with a LES thanks to an extension of subgrid scale model to anisotropic turbulence. Many geophysical flows are driven by thermal convection as illustrated here in the case of atmosphere. Compressible turbulent flows are investigated here by making extended use of databases from DNS of compressible mixing layers. Simulations of wall flows are run in order to characterize the growth of instabilities and the vortex structures associated with the transition to turbulence. These analyses are motivated by a need for better knowledge of fundamental mechanisms but also by future prospects of manipulating flows, for instance through wall oscillations. Rotation is also observed to affect the eddy structure of a wall flow. Complex flows develop because of geometry such as here in impinging jets or curved pipes. DNS or LES allow us now to obtain results of practical use. Configurations may be classical but the numerical simulations are made more difficult by the inclusion of mechanisms such as suspended particles or gas diffusion.

There is an increasing interest in the DNS and LES of combustion and reacting flows, with six contributions to be compared to only one at the first workshop held in Guildford. The challenge of reacting flows is that they require us to model accurately scalar mixing from large to very small scales at which chemistry operates. There is a triple challenge when addressing the large-eddy simulation of reacting flows with the need of modelling subgrid-scales for the velocity field, for mixing and for source terms. The latter terms differ radically from the others and cannot be modelled through eddy quantities. DNS and LES provide a good description of scalar mixing much better than any turbulent closure model. However, organized features which develop at very small scales have to be modelled in appropriate way, for instance through flamelet schemes. Modelling of turbulent combustion is often based on p.d.f. formulations which are extended to define subgrid-scales models. Because of the heavy computation for reacting flows, at least when detailed chemical schemes are used in DNS, two dimensional approximations are still useful and may take advantage of parallel computers. When attempting to better represent mixing and flow features, three dimensionality is required which makes necessary to use cruder models for the chemical process.

The flow past a square cylinder is considered in the last chapter, through the analysis of test case LES2. The work reported is a continuation of that carried out for a workshop organised by W.Rodi and J. Ferziger in 1995, and mostly involves the same groups. Further mesh refinement has been carried out by the participants for this test case but there is still considerable disagreement both between the LES computations and with the available data from experiment. Systematic studies of the effects of meshing, models, and numerical methods is reported in this chapter, but it appears that further work is still required to arrive at a definitive LES of the square cylinder flow.

The Second ERCOFTAC Workshop reported in this volume bears witness to continuing progress in the methods and application of Direct and Large-Eddy Simulation. The innovations described here may encourage us to persist in widening the scope of application while also strengthening the foundations of these increasingly important methods.

*J.P. Chollet*
*P.R. Voke*
*L. Kleiser*

# DIRECT NUMERICAL SIMULATIONS OF THE FREE, UNSTEADY, ROUND, UNFORCED JET AT LOW REYNOLDS NUMBERS

IONUT DANAILA
*I.N.E.R.I.S., Parc Tech. Alata, 60550 Verneuil en Halatte*
*I.R.P.H.E., 12 Av. Général Leclerc, 13003 Marseille*

JAN DUŠEK
*I.M.F., 2 rue Boussingault, 67000 Strasbourg*

AND

FABIEN ANSELMET
*I.R.P.H.E., 12 Av. Général Leclerc, 13003 Marseille*

**Abstract.** Three–dimensional direct numerical simulations of unforced, incompressible, free, spatially evolving round jets are used to investigate the onset of instability at low diametral Reynolds numbers ($Re \leq 500$). Compact, coherent structures are identified by means of iso–surfaces of vorticity and pressure fields. The Reynolds number proves to be the decisive parameter for the selection of the most amplified unstable mode once the inflow velocity profile is given. (A 'top–hat' profile is used). At the upper limit of the investigated range of Reynolds numbers, the present simulations are consistent with the widely accepted scenario of the space and time development of a round jet instability. For lower Reynolds numbers, a superposition of symmetry–breaking (helical) modes is shown to characterize the instability of the round jet. The Fourier decomposition of the fluctuating flow field allows to extract the helical modes and to identify them as the modes predicted already by Batchelor as possible linearly unstable modes in an axisymmetric parallel and inviscid jet. The dynamics of the unstable axisymmetric jet present some particular features that might be characteristic of axisymmetric flows in general: absence of a limit cycle and sensitivity of the asymptotic state to initial conditions.



*J.-P. Chollet et al. (eds.), Direct and Large-Eddy Simulation II,* 1-10.

## 1. Introduction

Since Rayleigh introduced stroboscopic illumination (1884) for flow visualizations and analyzed the jet instability problem (1879), the multitude of experimental, analytical and numerical studies concerning the free, homogeneous, axisymmetric jet instability have considered two distinct types (modes) of instabilities: varicose instability (characterized by axisymmetric modes in which waves travel as a succession of symmetrical swellings and contractions) and sinuous instability (characterized by helical modes in which the waves appear as a rhythmic undulation or twisting of the jet).

At the present time, the onset of one type of instability or another is not well defined in terms of Reynolds number range. In the literature, most of the studies deal (because of engineering applications) with high Reynolds number jets where the axisymmetric modes are the most amplified. The presence of the helical (symmetry–breaking) modes in the jet flow is reported, at low Reynolds numbers, by both linear stability analysis (Batchelor & Gill, 1962), (Mattingly & Chang, 1974), (Morris, 1976), (Mollendorf & Gebhart, 1973) and experimental studies (Reynolds, 1962), (Crow & Champagne, 1971), (Mollendorf & Gebhart, 1973), but reliable experimental and numerical data are not available for these Reynolds number values.

## 2. Numerical implementation

Direct numerical simulations of three–dimensional, spatially evolving jets at low Reynolds numbers ($10 \leq Re \leq 500$), were performed using the NEKTON code based on a spectral spatial discretization. The NEKTON code was successfully used in various studies concerning flow instabilities (Dusek *et al.*, 1994), (Neitzel *et al.*, 1995). A detailed description of the numerical procedures used by NEKTON code (domain design, computation of the time–dependent solution and data post–processing) is given by Neitzel *et al.*. Our three-dimensional calculations are performed in Cartesian coordinates in a cylindrical computational domain (fig. 1). The spectral spatial accuracy and the possibility to cluster collocation points in the jet mixing layer allowed us to capture all scales of the flow field in the considered Reynolds number interval. The overall temporal accuracy was only of first order $O(\Delta t)$. This was sufficient due to the fact that the numerical stability restrictions imposed a rather fine time step. About 250 time steps corresponded to the shortest time scale in the flow.

An important computational effort has been made to optimize the geometric domain parameters in order to capture all possible states of the jet flow in the indicated Reynolds numbers range. The grid mesh shown in Fig. 1 was selected for the present calculations: it has 65 spectral elements, slowly diverging in the downstream direction, covering an axisymmetric computational domain of radial size $D_{max} = 10.66$ and longitudinal length $L_z = 20$. The initial velocity profile is a 'top–hat' one ($U(r) = 1, r < 0.5$ and $U(r) = 0, r > 0.5$). All the variables are non–dimensionalized using the nozzle diameter ($D = 1$) as length scale, the inlet velocity ($U_0 = 1$) as velocity scale and the resulting time scale ($t_0 = D/U_0 = 1$). Non-reflecting boundary conditions (Outflow) are set everywhere, excepting at the inflow boundary where the previously specified initial velocity profile is imposed as the inflow boundary condition. Consequently, the numerical calculations simulate the evolution of a jet issuing from a circular opening in a wall. An

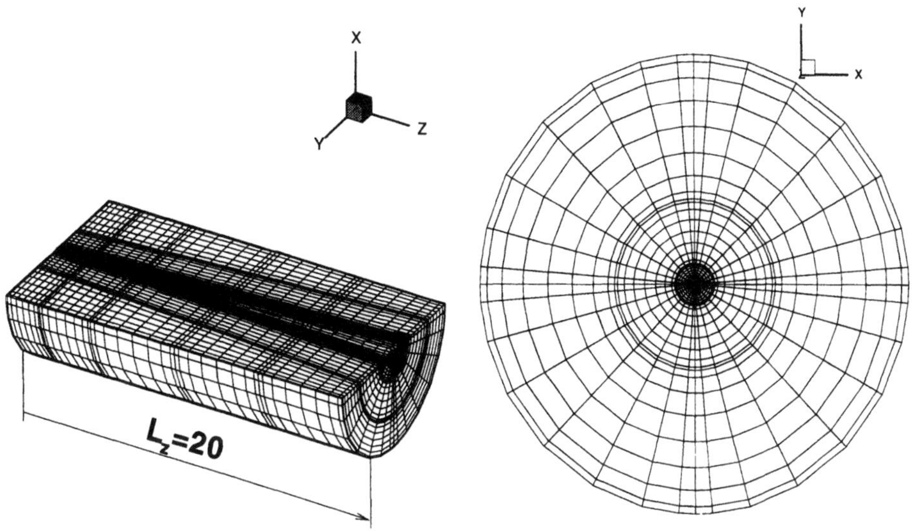


*Figure 1.* **Computational domain and spectral element mesh**

important parameter in the computational domain design is the order of the interpolation polynomials (NORDER), knowing that it has a greater effect solution accuracy than the number of spectral elements. In Sec. 3., where a relatively short time interval was sufficient for the undertaken investigation and the increased numerical cost related to the global spatial refinement (61 965 nodes) was easily tolerable, the NORDER parameter was set to 9. The effective momentum thickness achieved in this case was $\theta_0/D = 0.019$, i.e. $R/\theta_0 = 26.3$, a value close to that often used in numerical simulations ($R/\theta_0 = 22.6$) (Martin & Meiburg, 1991).

The absence of spurious reflections at the boundaries, which can affect the solution and even trigger global instabilities (Buelle & Huerre, 1988) was tested. The first numerical test verified that the vortex structures leave the computational domain without reflections. In the second numerical test the computational domain was shortened by the last layer of spectral elements, yielding a domain of $L_z = 15$. The highest Reynolds number considered in these simulations was chosen to put to the test the influence of the stream–wise length of the computational domain. The vorticity fields are identical for the two calculations for a time period equal to 2.5 times the residence time of the vortex structures within the domain considered ($t \approx 75$), corresponding to the onset of turbulence (figures not shown).

## 3. Low Reynolds number turbulence

For a large range of Reynolds numbers ($5500 < Re < 10^6$), the varicose instability was observed (see Gutmark & Ho 1983 for references) and the scenario of the early stages of evolution of the round jet seems to be universal: the shear layer originating from the nozzle lip is inviscidly unstable via the Kelvin–Helmholtz primary instability mechanism (Ho & Huerre, 1984); the instability waves grow downstream and roll up into coherent vortex rings (Yule, 1978); the structures merge as they are convected downstream and determine the shear layer spread (Ho & Huang, 1982); stream-wise vortex structures develop through a secondary three–dimensional instability of the thin vorticity layer (*braid*) between two neighboring vortex rings (Liepmann & Gharib, 1992).

For a Reynolds number of 500, the unforced jet flow is rapidly destabilized by the 'numerical noise' (given by the truncation errors) and becomes turbulent. Our numerical simulations provide a qualitative assessment of the formation and dynamics of large scale vortical structures described in the widely accepted scenario for high Reynolds numbers. The most important stages of the flow evolution from a linearly unstable mode to the turbulent state are shown in Fig. 2. The simulations use as initial conditions the flow field calculated for Re=200, once a steady state was obtained.

## 4. Primary instability

The basic difficulty in investigating the instability threshold and the instability development for only slightly super–critical Reynolds numbers is the necessity to simulate very long transients. For this reason the spatial discretization accuracy had to be decreased to 7 collocation points per spatial direction for most simulations.

The instability threshold was determined by following the dampening or amplification of the fluctuations in the time signals of the velocity at different collocation points placed in the mixing layer and on the jet axis. The already unstable flow field calculated for a higher Reynolds number was used as initial condition. At a sub–critical Reynolds number (Re=210), the instability fluctuations settled to a level of less than $10^{-5}$ of the inflow velocity value. Then the Reynolds number was progressively increased until a low amplification rate became perceptible in the velocity signals (Re=222). This allowed us to obtain rather rapidly an acceptable approximation of a slightly perturbed basic flow at nearly critical Reynolds number, suitable for the investigation of the instability onset and development.

At a slightly super–critical Reynolds number $Re > 220$ the instability was found to be characterized by oscillations with an non–dimensionalized frequency of 0.169, i.e. roughly half the Kelvin-Helmholtz frequency found at $Re = 500$. The amplification rate was very weak ($7 \cdot 10^{-4}$) and the maintained low level of the amplitudes ($\approx 10^{-4}$) suggested that we were close to the instability threshold. For different geometric discretizations the critical Reynolds number was found very close to the same value $Re_{crit} = 220$. The oscillating character of the instability shows that the latter is of Hopf type.

### 4.1. PRIMARY INSTABILITY DEVELOPMENT AND STRUCTURE

The instability so obtained was allowed to develop to see the nature of attractor that is reached. In spite of the immediate proximity of the instability threshold and an extremely low level of the instability oscillations a limit cycle has never been observed. Instead, beating, witnessing rather a limit torus dynamics, has been evidenced.

To understand better the reason for such a behavior we analyzed the flow by extracting the 'temporal' Fourier components of the instability, see (Dusek *et al.*, 1994), (Carte *et al.*, 1995), by the method described in Dusek (1996). The flow field decomposition has been added to the original code and yields the Fourier components of all velocity components and of the pressure at each node of the discretization. Considering that at each node of the computational domain the signals to have a periodicity $T = 2\pi/\omega$ in time $t$ domain, the flow field $\mathrm{v} = (v_r, v_\theta, v_z, p)$ can be expressed as Fourier series:

$$\mathrm{v}(r,\theta,z,t) = \sum_{n=-\infty}^{+\infty} \mathrm{c}_n(r,\theta,z)\, e^{in\omega t}, \quad \mathrm{c}_n(r,\theta,z) = \frac{1}{T}\int_0^T \mathrm{v}(r,\theta,z,t)\, dt \quad (1)$$

The axisymmetry breaking can be expected to occur via an onset of harmonics in the azimuthal direction $\theta$. The Fourier modes obtained above have therefore been further decomposed in the azimuthal direction:

$$\mathrm{c}_n(r,\theta,z) = \sum_{m=-\infty}^{+\infty} \mathrm{c}_{n,m}(r,z)\, e^{-im\theta} \quad (2)$$

with the coefficients $c_{n,m}$ calculated as

$$\mathrm{c}_{n,m}(r,z) = \frac{1}{2\pi}\int_0^{2\pi} \mathrm{c}_n(r,\theta,z) e^{im\theta}\, d\theta \quad (3)$$

The final decomposition of the flow field can be expressed as

$$\mathrm{v}(r,\theta,z,t) = \sum_{n,m=+\infty}^{-\infty} \mathrm{c}_{n,m}(r,z)e^{i(n\omega t - m\theta)} \tag{4}$$

The fundamental component of the signals (determined by letting $n = 1$) accurately describes the linear unstable mode and its analysis shows that it can be identified as a superposition of two counter–rotating helical modes ($m = 1$ and $m = -1$) quantified by the coefficients $c_{1,1}$ and $c_{1,-1}$. The 'raw' fluctuating flow field fundamental representing practically the total fluctuation field and the individual modes $c_{1,1}$ and $c_{1,-1}$ are shown in Fig. 3 in terms of iso–surfaces of stream–wise vorticity. We notice that the two modes have different magnitudes.

The linear theory shows that both modes are eigensolutions of the linear eigenvalue problem associated to the same unstable eigenvalue. The phase difference and the relative magnitude of the two counter–rotating modes, at least in the linear regime when the non-linear interactions are negligible, result thus from the initial conditions. This is easily confirmed by studying the fluctuating field structure for varying initial conditions using the flow field decomposition (4).

Two different initial conditions were created, the first containing only the first helical mode (the terms $c_{0,0}$, representing approximately the basic flow, and $c_{1,1}$, representing the initial perturbation, were retained in the sum (4)) and the second containing only his counter–rotating counterpart (terms $c_{0,0}$ and $c_{1,-1}$). For each analyzed period $T$, the three components of the mean kinetic energy of the helical modes can be evaluated in the spectral space as

$$K_m(\mathrm{v}) = \int_\Omega |c_{1,m}|^2 \, d\Omega \tag{5}$$

where $\Omega$ is the computational domain, $m = \pm 1$, and $\mathrm{v} = (v_r, v_\theta, v_z)$. Applying this analysis for different periods considered at different times of the simulation, we can draw the time evolution of the kinetic energy of the two helical modes. The two calculation cases displayed similar evolutions, shown in Fig. 4 for an initial condition containing only the mode $m = 1$. In both cases the initial helical mode grows linearly and reaches saturation after comparable times. The decay of its energy is related to the emergence of its counter–rotative helical counterpart. A rather surprising result was to find that the second helical mode never reaches the same strength as the mode dominating the initial condition. Higher order non–linear effects probably explain the final increase of the energy of the initial helical mode visible in Fig. 4. The net result is that the asymptotic state appears to be sensitive to initial conditions.

## 5. Conclusions and discussion

The main purpose of this paper is to examine the coherent structures dominating the flow at the onset of instability in an axisymmetric jet. There is no basic contradiction to find that instead of the inflow momentum thickness the relevant parameter appears to be the Reynolds number. In fully 3D simulations the momentum thickness increases stream-wise and its local value is the result of diffusion. The instability being a global phenomenon integrating the properties of the whole flow-field, it is not surprising to find the same trend for the increasing Reynolds number as that found in inviscid parallel flow theory for a decreasing momentum thickness. We focused, primarily, on low Reynolds numbers jet flows ($Re < 500$), which are of basic theoretical importance for the development of instabilities and for which the available experimental data appear to be incomplete.

Direct numerical simulations of unforced, spatially evolving round jets offered an ideal framework for this study; random initial perturbations occurring in experiments are avoided and information concerning the space evolution of the jet flow is available, in contrast to frequently used 'temporal' simulations, depleted of any *spatial* information (as flow spreading).

The spatial approach allowed us to obtain, for a given geometry, only the Reynolds number as a parameter and to investigate the *spontaneous* rise of the most unstable coherent structures. For low turbulent Reynolds number (Re=500) the spontaneous destabilization of an axisymmetric flow was investigated. This study represents a verification of the linear instability theory (Michalke, 1965) for axisymmetric flows characterized by this Reynolds number or by the same characteristic momentum thickness.

The coherent structures identified in the flow field are different for low super–critical Reynolds numbers in the immediate neighbourhood of the instability threshold. The linear instability is characterized by two counter-rotating helical modes both being equally amplified and resulting in an immediate symmetry breaking at the primary instability. A Fourier mode analysis of the fluctuating field allowed us to clearly identify the helical modes as responsible for the observed flow pattern. It was not surprising to find that the combination of both modes was determined by initial conditions throughout the whole linear regime. However, the fact that the asymptotic state remains sensitive to initial conditions has to be further verified and elucidated. This feature might be characteristic for all axisymmetric flows and is to be related to the degeneration of the spectrum of the linear Navier-Stokes operator giving rise to two different unstable modes. A numerical and theoretical study is under way to clarify this point.

## Aknowledgements

This work is funded by l'Institut National de l'Environnement Industriel et des Risques (I.N.E.R.I.S.). We are grateful for their support. Nekton is a registered trademark of Nektronics, Inc. and the Massachusetts Institute of Technology. ©1991 by creare.x, Inc., Hanover, New Hampshire, USA.

## References

Batchelor, G. K., Gill, A. E. (1962) Analysis of the stability of axisymmetric jets, *J. Fluid Mech.* **Vol. 14**, pp. 529–551

Becker, H. A., Massaro, T. A. (1968) Vortex evolution in a round jet, *J. Fluid Mech.* **Vol. 31**, pp. 405–448

Buelle, J. C., Huerre, P. (1988) Inflow/outflow boundary conditions and global dynamics of spatial mixing layers, *Summer Programm* **Rep. No. CTR–S88**, pp. 19–27

Carte, G., Dušek, J. & Fraunié, Ph., (1995) A Spectral Time Discretization for Flows with Dominant Periodicity, *J. Comp. Phys.* **Vol. 120**, pp. 171–183

Crow, S. C., Champagne, F. H. (1971) Orderly structure in jet turbulence, *J. Fluid Mech.* **Vol. 48**, pp. 547–591

Dusek, J., Fraunie, Ph., Le Gal, P (1994) A numerical and theoretical study of the first Hopf bifurcation in a cylinder wake, *J. Fluid Mech.* **Vol. 264**, pp. 59–80

Dusek, J. (1996) Spatial structure of the Bénard - von Kármán instability, *European Journal of Mechanics, B/Fluids.*, **Vol. 15**, No. 3

Gutmark, G., Ho, C. M. (1983) Preferred modes and the spreading rates of jets, *Phys. Fluids* **Vol. 26 no. 10**, pp. 2932–2938

Ho, C. M., Huerre, P. (1984) Perturbed free shear layers, *Ann. Rev. Fluid Mech.* **Vol. 16**, pp. 365–424

Ho, C. M., Huang, L. S. (1982) Subharmonics and vortex merging in mixing layers, *J. of Fluid Mech.* **Vol. 118**, pp. 443–473

Huerre, P., Monkewitz, P. A. (1985) Absolute and convective instabilities in free shear flows, *J. Fluid Mech.* **Vol. 159**, pp. 151–168

Liepmann, D., Gharib, M. (1992) The role of streamwise vorticity in the near–field entrainment of round jets, *J. Fluid Mech.* **Vol. 245**, pp. 643–668

Mattingly, G. E., Chang, C. C. (1974) Unstable waves on a axisymmetric jet column, *J. Fluid Mech.* **Vol. 65**, pp. 541–560

Martin, J. E., Meiburg, E. (1991) Numerical investigation of the three–dimensionally evolving jets subject to axisymmetric and azimutal perturbations, *J. Fluid Mech.* **Vol. 230**, pp. 271–318

Michalke, A. (1965) On spatially growing disturbances in an inviscid shear layer, *J. Fluid Mech.* **Vol. 23**, pp. 521–544

Mollendorf, J. C., Gebhart, B. (1973) An experimental and numerical study of the viscous stability of a round laminar vertical jet with and without buoyancy for symmetric and asymmetric disturbance, *J. Fluid Mech.* **Vol. 61**, pp. 367–399

Morris, P. J. (1976) The spatial viscous instability of axisymmetric jets, *J. Fluid Mech.* **Vol. 77**, pp. 511–529

Neitzel, G. P., Kirkconnell, C. S., Little L. J.(1995) Transient, nonaxisymmetric modes in the instability of unsteady circular Couette flow. Laboratory and numerical experiments, *Phys. Fluids* **Vol. 7 no. 2**, pp. 324–334

Reynolds, A. J. (1962) Observation of a liquid–into–liquid jet, *J. Fluid Mech.* **Vol. 14**, pp. 552–556

Yule, A. J. (1978) Large structures in the mixing layer of a round jet, *J. Fluid Mech.* **Vol. 89**, pp. 413–443

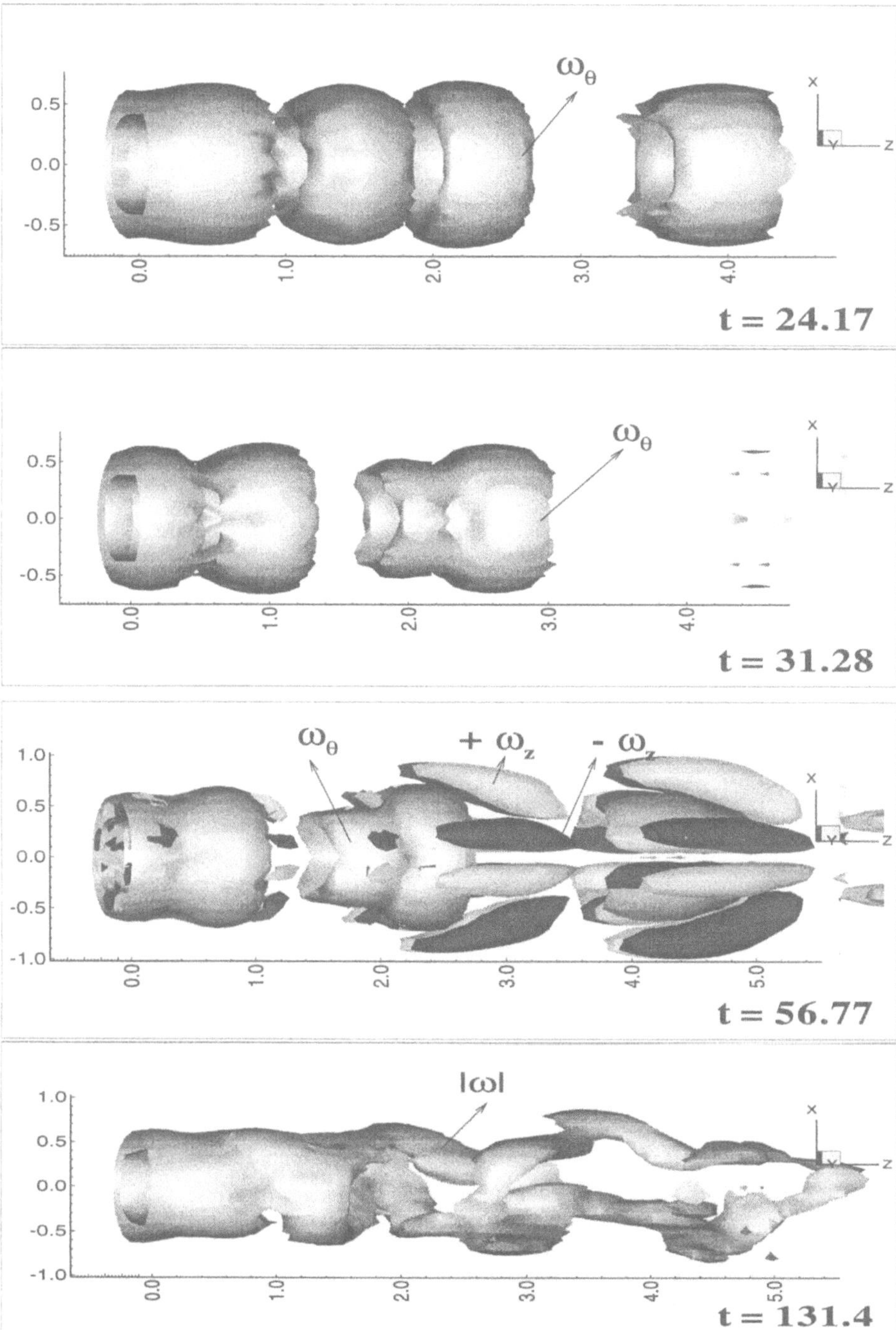


*Figure 2.* Re=500. The early stages of the jet flow evolution. From top to bottom: vortex ring roll-up, vortex pairing, axisymmetry breaking and stream–wise filaments, vortex reconnection. Instantaneous surfaces of vorticity components at a threshold corresponding to 35% of the the respective maxima and minima.

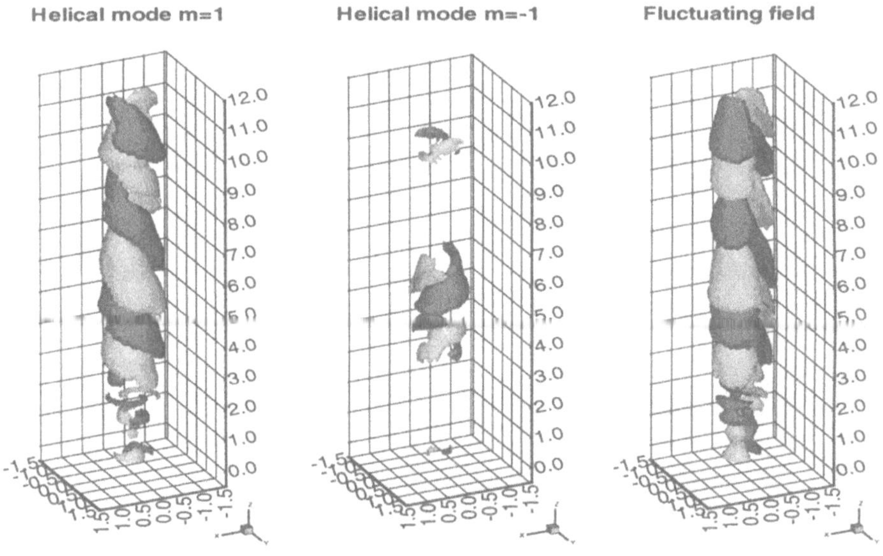


*Figure 3.* Re=225. Decomposition of the fluctuating field in helical modes. From left to right: mode $m = 1$, mode $m = -1$ and the fluctuating field. Surfaces of constant stream-wise vorticity (white for positive and black for negative) at a threshold corresponding to 35% of the the respective maxima and minima.

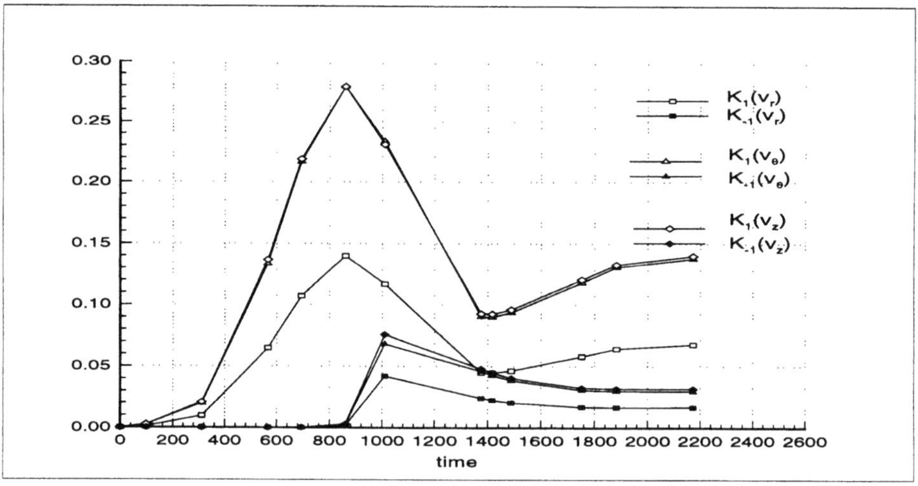


*Figure 4.* Re=225. Initial condition containing only the first helical mode $m = 1$. Temporal evolution of the components of the kinetic energy evaluated in the spectral space. The operator (K) is given by equation (5).

# LARGE EDDY SIMULATION OF A PLANAR CO-FLOWING JET

GUY HOFFMANN & CARLO BENOCCI
*von Karman Institute for Fluid Dynamics*
*B-1640 Rhode-St-Genèse, Belgium*

**Abstract.** The flowfield of a planar co-flowing jet is investigated numerically by means of the Large Eddy Simulation (LES) technique. The fundamental instantaneous flow features are correctly captured. The results obtained for the time-averaged flow quantities agree well qualitatively with the experimental reference data. Quantitatively, the largest disagreement concerns the numerical overprediction of the turbulence intensities and the Reynolds shear-stress.

## 1. Introduction

Turbulent jet flows are of considerable interest both for fundamental research and practical applications. However, in spite of extensive previous research, major deficiencies in the knowledge and understanding of this flow field still exist. This situation is mainly due to two reasons, namely, the technical limitations of the measurement techniques and the sensitivity of the jet to the initial conditions of the test facilities. Concerning the second point, it was shown by Gutmark & Ho (1983) that even where laminar initial conditions are used, the presence of a minute amount of *spatially coherent* background perturbation is sufficient to affect the initial instability frequency and the subsequent development and spreading rate of the jet. This effect explains the wide scattering of available data.

On the other hand, time-accurate computations, through direct or large eddy simulation, have become increasingly viable as a means of investigating turbulent flow problems. In view of the difficulties discussed above, the numerical simulation of the jet flow problem offers the advantages of accurately controlled initial conditions as well as direct access to instantaneous flow quantities, allowing both flow visualizations and calculation of



*J.-P. Chollet et al. (eds.), Direct and Large-Eddy Simulation II*, 11-22.

any desired correlation. While direct and large eddy simulation have been extensively applied to homogeneous turbulence and wall-bounded shear layers, where *equilibrium* or *near-equilibrium* boundary conditions can be imposed in the streamwise direction, Schumann (1991), the extension of these techniques to spatially-developing free shear layers has encountered greater difficulties. These problems are mainly related to the imposition of boundary conditions at the open portions of the domain. In the case of the free jet, the modelling of the open boundaries becomes crucial for the stability of the computation and the quality of results. In rather preliminary studies by Tsai,Voke & Leslie (1987) and Batcho,Karniadakis & Orszag (1991), periodic boundary conditions were used in the direction normal to the flow, thus effectively simulating a periodic array of jets. With the use of periodic boundary conditions, the average entrainment of fluid over these boundaries is necessarily zero. In a more recent study, Lo & Voke (1988) avoid the use of periodic boundary conditions by imposing the entrainment velocities at the transversal boundaries. The magnitudes of the imposed velocity profiles were adjusted in order to minimize the adverse pressure gradient and the backflow at the outflow boundary. In the present case it was found that, in order to ensure global stability of the flow field, a small co-flowing stream had to be introduced. As a preliminary step towards the application of time-accurate numerical simulation to jet flows, the LES of a co-flowing jet is proposed at $Re = 30000$ (based on the uniform jet velocity and the nozzle width $h$) and a ratio of free-stream velocity $U_1$ versus jet velocity $U_J$ of 0.16, reproducing the conditions of the experimental study by Bradbury (1965) (see Figure 1).

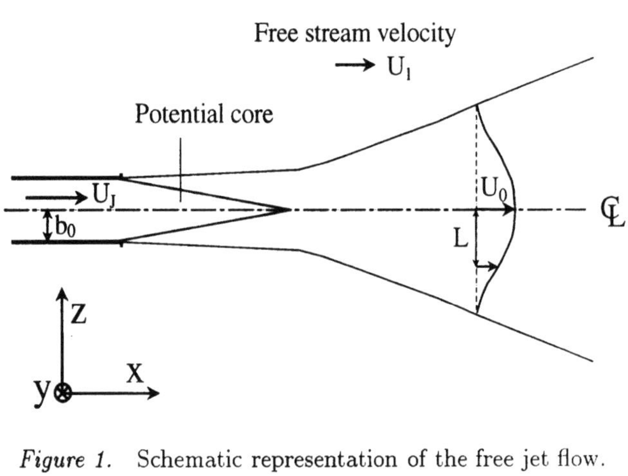


*Figure 1.* Schematic representation of the free jet flow.

## 2. Large Eddy Model

The LES approach is based upon the application of a filtering operation to the three-dimensional unsteady Navier-Stokes equations. The large scale velocity field is then obtained by the direct numerical solution of the filtered Navier-Stokes equations. The contribution of the scales smaller than the filter size appears as an unknown term which has to be modelled. In the present finite difference formulation, the filtering operation (denoted by an overbar) is a top-hat filter in physical space, applied implicitly by the finite difference operators. The resulting incompressible conservation equations of mass and momentum written for the large scale variables ($\overline{u}_i$, $P$) are:

$$\frac{\partial \overline{u}_i}{\partial x_i} = 0 \tag{1}$$

$$\frac{\partial \overline{u}_i}{\partial t} + \frac{\partial \overline{u}_i \overline{u}_j}{\partial x_j} = -\frac{\partial P}{\partial x_i} + \frac{1}{Re}\frac{\partial^2 \overline{u}_i}{\partial x_j \partial x_j} - \frac{\partial \tau_{ij}}{\partial x_j}. \tag{2}$$

Equation (2) is discretized over a staggered grid using $2^{nd}$ order finite differences, apart from the advection term, where a $4^{th}$ order central discretization is used together with a $4^{th}$ order dissipation term given here for an advection velocity $c$ and a generic transported variable $f$:

$$\begin{aligned}\left(c\frac{\partial f}{\partial x}\right)_i = {} & c_i(\alpha_1 f_{i-2} + \beta_1 f_{i-1} + \gamma_1 f_i + \delta_1 f_{i+1} + \epsilon_1 f_{i+2}) + \\ & \alpha|c_i|(\alpha_2 f_{i-2} + \beta_2 f_{i-1} + \gamma_2 f_i + \delta_2 f_{i+1} + \epsilon_2 f_{i+2})\end{aligned} \tag{3}$$

The coefficients $\alpha_1$ to $\epsilon_1$ and $\alpha_2$ to $\epsilon_2$ are obtained using Lagrangian interpolation formulas taking into account the grid stretching. The blending parameter $\alpha$ controls the magnitude of the $4^{th}$ order dissipation term. The present results were obtained using a value of $\alpha = 0.125$ in $x$ and $z$-direction, and a $4^{th}$ order central scheme ($\alpha = 0.$) in the periodic $y$-direction. These values of $\alpha$ were found sufficient to control *aliasing errors* while limiting the influence of the numerical dissipation on the resolved fluctuations even with relatively coarse grids. This conclusion was validated in turbulent channel simulations using values of $\alpha$ up to 0.5, as shown in Hoffmann & Benocci (1994), and was confirmed by a jet simulation with a 60% increased resolution on a smaller computational domain. In the latter case, the behaviour and the values of the resolved fluctuations were found to be similar to the results presented here.

The momentum equation is advanced in time using a three-step low-storage Runge-Kutta scheme in conjunction with a fractional step method. Continuity is satisfied by solving a Poisson equation with a direct Poisson solver.

The contribution of the unresolved small scales appears in equation (2) as the subgrid-scale stress (SGS) term $\tau_{ij}$, and is modelled using the concept of an eddy viscosity $\nu_t$, which relates the subgrid-scale stresses $\tau_{ij}$ to the large scale strain-rate tensor $\overline{S}_{ij}$:

$$\tau_{ij} - \frac{1}{3}\tau_{kk}\delta_{ij} = -2\nu_t\overline{S}_{ij}, \tag{4}$$

The presence of mixed laminar, turbulent and transitional flow regions in the initial development of the jet requires the use of a turbulence model which is entirely based on local quantities and predicts vanishing turbulent viscosities in laminar flow regions. These requirements can be fulfilled using the **filtered** structure function model (FSF) recently proposed by Ducros & Comte (1995):

$$\nu_t = C_{fsf}\overline{\Delta}\left[\overline{F}_2^{(3)}(\mathbf{x},\overline{\Delta},t)\right]^{\frac{1}{2}}, \tag{5}$$

where $C_{fsf} = 0.0014\ C_K^{-3/2}$, $C_K = 1.4$ being the Kolmogorov constant. The filtered structure function $\overline{F}_2^{(3)}$ is computed from the filtered velocity field, obtained by 3 iterations on a discrete Laplacian filter. The suitability of the FSF model for flows including mixed laminar/turbulent regions and for transitional flows is shown by its asymptotic long-wave limit ($\omega \to 0$) response to a discrete sine wave $\overline{u} = U\cos(\omega i + \phi)$ of pulsation $\omega$ and phase $\phi$. The response of the FSF model scales like $\omega^7$, while the unfiltered structure function scales like $\omega$, leading to a much smaller turbulent viscosity in well-resolved regions of the flow compared to the structure function model. Details of the implementation of this model and its validation can be found in Hoffmann (1996).

The main problem of the free jet simulation is related to the choice of boundary conditions at the open portions of the domain, which are crucial for the stability of the computation and the quality of results. The open boundary conditions used in the present study, Hoffmann (1996), allow fluid to enter the computational domain and impose the mean symmetry of the flow. The normal open boundary velocities are computed from a zero order derivative using the predicted velocity field, and corrected to ensure global mass conservation. The boundary conditions for the tangential velocity components at the fictitious cells assume a zero derivative of vorticity normal to the boundary. Unlike other alternative formulations, this condition proved stable in areas where turbulent structures are present and was used for the tangential velocity components on all open boundaries including the outflow area. As an example, Figure 2 illustrates the computation of the velocities at the fictitious grid points in the outflow area. The following expression is evaluated at the corner of the computational cell (*cf.*

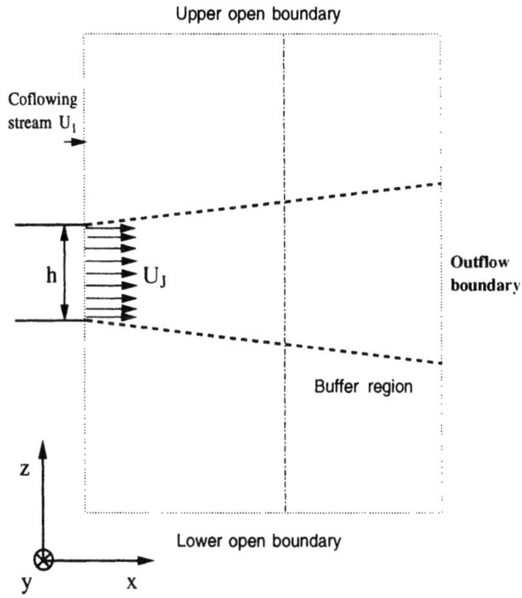


*Figure 2.* Free jet geometry.

Figure 3):

$$\left.\frac{\delta \overline{u}_i}{\delta x_j}\right|_1 - \left.\frac{\delta \overline{u}_j}{\delta x_i}\right|_1 = \left.\frac{\delta \overline{u}_i}{\delta x_j}\right|_2 - \left.\frac{\delta \overline{u}_j}{\delta x_i}\right|_2 , \tag{6}$$

where the discrete velocity derivatives are obtained using second order central differences. The index $i$ denotes the normal direction at the boundary while $j$ stands for either of the two remaining coordinate directions. The subscript 1 denotes a corner location on the open boundary, while position 2 is the corresponding inner location shifted by one cell along the normal direction $i$ into the domain. The boundary conditions described here must be supplemented with a symmetry condition. The solution adopted here is to adjust instantaneously the average pressure over the upper and lower open boundaries to the same level. The mean pressure difference existing between upper and lower open boundaries $\Delta\overline{p}$ is reduced to zero by increasing (respectively decreasing) the values of the instantaneous pressure field at the upper and lower open boundaries by a constant pressure value $\Delta\overline{p}/2$. Inside the domain a linear variation of the pressure modification is assumed between upper and lower open boundaries. The boundary condition for the normal velocity at the outflow boundary is based on a more complete version of the Sommerfeld radiation condition, which includes advective and dissipative terms in the direction tangential to the open boundary. In combination with a buffer domain as proposed by Street & Macaraeg (1989),

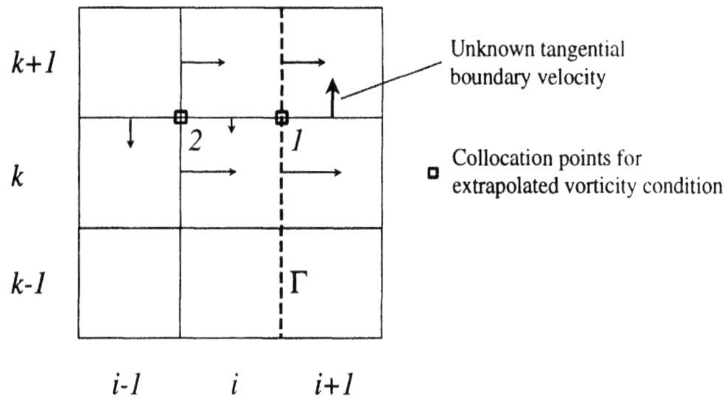


*Figure 3.* Extrapolated vorticity free stream boundary condition.

appended at the end of the computational domain (see Figure 2), it takes in account ingoing velocities, and allow developing vortical structures to leave the computational domain. However, in the case of the free jet, the present entrainment condition was unable to impose the long term symmetry of the flow. Indeed, the set of boundary conditions employed here avoids local instabilities at the boundaries, but cannot guarantee the global stability of the computation. This difficulty arises from the particular configuration of the jet problem, where the open boundaries exhibit unknown in- and outflowing velocities, and consequently, the total in- and outgoing mass flow over these boundaries is unknown. Although the fractional step method used to solve the conservation equations for mass and momentum ensures the satisfaction of the incompressibility constraint, and therefore of global conservation of mass, there is no similar global constraint on the momentum equation. Due to the approximations involved at the open boundaries of the domain, errors in the conservation of momentum cannot be avoided and accumulate from time step to time step. These errors are particularly important in the outflow region of the open domain, where large vortical structures cross the boundary and the relatively slow advection speed of these structures leads to the formation of large regions of ingoing velocities. In order to stabilize the computation it is possible to adjust the amount of entrainment, as done in Lo & Voke (1988), or fix the pressure to a constant reference value at the entrainment boundaries. For the present results the global stability problems were removed introducing a weak co-flowing stream.

## 3. Results

The computational domain of dimensions $30h \times 5h \times 50h$ is subdivided by $127 \times 24 \times 128$ grid points in the streamwise $x$-direction, the periodic $y$-direction and the normal $z$-direction. The jet was defined by the uniform inlet velocity $U_J$. Additionally, at each time step, random perturbations were superimposed on the three velocity components at the inflow plane. The magnitude of these perturbations was chosen equal to 3% of $U_J$. A variation of the magnitude of the random perturbations within the range of 0 to 10% did not change the results of the simulation.[1]

Figure 4 shows a 'snapshot' of the isovorticity contours, including the buffer domain which covers the range from $x/h = 25$ to $x/h = 30$. The flow field is characterized by the growth of large vortices in the jet shear layers, initially separated by a vorticity-free potential core. Through the initial phase of spatial development of flow the large scale vortices are located symmetrically in the upper and lower shear layers. The transition from the symmetrical to the asymmetrical mode of vortex formation in the upper and lower jet shear layers can be situated at $x/h \approx 8-9$. This result is consistent with the value of $x/h \approx 7$ found experimentally by Thomas (1986), in the case of a free jet without the confining influence of a coflowing stream.

The end of the potential core is located at $x/h \approx 8-9$, which is close to the experimental result ( Bradbury 1965) . Approximately the first third of the simulated domain (up to a streamwise distance of $x/h \approx 10$), is mainly characterized by two-dimensional flow. At $x/h \approx 10$ the 2-d vortices break down into 3-d structures. The 3-d structure of the flow is illustrated by Figures 5 and 6, which present the flow behaviour on the $x-y$ centre-plane ( $z/h = 25$). In Figure 5, the instantaneous fluctuating velocity vectors, obtained subtracting the mean velocity $U$ from $\overline{u_i}$, are shown: the lines created by velocity vectors pointing in opposite directions, mark the borders between two vortex rolls, and show the downstream growth of these structures. Figure 6 plots isovorticity contours of $\omega_z = \frac{\partial \overline{v}}{\partial x} - \frac{\partial \overline{u}}{\partial y}$, putting in evidence the initial potential behaviour of centerline and the development of turbulent structures after the break down.

A further indication of the streamwise evolution of the vortical structures can be obtained from the evolution of the turbulence intensities in the jet. Figure 7 shows the downstream development of the centre-line and peak value of the variance of the velocity fluctuations in each downstream section. The local changes of gradient in the evolution of the turbulent in-

[1]The encountered insensitivity to the magnitude of the random perturbations does not exclude a strong dependence on initial conditions. Different kinds of initial perturbations (like sine waves with random phases) or the forcing of selected frequencies, could have an important effect and have to be further investigated.

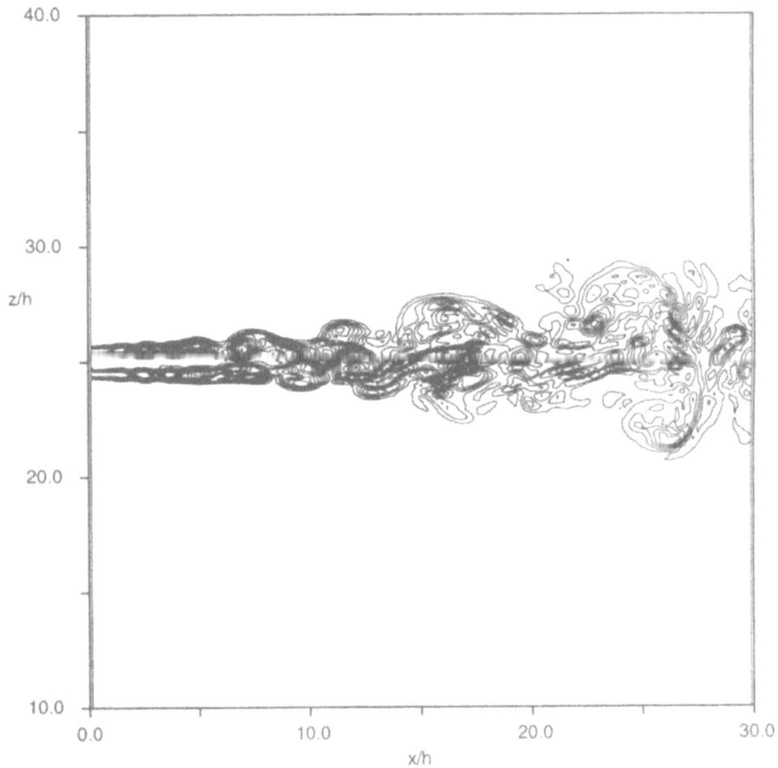


*Figure 4.* Instantaneous vorticity contours in $x - z$ plane at $y/h = 2.5$.

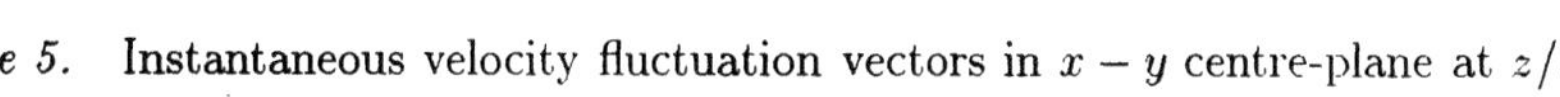


*Figure 5.* Instantaneous velocity fluctuation vectors in $x - y$ centre-plane at $z/h = 25$.

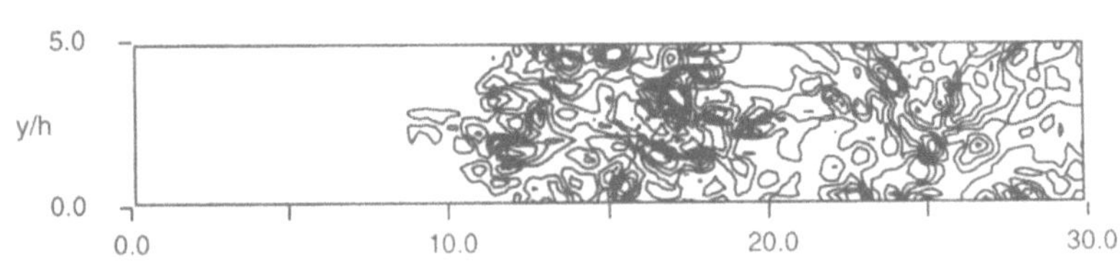


*Figure 6.* Instantaneous $\omega_z$-vorticity contours in $x - y$ centre-plane at $z/h = 25$.

tensities are caused by the formation and the change of regime of large scale vortices *cf.* Hsiao & Huang (1990). At $x/h > 20$ the growth of the turbulence intensities levels off, which is also in good agreement with experimental observations. From Figure 7, the location of vortex formation can be estimated as $x/h \approx 4$, while vortex merging occurs around $x/h \approx 6-10$, as it can also be seen in Figure 4 .

An important characteristic for the initial development of the jet is the most amplified frequency at the end of the potential core. Figure 8 shows the Fast Fourier Transformation (FFT) of the velocity fluctuations recorded on the centreline of the jet at the downstream position $x/h = 7$. The most amplified frequency $f = 0.21$ (*cf.* Figure 8) gives a Strouhal number $St = fh/(U_J - U_1) = 0.25$, which lies well within the experimentally obtained range of 0.15 to 0.42 reported in literature and compares well to the value of 0.27 found by Namer & Ötügen (1988) for jet Reynolds numbers ranging between 1000 and 7000.

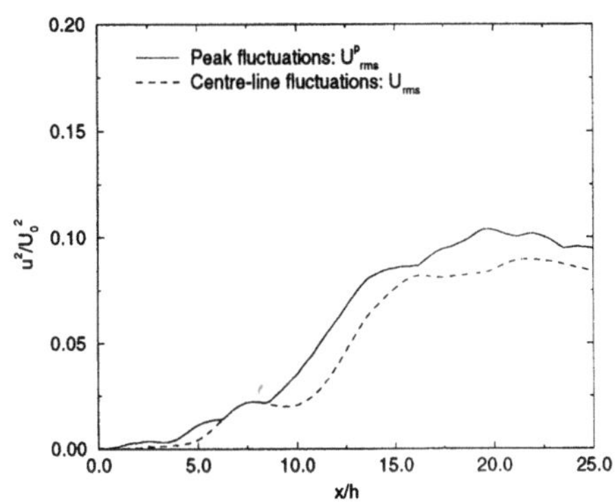


*Figure 7.* Turbulence intensity on the jet axis and peak intensity.

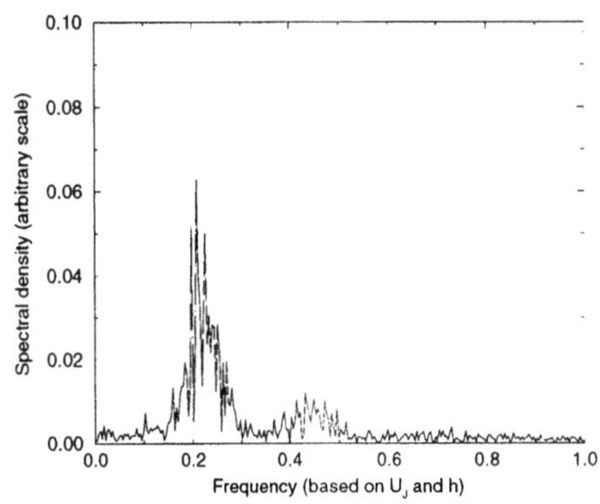


*Figure 8.* Energy spectrum of the $u$-velocity fluctuations detected at $x/h = 7$.

The results obtained for the time-averaged flow quantities agree well qualitatively with the results by Bradbury (1965). Quantitatively, the predicted evolution of the centerline velocity (*cf* Figure 9) is close to the curve proposed by Bradbury within the region $10 < x/h < 15$ . Beyond this threshold, which corrisponds to the establishment of developed turbulence, the prediction falls below the experimental trend and, correspondingly, the spread of the jet is higher than expected. This behaviour is due to the fact that, while the mean velocity profile is well predicted ( *cf.* Figure 10), the turbulence intensities and the Reynolds shear-stress are overpredicted by a significant amount with respect to the experimental values found for

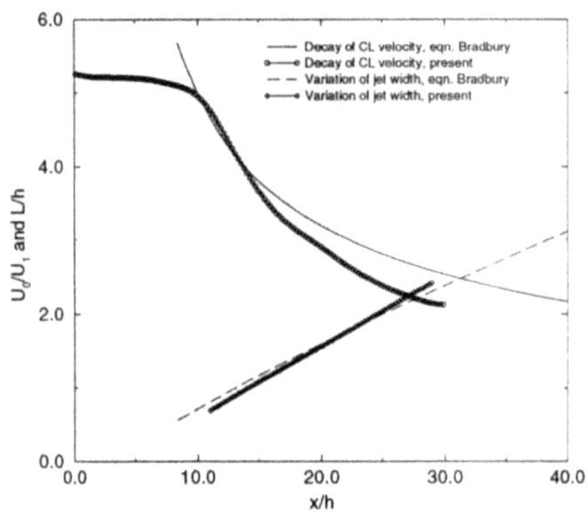


*Figure 9.* Variation of mean velocity $U_0/U_1$ on the centreline, and mean jet width $L/h$.

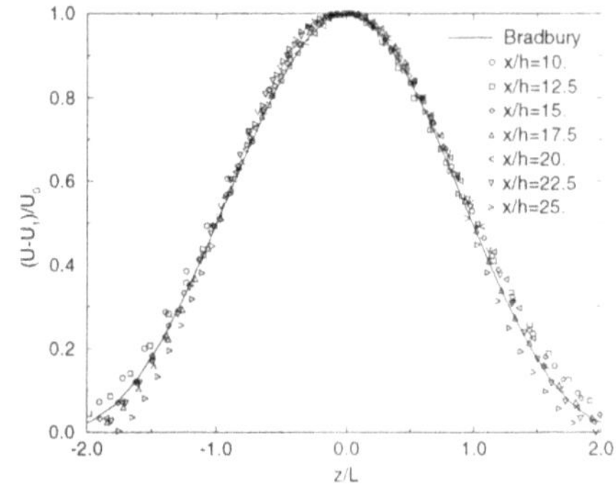


*Figure 10.* Time-averaged mean velocity profiles ($U_0 = U_{centreline} - U_1$).

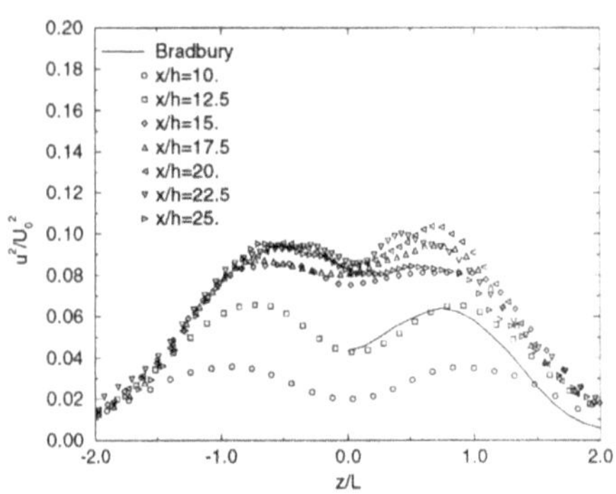


*Figure 11.* Turbulence intensities in the streamwise $x$-direction ($u$ = rms turbulence intensity).

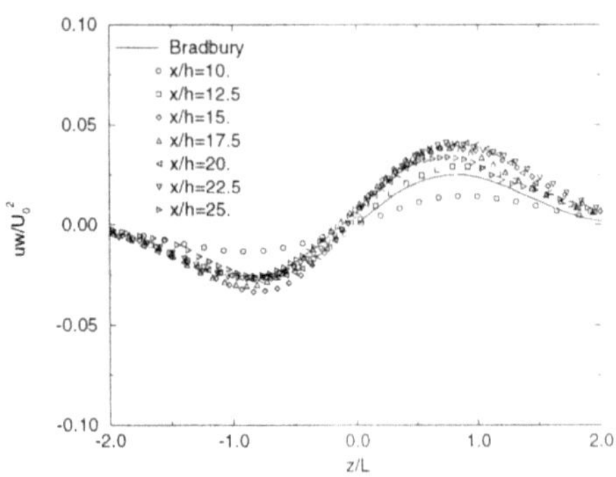


*Figure 12.* Profiles of the Reynolds shear-stress.

the self-similar region ( Figure 11 and Figure 12) and produce a turbulent diffusion stronger than the reality.

A moderate increase in the resolution of the jet or a doubling of the domain dimension in the periodic $y$-direction did not improve the prediction of the turbulence intensities and the mean shear-stress.

The lack of small scale turbulence might be the reason for the disagreement with experimental data found in the present simulation. Support for this conclusion comes from the experimental investigation of the plane jet at moderate Reynolds numbers by Namer & Ötügen (1988). They studied the effect of the jet Reynolds number on the turbulent structures in the flow. The Strouhal number measurements in the near field and the integral scale measurements in the far field showed that the size and number of turbulent large scale eddies is unaffected by the Reynolds number. However, higher Reynolds numbers led to the presence of smaller dissipating scales and a faster break-down of the initial shear layer. This broadening of the turbulent spectrum resulted in the decrease of the relative energy available to the large scales of turbulence with increasing Reynolds number. The consequences of the decrease in strength of the large scale vortices were: a weakening of turbulent mixing, a decrease in the jet spread rate, and a decrease in the turbulence level.

## 4. Conclusions

The flow field of the co-flowing jet is highly complex. After the initial transition to turbulence, the developed flow region exhibits an intermittent behaviour, characterized by the interaction of vortices, small scale turbulence and entrained outer fluid. Large eddy simulation of this complicated flow field was found to be particularly problematic in several areas. Due to the high velocity gradients and the coarse grid resolution, the use of an upwind-biased discretization of the advective term is necessary to preserve numerical stability. The filtered structure function turbulence model selected for the present simulations adapts itself to the changes in regime between laminar transitional and turbulent flow regions, and, in addition, is entirely based on local quantities. The most serious difficulties encountered stem from the necessary truncation of the infinite physical domain to a finite computational domain. The open boundary conditions used here allow a mean symmetric and stable computation while reducing the influence of the numerical boundaries on the solution in the interior of the computational domain.

Analysis of the instantaneous flow features reveals that the numerical simulation captures correctly the main primary instability mechanisms. The results for the time-averaged flow quantities show a qualitatively good agreement with the experimental reference data. The main flow features are correctly captured, but the turbulence intensities and Reynolds shear-stress are significantly overpredicted. A possible cause for the shortcomings of the numerical simulation might be that, due to the limited resolution, the break-up of the primary vortices is delayed, and the flow excessively retains a 2-d character.

## References

Batcho, P.F., Karniadakis, G.E. & Orszag, S.A. 1991 Numerical investigation of the spreading of self-excited stratified jets. *J. Fluids and Structures*, 5, 681–700.

Bradbury, L.J.S. 1965 The structure of a self-preserving turbulent plane jet. *J. Fluid Mech.*, **23**, 31–64.

Ducros, F. & Comte, P. 1995 Large-eddy simulation of spatially-developing boundary layer. In *Advances in Turbulence V*, Roberto Benzi ed., Kluwer.

Gutmark, E. & Ho, C. 1983 Preferred modes and the spreading rates of jets. *Phys. Fluids*, **26(10)**, 2932–2938.

Hoffmann, G. 1996 Engineering application of large eddy simulation to turbulent free and wall-bounded shear layers. PhD thesis, TU München.

Hoffmann, G. & Benocci, C. 1994 Numerical simulation of spatially-developing planar jets. In *74th AGARD FDP Symposium*, Chania, Greece, April 18-21.

Hsiao, F. & Huang, J. 1990 On the evolution of instabilities in the near field of a plane jet. *Phys. Fluids A*, **2(3)**, 400–412.

Lo, S.H. & Voke, P.R. 1993 Pattern recognition of eddy structures in a simulated turbulent jet. Technical Report ME-FD/93.11, University of Surrey.

Namer, I. & Ötügen, M.V. 1988 Velocity measurements in a plane turbulent air jet at moderate Reynolds numbers. *Exp. in Fluids*, **6**, 387–399.

Schumann, U. 1991 Direct and large-eddy simulation of turbulence. Summary of the state-of-the-art 1991. VKI LS 1991-02: Introduction to the modeling of turbulence.

Street, C.L. & Macaraeg, M.G. 1989/90 Spectral multi-domain for large-scale fluid dynamic simulations. *Applied Numerical Mathematics*, **6**, 123–139.

Thomas, F.O. & Goldschmidt V.W. 1986 Structural characteristics of a developing turbulent planar jet. *J. Fluid Mech.*, **163**, 227–256.

Tsai, H.M., Voke, P.R. & Leslie, D.C. 1987 Numerical Methods in Laminar and Turbulent Flow. In *Numerical Methods in Laminar and Turbulent Flow*, Montreal, Springer.

# AMPLIFIED FREQUENCIES IN THE PROXIMAL REGION OF A CIRCULAR JET

M. OLSSON AND L. FUCHS

*Department of Mechanics, Royal Institute of Technology,*
*S-100 44 Stockholm, Sweden*

**Abstract.** Earlier Large Eddy Simulations (LES) of the proximal region of the spatially developing jet, using a dynamic subgrid scale (SGS) model, showed that it was possible to capture the transition from laminar to turbulent flow, Olsson and Fuchs (1996). This work is focused on the amplification of certain frequencies in the proximal region of the jet. Frequency amplification is studied for different inlet disturbances and Reynolds numbers using a dynamic SGS-model. The disturbances correspond to perturbations used in experimental studies by Crow and Champagne (1971) and Longmire and Eaton (1992). The results of the dynamic models are compared with the results of a simulation without any explicit SGS-model. The numerical accuracy is studied using LES with different spatial resolutions. The validity of the LES assumption is assured by resolving the Taylor micro-length scales which implies spatial resolution within the inertial sub-range. The dynamic SGS-model employs artificial bounds of the SGS-model parameter instead of the commonly used spatial averaging. This is done to exclude unphysical values of the model parameter and to enhance numerical stability.

## 1. Introduction

The circular jet is of fundamental scientific and engineering interest, with applications in mixing, heat-transfer, combustion and aero acoustical flows. The study of the proximal region of the circular jet involves flow instability and transition to turbulence. The circular jet is initially dominated by axisymmetric vortex rings generated by a Kelvin-Helmholtz type of instability, Bradshaw et al. (1964), Crow and Champagne (1971). Crow and



*J.-P. Chollet et al. (eds.), Direct and Large-Eddy Simulation II, 23-34.*

Champagne (1971) showed in flow visualizations that smoke puffs occurred at a natural Strouhal number (St) of 0.30. The Strouhal number is defined as the frequency scaled with the inlet diameter and velocity . It was also shown that forcing at a frequency corresponding to a Strouhal number of 0.30 resulted in the strongest and fastest amplification of the turbulent intensity. It was reported that orderly axisymmetric smoke puffs were evident for Reynolds numbers between several hundred and about $7 \cdot 10^4$. The smoke puffs were found to be less frequent and less striking above this Reynolds number. Inviscid linear stability theory has been applied to the circular jet, Batchelor and Grill (1962), Michalke (1971). This theory shows the existence of unstable helical and axisymmetric modes in the proximal region of the jet. Further downstream, only the helical mode was found to be unstable. LES could be a powerful tool for studies of the physical processes occurring in the circular jet. Earlier LES of the proximal region of the spatially developing jet, using a dynamic SGS-model, showed that it was possible to capture the transition from laminar to turbulent flow, in the sense that the turbulent intensity increases rapidly in the proximal region of the jet, Olsson and Fuchs (1996). Linked to the prediction of the increase in turbulent intensity is the question of the ability of the LES to correctly capture the amplification of different frequencies of the jet. This work aims to investigate the ability to predict the amplified frequencies in a circular jet using LES. Also, the effect of inlet forcing on the amplified frequencies will be studied. The simulated cases are chosen to correspond to experimental studies by Crow and Champagne (1971) and Longmire and Eaton (1992). The Reynolds number was chosen to be $1 \cdot 10^4$ in most cases. This is not the same Reynolds number that was used in the experimental studies. However, it has been shown experimentally that the Reynolds number effects are marginal in the Reynolds number range from $1 \cdot 10^4$ to $7 \cdot 10^4$, Crow and Champagne (1971).

## 2. Mathematical formulation

The space filtered Navier-Stokes (NS) equations for incompressible flows can be written in the following non-dimensional form:

$$\frac{\partial \overline{u_i}}{\partial x_i} = 0, \tag{1}$$

$$\frac{\partial \overline{u_i}}{\partial t} + \frac{\partial \overline{u_i}\,\overline{u_j}}{\partial x_j} = -\frac{\partial \overline{p}}{\partial x_i} + \frac{1}{Re}\frac{\partial}{\partial x_j}\frac{\partial \overline{u_i}}{\partial x_j} - \frac{\partial \tau_{ij}}{\partial x_j}. \tag{2}$$

The spatial filtering is denoted by $(^{-})$. Equations (1) and (2) govern the large scale motion. The large scales are influenced by the small scales via the SGS-stress:

$$\tau_{ij} = \overline{u_i u_j} - \overline{u_i}\,\overline{u_j}. \tag{3}$$

Assume that the SGS-stress can be expressed as a local tensor,

$$\tau_{ij} - \frac{1}{3}\delta_{ij}\tau_{kk} = C\alpha_{ij}. \tag{4}$$

Using a second filter, usually named the test filter, gives a corresponding subtest scale (STS) stress. The STS-filter is denoted by ($\hat{}$) and it is assumed to have the same asymptotic behaviour as the SGS-filter ($\bar{}$). Germano et al. (1991) observed that the SGS-stress and the STS-stress were related. Following Lilly (1992) this relation can be solved in the least square sense assuming a slow spatial variation of $C$:

$$C\Delta^2 = -\frac{L_{mn}M_{mn}}{M_{kl}M_{kl}} \tag{5}$$

where

$$M_{ij} = \frac{1}{\Delta^2}(\widehat{\alpha_{ij}} - \beta_{ij}), L_{ij} = \widehat{\overline{u_i}\,\overline{u_j}} - \widehat{\overline{u_i}}\,\widehat{\overline{u_j}}.$$

In this work the SGS-stress and STS-stress are modeled by the Smagorinsky model (1963):

$$\alpha_{ij} = -2\Delta^2(2\overline{S_{kl}}\,\overline{S_{kl}})^{\frac{1}{2}}\overline{S_{ij}}, \tag{6}$$

$$\beta_{ij} = -2\hat{\Delta}^2(2\widehat{\overline{S_{kl}}}\,\widehat{\overline{S_{kl}}})^{\frac{1}{2}}\widehat{\overline{S_{ij}}}. \tag{7}$$

Where $\overline{S_{ij}} = \frac{1}{2}(\frac{\partial \overline{u_i}}{\partial x_j} + \frac{\partial \overline{u_j}}{\partial x_i})$. It is observed that the denominator of (5) is positive semi definite and has been reported, by Lilly (1992), to have better numerical properties than the original model proposed by Germano et al. (1991). In this work the filter ratio $\frac{\hat{\Delta}}{\Delta}$ was equal to two and a top-hat filter was used for filtering. As pointed out by Lilly (1992), equation (5) can be used locally or it can be averaged over larger regions. If $C$ is averaged over larger regions, certain local information is lost, but numerical stability is enhanced. On the other hand, if $C$ is used locally, numerical instability may be caused by occasional large positive or negative values of $C$. The instability can be avoided by the usage of local limits on $C$. An upper bound for $C$ can be found by the total viscous stability condition, taking into account molecular and SGS related viscosity. From the balance between the time derivative and the viscous and SGS-terms in the NS equations:

$$S = (\frac{1}{Re} + C\Delta^2(2\overline{S_{kl}}\,\overline{S_{kl}})^{\frac{1}{2}})\frac{D\,\Delta t}{h^2}. \tag{8}$$

$D$ is a constant given by the diagonal term in the numerical scheme, i.e. for an $O(h^2)$ scheme $\Rightarrow D = 6$ and for an $O(h^4)$ scheme $\Rightarrow D = \frac{90}{12}$. Assuming $S$ to be less than unity gives an upper bound on $C$,

$$C \leq \frac{1}{\Delta^2(2\overline{S_{kl}}\ \overline{S_{kl}})^{\frac{1}{2}}}(\frac{h^2}{D\Delta t} - \frac{1}{Re}). \tag{9}$$

The upper bound on $C$ can be seen as a bound for the eddy-turnover time. Negative values of $C$ imply inherent instability. This instability can be avoided by bounding $C$. Equation (5) may give rise to negative $C$, i.e. backscatter. From a physical point of view it can be reasonable to limit $C$ from below. Consider the total dissipation, consisting of the viscous-, the SGS- and numerical-dissipations:

$$\varepsilon_v = \frac{1}{Re}2\overline{S_{ij}}\frac{\partial \overline{u_i}}{\partial x_j}, \varepsilon_{sgs} = -\tau_{ij}\frac{\partial \overline{u_i}}{\partial x_j}.$$

The numerical dissipation is given by the numerical scheme as the dissipative part of the leading order truncation error term. For the scheme used here:

$$\varepsilon_n = \frac{1}{12}(\mid \overline{u_j} \mid h^3)\frac{\partial^2 \overline{u_i}}{\partial x_k^2}\frac{\partial^2 \overline{u_i}}{\partial x_l^2}\delta_{jk}\delta_{jl}. \tag{10}$$

The total dissipation should satisfy an entropy condition which may be expressed as, $\varepsilon_v + \varepsilon_{sgs} + \varepsilon_n \geq 0$. This gives the lower bound on $C$ used in the present Cases $V - VII$, which allows for negative values of $C$,

$$C \geq -\frac{1}{\Delta^2(2\overline{S_{kl}}\ \overline{S_{kl}})^{\frac{1}{2}}}(\frac{1}{Re} + \frac{1}{(2\overline{S_{ij}}\ \overline{S_{ij}})}\varepsilon_n). \tag{11}$$

The effects of the SGS-model are discussed below.

## 3. Problem description

The LES were performed at two different Reynolds numbers (Re), $1 \cdot 10^4$ and $50 \cdot 10^4$ respectively. The Reynolds number was based on the inlet diameter of the jet and the mean inlet velocity. The time was scaled by the ratio of the inlet diameter and the mean inlet velocity. The flat inflow profile was disturbed by different perturbations. Overall mass conservation was ensured. This was done to trigger the unstable axisymmetric and helical modes of the jet. The jet was either unperturbed, perturbed with a random disturbance or perturbed with a sinusoidal disturbance in time. The Strouhal numbers and amplitude of the inlet perturbations are shown for the different Cases in Table 1. The sinusoidal perturbations of Cases $VI$

TABLE 1. Computed Cases.

| Case | Re | $h$ | data at (z) | SGS-model | St | Amp(x,y,z) |
|---|---|---|---|---|---|---|
| $I$ | $1 \cdot 10^4$ | 0.0625 | 3,6.50,7.75 | *dynamic* | random | 0.025,0.025,0.05 |
| $II$ | $1 \cdot 10^4$ | 0.1000 | 3,6.50,7.75 | *none* | random | 0.025,0.025, 0.05 |
| $III$ | $1 \cdot 10^4$ | 0.1000 | 3,6.50,7.75 | *dynamic* | random | 0.025,0.025, 0.05 |
| $IV$ | $50 \cdot 10^4$ | 0.1000 | 3,6.50,7.75 | *dynamic* | random | 0.025,0.025, 0.05 |
| $V$ | $1 \cdot 10^4$ | 0.1000 | 1,2,3,4 | *dynamic* | - | |
| $VI$ | $1 \cdot 10^4$ | 0.1000 | 1,2,3,4 | *dynamic* | 0.28 | 0,0,0.028 |
| $VII$ | $1 \cdot 10^4$ | 0.1000 | 1,2,3,4 | *dynamic* | 0.43 | 0,0,0.37 |

and $VII$ corresponds to root-mean-square perturbation velocities of 2% and 26% of the mean inflow velocity, respectively. The computational domain consists of a rectangular box of dimension $8 \times 8 \times 12$. The jet enters at the center of the inlet plane. The long-sides are solid walls. The coordinate system was chosen so that the center of the jet inlet corresponded with the origin and the jet was developing along the z-coordinate. At the outflow boundary the axial derivative of the velocity vector vanishes. The effect of the outflow boundary condition was studied by considering the instantaneous vorticity field near the outflow boundary. Also, a simulation with an extended domain showed no indication of outflow boundary condition dependence for $z \leq 9$.

## 4. Numerical methods

Equations (1) and (2) were discretised on a staggered uniform Cartesian grid. The derivatives were approximated by $O(h^4)$ central finite differences, except for the advective terms. The advective terms were discretised using an $O(h^3)$ upwind biased scheme proposed by Rai and Moin (1991). Near the long-side boundaries, central $O(h^2)$ finite differences were used. This lower order approximation does not affect the final results since at that distance all field derivatives were small. The equations were integrated in time using a time splitting method. First, the advective parts of the NS equations were updated using an $O(\Delta t^4)$ multi-stage Runge-Kutta method. Then, the viscous and pressure parts together with the continuity equation were advanced in time using an $O(\Delta t^2)$ Crank-Nicholson (CN) implicit scheme. Each CN step is solved by using a single Multi-Grid (MG) cycle, Brandt (1977). The MG solver uses point Gauss-Seidel relaxations as smoother, flux averaged restriction and trilinear interpolations for prolongation. Four or five MG levels were used for the simulation, depending on the resolution. The three coarsest levels were global and covered the whole computational domain. The finer MG levels were local refinements covering only parts

TABLE 2. Numerical accuracy.

| z | LES | $h$ | $\lambda$ | $< w'^2 >^{0.5}$ | $Re_\lambda$ |
|---|---|---|---|---|---|
| 6.50 | $I$ | 0.1 | 0.25 | 0.17 | 420 |
| | $III$ | 0.0625 | 0.24 | 0.18 | 430 |
| 7.75 | $I$ | 0.1 | 0.27 | 0.21 | 570 |
| | $III$ | 0.0625 | 0.23 | 0.24 | 560 |

of the domain. It should be pointed out that the local refinements were formed by locally halving the grid size in each direction. Thus, all of the grids were uniform and hence maintained the same accuracy as the formal accuracy on uniform grids, i.e. $O(h^3)$. The extent of the local refinements can be controlled by minimizing the local truncation error. This approach has been exploit in the past by Fuchs (1986). The solution field of the finest local refinements were used to calculate the turbulent statistics. The time step was chosen to ensure adequate resolution in time and was taken to be 0.005 dimensionless time units in all the Cases.

## 5. Results

Instantaneous velocity and pressure fields were stored at certain locations in the streamwise direction (z) for every time step, as the numerical solution was integrated in time. This allowed the statistics to be calculated at a later occasion and the solution fields to be used as a future data base. Data were stored at different locations for the different Cases given in Table 1. At each station, data were stored along the positive and negative x-,y- and z-directions. The amount of stored data allowed the calculation of all spatial first and second derivatives. Turbulent statistics, symmetrical with respect to the jet center line have been computed by averaging in the four mutually perpendicular positive and negative x-,y-directions. No other radial directions were used, to avoid the need for interpolation. The turbulent statistics were based on data which was independent of the initial state. The initial transients of the developing jet were allowed to develop before the turbulent statistics were calculated. Cases $V - VII$ are simulations which have been performed with the aim to increase the understanding of the roll of the inflow perturbation in the development of the jet in the proximal transition region.

### 5.1. NUMERICAL ACCURACY

To study the effects of the spatial resolution and different sizes of the local refinement on the numerical solution, two of the Cases, $I$ and $III$, were simulated using different spatial resolutions and different sizes of the local refinements. The Taylor micro-length scale $\lambda^2 = 2 < w'^2 > / (\frac{\partial < w'^2 >^{\frac{1}{2}}}{\partial z})^2$, the turbulent intensities and the corresponding Reynolds number based on those scales $Re_\lambda$ are compared in Table 2 for two locations on the centerline. The Taylor micro-length scale was resolved for both the fine grid Case $III$ and the coarse grid Case $I$. This can be compared with $\lambda = 0.24$ at $z = 50$ and $\lambda = 0.17$ at $z = 20$ found at the centerline by Wygnanski and Fiedler (1969) and Corrsin and Uberoi (1950), respectively. These results and a more detailed comparisons of the first and second moments, not included, indicate that the spatial resolution, the extent of the local refinements and the numerical methods, yield accurate and reliable results. Also, the high correlation found below for the mean velocity between the present LES and the experimental results by Crow and Champagne (1971) indicates this.

### 5.2. MEAN VELOCITY

Figure 1 displays the streamwise mean velocities at four stations along the jet, at $z = 1.00, 2.00, 3.00$ and $4.00$, for the unforced and forced (St=0.28) Cases, respectively. The profiles are staggered according to their streamwise location, to indicate the spatial development of the jets. Case $V$ and $VI$

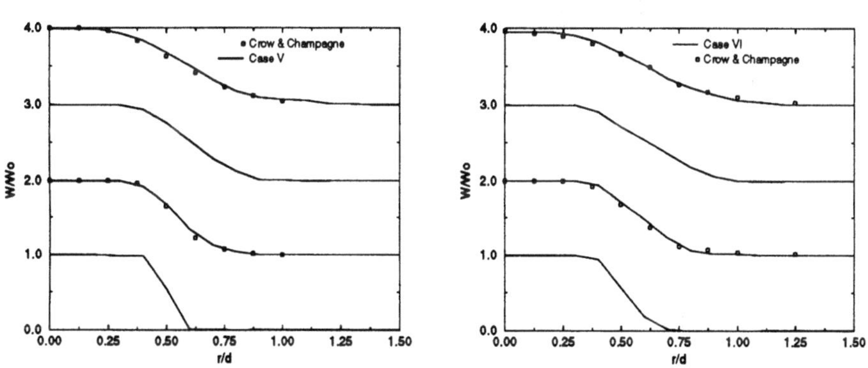


*Figure 1.* The mean streamwise velocity vs. the radius. The left plot shows the unforced Case $V$ and the right plot shows the forced $St = 0.28$ Case $VI$.

correspond to experiments done by Crow and Champagne (1971). As can be seen the LES correlates well with the experimental results at $z = 2$ and 4. It can also be seen that the forcing has only small effects on the mean velocity. Case $VII$ corresponds to experiments done by Longmire and Eaton (1992) with a higher forcing amplitude of 0.37 and St=0.43. Unfortunately, there are no experimental results available for the mean velocity to compare with the LES.

### 5.3. FREQUENCY SPECTRUM

Figures 2, 3 and 4 display the energy spectrum along the centerline, at $z = 1, 2, 3$ and 4, for the unforced, forced St=0.28 and St=0.43 Cases $V - VII$, respectively. It should be pointed out that the scaling differs between the figures. In Figure 2, it can be seen that the unforced jet is initially dominated by the fundamental frequency of St=0.27 and the three corresponding harmonics, 0.54, 0.81 and 1.08. The fundamental frequency is the same as was found to be the naturally occurring frequency of smoke-puffs in flow visualizations by Crow and Champagne (1971). This frequency was also found in the jet-noise spectrum measured outside the jet. Longmire and Eaton (1992) have performed experimental investigations on harmonicly forced jets. The minimum forcing amplitude required to induce vortices was monitored for different frequencies. Large differences were found in

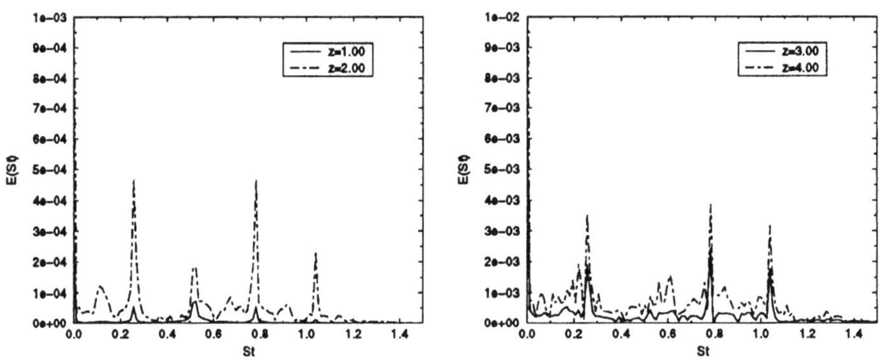


*Figure 2.* The energy spectrum vs. the Strouhal number for the unforced Case $V$, at $z = 1$, 2 left and 3, 4 right.

the minimum required forcing for different frequencies. A comparison with the present LES shows that frequencies close to the naturally occurring frequencies found above required less forcing. From Figure 2 it can be observed that all but the first harmonic persists in the studied region. The first harmonic can be seen to lose energy relatively faster than the other frequencies. At $z = 1$ the first harmonic contains as much energy as the fundamental frequency. But, as the jet develops the first harmonic can not be distinguished from the background spectrum. As the jet develops a broadening of the spectra can be observed, until the jet becomes fully turbulent for $z > 4$. In Case $VI$ the jet was forced close to the naturally occurring fundamental frequency, at St=0.28 with an amplitude of 0.028. This Case is also in close resemblance to the measurements by Crow and Champagne (1971), at St=0.30 and an amplitude of 0.028. From Figure 3 it can be seen that the jet forced with a frequency of St=0.28 is dominated by this fundamental frequency and the corresponding first four harmonics. If this Case is compared with the above Case $V$ several differences can be noticed. First, the first harmonic which was damped in the unforced Case $V$ persists in the studied region. Also, the forced Case $VI$ shows a spectrum where the energy at the studied stations z=1,2 and 3 is more concentrated to the fundamental frequency and the harmonics. This is not the case for the station z=4. At z=4 the energy in the frequencies surrounding the fundamental frequency and the harmonics are closer to the levels found in the unforced Case $V$. Also, the energy growth in the fundamental and the harmonics is higher from z=3 to z=4 for the forced case than in the unforced case. This difference between the forced case and unforced case is not as pronounced for the earlier stations. Case $VII$ corresponds to one of the experiments performed by Longmire and Eaton (1992). The forcing was at St=0.43 and with a high amplitude of 0.37. The amplitude was chosen to be the minimum forcing amplitude required to induce formation of vortices. It is clear from the results of Case $V$ and $VI$ that this high amplitude was needed since the forcing was applied far from any of the naturally amplified frequencies of the jet. From Figure 4 it can be seen that the energy levels of the energy spectra in Case $VII$ are higher than in Cases $V$ and $VI$ and that the energy spectrum is not as dominated by certain frequencies. Even so some frequencies with more energy can be identified. First, the fundamental forcing frequency and the corresponding harmonics, St=0.43, 0.86 and 1.29. These frequencies are clearly identifiable and the first two persist as the jet develops downstream while the third is damped. Also clearly seen are two frequencies at St=0.81 and 1.35. These are the

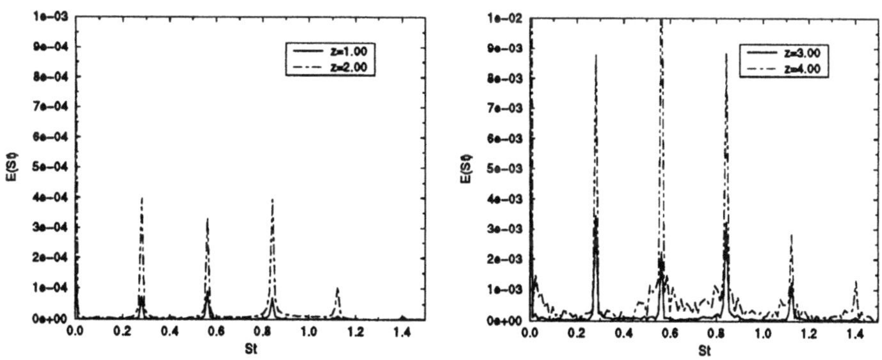


*Figure 3.* The energy spectrum vs. the Strouhal number for the forced, St=0.28 Case $VI$, at $z = 1$, 2 left and 3, 4 right.

second and fourth harmonics of the naturally amplified frequencies found in Case $V$. These naturally amplified frequencies are probably triggered by the closely located first and second harmonics of the forcing frequency, located at 0.86 and 1.29 respectively. Peak frequencies can also be found at St=0.40, 0.60 and 0.63. But the origin of these frequencies are not as clear. Cases *III* and *IV* were simulated using the same grid, SGS-model, and random perturbation, but at different Reynolds numbers. The Cases were simulated at Re=$1 \cdot 10^4$ and Re=$50 \cdot 10^4$, respectively. This was done to study the effects of high Reynolds numbers on the energy spectrum. Case *III* shows a spectrum with the same fundamentals and harmonics as Case *V*, but with higher energy levels. This is expected due to the random forcing. For the high Reynolds number Case *IV*, the energy spectrum showed no clear peak frequencies. Even at the most upstream location, z=1, the energy spectrum was continuous and without clear peaks. This is consistent with measurements by Crow and Champagne (1971). They observed in smoke visualizations that the regularly occurring smoke puffs, which were observed below Reynolds numbers of about $7 \cdot 10^4$, were less frequent and striking above this Reynolds number.

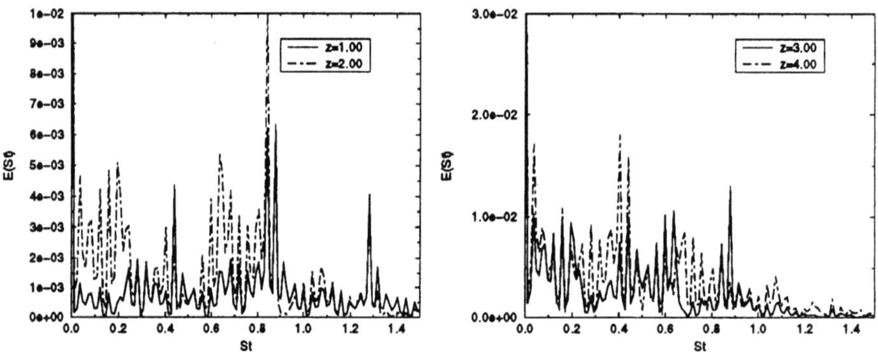


*Figure 4.* The energy spectrum vs. the Strouhal number for the forced, St=0.43 Case *VII*, at $z = 1$ and 2 in the left plot and 3 and 4 in the right plot.

## 5.4. EFFECTS OF THE SGS-MODEL

Cases *I* and *II* were used for *a posteriori* tests of the effect of the SGS-model. The Cases are identical in all aspects except for the use of the dynamic SGS-model. Case *I* was simulated with and Case *II* without SGS-model. Earlier studies by Olsson and Fuchs (1996) using the data bases of Cases *I* and *II* showed that the main effect of the dynamic SGS-model on the turbulence intensity, is a prediction of higher values along the centerline. This is caused by the fact that the SGS-model reduces the effective viscosity, by negative values of the dynamic model parameter. The reduced dissipation when the SGS-model is used, is noticed in a higher turbulence intensity along the centerline for $z > 4$. The energy spectrum of the two

Cases *I* and *II* was compared to study the effects of the SGS-model on the energy spectrum. In the studied region, $z < 4$, the spectrum was found to be nearly identical for the two Cases. The only difference found was in a slightly lower energy level for Case *II* without SGS-model. The maximum difference was found to be less than 0.1%.

## 6. CONCLUDING REMARKS

LES were performed to study the amplified frequencies in the proximal region of a circular jet. The inflow perturbations were chosen to correspond with perturbations used in experimental studies by Crow and Champagne (1971) and Longmire and Eaton (1992). Several LES were used to verify the accuracy, to study the Reynolds number effects and the effects of the SGS-model. The spatial resolution was found to be sufficient by comparing simulations on different grids, by computing the Taylor micro-length scales on grids with different resolution, as well as comparing with experimental data in the near entrance region. Case *V* without inlet perturbation and Case *VI* with a forced inlet perturbation at St=0.28, yield comparable results to experimental data by Crow and Champagne (1971). The mean velocity of these Cases correlated well with the experimental results. The major difference in the mean velocity for the studied region was found closest to the inlet. At z=1 the strongly forced Case *VII* showed a more disturbed velocity profile. Studies of the instantaneous velocity field showed that the strong forcing of this Case generates a vortex ring type of flow. The unforced jet, Case *V*, was found to be initially dominated by the fundamental frequency of St=0.27 and the three harmonics, 0.54, 0.81 and 1.08. This naturally occurring fundamental frequency agrees with what was found experimentally by Crow and Champagne (1971). As the jet was forced at the fundamental frequency, Case *VI*, it was found that most of the energy was initially concentrated to the fundamental mode and the corresponding four harmonics. But, as the jet developed more energy could be found in the surrounding frequencies. In the strongly forced jet Case *VII* higher energy levels were found at several frequencies. Energy peaks could be found at the forcing frequency St=0.43 and the first two harmonics, St=0.86 and 1.29. Also, energy peaks were found at the second and forth harmonics of the naturally fundamental frequency. These frequencies are close to the frequencies of the multiples of the forcing frequency St=0.43. In this way the second and forth harmonics of the naturally occurring frequencies were triggered by the forcing at St=0.43. For Re=$50 \cdot 10^4$ no dominating frequencies were found in the considered region. A similar observation was also done by Crow and Champagne (1971). Orderly structures and a dominating mode were found for Re in the range between several hundred and about $7 \cdot 10^4$, whereas for higher Re no particular structure or dominating mode was naturally occurring. The effect of the SGS-model was found to be marginal in the region of major interest $z < 4$. The SGS-model has been found, Olsson and Fuchs (1996), to have a larger effect further downstream, $z > 4$.

## Acknowledgment

This work was supported by TFR, the Swedish Research Council for Engineering Sciences.

## References

Batchelor, G. K. & Gill, A. E. 1962. Analysis of the stability of axisymmetric jets. *J. Fluid Mech.* **14**, 529.

Bradshaw, P., Ferriss, D. H. & Johnson, R. F. 1964. Turbulence in the noise-producing region of a circular jet. *J. Fluid Mech.* **19**, 591.

Brandt, A. 1977. Multi-level adaptive solutions to boundary-value problems. *Math. Comput.* **31**, 333.

Corrsin, S. & Uberoi, M. S. 1950. Further experiments on the flow and heat transfer in a heated turbulent air jet. Technical Report Report 998, NACA.

Crow, S. C. & Champagne, F. H. 1971. Orderly structure in jet turbulence. *J. Fluid Mech.* **48**, 547.

Fuchs, L. 1986. A local mesh-refinement technique for incompressible flows. *Comput. Fluids.* **14**, 69.

Germano, M., Piomelli, U., Moin, P. & Cabot, W. 1991. A dynamic subgrid-scale eddy viscosity model. *Phys. Fluids A.* **3**, 1790.

Lilly, D. K. 1992. A proposed modification of the germano subgrid-scale closure method. *Phys. Fluids A.* **4**, 633.

Longmire, E. K. & Eaton, J. K. 1992. Structure of a particle-laden round jet. *J. Fluid Mech.* **236**, 217.

Michalke, A. 1971. Instabilität eines kompressiblen runden freistrahls unter berücksichtigung des einflusses der strahlgrenzschichtdicke. *Z. Flugwiss.* **19**, 319.

Olsson, M. & Fuchs, L. 1996. Large eddy simulation of the proximal region of a spatially developing circular jet. *Phys. Fluids.* **8**, 2125.

Rai, M. M. & Moin, P. 1991. Direct simulations of turbulent flow using finite-difference schemes. *J. Comput. Phys.* **96**, 15.

Smagorinsky, J. 1963. General circulation experiments with the primitive equations. *Mon. Weather Rev.* **91**, 99.

Wygnanski, I. & Fiedler, H. 1969. Some measurements in the self-preserving jet. *J. Fluid Mech.* **38**, 577.

# LARGE-EDDY SIMULATIONS OF THREE-DIMENSIONAL SPATIALLY-DEVELOPING ROUND JETS

GÉRALD URBIN
*CEA DRN/DTP/STR/LMTL, 17, rue des Martyrs, 38050 Grenoble cedex 09, FRANCE*

AND

OLIVIER MÉTAIS
*LEGI/IMG, BP 53, 38041 Grenoble cedex 9, FRANCE*

**Abstract.**

We present a statistical and topological study of the spatial growth of a round jet up to twelve diameters downstream from the nozzle. The use of large-eddy simulations allow us to reach high Reynolds number values: here, $Re = 25000$. It is shown that, at a reasonable computational cost, good comparisons with experimental data can be achieved. We successively consider the case of the "natural" unexcited jet and the case of the jet excited with specified inflow perturbations at the nozzle. The natural jet alternatively exhibits axisymmetric (rings) and helicoidal vortex structures. Further downstream, we observe that the alternate inclination of the rings yields localized alternated pairings. We have showed that this structure can be forced to appear from the nozzle with an adhoc excitation leading to the preferential development of the jet in one particular direction. When axisymmetric excitation is applied, after vortex rings have formed, pairs of counter-rotating longitudinal vortices appear linked with primary rings and these create horizontal side jets. Longitudinal vortices are still present when a helicoidal excitation is imposed.

## 1. Introduction

Because of their numerous engineering applications, round jets have been the subject of extensive research since the experimental and theoretical



*J.-P. Chollet et al. (eds.), Direct and Large-Eddy Simulation II*, 35-46.

early work by Abramovich (1958). More recent works have brought to light the presence of complex coherent vortex structures imbedded within this flow. For example, Lasheras, Lecuona and Rodriguez (1991), Monkewitz and Pfitzenmaier (1991), Liepmann and Gharib (1992) have mentioned the appearance of streamwise vortices and side jets which appear to play a prominent role in the flow development. Axisymmetric jets are a prototype of free shear flows of vital importance from both a fundamental as well as a more applied point of view. Indeed, a better understanding of the jet vortex structures should make possible the active control of the jet (spreading rate, mixing enhancement...) for engineering applications (Zaman and al, 1994; Reeder and Samimy, 1996). We here present numerical simulations investigating the three-dimensional coherent vortices in the near region of a round jet (less than 15 diameters away from the exit), the role of the inflow perturbation on these structures, and the jet statistics. Two "classical" approaches exist in literature to numerically simulate jet flows, and more precisely round jets. The first one, based upon the axisymmetry hypothesis, considers the flow only in a two-dimensional radial plan, and calculates its spatial development (spatial box) (Gamet and Estivalezes, 1995; Reynier, Kourta et Ha-Minh 1995). The second one considers the three-dimensional aspects, but restricts itself to a temporal evolution (temporal box) in order to be able to conserve a fine enough discretisation grid (Martin and Meiburg, 1991; Abid and Brachet, 1992; Comte *et al.*, 1992; Verzicco and Orlandi, 1994; Brancher, 1996). Our purpose here is to demonstrate that a simulation, both three-dimensional and in a spatial box, is feasible with reasonable computational cost for engineering applications. Through Large Eddy Simulations (LES), we first describe the different coherent structures growing in a free round jet at high Reynolds number $Re = 25000$ based upon the jet diameter $D$ and its bulk velocity $W_o$. In particular, we will described an original mode, called "alternated pairing" resulting from the late evolution of the varicose mode. We then show how the jet dynamics can be controlled by imposing the inflow excitation.

## 2. Subgrid-scale model and numerical methods

For the simulation of a turbulent jet at $Re = 25000$, the use of LES techniques is compulsory. We have chosen the *structure function model* proposed by Métais and Lesieur (1992) in its selective version (see Lesieur and Métais, 1996), which is well adapted for transitional flows and accepts non uniform grids (Fallon, 1994). Its main characteristic is that it switches off the eddy-viscosity when the flow is not three-dimensional enough. The LES filtered Navier-Stokes equations are solved using the TRIO-VF code. This is an industrial software developed for thermal-hydraulics applications at the Commissariat a l'Energie Atomique de Grenoble by Grand *et al.* (1988). It has been thoroughly validated in many LES of various flows (see e.g. Silveira *et al.*, 1993, for the backward facing step). It uses the finite volume element method on a structured mesh. The variables are located on a staggered grid, with pressure and scalar quantities located at the centre of the control volume and the velocity components on the side. Temporal discretisation is a first order Euler scheme; spatial discretisation is a third order Quick Sharp scheme. The Poisson equation for pressure uses an iterative method of conjugated gradient.

## 3. Flow configuration and boundary conditions

### 3.1. LATERAL BOUNDARIES

We here consider the domain starting at the nozzle and extending up to a dozen diameters downstream. This domain, including the potential core, is characterized by strong vorticity and coherent vortex structures generation. Although we would like to simulate a free jet, radial limitations exist. We must therefore ensure that the presence of lateral artificial boundaries does not modify the numerical solution in a significant way. It is important to stress that recirculating flows could appear and result in an excitation of the jet. For instance, Melander *et al.* (1991) showed, for their Direct Numerical Simulation of a temporally evolving round jet, that the lateral periodicity conditions create an unnatural development of longitudinal vortices in a 4th order azimuthal mode. Villermaux and Hopfinger (1994) in an experimental work on confined jets, evaluate these effects. Their results support the idea that the round jet can be considered as free, when lateral limits are pushed further than five diameters away. This has been confirmed by the recent experiment by Djeridane (1994) of a round jet at $Re = 21000$. For our numerical jet, lateral boundary conditions are therefore imposed at $5D$ and free-slip boundary conditions are applied, without taking into account lateral fluid entrainment. Since a grid with cartesian coordinates is here used the section perpendicular to the jet axis consists of a square. Cartesian coordinates were preferred to cylindrical coordinates since the former allow to consider more complex geometries of industrial interest like multi-jets for instance. Furthermore, phenomena like side jets, experimentally visualized by Liepmann and Gharib (1992) at $Z/D = 3$ downstream of the nozzle, are difficult to adequately reproduce with cylindrical coordinates: these could indeed develop in regions far from the jet axis where the resolution is very coarse.

### 3.2. INFLOW

The nature of the first instabilities is mainly conditioned by the mean velocity profile shape at the nozzle (Morris, 1976; Michalke and Hermann, 1982; Cohen and Wygnanski, 1987 among others). So it is essential to correctly reproduce the initial vorticity layer thickness. Here, we did not simulate the flow inside the nozzle but we imposed the mean axial velocity profile in accordance with the experimental measurements by Moore (1977) and Cohen and Wygnanski (1987). The same type of profile has been used in the numerical simulation by Martin and Meiburg (1991), Abid and Brachet (1992) and Brancher (1996). It writes:

$$W(r) = \frac{1}{2} W_o \left[ 1 - tanh \left( \frac{1}{4} \frac{R}{\theta} \left( \frac{r}{R} - \frac{R}{r} \right) \right) \right] \tag{1}$$

where $W_o$ is the axial bulk velocity, $R$ the jet radius defined with $W(R) = W_o/2$, and $\theta$ the width of the shear layer

$$\theta = \int_0^\infty \frac{W(r)}{W_o} \left( 1 - \frac{W(r)}{W_o} \right) dr \tag{2}$$

Notice that the parameter $R/\theta$ depends on the experimental conditions (nozzle striction, Reynolds number, roughness...) and may vary from 10 (Becker and Massaro, 1968) up to 300 (Zaman and Hussain, 1980). We have checked that it is crucial to correctly resolve the shear zone at the border of the nozzle (at least with five computational cells): we are therefore limited to low values of the parameter $R/\theta$. Here, $R/\theta = 10$. The question is to know if this type of jet represents an experimental one exhibiting a more abrupt inflow velocity profile. In the case of a thin vorticity layer, the fastest growing mode predicted by the linear stability analysis corresponds to a varicose mode leading to vortex rings. Consequently, the resulting structures, whose typical size is of the order of seven times the vorticity thickness, will grow in a pairing cascade mechanism (Zaman and Hussain, 1980). At each step, their sizes are doubled and their frequencies are reduced by half, to the end of the potential core where they reach a Strouhal number $Str_D = 0.3$ to $0.5$ (Hussain and Zaman, 1981). For $R/\theta = 10$ the most amplified mode is still of varicose type. However, the pairing cascade will not be represented since the associated vortex-shedding frequency is already low. Consequently, our simulation does not represent the abrupt experimental jets in the domain between the nozzle and two diameters downstream (Zaman and Hussain, 1980). Nevertheless, it is reasonable to assume that the further flow development is well represented according to Moore's results (1977).

### 3.3. GRID CHARACTERISTICS

The grid is three-dimensional, cartesian, non-uniform, with 204000 cells. The $Z$ axis is along the jet axis, $X = 0, Y = 0$ corresponds to the nozzle center. The minimal grid spacing ($Dy_{min} = D/25 = Dx_{min}$) is sufficient to resolve the vorticity layer with $R/\theta = 10$. This relatively low number of cells makes possible long time integration, which is compulsory to reach statistical convergence: over one hundred adimensional times ($D/W_o$) are here represented. The main problem arising associated with the use of a cartesian grid is the difficulty to code a round jet. In order to prevent a preferential spreading along the $X$ and $Y$ axis, we have found that it is important to try to correctly reproduce both the velocity field and momentum flux axisymmetry. Another dominant aspect is the correct representation of the inflow noise since the nozzle boundary layer has a great influence on the downstream flow development As mentioned above, the flow inside the nozzle is not here simulated in order to limit the computer cost and some white noise is imposed at the nozzle exit. Nevertheless, the intensity of the noise mimics experimental measurements performed at the nozzle: we choose 1% for the rate of turbulence level at the center and 5% near the border. If another type of excitation is imposed, this will be clearly explicited.

### 3.4. OUTFLOW

A nice aspect of the jet is that it is convectively unstable for the Kelvin-Helmholtz waves, referring to Monkewitz and Sohn (1988), Jendoubi and Strykowski (1994), Brancher (1996). This means that errors introduced by the numerics at the outflow boundary conditions will have a limited effect and should not contaminate the whole jet. At the outflow, we here impose a zero normal pressure gradient. Besides, the stretching in the $Z$ direction of the last mesh improves the vortices exit. We have checked with a longer box ($Z_{max} = 16D$) that only the last two diameters are susceptible to be affected by the outflow conditions.

## 4. Numerical experiments

We recall that the Reynolds number is set to $Re = 25000$ and that $R/\theta = 10$. The results are divided into two distinct sections. The first one presents the various coherent structures present within the jet with a white noise inflow perturbation, which will be called the natural jet. In the second section, we study the effects of specific inflow excitations and will show how the jet can be controlled.

### 4.1. THE NATURAL JET

In order to validate the numerical approach, we first compare the computed statistics with experimental results. Figure 1a shows the donwstream evolution of the mean axial velocity on the jet axis. The fall around $Z/D = 5.5$ indicates the end of the potential core. One may notice the big dispersion in the experimental results. The numerical value falls within the experimental range. The downstream evolution of the r.m.s. fluctuating axial velocity on the jet axis is displayed on Figure 1b. The abrupt rise of the fluctuation experimentally observed just after the nozzle, results from the pairing cascade previously mentioned. It is not reproduced by the numerical simulation because of the lower value of $R/\theta$. Nevertheless, for $Z/D > 3$, the agreement is good. The characteristic peak is well reproduced both in terms of its maximal value and its axial position. In order to investigate the role played by the turbulent viscosity as compared with the numerical viscosity,

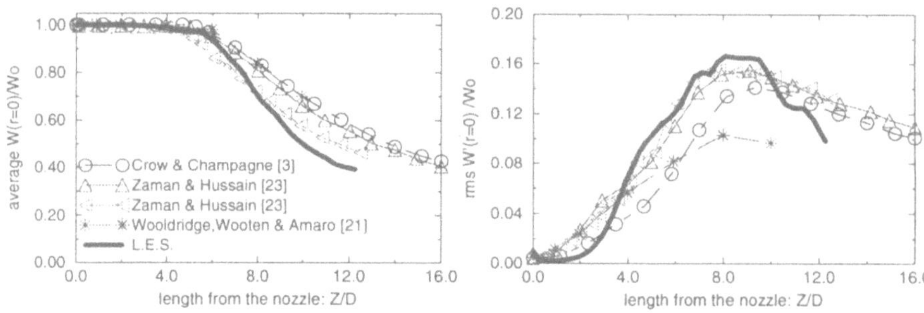


*Figure 1.* Numerical and experimental results comparisons. Centerline axial velocity: a) mean ; b) r.m.s. fluctuations

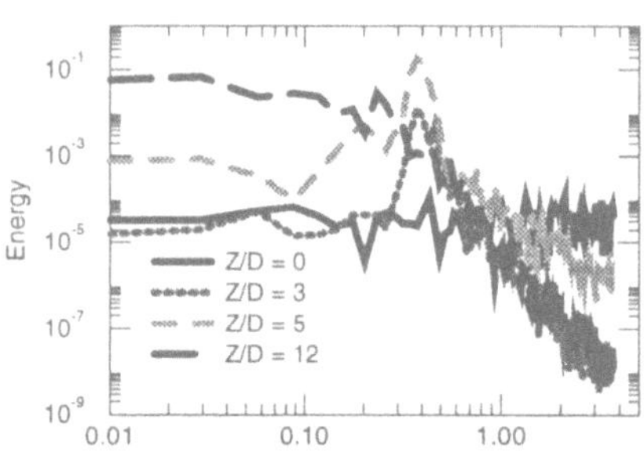


*Figure 2.* Frequency spectra of the fluctuating axial velocity at different dowstream locations

we have performed a simulation without any subgrid model. It exhibits a large discrepancy with the laboratory data: the centerline axial velocity r.m.s. fluctuations (corresponding to Figure 1b) indeed reach 0.20 at $Z/D = 4$.

We then turn to the frequential aspect by considering the frequency spectra (see Figure 2). From the white noise ($Z = 0$) emerges a predominant vortex-shedding Strouhal number,

$Str_D = 0.35$ (based on the diameter) in good correspondance with the experimental value ($Str_D$ between 0.3 and 0.5; see Hussain and Zaman, 1981). This demonstrates that the inflow vorticity layer is thin enough to reproduce the big structures.

The Kelvin-Helmholtz instability along the border of the jet yields, further downstream, vortex structures mainly consisting in axisymmetric toroidal shape. However, these structures are not always present and alternate with vortices of helicoidal shape (see figures 3a and 3b taken at two different times). It has to be stressed that temporal linear stability analysis performed on the inlet jet profile given by (1) (with $R/\theta = 10$) predicts a higher amplification rate for the axisymmetric (varicose) mode than for the helicoidal mode. The spatial jet variations of the present jet could be at the origin of the alternance between the two structures. We then follow the structures further downstream. The 3D visualisation exhibits an original vortex arrangement subsequent to the varicose mode growth: the "alternated pairing". Such a structure was previously observed by Comte *et al.* (1992) in a direct simulation of a temporally evolving round jet at low Reynolds number ($Re = 2000$). We have checked that the direction normal to the toroidal vortices symmetry plane, during their advection downstream, tends to differ from the jet axis. The inclination angle of two consecutive vortices appears to be of opposite sign eventually leading to a local pairing with an alternate arrangement. Note that vortex loop's inclination at the end of the potential core was experimentally observed by Petersen (1978). Experimental evidence of "alternated pairing" was recently shown by Broze and Hussain (1996). This mode will be further discussed in the next section.

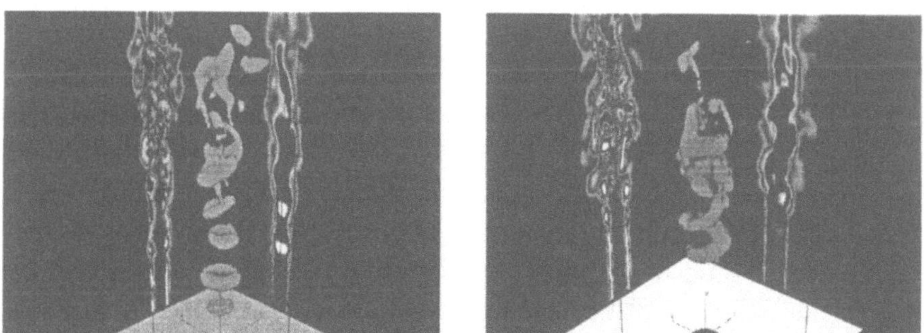

*Figure 3.* Natural jet three-dimensional visualization at two different times. Low pressure isosurface in light gray; cross-sections at $X = 0$ of the vorticity modulus and at $Y = 0$ of the velocity modulus.

### 4.2. THE EXCITED JET

Numerous experimental visualizations have been performed with an axisymmetric excitation of the jet. Crow and Champagne (1971) first have noticed that the jet response is maximal with a preferred mode frequency corresponding to $Str_D$ between 0.3 and 0.5. We here reproduce such an excitation of the varicose mode by imposing a periodic perturbation (alternatively low-speed and high speed) to the axial velocity at the nozzle:

$$W(r) + \epsilon\ W_o\ sin\left(\frac{Str_D W_o}{D}t\right) \quad (3)$$

This is superposed to the white random perturbation of the previous case. The periodic fluctuation level is $\epsilon = 1\%$, and its imposed frequency corresponds to $Str_D = 0.35$. The simulation shows that the varicose mode is now present at every instant at the beginning of the jet $Z/D < 6$. The vortex structures are more intense than in the natural case with well marked and organized pressure trough (Figure 5a below).

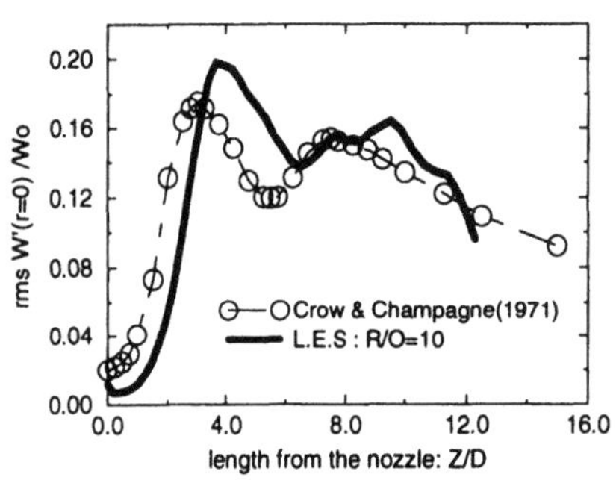


*Figure 4.* Similar to Figure 1b for the jet with varicose mode excitation

Comparisons of the axial velocity fluctuations with experimental results, show that, as opposed to the unexcited jet, a strong and fast amplification of the instability appears, resulting in a peak at $Z/D = 4$ (Figure 4). This saturation is followed by a decrease until a second peak is reached. Notice that the jet excitation modifies mean velocity too by reducing the potential core length.

The Figure 5b shows that the rings resulting from the varicose mode are linked together with longitudinal vortices. These have been observed experimentally by Lasheras, Lecuona and Rodriguez (1991), Monkewitz and Pfitzenmaier (1991) and Liepmann and Gharib(1992) at moderate Reynolds number flow. These longitudinal vortices are known to entrain and eject fluid outside, thus creating transverse side jets. The latter were studied numerically, in temporal simulations of Martin and Meiburg (1991), Abid and Brachet (1992), Verzicco and Orlandi (1994), Brancher (1996) at low Reynolds number. Notice that, in these works, they were directly excited with a specified azimuthal excitation of the vortex rings: this is at variance with the present study. Here, we evaluate their number to be between 3 to 7 at the end of the potential core. The maximum vorticity magnitude within the longitudinal structures is about 40% of the vorticity of the associated rings.

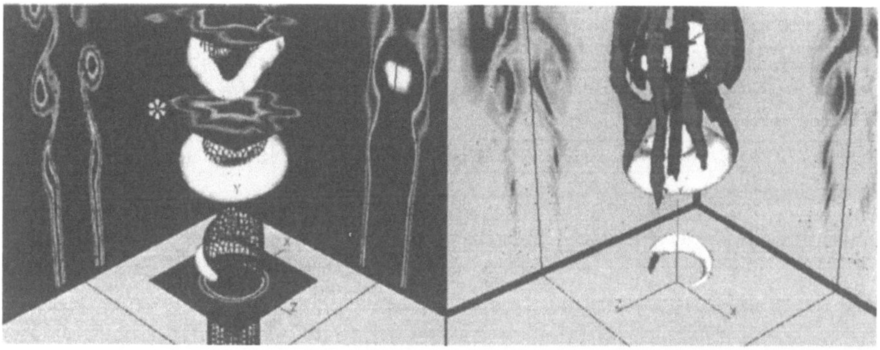

*Figure 5.* Jet with varicose mode excitation: low-pressure isosurfaces in white; cross-sections at $X = 0$ and at $Y = 0$, a) similar to Figure 3a, b) longitudinal vorticity component. On a), horizontal cuts of the axial velocity component show side jets (see the star). On b), positive and negative longitudinal vorticity isosurfaces corresponding to $\omega_z = \pm 2/3 \omega_z^{max}$ are represented in black and grey

As mentioned above, when the toroidal structures are advected downstream, they display alternated localized pairings. Since this vortex arrangement seems to be characteristic of the jet topology, we next propose to preferentially excite it from the nozzle. The excitation method is based on the same principle as previously described. The perturbation intensity is here $\epsilon = 5\%$. Its frequency is the same as before except that now half of the jet presents a speed excess, while a speed defect is imposed on the other half, and this alternatively. Note that this perturbation has a preferred direction, chosen along the $Y$ axis. The resulting structures are shown on Figure 6.

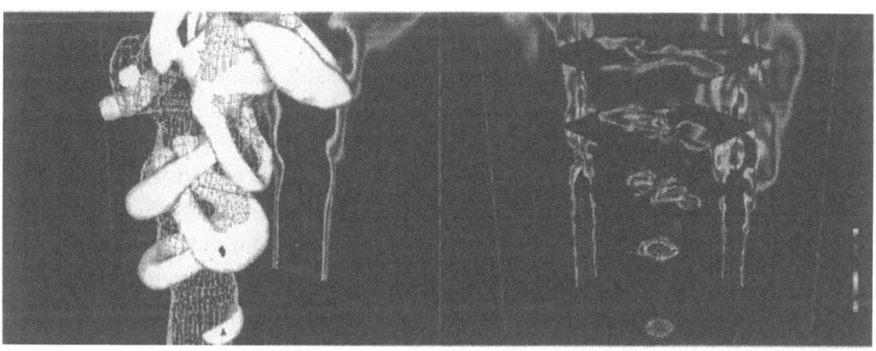

*Figure 6.* Jet with "alternate pairing" excitation: a) low-pressure isosurface in white, wire isosurface of the axial velocity $W = W_o/2$; b) vertical and horizontal cross-sections of $W$

Inclined vortex rings exhibiting localized pairing persist far downstream till $Z/D = 10$. The striking fact is the very distinct spreading rates in the $Y$ and $X$ directions. In the $YZ$ plane, the spreading rate reaches nearly $25^o$ and even higher values near the exit. Conversely, it is close to zero in the $XZ$ plane. Recently, the experiment performed by Longmire and Duong (1996) displays similar results by using a special geometry of the nozzle (made with two half nozzles). One important technological application of this excitation resides in the ability to polarize the jet in a preferential direction.

Finally, the next excitation is designed to trigger the first helicoidal mode by imposing some white noise superposed on the following inflow velocity profile

$$W(r) + \epsilon\ W_o\ sin\left(\Theta - 2\pi \frac{Str_D W_o}{D} t\right) \tag{4}$$

where $W(r)$ is given by equation (1), and $\Theta$ stands for the azimuthal angle. The same intensity and frequency are used as before.

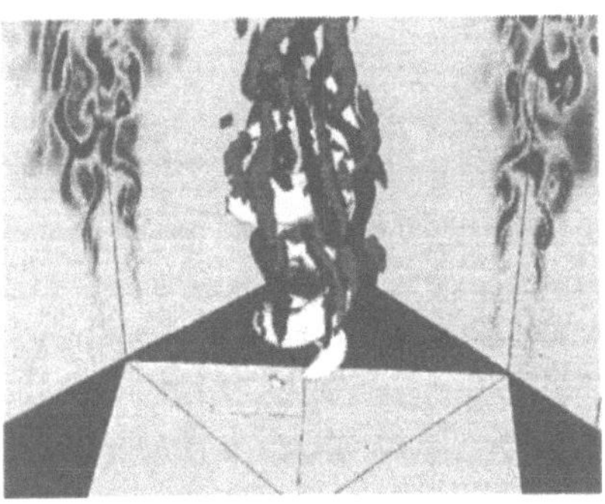

*Figure 7.* Jet with helicoidal mode excitation: similar to Figure 5b

The response of the jet indeed consists in the development of an helicoidal coherent vortex structure, as found experimentally by Kusek, Corke and Reisenthel (1990). Furthermore, the Figure 7 shows that longitudinal vortices are present both of positive and negative $Z$ component of vorticity. These secondary structures are wrapped around the jet axis, in opposite direction with respect to the primary helix.

## 5. Conclusion

With the aid of Large-Eddy Simulations based upon the *selective structure function model* (see Lesieur and Métais, 1996), the statistical and topological behaviour of the round jet spatial evolution has been studied. For the simulated jet, we have chosen a velocity profile close to a top-hat shape. Due to numerical constraints, the simulated profile is however shallower than the experimentally measured profiles: here $R/\theta = 10$. Despite this difference, a good agreement with the experimental data is found for the mean and fluctuating axial velocities. As expected, the growth of the axisymmetric instability mode leads to axisymmetric vortices (rings). This preferred mode appears at a Strouhal number $Str_D = 0.35$ in good correspondance with the experimental predictions. In alternance with rings, we also observe helicoidal vortices. Further downstream, the rings also give rise to a vortex structure consisting in alternated localized pairings between vortex rings (see also Comte *et al.*, 1992). We have designed a specific excitation of the jet aimed at triggering this "alternated-pairing" mode. The jet behaviour is dramatically modified in that case: in one radial plane, the jet expansion rate is greatly enhanced as compared to the natural jet, while in the plane perpendicular it is totally inhibited. One important technological application of this excitation resides in the ability to control the jet by polarizing in a preferential direction. When axisymmetric excitation is applied, after vortex rings have formed, pairs of counter-rotating longitudinal vortices appear linked with primary rings and these create horizontal side jets. Longitudinal vortices are still present when helicoidal excitation is imposed. In concordance with several experimental results, we have thus clearly showed how a specific excitation (axisymmetric, alternate, helicoidal, etc) of the jet at the nozzle renders possible the control of the jet vortical structures and a strong modification of its statistics (spreading rate, turbulence level, etc).

## References


Abid, M. and Brachet, M.E. (1992) Mécanisme de génération des jets latéraux dans les jets axisymetriques forcés, *C.R.Acad.Sci.Paris.*

Abramovich, G.N. (1957) The turbulent jet in a moving fluid. *Library translation n° 28.*

Becker, H.A. and Massaro, T.A. (1968) Vortex evolution in a round jet. *J.Fluid Mech*, **31**, pp 435–448.

Brancher, P. (1996) Etude numérique des instabilités secondaires de jets. *Thèse de l'École polytechnique*, Paris, France.

Broze, G. and Hussain, F. (1996) Transition to chaos in a forced jet: intermittency, tangent bifurcations and hysteresis. *J.Fluid Mech*, **311**, pp 37–71.

Cohen,J. and Wygnansky I. (1987) The evolution of instabilities in the axisymmetric jet, part1: the linear growth of disturbances near the nozzle. *J.Fluid Mech.*, **176**, pp 191–219.

Comte, P., Fouillet, Y. and Lesieur, M., (1992) Simulation numérique des zones de mélange compressibles. *Revue scientifique et technique de la defense*, 3ème trimestre.

Crow, S.C. and Champagne, F.H. (1971) Orderly structure in jet turbulence. *J.Fluid Mech*, **48**, pp 547–591.

Djeridane,T. (1996) Contribution à l'étude experimentale de jets turbulents axisymétriques à densité variable. *Thèse de l'Université d'Aix-Marseille II.*

Fallon, B. (1994) Simulation des grandes échelles d'écoulements turbulents stratifiés en densité. *Thèse de l'Institut National Polytechnique de Grenoble.*

Gamet L. and Estivalezes J.L (1995) Simulation numérique de jets isothermes et chauffés en régime transonique. Société française des thermiciens. Journée d'études du 15 mars.

Grand, D., Coulon, N., Magnaud, J.P. and Villand, M. (1988) Computation of flow with distributed resistance and heat sources. In *Proc. Third Intl. Symp. on Refined Flow Modelling and Turbulence Measurements, Nipon Toshi Center Tokyo* (ed. Y. Iwasa), pp 487–494.

Hussain, F. and Zaman K.B.M.Q. (1981) The "preferred mode" of the axisymmetric jet *J.Fluid Mech*, **110**, pp. 39–71.

Jendoubi, S. and Strykowski, P.J. (1994) Absolute and convective instabilitiy of axisymmetric jets with external flow, *Phys. Fluids* **6** (9), pp 3000-9.

Kusek, Corke and Reisenthel, Seeding of helical modes in the initial region of an axisymmetric jet (1990), *Experiments in Fluids*, **10**, pp 116–124.

Lasheras, J.C., Lecuona A. and Rodriguez, P. (1991) Three dimensionnal structure of the vorticity field in the near region of laminar co-flowing forced jets. In *The Global Geometry of Turbulence*, edited by J.Jimenez (Plenum Press, New-York).

Lesieur, M. and Métais, O. (1996) New trends in large-eddy simulations of turbulence. *Annu. Rev. Fluid. Mech.*, **28**, pp 45–82.

Liepmann, D. and Gharib, M. (1992) The role of streamwise vorticity in the near-field entrainment of round jets. *J.Fluid Mech*, **245**, pp 643–668.

Longmire, E.K. and Duong, L.H. (1996) Bifurcating jets generated with stepped and sawtooth nozzles. *Phys. Fluids*, **8** (4), pp 978–992.

Martin, J.E. and Meiburg, E. (1991) Numerical investigation of three-dimensionally evolving jets subject to axisymmetric and azimuthal perturbations. *J.Fluid Mech*, **230**, pp 271–318.

Melander, M.R., Hussain, F. and Basu, E. (1991) Breakdown of a circular jet into turbulence. In *Proc. of eighth Symposium on Turbulent Shear Flow*, Münich.

Métais, O. and Lesieur, M. (1992) Spectral large-eddy simulation of isotropic and stably stratified turbulence. *J.Fluid Mech*, **239**, pp 157–194.

Michalke, A. and Hermann, G. (1982) On the inviscid instability of a circular jet with external flow. *J.Fluid Mech*, **114**, pp 343–359.

Monkewitz, P.A. and Pfizenmaier, E. (1991) Mixing by 'side jets' in strongly forced and self-excited round jets. *Phys. Fluis A*, **3** (5), pp 1356–1361.

Monkewitz, P.A. and Sohn, K.D. (1988) Absolute instability in hot jets. *A.I.A.A. J.*, **26**, pp 911–916.

Morris, P.J. (1976) The spatial viscous instability of axisymmetric jets. *J.Fluid Mech*, **77**, pp 511–529.

Moore, C.J. (1977) The role of shear-layer instability waves in jet exhaust noise. *J.Fluid Mech*, **80**, pp 321–367.

Orlanski, I. (1976) A simple boundary condition for unbounded hyperbolic flows. *J.Comp.Phys.*, **21**, pp 251–269.

Petersen, R.A. (1978) Influence of wave dispersion on vortex pairing in a jet. *J.Fluid Mech*, **89**, pp 469–495.

Reeder, M.F. and Samimy, M. (1996) The evolution of a jet with vortex-generating tabs: real-time visualization and quantitative measurements. *J.Fluid Mech*, **311**, pp 73–118.

Reynier, P., Kourta, A. and Ha-Minh, H. (1995) Simulation numérique de jets ronds turbulents, compressibles et instationnaires. Société française des thermiciens. Journée d'études du 15 mars.

Silveira-Neto, A., Grand, D. Métais, O. Lesieur, M., A numerical investigation of the coherent structures of turbulence behind a backward-facing step. *J.Fluid Mech*, **256** (1993), pp. 1–55.

Villermaux, E. and Hopfinger,E. (1994) Self-sustained oscillations of a confined jet: a case study for the non-linear delayed saturation model. *Physica D*, **72**, pp 230–243.

Verzicco, R. and Orlandi, P. (1994) Direct simulations of the transitional regime of a circular jet. *Phys.Fluids*, **6** (2), pp 751–759.

Wooldridge, C.E., Wooten, D.C., and Amaro, A.J. (1971) The structure of jet turbulence producing noise. NASA contractor Rep n°126483.

Zaman, K.B.M.Q., Reeder, M.F. and Samimy, M. (1994) Control of an axisymmetric jet using vortex generators. *Phys.Fluids*, **6** (2), pp 778–794.

Zaman, K.B.M.Q. and Hussain, A.K.M.F. (1980) Vortex pairing in a circular jet under controlled excitation, part 1, general jet response. *J.Fluid Mech*, **101**, pp 449–491.

# NONCOMMUTING FILTERS AND DYNAMIC MODELLING FOR LES OF TURBULENT COMPRESSIBLE FLOW IN 3D SHEAR LAYERS

B. GEURTS, B. VREMAN, H. KUERTEN AND R. VAN BUUREN
*Department of Applied Mathematics, University of Twente*
*P.O. Box 217, 7500 AE Enschede, The Netherlands*

**Abstract.** Large-eddy simulation of complex turbulent flows involves the filtering and modelling of small scale flow structures whose intensity shows large spatial variations in the flow domain. This suggests the use of filters with nonuniform filterwidth. Such filters fail to commute with spatial derivatives and give rise to additional 'noncommutation' terms in LES. We construct higher order filters and show that the subgrid terms and the new noncommutation terms are *a priori* of comparable magnitude. We apply these filters to DNS data of the temporal mixing layer. The magnitude of the noncommutation terms and their contribution to the kinetic energy dynamics is determined. Finally, we show that LES predictions significantly depend on the specific explicit filter used in dynamic subgrid modelling.

## 1. Introduction

The numerical simulation of transitional and turbulent flow forms a field of considerable interest (Kleiser and Zang, 1991). Through continuous advances in computing capabilities, direct numerical simulation (DNS) of transitional and low-Reynolds turbulent flow in simple geometries has become feasible. This requires the detailed spatial and temporal resolution of all relevant scales of motion. Restrictions arising from available computing capabilities render DNS impossible for practically relevant configurations at high Reynolds numbers. Thus, modelling of the governing equations, which reduces the degrees of freedom of the dynamical system, is required. For this purpose the central first step consists of an averaging of the Navier-Stokes equations. A detailed modelling is obtained in Large-Eddy Simulation (LES) which can be obtained from the Navier-Stokes equations through



*J.-P. Chollet et al. (eds.), Direct and Large-Eddy Simulation II*, 47-56.

the application of a spatial filter. In the derivation of the LES equations one commonly adopts a convolution filter which has the property that filtering and partial differentiation can be interchanged. In that case the effect of the filtering is restricted to the appearence of so called subgrid scale terms and the resulting equations can be written in conservative form. A large drawback of convolution filters, however, is the fact that the filterwidth is constant. This complicates the extension of the LES approach to complex geometries for which a varying filterwidth is required in view of the strong spatial variations in the small scale turbulence intensities. However, a nonuniform filterwidth gives rise to additional 'noncommutation' terms in the filtered equations (Geurts *et al.*, 1994; Ghosal and Moin, 1995; van der Ven, 1995).

The higher order compact support filters developed in this paper are obtained by requiring invariance of polynomials up to a certain order. The 'noncommutation' properties of these filters will be investigated and it will be shown that the common subgrid-terms and the noncommutation terms are in general of comparable magnitude. Apart from a varying filterwidth, these filters are characterized by their 'skewness'. Through an application of these filters to DNS-data for the compressible mixing layer *a priori* estimates are obtained for the different contributions to the LES equations for varying grid nonuniformity. Whereas the turbulent stress tensor has a predominant dissipative contribution to the dynamics of the kinetic energy the noncommutation terms also contribute to backscatter. Finally, we consider the influence of the specific realization of the explicit filtering in dynamic subgrid modelling and compare simulation results obtained with the dynamic eddy viscosity model and the dynamic mixed model.

The organization of this paper is as follows. In section 2 we introduce higher order compact support filters and study their noncommutation properties. Section 3 is devoted to *a priori* evaluation of the magnitude and effects of the noncommutation terms compared to the common subgrid terms for the mixing layer. In section 4 we present LES results for the mixing layer showing the influence of a higher order realization of the explicit filtering in the dynamic modelling. We summarize our findings in section 5.

## 2. Higher order non-commuting filters

In this section we introduce higher order filters and study their commutation properties in one spatial dimension. The consequences of the realizability conditions on the filters are established and the effects related to the 'skewness' of the filter are illustrated with a Fourier analysis.

We consider a signal $f : I\!R \to I\!R$ and define the filter operation $f \to \overline{f}$

by:

$$\overline{f}(x) = \int_{x-\Delta_-(x)}^{x+\Delta_+(x)} \frac{\mathcal{H}(x,\xi)}{\Delta(x)} f(\xi)d\xi \tag{1}$$

in which $\Delta_+, \Delta_- \geq 0$ denote the $x$-dependent upper- and lower filterwidths respectively, $\Delta = \Delta_+ + \Delta_- > 0$ is the filterwidth and $\mathcal{H}/\Delta$ the filter-kernel. We arrive at a more convenient formulation after a change of coordinates in which we put $y = (\xi - x)/\Delta(x)$. This leads to:

$$\overline{f}(x) = \int_{I_x} \mathcal{H}(x, x+\Delta(x)y) f(x+\Delta(x)y)dy \tag{2}$$

where $I_x = [(s(x)-1)/2, (s(x)+1)/2]$. Here we introduced the 'normalized skewness' $s(x) = S(x)/\Delta(x)$ in terms of the skewness $S(x) = \Delta_+(x) - \Delta_-(x)$. We introduce $N$-th order filters by requiring:

$$\int_{I_x} \mathcal{H}(x, x+\Delta(x)y)y^k dy = \delta_{k0} \;\; ; \;\; k = 0, 1, .., N-1 \tag{3}$$

in which $\delta_{ij}$ denotes the Kronecker delta. These filters have the property that $\overline{P_{N-1}(x)} = P_{N-1}(x)$ for any polynomial $P$ of order $N-1$. Application of this filter to the $M$-th order Taylor expansion of $f$ around $x$ yields:

$$\overline{f}(x) = f(x) + \sum_{k=N}^{M-1} \left(\Delta^k(x)M_k(x)\right) f^{(k)}(x) + \Delta^M(x)R_M(f) \tag{4}$$

where $f^{(k)}$ denotes the $k$-th derivative of $f$ and we introduced

$$M_k(x) = \frac{1}{k!}\int_{I_x} \mathcal{H}(x, x+\Delta(x)y)y^k dy \tag{5}$$

Moreover, $R_M(f)$ denote the rest-term. Hence, if $M > N$ the leading order term of $\overline{f} - f$ scales with $\Delta^N$. For notational convenience we will not denote the explicit $x$-dependence in the sequel and we ignore the rest-terms.

The effect of the filter-operation on the signal $f$ as expressed in (4) can straightforwardly be extended to derivatives of $f$ and to nonlinear operations $g$ on the signal. These expressions can be used in order to derive the basic noncommutation properties. For $N$-th order filters the commutator with differentiation can be written as:

$$\overline{f'} - \overline{f}' = -\sum_{k=N}^{M-1} \left(\Delta^k M_k\right)' f^{(k)} + ... \tag{6}$$

in which the dots denote higher order terms in $\Delta$. This commutator is written in the usual way in terms of higher order derivatives of $f$ where now a factor $(\Delta^k M_k)'$ appears. In general it is quite complicated to obtain detailed estimates for this term. Using the definition of the filter (1) and some partial integration one may distinguish two contributions, one due to $x$-dependence of $\Delta_\pm$ and one from the fact that the filter-kernel is not of convolution type which results in a contribution containing $\partial_x \mathcal{H} + \partial_\xi \mathcal{H}$.

Another commutator which is relevant for filtering nonlinear terms arises when the filter-operation is combined with an algebraic operation on the signal: $\overline{(g \circ f)} - (g \circ \overline{f})$. An important example of such a nonlinearity is $g(z) = z^2$. For this particular example we obtain:

$$\overline{f^2} - \overline{f}^2 = \sum_{k=N}^{M-1} (\Delta^k M_k) \left( (f^2)^{(k)} - 2ff^{(k)} \right) - (\overline{f} - f)^2 + \tag{7}$$

The scaling with $\Delta^N$ is readily verified for $N > 1$. In case $N = 1$ the lowest order term in the summation equals 0 since $(f^2)' = 2ff'$ and the commutator scales with $\Delta^2$ with contributions from the term $k = 2$ in the summation and an additional contribution from $(\overline{f} - f)^2$. In the Navier-Stokes equations the latter commutator arises inside a divergence operator which gives rise to a term $\sim \Delta^{N-1}$, i.e. comparable to the commutator with first order derivatives. Hence, there is *a priori* no justification to ignore the latter terms while retaining the common subgrid-terms, which is in contrast with the findings in (van der Ven, 1995). Of course the actual magnitude of the noncommutation with differentiation also depends on the spatial variation of $\Delta(x)$ and $s(x)$ which may reduce the magnitude considerably. Moreover, the effect on the evolution of the solution arising from the various noncommutation terms can be quite different.

The construction of specific higher order filters relies on a Taylor expansion of the filter-function $\mathcal{H}$. In the definition of the filter-operation (1) the filter-function $\mathcal{H}$ is required for $\xi$ in a neighborhood of $x$:

$$\mathcal{H}(x, \xi) = \mathcal{H}(x, x) + \sum_{m=1}^{N-1} \frac{\partial_\xi^{(m)} \mathcal{H}(x, x)}{m!} (\xi - x)^m + \ldots \tag{8}$$

in which $\partial_\xi^{(m)} \mathcal{H}(x, x)$ denotes the $m$-th partial derivative of $\mathcal{H}(x, \xi)$ with respect to $\xi$ evaluated at $\xi = x$. With this filter we arrive at

$$(\Delta^k M_k) = \frac{\Delta^k}{k!} \sum_{m=0}^{L-1} \frac{\Delta^m \partial_\xi^{(m)} \mathcal{H}(x, x)}{m!} \int_{I_x} y^{m+k} dy + \ldots \tag{9}$$

The definition of $N$-th order filters as given in (3) can also be expressed as $(\Delta^k M_k) = \delta_{k0}$ for $k = 0, 1, .., N - 1$. Thus the truncated polynomial representation of $N$-th order filter satisfies a linear system of equations from which the required $\Delta^m \partial_\xi^{(m)} \mathcal{H}$ follow. We can specify the form of the filter-function conveniently as $\mathcal{H}(x,\xi) = \mathcal{G}(s,y)$. In view of the symmetries it appears that $\mathcal{G}(-s,y) = \mathcal{G}(s,-y)$ for $y \in I_x$ and in particular that $\mathcal{G}(0,y)$ is an even function of $y$. This implies that the $(2N-1)$-th and the $(2N)$-th order filters coincide if $s = 0$.

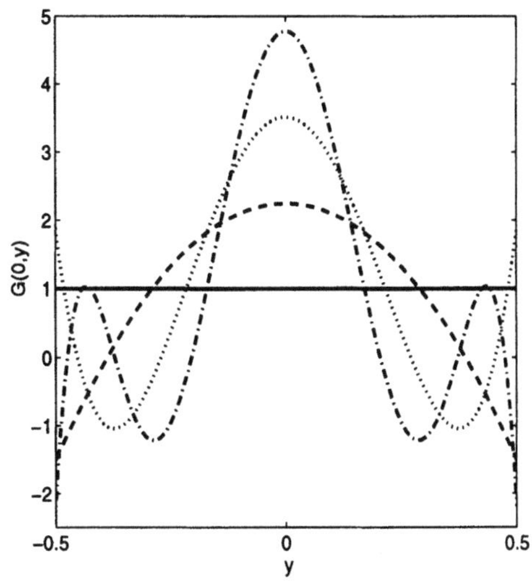


*Figure 1.* The filter-kernels $\mathcal{G}_N$ for symmetric higher-order filters, i.e. $s = 0$. The results for $N = 1$ (solid); $N = 3$ (dashed); $N = 5$ (dotted) and $N = 7$ (dash-dotted) are shown. Notice $\mathcal{G}_{2N} = \mathcal{G}_{2N-1}$ for symmetric filters.

In figure 1 we plotted the filter-kernels $\mathcal{G}_N$ at $s = 0$. In (Vreman *et al.*, 1994a) it was shown that the realizability conditions for the turbulent stress tensor are satisfied if and only if the kernel is positive. The explicit filters plotted above, however, show filter-kernels which are not strictly positive in case $N \geq 3$. Also, the $N = 2$ kernel is positive only if $|s| < 1/3$. Hence, only first or second order filters are allowed if the turbulent stress tensor should be realizable and the normalized skewness may not exceed $1/3$ for second order filters.

So far we considered the effect of applying an $N$-th order filter to a signal which can be represented by a Taylor-expansion. As long as the higher order contributions to $\overline{f} - f$ are small we obtain an accurate representation of $\overline{f}$. However, for very rapidly fluctuating signals on a scale comparable to $\Delta$ this approximation is no longer adequate. In order to analyze this we focus on the filtering of $\sin(kx)$ for which

$$\overline{\sin}(kx) = F_1(s,k\Delta)\sin(kx) + F_2(s,k\Delta)\cos(kx) \qquad (10)$$

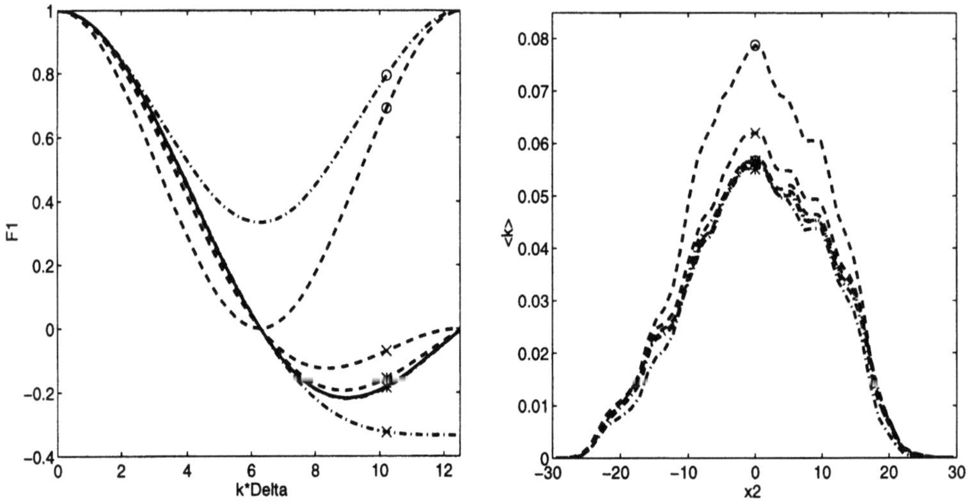


*Figure 2.* The numerical approximation to $F_1$ of the symmetric top-hat filter (solid) with the composite trapezoidal rule (dashed) and the composite Simpson rule (dash-dotted) (left). The turbulent kinetic energy profiles $< k >$ (right) evaluated at $t = 100$. We used $\Delta = 2h$, $\Delta = 4h$ and $\Delta = 8h$ marked with 'o', 'x' and '*' respectively.

where we introduced the 'characteristic filter-functions':

$$F_1(s, k\Delta) = \int_{I_x} G(s,y)\cos(k\Delta y)dy \quad ; \quad F_2(s, k\Delta) = \int_{I_x} G(s,y)\sin(k\Delta y)dy \tag{11}$$

Notice that $F_2(0, k\Delta) = 0$ and the only effect of symmetric filters is an amplitude change. This also shows that skewness of the filter contributes to a phase-shift in the filtered signal. Further analysis of $F_1$ and $F_2$ for different filters and skewness shows that the higher order filters can even increase the amplitude of rapidly oscillating signals on a length-scale comparable to $\Delta$, and only effectively reduce the signal as $k\Delta \gg 1$. An analysis of the effect of $N$-th order filters on stochastic signals could be interesting in the context of LES. Moreover, the influence of the skewness on the models for the subgrid-terms requires further attention.

## 3. A priori evaluation of noncommutation terms

In this section we formulate the higher order filters in a numerically consistent way using Newton-Cotes integration and apply these filters to DNS data obtained for the temporal compressible mixing layer at convective Mach-number $M = 0.2$ and Reynolds number $Re = 100$ based on the initial vorticity thickness (Vreman *et al.*, 1995).

In order to arrive at a numerically consistent representation of the filters introduced in the previous section the numerical integration should be sufficiently accurate in order to maintain the basic invariance properties of these filters. The discrete data $\{f_j\}$ are assumed to be represented on a grid $\{x_j\}$. In order to define the filter we adopt the following definition for the filterwidth and the skewness: $\Delta(x_i) = x_{i+n_+} - x_{i-n_-}$; $S(x_i) = x_{i+n_+} - 2x_i + x_{i-n_-}$ where $n_\pm \geq 0$ specify the upper and lower filterwidth distributions on the grid. The numerical filtering should be such that invariance of $x^\alpha$, $\alpha = 0, .., N-1$ is maintained. Within the framework of Newton-Cotes integration such integration rules can readily be specified, also in composite form. If $n_\pm$ are small the numerical filtering is quite different from the analytical filtering. Numerical filtering of $\sin(kx)$ using the symmetric top-hat filter and the (composite) trapezoidal and Simpson rules at different values of $n_\pm$ is shown in figure 2. As $n_\pm = 1$, the numerical filtering of the small scale structures differs considerably from the analytic filtering. The use of Simpson integration instead of the trapezoidal rule leads to less reduction of the small scale structures which is also illustrated in the profiles of the turbulent kinetic energy $< k >$ shown in figure 2 which are lower in case Simpson integration is adopted.

We proceed with the filtering of the Navier-Stokes equations and use DNS-data obtained from the temporal compressible mixing layer on a uniform grid with $192^3$ grid-cells. The evolution of the flow displays four large rollers at $t = 20$ which subsequently interact and give rise to two spanwise rollers at $t = 40$ and one large roller with many small-scale structure at $t = 80$. A mixing transition to turbulence arises in this flow.

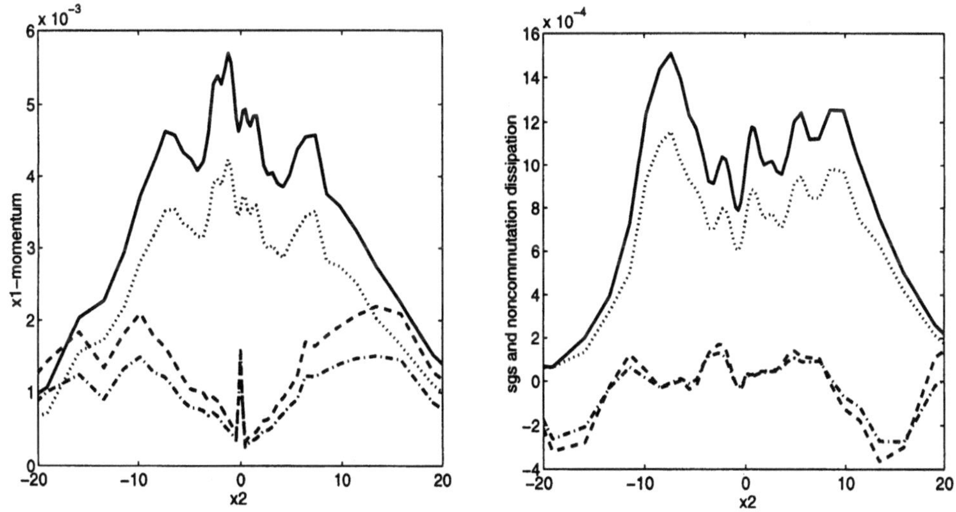


*Figure 3.* Discrete $L_2$-norm of the subgrid (T: solid, S: dotted) and noncommutation term (T: dashed, S: dash-dotted) in the $x_1$-momentum equation (left). Contribution to the kinetic energy dynamics from the turbulent stress tensor (T: solid, S: dotted) and due to the noncommutation term (T: dashed, S: dash-dotted). The data at $t = 100$ and $a = 0.2$ are shown on the $48^3$ grid after averaging over the homogeneous directions.

In order to evaluate the effects of the filtering the data have been transferred to a grid which is nonuniform in the (normal) $x_2$ direction only. For the nonuniform grid we use the mapping

$$x_2 = \frac{1}{2} L_2 \frac{a\eta}{1 + a - \eta} \tag{12}$$

to generate the positive $x_2$-nodes and we complete this grid by reflection in $x_2 = 0$. Here $\eta$ is in $[0, 1]$, $L_2$ denotes the extent in the $x_2$ direction and the parameter $a$ controls the grid-nonuniformity. The original data on the uniform grid have been transferred to the nonuniform grid by using fourth order accurate interpolation. In the sequel we consider a coarse grid with $48^3$ grid-cells, an intermediate grid with $96^3$ grid-cells and a fine grid with $192^3$ grid-cells. On the coarse, intermediate and fine grids we use $n_\pm = 1$, $n_\pm = 2$ and $n_\pm = 4$ respectively. For the filtering in 3D we adopt a 'product-filter' in which the filtering is performed independently in each direction with a 1D filter as described in the previous section.

The effect of the filter is expressed by the following decomposition. Consider a typical convective term in the momentum equations given by:

$$\begin{aligned} \overline{\partial_j(\rho u_i u_j)} &= \partial_j(\overline{\rho}\tilde{u}_i\tilde{u}_j) \\ &+ [\partial_j(\overline{\rho}(\widetilde{u_i u_j} - \tilde{u}_i\tilde{u}_j))] + [\overline{\partial_j(\rho u_i u_j)} - \partial_j(\overline{\rho}\widetilde{u_i u_j})] \end{aligned} \tag{13}$$

where $\partial_j$ denotes the partial derivative with respect to $x_j$, $\rho$ the density, $u_i$ the velocity in the $x_i$ direction and $(\tilde{\cdot})$ is the Favre filter. The first term on the right hand side corresponds to the mean term, the second is the common turbulent stress term and the last denotes the noncommutation term. Since the grid is nonuniform only in the $x_2$ direction the noncommutation term is nonzero only if derivatives with respect to $x_2$ are concerned. As a typical example we show the subgrid and noncommutation terms for $\partial_2(\rho u_1 u_2)$ in figure 3. Use was made of the top-hat filter in combination with trapezoidal (T) and Simpson (S) integration. The noncommutation term is comparable with the turbulent stress term in large parts of the flow domain. Moreover, a definite 'spike' can be seen near $x_2 = 0$ due to a strong local variation of $\Delta$. In order to assess *a priori* some of the effects of these terms on the dynamics of the kinetic energy we consider the turbulent stress contribution given by $\overline{\rho}\tau_{ij}\partial_j\tilde{u}_i$ where $\tau_{ij} = \widetilde{u_i u_j} - \tilde{u}_i\tilde{u}_j$ and compare this with the contribution arising from the noncommutation term $\tilde{u}_i R_i$ where $R_i = \overline{\partial_j(\rho u_i u_j)} - \partial_j(\overline{\rho}\widetilde{u_i u_j})$ (the summation convention is adopted here). The result is shown in figure 3 from which it becomes clear that the turbulent stress tensor has a dissipative influence. The noncommutation term is considerably lower and can give rise to backscatter.

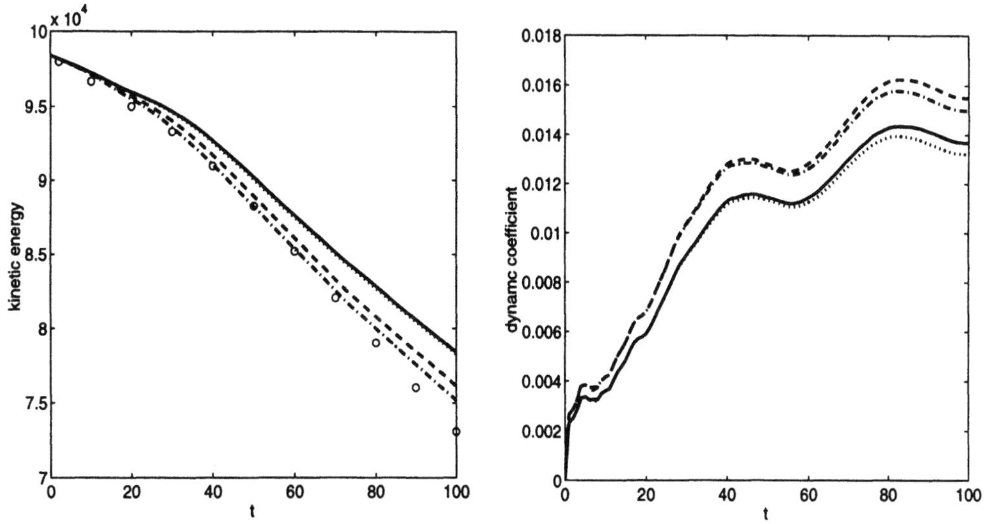


*Figure 4.* The evolution of the kinetic energy (left): ES (solid); ET (dotted); MS (dashed) and MT (dash-dotted) and the evolution of the dynamic coefficient $C_d$ for the eddy viscosity model (right): SS (solid); ST (dashed); TS (dotted) and TT (dash-dotted). The markers denote the filtered DNS-data.

## 4. Higher order explicit filtering and dynamic modelling

The formulation of dynamic subgrid models for the turbulent stress tensor implies an explicit filtering of a LES solution. In the previous section we showed that if $\Delta/h$ is relatively small the numerical realization of the filter differs significantly from the analytic filter. In this section we show that these differences can contribute also to the simulation results. We concentrate on the top-hat filter and for the numerical realization we compare the use of the trapezoidal rule with Simpson integration using $\Delta = 2h$ in the simulation with a test filter $\Delta = 4h$. The LES results were obtained on a uniform grid with $32^3$ grid-cells adopting the dynamic eddy viscosity model (Germano *et al.*, 1991) or the dynamic mixed model (Vreman *et al.*, 1994b). We use the abbreviations E, M for the dynamic eddy viscosity and mixed models. Combination with trapezoidal or Simpson integration is denoted with either T or S.

The formulation of the dynamic models is based on the Germano identity. In this formulation a model coefficient is determined in accordance with the local flow structure. In figure 4 we compare the predictions for the kinetic energy. The influence of the explicit filtering is much larger for the dynamic mixed model compared to the dynamic eddy viscosity model. This directly corresponds to the occurrence of both filters in the formulation of the dynamic mixed model whereas only the $\Delta = 4h$ filter arises in the dynamic eddy viscosity model for which the difference between T and S integration is much smaller. Moreover, in figure 4 the model-coefficient arising in the dynamic model is shown. Two simulations have been performed; one in which the model is calculated with the trapezoidal rule and one with Simpson integration. During each of these runs the coefficient was also evaluated with the complementary integration rule although this result was not used in the flux-calculation. In the figure these four combinations are labeled AB where A denotes the rule used for the flux and B the rule for the additional evaluation. The result for the model-coefficient $C_s$ shows a decrease in $C_s$ in case Simpson integration is used in the explicit filter. The evaluation of $C_s$ with either the trapezoidal or the Simpson rule is more important than the specific dynamic model used for the flux calculation.

## 5. Concluding remarks

We presented the construction of higher order noncommuting filters in LES and showed that the contribution of the additional subgrid terms is in general comparable to that of the common subgrid-terms. A numerically consistent formulation of the higher order filters was applied to DNS data of the mixing layer. It was shown that the noncommutation terms can be comparable to the common subgrid terms and the contribution to e.g. the kinetic energy evolution is quite small and contributes to backscatter. The numerical filtering was shown to differ significantly from the analytic filtering in case $\Delta = 2h$. Moreover, the differences between e.g. the trapezoidal and Simpson realization of the filtering were found to be considerable at this $\Delta/h$ ratio. In LES the use of dynamic subgrid models for the turbulent stress tensor has given rise to accurate LES predictions for the mixing layer. The specific realization of the explicit filtering which is required in this approach has a large influence on the simulation results for the dynamic mixed model at $\Delta = 2h$.

## References

L. Kleiser and T.A. Zang, 'Numerical simulation of transition in wall-bounded shear flows', *Ann. Rev. Fluid Mech.*, **23**, 495, (1991)

B.J. Geurts, A.W. Vreman, J.G.M. Kuerten, 'Comparison of DNS and LES of transitional and turbulent compressible flow: flat plate and mixing layer', AGARD Conf. Proc. 551, 5.1-5-14, Chania, Crete, Greece, (1994)

S. Ghosal, P. Moin, 'The basic equations for the Large Eddy simulation of turbulent flows in complex geometry', *J. Comp. Phys.*, **118**, 24, (1995)

H. van der Ven, 'A family of LES filters with nonuniform filter widths', *Phys. Fluids A*, **7**, 1171, (1995)

A.W. Vreman, B.J. Geurts, J.G.M. Kuerten, 'Realizability conditions for the turbulent stress tensor in Large Eddy Simulation', *JFM*, **278** , 351, (1994a)

A.W. Vreman, B.J. Geurts, J.G.M. Kuerten, 'A priori tests of large eddy simulation of the compressible plane mixing layer', *J.Engg.Math.*, **29**, 299, (1995)

M. Germano, U. Piomelli, P. Moin and W.H. Cabot, 'A dynamic subgrid-scale eddy viscosity model', *Phys. Fluids A*, **3**, 1760, (1991)

A.W. Vreman, B.J. Geurts, J.G.M. Kuerten, 'On the formulation of the dynamic mixed subgrid-scale model', *Phys. Fluids A*, **6**, 4057, (1994b)

# DEVELOPMENT OF A FILTERING OPERATION FOR DYNAMIC SGS MODEL USING THE FINITE ELEMENT METHOD

MARIE OSHIMA, TOSHIO KOBAYASHI, NOBUYUKI TANIGUCHI AND MAKOTO TSUBOKURA
*Institute of Industrial Science, University of Tokyo*
*7-22-1 Roppongi Minato-ku, Tokyo 107, Japan*

**Abstract.** The main focus of this paper is to investigate the filtering operation for the finite element method in a LES (Large Eddy Simulation) with the dynamic sub-grid scale (SGS) model. A turbulent channel flow has been calculated using the present filtering operation based on characteristics of unstructured grids. The results have been compared with those of a finite difference method (FDM), and have verified applicability of the method.

## 1. Introduction

Large eddy simulation has been considered to be more practical numerical tool than DNS (Direct Numerical Simulation) or RANS (Reynolds Averaged Numerical Simulation) to solve a turbulent flow in an engineering problem. An application, however, is still limited to a relatively simple flow geometry such as a channel flow using structured grids. Structured grids are more efficient than unstructured grids in terms of memory usage and CPU. But they have difficulty representing complex domains and capturing complicated flow features, and also become inefficient for flows at high Reynolds numbers, which require fine grids to resolve the viscous sublayer. Since the structured grids do not allow local mesh refinement grids, they need to carry the fine resolution throughout the domain even in the region away from the wall to obtain reasonable accuracy. On the contrary, the unstructured grids have the ability to refine only the near-wall region and also to allocate grids depending on geometries and flow characteristics. This flexibility will contribute not only to accuracy but also to reduction of an overall number of grid points.



*J.-P. Chollet et al. (eds.), Direct and Large-Eddy Simulation II,* 57-68.

The authors have conducted LES using a FEM (Finite Element Methods) (Tsubokura, et. al 1995), and have verified capability of the FEM in the LES of a channel flow. However, the Smagorinsky model has intrinsic drawbacks such as usage of a constant model coefficient throughout a region or usage of a damping function to account for near-wall effects. Since the dynamic SGS model dynamically calculates the model coefficient based on an algebraic identity between two SGS stresses at two different filter levels(Germano, et al. 1991), this model has been applied to the channel flow in the paper.

Unlike the structured grids, the filtering operation cannot be conducted in a recursive manner for the unstructured grids because the location of adjacent grids cannot be systematically determined. Thus, the main focus of this paper is to investigate an effective filtering operation for the finite element method. The filtering operation suggested by Taniguchi (1995) is used and is applied to the channel flow. The results have been compared with those of FDM to verify applicability of the filtering operation.

## 2. Mathematical Formulation

### 2.1. GOVERNING EQUATIONS OF LES AND DYNAMIC SGS MODEL

For the LES of an incompressible flow, the governing equations consist of the filtered continuity and Navier-Stokes equations, and are given by

$$\frac{\partial \overline{u_i}}{\partial x_i} = 0 \quad , \tag{1}$$

$$\frac{\partial \overline{u_i}}{\partial t} + \frac{\partial \left(\overline{u_i}\,\overline{u_j} + \tau_{ij}\right)}{\partial x_j} = -\frac{\partial \overline{p}}{\partial x_j} + \frac{1}{\mathrm{R_e}}\frac{\partial^2 \overline{\mathrm{u_i}}}{\partial x_j^2} \quad , \tag{2}$$

where, u and p are velocity, pressure, and $R_e = UL/\nu$ is Reynolds number($U$: characteristic velocity, $L$: characteristic length, and $\nu$: viscosity). The variable $\overline{x}$ denotes the grid scale (GS) filtered value x. In Equation (2), the effect of the subgrid scale fluctuations is expressed only by a SGS stress $\tau_{ij}$:

$$\tau_{ij} = \overline{u_i u_j} - \overline{u_i}\,\overline{u_j} \quad . \tag{3}$$

The trace of the SGS stress can be included in the pressure, but the rest needs to be approximated for example by a Smagorinsky model as follows:

$$\tau_{ij}^{*} = \tau_{ij} - 1/3\delta_{ij}\tau_{ij} = -2(C_s\overline{\Delta})^2 \left|\overline{S}\right| \overline{S_{ij}} \quad , \tag{4}$$

$$\overline{S_{ij}} = \frac{1}{2}\left(\frac{\partial \overline{u_i}}{\partial x_j} + \frac{\partial \overline{u_j}}{\partial x_i}\right) \quad , S = \sqrt{2S_{ij}S_{ij}} \quad .$$

According to the dynamic model proposed by Germano, et. al (1991), after a test scale (TS) filter is operated in Equation (2), the TS filtered SGS stress yields

$$T_{ij} = \widetilde{\overline{u_i u_j}} - \widetilde{\overline{u_i}}\,\widetilde{\overline{u_j}} \quad , \tag{5}$$

where the variable $\tilde{\bar{x}}$ denotes the TS filtered value.

Now, if the same model such as the Smagorinsky model is applied to the two SGS stresses ($T_{ij}$ and $\tau_{ij}$), unknown values of the Smagorinsky model ($Cs\Delta$) can be 'dynamically' determined. Using the least square method, Lilly (1991) proposed the following equations:

$$(Cs\ \Delta)^2 = -\frac{\left\langle L^*_{ij}\ M_{ij}\right\rangle}{\langle M_{ij}\ M_{ij}\rangle} \quad , \tag{6}$$

$$L^*_{ij} = L_{ij} - \frac{1}{3}\delta_{ij}L_{ij} \quad , \tag{7}$$

$$L_{ij} = T_{ij} - \widetilde{\tau_{ij}} = \widetilde{\overline{u_i}\,\overline{u_j}} - \widetilde{\overline{u_i}}\,\widetilde{\overline{u_j}} \quad .$$

$$M_{ij} = \alpha^2 \left|\widetilde{\overline{S}}\right| \widetilde{\overline{S_{ij}}} - \widetilde{\left|\overline{S}\right| \overline{S_{ij}}} \quad , \tag{8}$$

$$\alpha = \frac{\widetilde{\overline{Cs}}\ \widetilde{\overline{\Delta}}}{\overline{Cs}\ \overline{\Delta}} \quad . \tag{9}$$

The variable $\langle x\rangle$ denotes an averaging operation in terms of x to prevent the value $(Cs\Delta)$ from becoming negative locally. For a channel flow simulation in this paper, the averaging operation is provided in the homogeneous plane parallel to the wall.

## 2.2. THE FILTERING OPERATION

The above model requires an explicit filter operation for a TS value. It has been numerically validated by the authors (Tsubokura, et.al 1995) that the size of a GS filter can be equal to a size of a FEM element in the same manner as a volume filter in the FDM grid. Similarly, a TS filter in the FEM can be formulated by a Taylor expansion with respect to a volume filter as follows:

$$\widetilde{\overline{u_i}} = \overline{u_i} + \frac{\tilde{\Delta}^2}{24}\frac{\partial^2\,\overline{u_i}}{\partial x_j \partial x_j} + O\left(\tilde{\Delta}^4\right) \quad . \tag{10}$$

The test filter width $\tilde{\Delta}$ can be expressed in terms of the width of grid $h$ as:

$$\gamma = \tilde{\Delta}\,/\,h \quad . \tag{11}$$

For unstructured grids like FEM elements, the width of grid $h$ is given by $h = (V_{el})^{1/3}$.

As shown above, the $Cs$ value in Equation (6) is determined from Equations (7)-(9) using the test filtered values from Equations (10) and (11). In this procedure, it is important to note two model coefficients: $\alpha$ and $\gamma$. Generally, $\tilde{\overline{\Delta}}$ cannot be determined explicitly, and also $\tilde{\overline{\Delta}} \neq \tilde{\Delta}$. Thus, the effective filter widths $\overline{\Delta}$, $\tilde{\Delta}$, and $\tilde{\overline{\Delta}}$ have the following relationship:

$$\tilde{\overline{\Delta}}^2 = \tilde{\Delta}^2 + \overline{\Delta}^2 \tag{12}$$

Equation (12) can be derived using both the Gaussian and the box filters to satisfy Equation (10).

If $\widetilde{\overline{Cs}} = \overline{Cs}$ is assumed, the following relationship between the variables $\alpha$ and $\gamma$ can be defined based on the relationship in Equations (11) and (12):

$$\alpha^2 \cong \left(\frac{\tilde{\overline{\Delta}}}{\overline{\Delta}}\right)^2 = 1 + \frac{\tilde{\Delta}^2}{\overline{\Delta}^2} = 1 + \frac{\gamma^2}{\left(\overline{\Delta}/h\right)^2} \quad . \tag{13}$$

Taking the above discussion into account, Equation (10) can be rewritten as

$$\tilde{\overline{u_i}} = \overline{u_i} + \frac{\gamma^2 (V_{el})^{2/3}}{24} \frac{\partial^2 \overline{u_i}}{\partial x_j \partial x_j} \quad . \tag{14}$$

A preliminary FDM calculation for a homogeneous turbulence decay by Tsubokura, et al (1996) indicates that a set of $\gamma^2 = 6$ ($\overline{\Delta} = \sqrt{2}h$) and $\alpha^2 = 4$ shows better agreement with the Smagorinsky model than a priori predicted set of $\gamma^2 = 3$ ($\overline{\Delta} = h$) and $\alpha^2 = 4$ when the second order central schemes are applied to all terms.

Next, let us consider types of test filters for grids with a large aspect ratio. Most researches using the FDM (Akselvoll, et al. 1993 and Zang, et al. 1993) have applied Equation (10) in the one-dimensional form to each coordinate as schematically illustrated in Figure 1-a). This method may not be suitable for a general FEM grid due to an unstructured geometry. An alternative method is to operate Equation (10) directly as shown in Figure 1-b), where the TS filter will be isotropic and independent of a grid. A preliminary calculation by the FDM (Taniguchi 1995) indicates that both types of filters work similarly at least for estimating an averaged value, $(Cs\Delta)$. Therefore, the second type is applied to this FEM research.

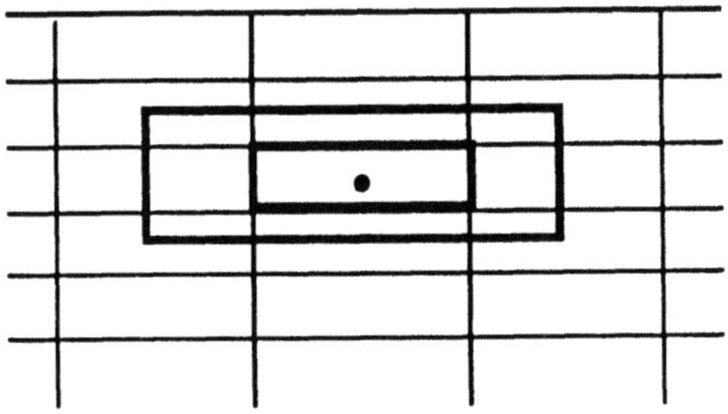

a) Type A

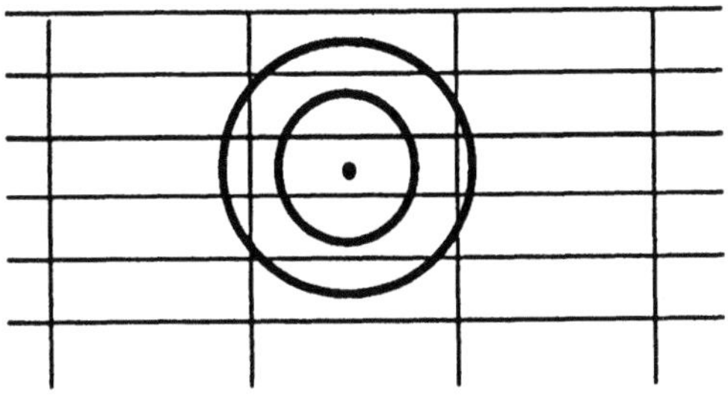

b) Type B

*Figure 1.* Schematic illustration of filtering operation

## 3. Finite Element Formulation

The FEM formulation of the LES governing equations with the dynamic SGS model can be given by:

$$\int_\Omega \frac{\partial\, N_E}{\partial\, x_i}\, \overline{u_{iE}}\; d\Omega = 0 \quad , \tag{15}$$

$$\int_\Omega N_A N_B \frac{\partial \overline{u_{iB}}}{\partial t} d\Omega = -\int_\Omega N_A \left( N_C \overline{u_{jC}} \frac{\partial N_B}{\partial x_j} \overline{u_{jB}} \right) d\Omega + \int_\Omega \frac{\partial\, N_A}{\partial x_j} \delta_{ij} P_A d\Omega$$

$$-\int_\Omega \left( \nu_t + \frac{1}{R_e} \right) \frac{\partial N_A}{\partial x_j} \left( \frac{\partial N_B}{\partial x_j} \overline{u_{iB}} + \frac{\partial N_B}{\partial x_i} \overline{u_{jB}} \right) d\Omega + \int_\Gamma N_A\, h_{iA} d\Gamma \quad , \tag{16}$$

$$\nu_t = (C_s \Delta)^2 \left( 2 \overline{S_{ij}} \overline{S_{ij}} \right)^{1/2} \quad ,$$

$$h_i = \left( -\delta_{ij} P + \left( \nu_t + \frac{1}{R_e} \right) \left( \frac{\partial u_i}{\partial x_j} + \frac{\partial u_j}{\partial x_i} \right) \right) n_j$$

where $N_A$, $N_B$, $N_C$, and $N_E$ denote the basis function, and the subscripts A, B and C are the nodes, and E is the element. Here, the above equations are derived from the weak forms of Equations (1) and (2), which are obtained by integrating the weighted residual equations of (1) and (2) by parts.

In the analysis, an eight-node brick element is used. The velocity is approximated by a trilinear basis function while the pressure is approximated by a piecewise constant.

The MAC method is used for the coupling algorithm to advance the velocity and the pressure in time. This approach leads to Poisson equation for the pressure, and is solved by the Conjugate Gradient method. A 2nd-order Adams-Bashforth method is used for the time integration.

In order to obtain a TS filtered value, a similar method to derive the FEM formulation of Equations (15) and (16) is applied to Equation (14). As a result, the following FEM formulation is derived:

$$\int_\Omega N_A N_B \widetilde{\overline{u_{iB}}} d\Omega = \int_\Omega N_A N_B \overline{u_{iB}} d\Omega - \int_\Omega \frac{\gamma^2 (V_{el})^{2/3}}{24} \frac{\partial N_A}{\partial x_j} \frac{\partial N_B}{\partial x_j} \overline{u_{iB}} d\Omega + \int_\Gamma N_A \frac{\partial N_B}{\partial x_j} \overline{u_{iB}} d\Gamma \quad . \tag{17}$$

The above equation is also used to obtain $\widetilde{|\overline{S}| \overline{S_{ij}}}$ and $\widetilde{\overline{u_i}\,\overline{u_j}}$. Since the second term on the left hand side of Equation (17) can be formulated like the diffusion term in Equation (16), extra coding is not necessary.

## 4. Numerical Results

### 4.1. ANALYSIS MODEL

The analysis has been conducted for a 3.2H x 1.0H x 1.6H channel flow for $R_{e\tau}(= u_\tau \delta/\nu : \delta$ is half the channel height)=180 as illustrated in Figure 2. The analysis region is divided into 32 x 40 x 32 elements with the uniform length in the streamwise (x) and the spanwise directions (z) and non-uniform length in the wall direction (y). The periodic boundary conditions are imposed on the boundaries of the streamwise and spanwise directions while a no slip condition is imposed on the boundaries of the wall direction. The computation result of a fully developed turbulent channel flow based on the Smagorinsky model is used as an initial condition.

### 4.2. NUMERICAL RESULTS

The results have been compared with those of the FDM analyses, in which the channel with 3.2H x 1.0H x 1.6H is divided into 32 x 64 x 32. The convection and the diffusion terms are discretized with a 2nd order central difference scheme, and a 2nd order Adams-Bashforth method. The coupling method is SMAC, and Poisson equation for the pressure is solved by ICCG method (Tsubokura, et al. 1996).

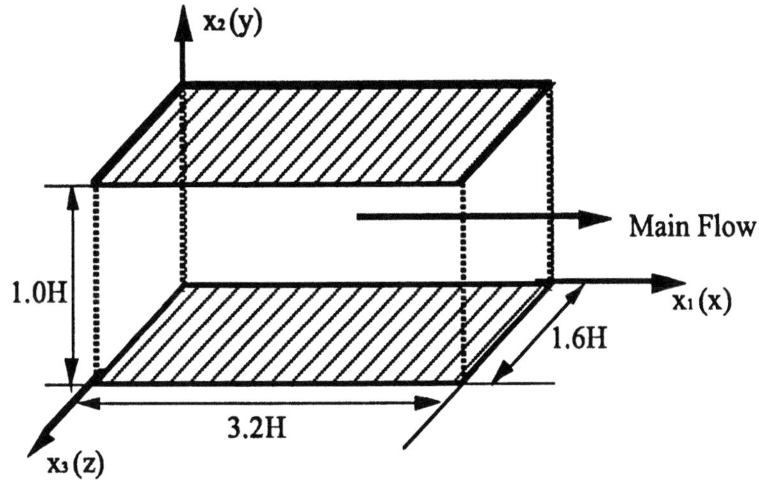


*Figure 2.* Analysis model

Table 1 summarizes types of FDM and FEM used in the paper. The

TABLE 1. Summary of the Filtering Operation

| Case | Discretization | Filter Type | Evaluation of $\widetilde{\overline{S_{ij}}}$ | $\alpha$ | $\gamma$ |
|---|---|---|---|---|---|
| DAT1 | FDM | A | Test Filter Space | 2 | $\sqrt{3}$ |
| DBT1 | FDM | B | Test Filter Space | 2 | $\sqrt{3}$ |
| DAT2 | FDM | A | Test Filter Space | 2 | $\sqrt{6}$ |
| DBT2 | FDM | B | Test Filter Space | 2 | $\sqrt{6}$ |
| DBG2 | FDM | B | Grid Filter Space | 2 | $\sqrt{6}$ |
| EBG2 | FEM | B | Grid Filter Space | 2 | $\sqrt{6}$ |

first letter of cases, D or E corresponds to FDM or FEM, the second letter, A or B corresponds to Type A or B, The third letter, T or G corresponds to Test Filter Space or Grid Space filter, and finally the last number, 1 or 2 corresponds to $\sqrt{3}$ or $\sqrt{6}$. The test filtered values $\widetilde{\overline{S_{ij}}}$ are evaluated either in the test filter space or in the grid filter space as schematically illustrated in Figure 3.

The statistics of the FDM (Cases DAT1, DBT1, DAT2, DBT2, and DBG2) is averaged over 20,000 time increments with a size of time step $\Delta t$=0.0005 after 40,000 time increments to omit effects of an initial conditions. The statistics of the FEM (Case EBG2) is averaged over 20,000 time increments with a size of time step $\Delta t$=0.0005 after the results of the Smagorinsky model is used as an initial condition.

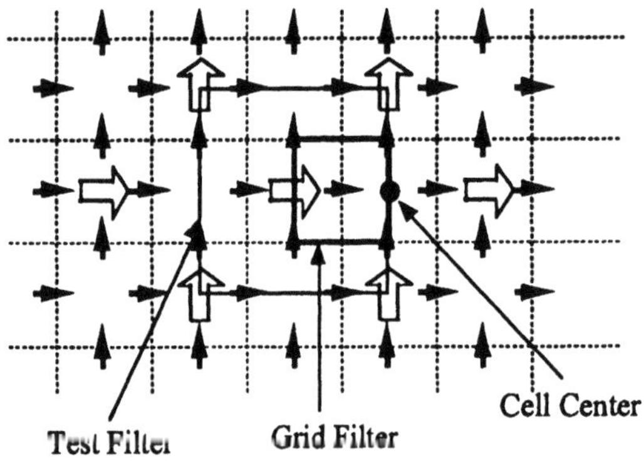


*Figure 3.* Definition of filtering space for the test filtered values

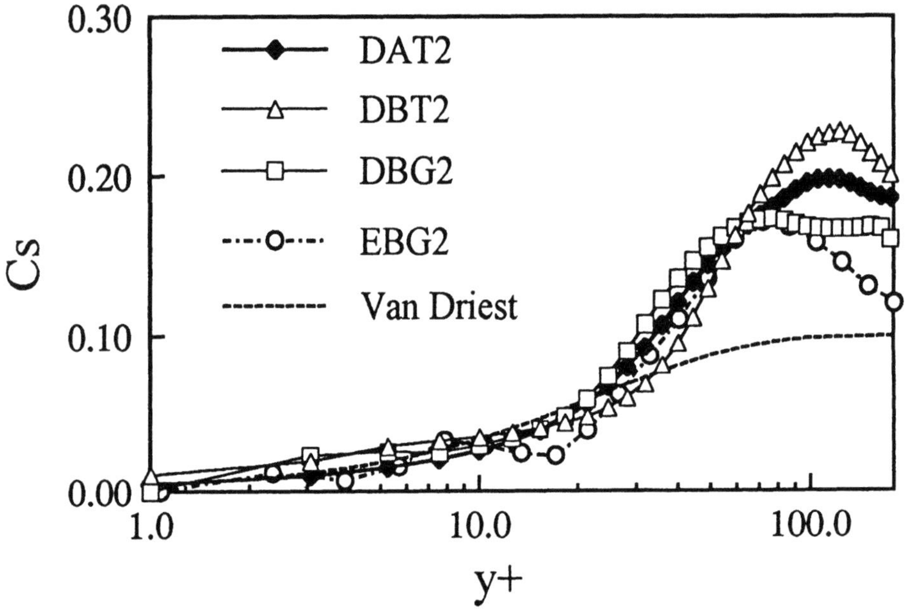


*Figure 4.* The profile of $Cs$ values for DAT2, DBT2, DBG2 and EBG2

Figures 4 and 5 describe the $Cs$ distribution compared with that of the FDM analyses and the Van Driest exponential damping function, which is given by

$$f = 1 - \exp\left(-\frac{y^+}{25}\right) \tag{18}$$

As shown in Figures 4 and 5, the $Cs$ profile of the FEM analysis (Case EBG2) correlates well with that of the FDM analyses with $\gamma = \sqrt{3}$ (Cases DAT1 and DBT1) rather than that with the same $\gamma = \sqrt{6}$ (Cases DAT2

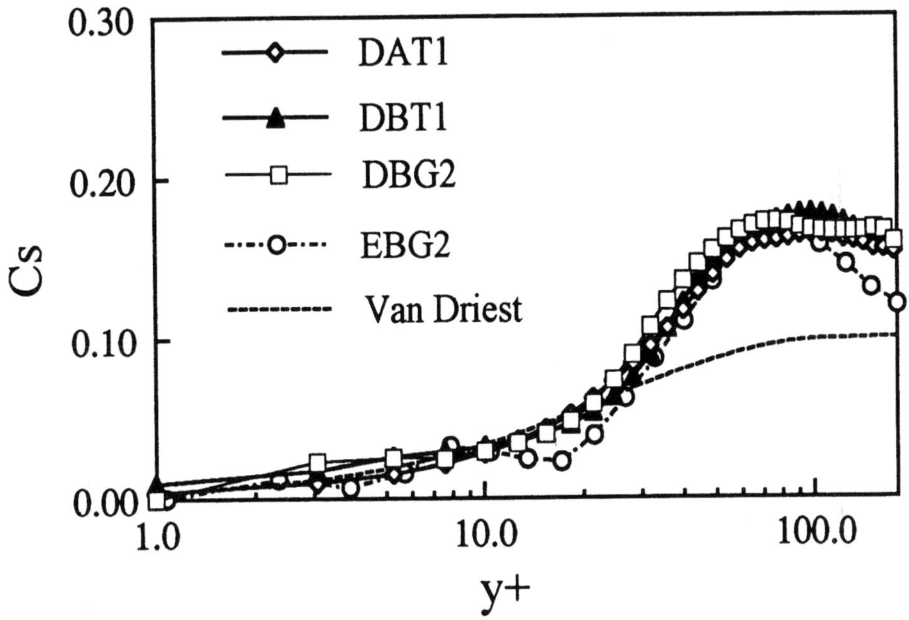


*Figure 5.* The profile of $Cs$ values for DAT1, DBT1, DBG2 and EBG2

and DBT2). On the other hand, under the same $\gamma = \sqrt{6}$, $Cs$ profile of Case DBG2 correlates well with that of Case EBG2. These discrepancies arise due to differences in evaluating test filtered values. As shown in Figure 3, the test filtered values of $\widetilde{\overline{S_{ij}}}$ in Cases DAT1, DBT1, DAT2, and DBT2 are evaluated using the derivative values in the test filter space while those of Cases DBG2 and EBG2 are evaluated using the derivative values in the grid space.

If the unstructured grids are used, it is difficult to evaluate the test filtered values of $\widetilde{\overline{S_{ij}}}$ in the test filter space. Thus, evaluation in the grid space is more suitable. However, it is important to note that the relationship between $\alpha$ and $\gamma$, which is defined in Equation (13) is assumed under $\widetilde{\overline{Cs}} = \overline{Cs}$. Strictly speaking, the assumption of $\widetilde{\overline{Cs}} = \overline{Cs}$ is satisfied if the test filtered values of $\widetilde{\overline{S_{ij}}}$ is evaluated in the test filter space with the structured grids. Thus, in case of the unstructured analysis, $\alpha$ and $\gamma$ may become independent, and both of them may require optimization.

Figure 6 describes the profile of the mean velocities in Case EBG2 compared with that of Case DBT2, DNS and the law of the wall (the linear law and the log law). Although the mean velocities are underestimated comparing to those of FDM analyses, the FEM result correlates well with the law of the wall and DNS. In general, the FDM analysis shows that the mean velocity profile are overestimated and is improved if the higher dicretization scheme is used or the finer refinement is used. On the contrary, the FEM analysis shows that despite coarse grid refinement, mean velocity profile is

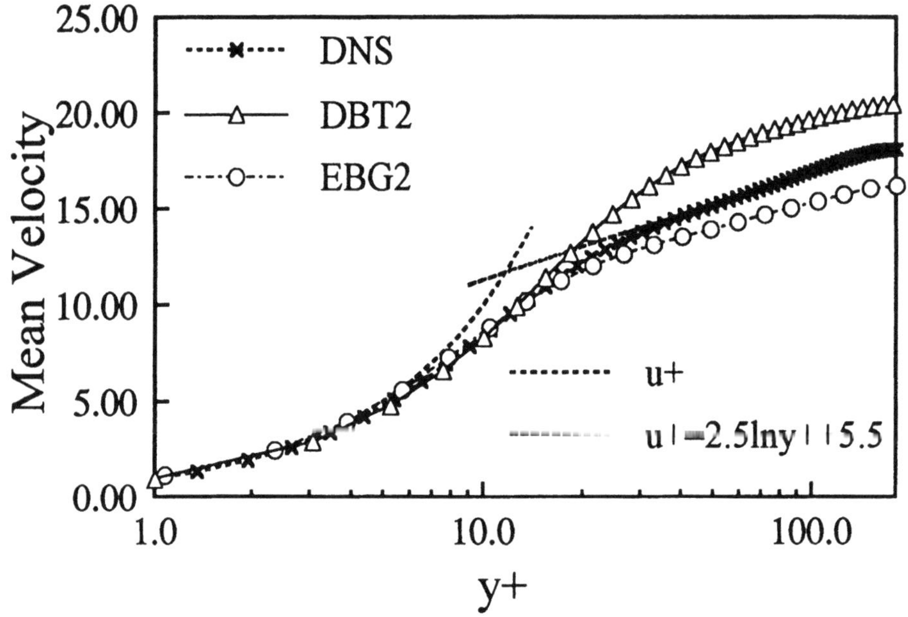


*Figure 6.* The profile of mean velocities

underestimated.

Figures 7 - 9 describe the profiles of the turbulent intensity in the streamwise, the normal to the wall and spanwise directions compared with those of Case DBT2 and DNS. Regarding the profile in the streamwise direction, a significant difference between FDM and FEM has not been observed. Since the resolution in the spanwise direction is not enough for both cases, both profiles of the turbulent intensity in the streamwise direction are overestimated. On the other hand, the turbulent intensity in the normal to the wall direction by the FEM analysis agree with that of DNS as shown in Figure 8, particularly near the wall region compared with that by FDM. The profile of FDM is smaller than that of DNS throughout the region while that of FDM started to deviate toward the center. This is caused again by coarse refinement in the direction normal to the wall such that 40 grids are used in the FEM analysis as opposed to 64 grids in the FDM analysis. Regarding the turbulent intensity in the spanwise direction, the profile of FEM is overestimated while that of FDM is underestimated as shown in Figure 9. In order to clarify the reasons of the discrepancies, the further investigation and calculations will be carried out.

## 5. Conclusions

The filtering operation for the FEM has been investigated in order to apply the finite element LES with the dynamic SGS model to a turbulent channel flow. The simulation has been conducted with the isotropic filter (Type B)

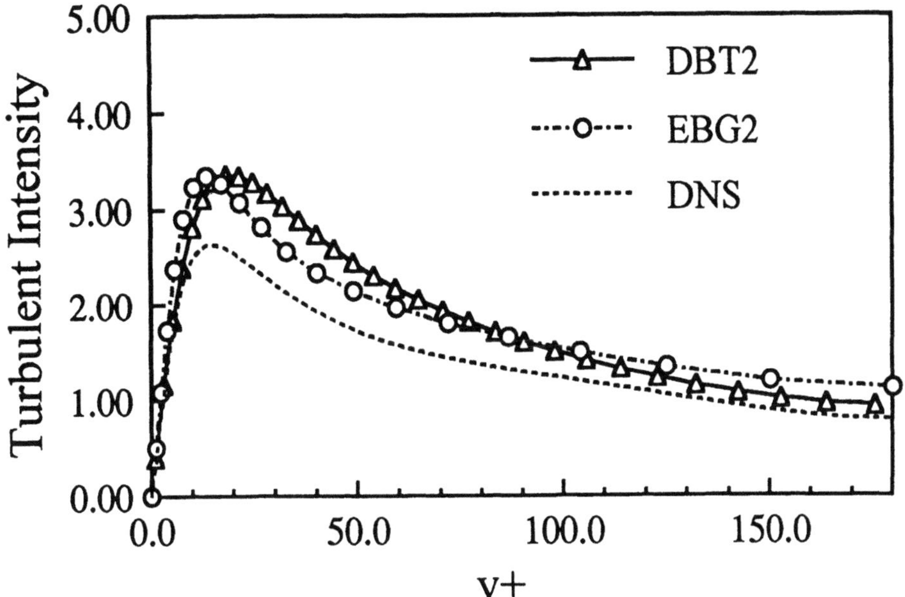


*Figure 7.* The profile of the turbulent intensity in the streamwise direction

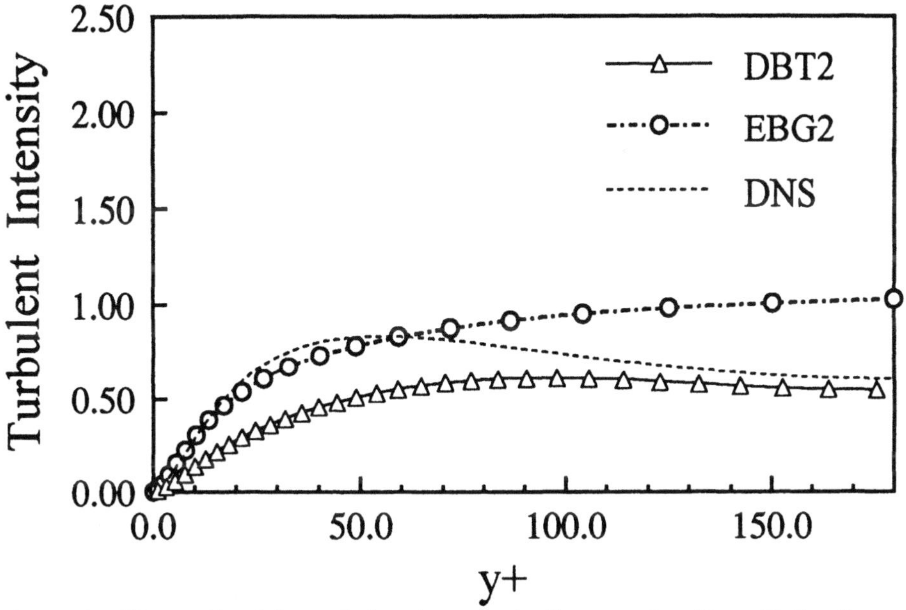


*Figure 8.* The profile of the turbulent intensity in the normal to the wall direction

and the parameters $\alpha^2 = 4$ and $\gamma^2 = 6$.

The results have been compared to those with different types of filtering operation using the FDM. It is found that if the test filtered values of $\widetilde{\overline{S_{ij}}}$ are evaluated in the grid space, the similar $Cs$ profiles are obtained under the same parameters $\alpha^2 = 4$ and $\gamma^2 = 6$ regardless of the discretization scheme. It is also observed that the $Cs$ profile by the FEM correlates well

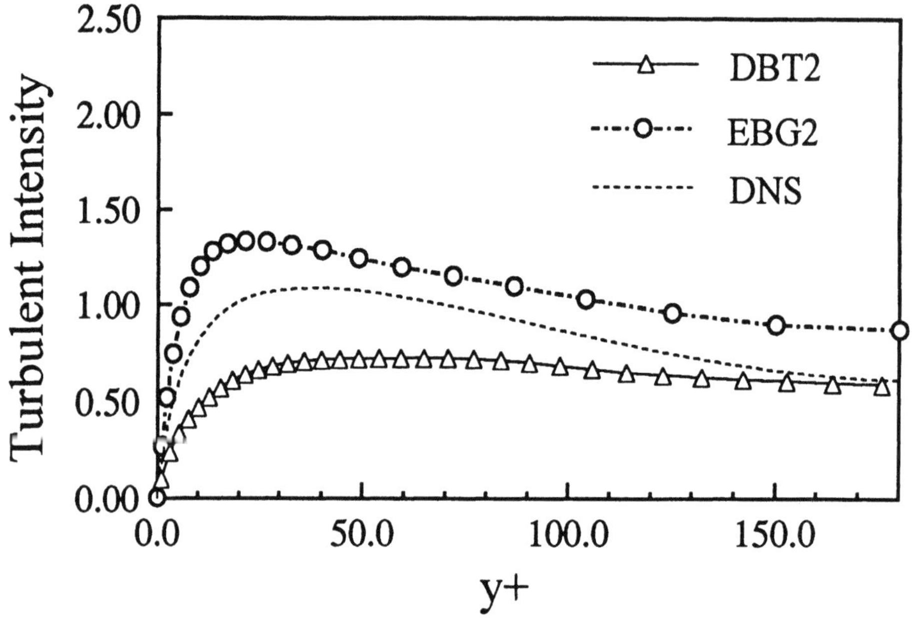


*Figure 9.* The profile of the turbulent intensity in the spanwise direction

with those by the FDM analyses, in which the test filtered values of $\widetilde{\overline{S_{ij}}}$ are evaluated in the test filter space with the parameters $\alpha^2 = 4$ and $\gamma^2 = 3$.

In the future, the unstructured analysis will be carried out to investigate the filtering operation, particularly to clarify the relationship among the grid space, the grid filter width and the effective test filter width.

## References


Akselvoll K. and Moin P. 1993 Large eddy simulation of a backward facing step flow, *Engineering Turbulence Modelling and Experiments 2, Elsevier Pub.*, 315-324.

Germano M., Piomelli U., Moin P. and Cabot W.H. 1991 A dynamic subgrid-scale eddy viscosity model,*phys. of Fluids* **A3(7)**, 1760-1765.

Lilly D.K. 1991 A proposed modification on the Germano subgrid-scale closure method, *Phys. of Fluids*, **A4(3)**, 633-635.

Taniguchi N.1995 Dynamic SGS model by finite difference method, *Seisan-kennkyu IIS report, University of Tokyo*, **47-2**,120-124 (in Japanese).

Tsubokura, M.,Kobayashi, T., Taniguchi, N. and Oshima, M. 1995 Large eddy simulation of turbulent channel flow using an unstructured mesh,*ASME, FED,Numerical Developments in CFD*, **215**, 81-87.

Tsubokura M., Kobayashi., T., and Taniguchi, N. 1996 An investigation of improved dynamic sub-grid scale models by simulating a turbulent channel flow, *JSME*, **62**, 3015-3022(in Japanese).

Zang Y., Street R.L. and Koseff J.R. 1993 A dynamic mixed subgrid-scale model and its application to turbulent recirculating flows, *Phys. of Fluids*, **A5(12)**, 3186-3196.

# INVESTIGATION OF THE SELF-CONSISTENCY OF THE SMAGORINSKY CONSTANT AND THE VALUE OF THE ROTTA PARAMETER AT HIGH REYNOLDS NUMBERS

K. ALVELIUS, M. HALLBÄCK AND A. V. JOHANSSON
*Department of Mechanics, Royal Institute of Technology,*
*S-100 44 Stockholm, Sweden*

**Abstract.** The aim of the first part of this work is to investigate if the value of the Kolmogorov constant, $\alpha$, predicted by LES is consistent with the assumed value of $\alpha$ that was used to specify the Smagorinsky constant, $C_s$, in the LES, and to determine the appropriate value of $C_s$ for decaying homogeneous turbulence at high Reynolds numbers. In the second part LES of axisymmetric decaying turbulence is used to calibrate the Rotta parameter, $C_1$, in the Rotta model for the slow pressure-strain rate.

## 1. The Smagorinsky Model

The filtered Navier-Stokes equations for incompressible flow are

$$\frac{\partial \bar{u}_i}{\partial t} + \bar{u}_j \frac{\partial \bar{u}_i}{\partial x_j} = -\frac{1}{\rho}\frac{\partial \bar{p}}{\partial x_i} + \nu \frac{\partial^2 \bar{u}_i}{\partial x_j \partial x_j} - \frac{\partial \tau_{ij}}{\partial x_j} \tag{1}$$

where an over-bar denotes a filtered quantity and $\tau_{ij} = \overline{u_i u_j} - \bar{u}_i \bar{u}_j$ is the sub-grid scale (SGS) stress which has to be modelled. The Smagorinsky model (1963) for the SGS stress reads

$$\tau_{ij} = \frac{1}{3}\tau_{kk}\delta_{ij} - 2\nu_T \bar{s}_{ij} \tag{2}$$

where $\nu_T = (C_s L)^2 (2\bar{s}_{pq}\bar{s}_{pq})^{1/2}$ and $\bar{S}_{ij} = (\bar{u}_{i,j} + \bar{u}_{j,i})/2$. $L$ is a filter width and in the case of spectral filtering we define $L = \pi/k_c$, where $k_c$ is the cut-off wavenumber in the Fourier representation. The trace $\tau_{kk}$ is put into



*J.-P. Chollet et al. (eds.), Direct and Large-Eddy Simulation II, 69-80.*

the pressure according to $\bar{q} \equiv \bar{p}/\rho + \tau_{kk}/3$. The modelled form of (1) is then given by

$$\frac{\partial \bar{u}_i}{\partial t} + \bar{u}_j \frac{\partial \bar{u}_i}{\partial x_j} = -\frac{\partial \bar{q}}{\partial x_i} + \nu \frac{\partial^2 \bar{u}_i}{\partial x_j \partial x_j} - \frac{\partial \tau^a_{ij}}{\partial x_j} \tag{3}$$

where $\tau^a_{ij} \equiv \tau_{ij} - \tau_{kk}\delta_{ij}/3 = -2\nu_T \bar{s}_{ij}$.

## 2. The Energy Spectrum

Let $\langle \cdot \rangle_V$ denote a volume average. The two-point-correlation spectrum tensor $\phi^f_{ij}$ is then defined through the relation

$$\langle \bar{u}_i(\mathbf{x})\bar{u}_j(\mathbf{x}+\mathbf{r})\rangle_V = \int \phi^f_{ij}(\mathbf{k}) e^{i\mathbf{k}\cdot\mathbf{r}}\, d^3k \tag{4}$$

The energy spectral function of the filtered field is defined as $E^f(k) \equiv \frac{1}{2}\iint_S \phi^f_{ii}(\mathbf{k})\, dS$, where $S$ is a spherical surface at distance $k$ from the origin. Equation (4) gives the relation $\phi^f_{ij} = \widehat{\bar{u}}_j(\mathbf{k})\widehat{\bar{u}}_i(\mathbf{k})^*$ where a caret, $\widehat{\cdot}$, denotes a Fourier transformed quantity. This yields that $E^f(k)$ for $k < k_c$, under the assumption that the filter function in spectral space, $\widehat{G}$, only is a function of $k$ in that range, can be written as

$$E^f(k) = \frac{1}{2}\int_S \widehat{\bar{u}}_i\widehat{\bar{u}}_i^*\, dS = \frac{1}{2}\int_S \widehat{G}(k)\widehat{G}(k)^*\widehat{u}_i\widehat{u}_i^*\, dS = \\ |\widehat{G}(k)|^2 \frac{1}{2}\int_S \widehat{u}_i\widehat{u}_i^*\, dS = |\widehat{G}(k)|^2 E(k) \tag{5}$$

where $E(k)$ is the energy spectrum of the complete (unfiltered) field. For the spectral cut-off filter

$$\widehat{G}(\mathbf{k}) = \begin{cases} 1 & \text{if } k_i \le k_c \;\; i = 1,2,3 \\ 0 & \text{otherwise} \end{cases} \tag{6}$$

$E^f(k) = E(k)$ if $k < k_c$. In the present LES code the average of $\widehat{\bar{u}}_i\widehat{\bar{u}}_i^*$ is calculated over a spherical shell with radius $k$. This average is then multiplied with the shell volume to give $E(k)\, dk$ from which the energy spectrum is obtained. This enables the spectrum in the corners of the spectral domain to be captured correctly.

## 3. The Self-Consistency of the Smagorinsky Constant

By integrating an appropriate model spectrum over the resolved spectral domain one may obtain an estimate of the Smagorinsky constant. For high

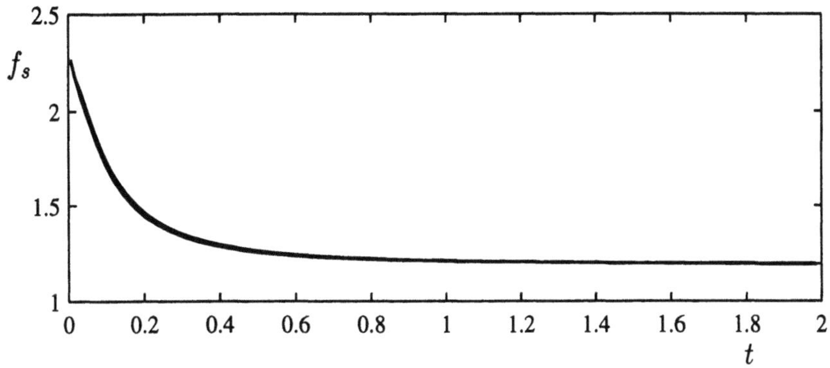


*Figure 1.* The development of $f_s$ with time for several different LES realizations

Reynolds numbers we have that the dissipation of the filtered field, $\epsilon^f$, is approximately equal to the dissipation by the sub-grid scales, $\epsilon^{sgs}$,

$$\epsilon^f = \epsilon^{res} + \epsilon^{sgs} \approx \epsilon^{sgs} \gg \epsilon^{res} \tag{7}$$

Here $\epsilon^{res} = \langle 2\nu \bar{s}_{pq}\bar{s}_{pq}\rangle$ is the resolved dissipation and $\langle\cdot\rangle$ denotes ensemble averaging. Using the Smagorinsky model the dissipation by the sub-grid scales is given by $\epsilon^{sgs} = \langle 2\nu_T \bar{s}_{pq}\bar{s}_{pq}\rangle$. With the filter width taken as $L = \pi/k_c$ one then obtains

$$\epsilon^f = C_s^2(\frac{\pi}{k_c})^2\langle(2\bar{s}_{pq}\bar{s}_{pq})^{3/2}\rangle = C_s^2(\frac{\pi}{k_c})^2 f_s\langle\bar{s}^2\rangle^{3/2} \tag{8}$$

where $\bar{s} \equiv (2\bar{s}_{pq}\bar{s}_{pq})^{1/2}$ and $f_s \equiv \langle\bar{s}^3\rangle/\langle\bar{s}^2\rangle^{3/2}$. In homogeneous turbulence

$$\langle\bar{s}^2\rangle = \int \widehat{G}(\mathbf{k})\widehat{G}(\mathbf{k})^*\langle\widehat{u}_i\widehat{u}_i^*\rangle k^2 d^3k \tag{9}$$

and isotropy gives

$$\langle\bar{s}^2\rangle = \int \widehat{G}(\mathbf{k})\widehat{G}(\mathbf{k})^*\frac{E(k)}{2\pi}d^3k \tag{10}$$

Equation (8) and (10) together with a model spectrum may be used to formulate the estimate of $C_s$. The crucial points now are how the integral is evaluated and the estimation of $f_s$. The usual way of estimating the Smagorinsky constant has been to integrate the model spectrum over a spherical volume and to let $f_s$ equal unity. LES then gives a value of $\alpha$ which is inconsistent with the one which was used to specify $C_s$.

### 3.1. SECOND ORDER MOMENTS VERSUS THIRD ORDER MOMENTS

The factor $f_s$ can be calculated in the LES and was found to deviate very little from 1.2. In figure 1 the development of $f_s$ with time for several different LES realizations at different Reynolds numbers are shown.

### 3.2. EVALUATION OF THE INTEGRAL

The evaluation of the integral determining $\langle \bar{s}^2 \rangle$ will be based on a high Reynolds number model spectrum of the form

$$E(k) = \begin{cases} E_1(k) = Ak^a & \text{if } k < k_p \\ E_2(k) = \alpha\epsilon^{2/3}k^{-5/3} & \text{if } k > k_p \end{cases} \tag{11}$$

where $A, a, \alpha$ are constants and $\epsilon, k_p$ may vary with time. $\epsilon$ is the energy transfer rate through the spectrum, which for high Reynolds numbers equals the total dissipation rate. The present spectral filter function gives that the

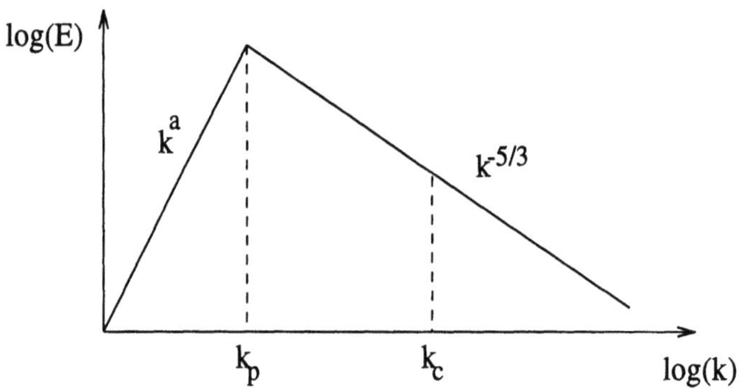


*Figure 2.* The model energy spectrum

integration is to be taken over a cubic volume and not a spherical volume. This is done by integrating the spectrum $E$ over the sphere $k < k_c$ and adding an extra term from the $E_2$ spectrum at the corners of the cube.

$$\langle \bar{s}^2 \rangle = \int_{k<k_c} \frac{E(k)}{2\pi} d^3k + \int_{corners} \frac{E_2(k)}{2\pi} d^3k \tag{12}$$

Matching the energy spectra at $k_p$ and introducing the non-dimensional wavenumber $\kappa = k/k_p$ gives that $E(k) = \alpha\epsilon^{2/3}k_p^{-5/3}E'(\kappa)$, where

$$E'(\kappa) = \begin{cases} \kappa^a & \text{if } \kappa < 1 \\ \kappa^{-5/3} & \text{if } \kappa > 1 \end{cases} \tag{13}$$

From this we obtain that

$$\int_{k<k_c} \frac{E(k)}{2\pi} d^3k = 2\alpha\epsilon^{2/3}k_p^{4/3} \int_0^{\kappa_c} \kappa^2 E'(\kappa) d\kappa = \frac{3}{2}\alpha\epsilon^{2/3}k_c^{4/3} f_1(\kappa_c) \tag{14}$$

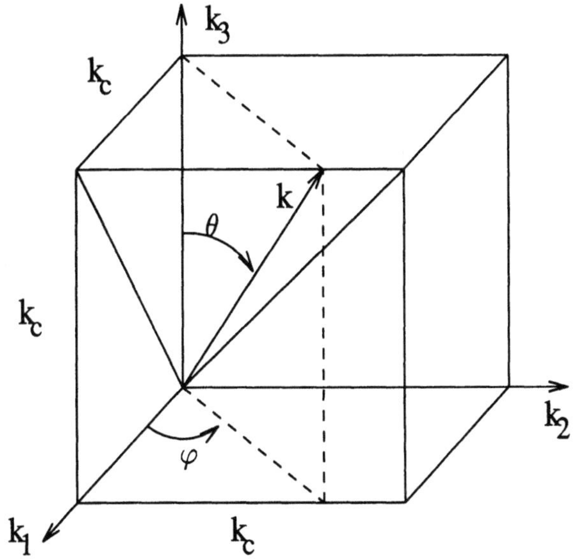


*Figure 3.* Upper integration limits

where $f_1(\kappa_c) = 1 - \kappa_c^{-4/3}(1 - \frac{4/3}{3+a})$. The integral over the eight corners is evaluated by use of spherical coordinates

$$\int_{corners} \frac{E_2(k)}{2\pi} d^3k = \frac{4}{\pi}\alpha\epsilon^{2/3} \int_\varphi \int_\theta \int_k k^{-5/3} k^2 \sin\theta dk d\theta d\varphi \tag{15}$$

where the integral on the right-hand side is to be evaluated over one corner. Here the spherical coordinates have been introduced as $k_1 = k \sin\theta\cos\varphi$, $k_2 = k \sin\theta\sin\varphi$, $k_3 = k \cos\theta$. Due to symmetry it is only necessary to integrate over one part out of six, which is formed by the vectors $\mathbf{k}$ that intersect the left triangle of the top surface, see figure 3. The integration limits are: $k \in [k_c, \frac{k_c}{\cos\theta}]$, $\tan\theta \in [0, \frac{1}{\cos\varphi}]$, $\varphi \in [0, \frac{\pi}{4}]$. This gives

$$\int_{corners} \frac{E_2(k)}{2\pi} d^3k \approx 0.465\alpha\epsilon^{2/3} k_c^{4/3} \tag{16}$$

where the last integration, with respect to $\varphi$, has to be carried out numerically. The total expression for the integral becomes

$$\langle \bar{s}^2 \rangle = (f_1(\kappa_c) + 0.310)\frac{3}{2}\alpha\epsilon^{2/3} k_c^{4/3} \tag{17}$$

Let $f_E = f_1(\kappa_c) + 0.310$ and use the expression (8) for the dissipation

$$\epsilon^f = C_s^2(\frac{\pi}{k_c})^2 f_s f_E^{3/2} (\frac{3\alpha}{2})^{3/2} \epsilon k_c^2 \tag{18}$$

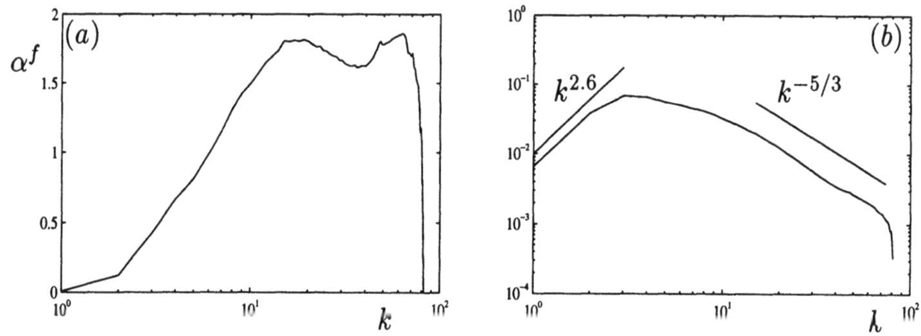


*Figure 4.* Left: the 'local Kolmogorov constant' $E(k)k^{5/3}/(\epsilon^f)^{2/3}$, $\alpha^f = 1.67$ at $k = 30$. Right: $E(k)/(\epsilon^f)^{2/3}$ versus the wavenumber, $k_p \approx 4$. The cut-off wavenumber is $k_c = 47$.

Introducing a new quantity $\alpha^f$ defined by

$$\alpha^f \equiv \alpha(\frac{\epsilon}{\epsilon^f})^{2/3} \tag{19}$$

one finally gets an expression for the Smagorinsky constant

$$C_s = \frac{1}{f_E^{3/4} f_s^{1/2}} \left[ (\frac{2}{3\alpha^f})^{3/4} \frac{1}{\pi} \right] \tag{20}$$

It is $\alpha^f$ that is obtained from the LES results if $\epsilon^f$ is used for the normalization of the energy spectrum. If $\alpha^f = \alpha$, the square brackets contain the original expression for the Smagorinsky constant suggested by Lilly (1967).

### 3.3. RESULTS

Initial isotropic turbulence fields were generated using random phases with the model energy spectrum described earlier. These velocity fields were then relaxed using LES. The LES is solved with a spectral method using FFT's to calculate the nonlinear terms. Results from two LES computations on a $96^3$ cube are shown in figures 4 and 5. For each case ten realizations were carried out in order to get sufficient statistics. The difference between them is the slope of the spectrum at low wavenumbers. The Smagorinsky constant has been chosen so that the energy spectrum gets a distinct inertial region with a $k^{-5/3}$ slope. In both runs the turbulent Reynolds number ($Re_T \equiv (2K^f)^2/\nu\epsilon^f$) varies between 200000 and 40000, where $K^f \equiv \langle \bar{u}_i \bar{u}_i \rangle /2$, and the value $C_s = 0.12$ has been used. Figure 4b gives that $\kappa_c = 12$ and $a = 2.6$. If we take $\alpha^f = 1.67$ from the resolved part of the inertial range of

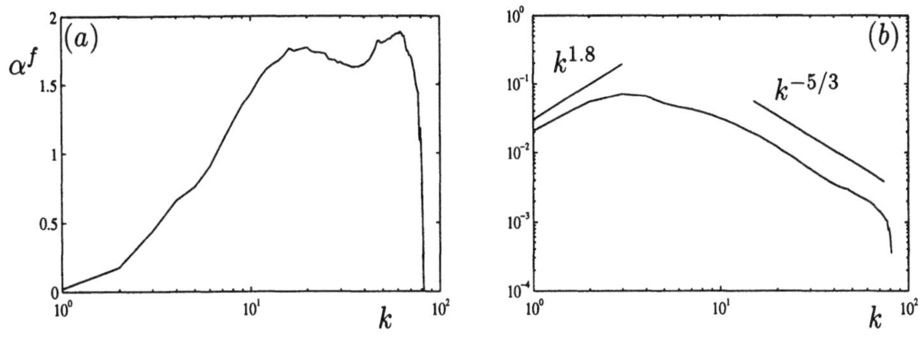


*Figure 5.* Left: the 'local Kolmogorov constant' $E(k)k^{5/3}/(\epsilon^f)^{2/3}$, $\alpha^f = 1.66$ at $k = 30$. Right: $E(k)/(\epsilon^f)^{2/3}$ versus the wavenumber, $k_p \approx 4$, $k_c = 47$.

the LES predicted spectrum in figure 4a we get $f_E = 1.282$ which in turn from equation (20) gives that $C_s = 0.121$. Figure 5b gives that $\kappa_c = 12$ and $a = 1.8$. The value $\alpha^f = 1.66$ (figure 5a) yields that $f_E = 1.284$ and hence $C_s = 0.122$. In both cases the value $f_s = 1.2$ was used for the estimation of $C_s$. The correction factor $f_E^{-3/4} f_s^{-1/2}$ to the Lilly expression is about 0.76 in both cases. For $\kappa_c = 12$ we get that $f_1(\kappa_c)$ differs from unity by a few per cent only. The main contribution to $f_E$'s deviation from unity is due to the difference in integration domain, which is in agreement with the results of Scotti et al. (1993).

To conclude, the agreement between equation (20) and the results of the LES is quite satisfactory. Hence self-consistency in this respect is demonstrated. The appropriate value of the Smagorinsky constant seems to be $C_s = 0.12$ when a cubic spectral cut-off filter is used, at least for isotropic decaying turbulence at high Reynolds numbers. Equation (19) gives the Kolmogorov constant $\alpha$ in terms of $\alpha^f$ and the ratio between the dissipation of the total and filtered fields. An estimate of this ratio is given by equations (23) and (35) below. The value $\alpha^f = 1.67$ in the case above gives $\alpha = 1.45$. However, since the Kolmogorov constant is evaluated from a spectrum whose shape depends strongly on the SGS model the results concerning its value should be viewed cautiously.

## 4. Evaluation of the Rotta Parameter

The Rotta model (1951) for the slow pressure-strain-rate correlation ($\Pi_{ij} \equiv 2\langle ps_{ij}\rangle/\rho$) reads

$$\Pi_{ij} = -C_1 \epsilon a_{ij} \tag{21}$$

where $a_{ij}$ is the Reynolds-stress-anisotropy tensor and $C_1$ is the Rotta parameter which has been found to be a function of the turbulence Reynolds number. The Reynolds number dependence has been suggested by Hallbäck et al. (1993) to be

$$C_1 = C_{1\infty}\left(\sqrt{\frac{2Re_T}{D_1} + \left(\frac{Re_T}{D_1}\right)^2} - \frac{Re_T}{D_1}\right) \tag{22}$$

The expression was calibrated at low Reynolds numbers ($Re_T < 300$) against DNS data which gave $D_1 = 1307.6$ and $C_{1\infty} = 2.58$. Wind tunnel experiments at higher Reynolds numbers by Sjögren and Johansson (1995) has supported the calibration, although possibly with a somewhat lower value of $C_{1\infty}$.

Here we wish to use LES to investigate this dependence. However, an LES only gives filtered quantities, hence it is only possible to calculate the pressure strain-rate from the filtered velocity and pressure fields, $\Pi^f_{ij} \equiv 2\langle \bar{p}\bar{s}_{ij}\rangle/\rho$, the dissipation of the filtered field, $\epsilon^f$, and the anisotropy of the resolved scales, $a^f_{ij}$.

### 4.1. CORRECTIONS TO THE DISSIPATION AND ANISOTROPY

In homogeneous decaying turbulence the rate of change of turbulent kinetic energy is set by the dissipation alone and the relations $\epsilon^f = -\dot{K}^f$ and $\epsilon = -\dot{K}$, where $K \equiv \langle u_i u_i\rangle/2$, are valid. To obtain the true dissipation the dissipation calculated from LES may be corrected according to

$$\epsilon = \frac{\dot{K}}{\dot{K}^f}\epsilon^f \tag{23}$$

provided that we have an estimate of $\dot{K}/\dot{K}^f$. For the SGS-stress tensor we have $\langle \tau_{ij}\rangle = \langle \overline{u_i u_j}\rangle - \langle \bar{u}_i\bar{u}_j\rangle$. Introducing $\tau^a_{ij}$ and noticing that the ensemble average of a filtered quantity in homogeneous turbulence equals the ensemble average of the quantity itself we obtain

$$\langle u_i u_j\rangle = \langle \bar{u}_i\bar{u}_j\rangle + \langle \tau^a_{ij}\rangle + \frac{2}{3}K^{sgs}\delta_{ij} \tag{24}$$

where $K^{sgs} \equiv \langle \tau_{kk}\rangle/2 = K - K^f$. The SGS-stress tensor represents the effects from the small scales which are expected to be close to isotropic. If $\langle \tau^a_{ij}\rangle$ is assumed to be zero we get the total anisotropy

$$a_{ij} \equiv \frac{\langle u_i u_j\rangle}{K} - \frac{2}{3}\delta_{ij} = \frac{\langle \bar{u}_i\bar{u}_j\rangle + \frac{2}{3}K^{sgs}\delta_{ij}}{K} - \frac{2}{3}\delta_{ij} \tag{25}$$

Using an equivalent definition of $a^f_{ij}$ as in (25) the resolved stresses can be written as $\langle \bar{u}_i \bar{u}_j \rangle = K^f a^f_{ij} + \frac{2}{3}\delta_{ij} K^f$. Inserted in the equation (25) this gives an estimation of the anisotropy tensor

$$a_{ij} = \frac{K^f}{K} a^f_{ij} \tag{26}$$

The ratios $K/K^f$ and $\dot{K}/\dot{K}^f$ will be estimated in section 4.3.

### 4.2. THE PRESSURE-STRAIN RATE

The pressure-strain rate can be decomposed as

$$\langle p s_{ij} \rangle = \langle \bar{p} \bar{s}_{ij} \rangle + \langle \bar{p} s'_{ij} \rangle + \langle p' \bar{s}_{ij} \rangle + \langle p' s'_{ij} \rangle \tag{27}$$

where a prime denotes an SGS quantity and $\langle \bar{p} \bar{s}_{ij} \rangle / \rho$ is the pressure-strain rate based on filtered quantities. For a spectral cut-off filter the terms $\langle p' \bar{s}_{ij} \rangle$ and $\langle \bar{p} s'_{ij} \rangle$ are identically zero. The term $\langle p' s'_{ij} \rangle$ will appear in the transport equation for $\langle u'_i u'_j \rangle$, which is identical to $\langle \tau_{ij} \rangle$ for spectral cut-off filters. Since we have already assumed $\langle \tau_{ij} \rangle$ to be close to isotropic we may to the same approximation assume $\langle p' s'_{ij} \rangle$ to be isotropic and hence zero. This gives that the total contribution to the pressure strain rate from the sub-grid scales should be small.

When the Smagorinsky model is used in an LES one obtains $\Pi^{smag}_{ij} \equiv 2\langle \bar{q} \bar{s}_{ij} \rangle$ which differs from $\Pi^f_{ij}$ by the term $2\langle \tau_{kk} \bar{s}_{ij} \rangle / 3$. This term should be small since the magnitude of the SGS-stress tensor is of the order of the energy in the sub-grid scales and also this small-scale quantity should be poorly correlated with the large-scale quantity $\bar{s}_{ij}$. Lacking means to quantitatively estimate the difference between $\Pi^{smag}_{ij}$ and $\Pi_{ij}$, but with the above reasoning making it plausible that the difference is smaller than for the dissipation and anisotropy, the pressure-strain rate is here estimated by

$$\Pi_{ij} = \Pi^{smag}_{ij} \tag{28}$$

We now get an expression, which only contains filtered quantities and the energy in the resolved and sub-grid scales, from which the Rotta parameter can be evaluated.

$$\frac{K}{K^f} \frac{\dot{K}^f}{\dot{K}} \Pi^{smag}_{ij} = -C_1 \epsilon^f a^f_{ij} \tag{29}$$

### 4.3. ESTIMATION OF THE RESOLVED AND SGS ENERGIES

The energy in the resolved and sub-grid scales is estimated by using the model spectrum in the previous chapter. The resolved and sub-grid scale kinetic energies are

$$K^f = \int_0^{k_p} E_1 dk + \int_{k_p}^{k_c} E_2 dk = \alpha\epsilon^{2/3}(\frac{3a+5}{2a+2}k_p^{-2/3} - \frac{3}{2}k_c^{-2/3}) \quad (30)$$

$$K^{sgs} = \int_{k_c}^{\infty} E_2 dk = \frac{3}{2}\alpha\epsilon^{2/3}k_c^{-2/3} \quad (31)$$

where matching of $E_1$ and $E_2$ at $k_p$ has been used to eliminate $A$. The total kinetic energy is then given by $K = K^f + K^{sgs}$ and the ratio between the filtered kinetic energy and total kinetic energy becomes

$$\frac{K^f}{K} = 1 - \frac{3a+3}{3a+5}(\frac{k_p}{k_c})^{2/3} \quad (32)$$

The expressions (30) and (31) together with $k_p = (\alpha/A)^{\frac{3}{3a+5}}\epsilon^{\frac{2}{3a+5}}$ gives

$$\dot{K}^f = \alpha\epsilon^{-1/3}(k_p^{-2/3} - k_c^{-2/3})\dot{\epsilon} \quad (33)$$

$$\dot{K}^{sgs} = \alpha\epsilon^{-1/3}k_c^{-2/3}\dot{\epsilon} \quad (34)$$

and the ratio between the time derivatives of the filtered and total kinetic energies

$$\frac{\dot{K}^f}{\dot{K}} = 1 - (\frac{k_p}{k_c})^{2/3} \quad (35)$$

Finally this yields that the total correction factor in equation (29) may be estimated as

$$\frac{K}{K^f}\frac{\dot{K}^f}{\dot{K}} = \frac{1 - (\frac{k_p}{k_c})^{2/3}}{1 - \frac{3a+3}{3a+5}(\frac{k_p}{k_c})^{2/3}} \approx 1 - \frac{2}{3a+5}(\frac{k_p}{k_c})^{2/3} \quad (36)$$

If $a = 4$ and $k_c = 12k_p$ the correction factor becomes 0.97.

### 4.4. RESULTS

Homogeneous axisymmetric turbulence fields were generated by straining isotropic fields, with a $k^4$ low wavenumber spectrum, using rapid distortion theory. The decay and relaxation of these velocity fields was then computed using LES. Calculations on a $96^3$ cube were made for different starting conditions with respect to the anisotropy and Reynolds number. For each case ten realizations were carried out in order to get sufficient statistics. In figures 6 and 7 time evolutions of $C_1$ versus $a_{11}$ and $Re_T$ are displayed. During an initial phase higher order moments develop and $\Pi_{ij}$ increases in strength. At the end of the simulations $C_1$ shows a more rapid decrease with

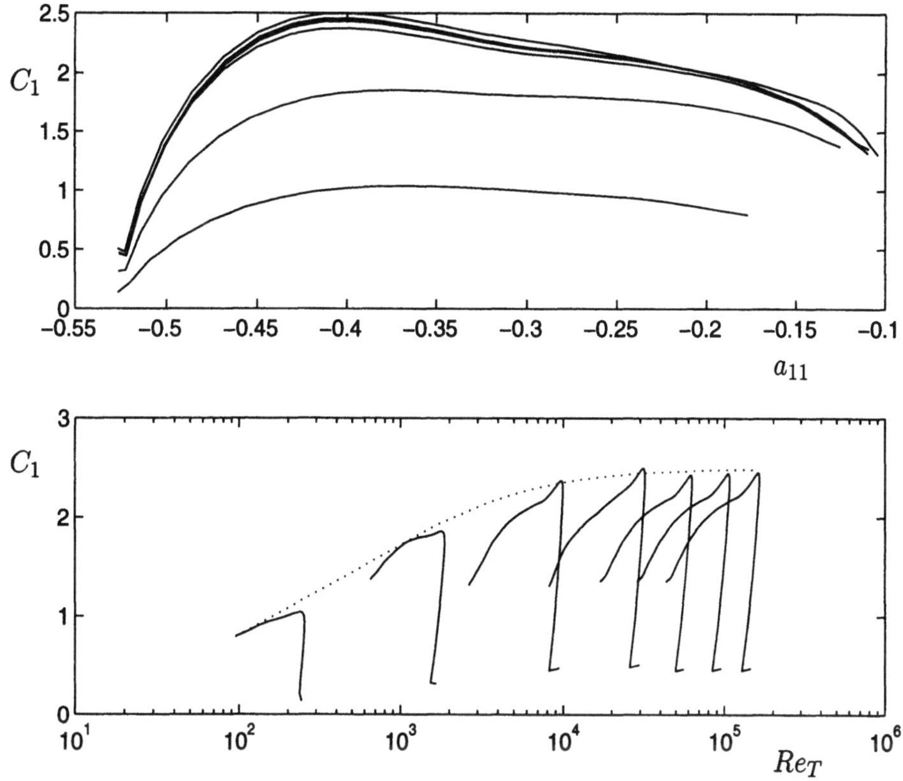


*Figure 6.* Top: the Rotta parameter versus the anisotropy component $a_{11}$. Bottom: the Rotta parameter versus the turbulence Reynolds number. Full lines: LES, dotted line: explicit model with $C_{1\infty} = 2.5$.

both $a_{11}$ and $Re_T$. This may be a significant feature at weak anisotropies but may also be a result of the large turbulence scales reaching a size comparable to that of the computational domain.

Figure 6 shows that the Rotta parameter is quite independent of $Re_T$ for large values of $Re_T$. LES predicts the value $C_{1\infty} = 2.5$ for an anisotropy value of $a_{11} = -0.4$. The asymptotic value 2.58 was originally suggested by Hallbäck et al. (1993), based on observations made for $a_{11}$ between $-0.38$ and $-0.30$. For $a_{11}$ in this interval LES predicts a $C_{1\infty}$ between 2.5 and 2.3. The results displayed in figures 6 and 7 indicate that there is a slight dependence of $C_1$ on $a_{11}$, and that this dependence seems to be smaller for lower values of $Re_T$ (see figure 6).

To conclude, the correction analysis shows that the filtered data obtained from an LES can be used to calibrate quantities such as $C_1$ in the asymptotic limit of high $Re_T$, where it is impossible to do a direct nu-

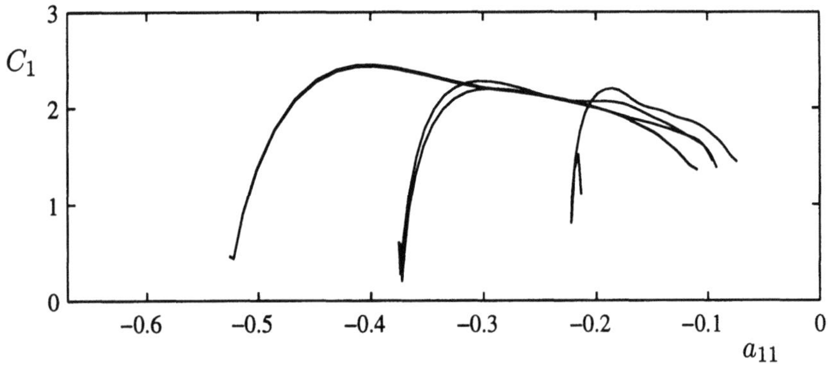


*Figure 7.* Time realizations of $C_1$ versus the anisotropy component $a_{11}$ at high Reynolds numbers for cases with different initial anisotropy levels.

merical simulation. Further work will be made to investigate the isotropy assumption of the small scales and the dependence of the special choice of SGS-stress model.

## References

Hallbäck, M. Sjögren, T. and Johansson, A.V. 1993 'Modeling of intercomponent transfer in Reynolds stress closures of homogeneous turbulence.'. In Proc. of Turbulent Shear Flows IX, eds. Durst, Kasagi, Launder and Whitelaw, pp. 23.4.1-6.

Lilly, D. K. 1967 'The representation of small-scale turbulence in numerical simulation experiments', in Proceedings of the IBM Scientific Computing Symposium on Environmental Science, IBM Form No. 320-1951, p. 195.

Rotta, J. 1951 'Statistische theorie nichthomogener turbulenz I'. Z. für Physik 129, pp. 547-572.

Scotti, A. Meneveau, C. and Lilly, D. K. 1993 'Generalized Smagorinsky model for anisotropic grids', Phys. Fluids A 5 (9), pp. 2306-2308.

Sjögren, T. and Johansson, A.V. 1995 'Measurement and modelling of anisotropic dissipation and slow pressure-strain in homogeneous turbulence'. In Advances in Turbulence V, pp. 486-490, ed. Benzi, Kluwer.

Smagorinsky, J 1963. 'General circulation experiments with the primitive equations'. Mon. Weather Rev. 91, p. 99.

# SOME INVESTIGATIONS ON THE SENSITIVITY OF LARGE EDDY SIMULATION

PIERRE SAGAUT, THIEN HIEP LÊ

*Aerodynamics Department, ONERA*
*29 Av. de la Division Leclerc 92320 Châtillon FRANCE*

**Abstract.** The sensitivity of the large eddy simulation of transitional shear flows is investigated for the cases of the flow past a backward facing step and the round jet. A strong influence of the following parameters of the simulations is reported: representation of the upstream turbulent flow, subgrid-scale model, subgrid term formulation. Two new formulations of the Mixed Scale Model, which account explicitly for the mesh anisotropy, are proposed. The observed dependencies of the results prove that further work is needed to be able to handle flows in complex industrial geometries on highly anisotropic grids.

## 1. Introduction

The suitability of Large-Eddy Simulation (LES) for the analysis of academic homogeneous flows has already been proven. However, the potential of LES for the prediction of more complex and inhomogeneous flows still remains to be investigated. That category of flows involves more problems than the previous one, concerning the following three topics: subgrid-scale modeling, numerics (numerical schemes, boundary conditions), and detailed data for the representation of the experimental facilities.

In this paper, the sensitivity of LES to these three problems is investigated. Two cases of transitional shear flows have been considered for that purpose: the three-dimensional flow past a backward facing step at Re=11 200 and the round jet flow at Re=23 000.



*J.-P. Chollet et al. (eds.), Direct and Large-Eddy Simulation II,* 81-92.

## 2. Governing Equations and Numerical Method

### 2.1. MATHEMATICAL MODEL

The PEGASE code, incompressible version, for DNS and LES, developed at ONERA, solves the unsteady Navier-Stokes equations written in velocity-pressure formulation (Sagaut *et al.*, 1995). The incompressibility constraint is enforced employing an approximate projection method. The pressure is calculated using a Poisson equation, leading to the following system:

$$\frac{\partial \overline{u}}{\partial t} + \frac{1}{2}(\nabla \cdot (\overline{u} \otimes \overline{u}) + \overline{u} \cdot \nabla \overline{u}) = -\nabla \Pi + \frac{1}{Re}\nabla^2 \overline{u} - \nabla \cdot m \qquad (1)$$

$$-\nabla^2 \Pi = \frac{\partial (\nabla \cdot \overline{u})}{\partial t} + \nabla \cdot \left(\frac{1}{2}(\nabla \cdot (\overline{u} \otimes \overline{u}) + \overline{u} \cdot \nabla \overline{u})\right) + \nabla \cdot \nabla \cdot m \qquad (2)$$

where $\bar{u}$ and $\Pi$ are respectively the filtered velocity and the modified filtered static pressure defined as

$$\Pi = \overline{p} + \frac{1}{3} m_{kk} I \qquad (3)$$

and $m$ is a model for the anisotropic part of the subgrid scale stress tensor.

The discretization is achieved using an hybrid Finite Elements/Finite Differences approach. The time integration is performed using an Adams-Bashforth/Crank-Nicolson scheme. The overall order of the method is 2 in space and time. A third-order upwind scheme is used for the non-conservative part of the convection term in order to improve the stability of the algorithm.

### 2.2. SUBGRID-SCALE TERMS FORMULATION

We are now focusing on eddy-viscosity type SGS models. These models are of the generic form:

$$m_{ij} = -2\nu_{SM}\overline{S}_{ij}, \qquad \overline{S}_{ij} = \frac{1}{2}\left(\frac{\partial \overline{u}_i}{\partial x_j} + \frac{\partial \overline{u}_j}{\partial x_i}\right) \qquad (4)$$

where $\nu_{SM}$ is the subgrid viscosity. Three formulations of the term $\nabla \cdot m$ have been considered:

- The conservative form: $-\nabla \cdot (\nu_{SM}(\nabla \overline{u} + \nabla^T \overline{u}))$
- The non-conservative form: $-\nabla \nu_{SM} \cdot (\nabla \overline{u} + \nabla^T \overline{u}) - \nu_{SM} \nabla^2 \overline{u}$
- The Laplacian form: $-\nu_{SM} \nabla^2 \overline{u}$

The conservative form is treated in a fully explicit way, while for the two other forms, the Laplacian contribution is taken into account with the Crank-Nicolson scheme. To save the second-order accuracy, a special linearization is applied to this term (Sagaut *et al.*, 1996).

## 3. Subgrid Scale Models

### 3.1. THE SMAGORINSKY MODEL

The subgrid scale tensor is formulated in a standard way by the Smagorinsky model:

$$\nu_{SM} = c_s \overline{\Delta}^2 \left| \overline{S} \right|, \qquad \left| \overline{S} \right| = \sqrt{2 \overline{S}_{ij} \overline{S}_{ij}} \tag{5}$$

where $\overline{\Delta}$ is evaluated at each grid point as $(\Delta x \Delta y \Delta z)^{1/3}$.

The main well-known drawbacks of this model come from the fact that it depends exclusively on large scales and its eddy viscosity does not vanish towards a wall and for laminar flow.

### 3.2. THE MIXED SCALE MODEL

In order to alleviate the difficulties mentioned above, the Mixed Scale Model (MSM), based on both large and small scale dependencies, was developed at ONERA in collaboration with LIMSI (CNRS) (Sagaut *et al.*, 1995). The subgrid-scale viscosity of this model is defined as follows:

$$\nu_{SM}(\alpha) = C \left( \left| \overline{S} \right| \right)^{\alpha} q^{\left( \frac{1-\alpha}{2} \right)} \overline{\Delta}^{(1+\alpha)} \tag{6}$$

where $0 \leq \alpha \leq 1$ and where $q$ is the fluctuating kinetic energy evaluated by the following formula:

$$q = \frac{1}{2} \overline{(u_i')} \, \overline{(u_i')} \tag{7}$$

The filtered characteristic subgrid-scale velocity $\overline{(u_i')}$ is calculated using Bardina's *similarity hypothesis* (Bardina *et al.*, 1983):

$$\overline{(u_i')} \approx (\overline{u}_i)' \tag{8}$$

The fluctuating resolved scale $(\overline{u}_i)'$ is extracted from the resolved velocity field by employing a *testfilter* with a characteristic cut-off lengthscale equal to $2 \times \overline{\Delta}$:

$$(\overline{u}_i)' = \overline{u}_i - \tilde{\overline{u}}_i \tag{9}$$

The small-scale dependency of the model ensures that it will adapt itself to the local state of the flow, and then will disappear in fully resolved regions of the flow and near solid boundaries.

### 3.3. THE SCALAR ANISOTROPIC MIXED SCALE MODEL

The Scalar Anisotropic Mixed Scale Model (SAMSM) is the first proposed modification of the MSM to account for the grid anisotropy. The evaluation of the characteristic cut-off lengthscale associated with the grid is evaluated using Scotti's formula (Scotti *et al.*, 1993):

$$\overline{\Delta} = \Delta_{iso} f(a_1, a_2) \tag{10}$$

where $a_1$ and $a_2$ are the mesh aspect ratio and

$$f(x,y) = 1 + \frac{2}{27}\left[(\ln x)^2 - \ln x \ln y + (\ln y)^2\right]$$

The length $\Delta_{iso}$ is the isotropic length scale defined by Deardorff:

$$\Delta_{iso} = (\Delta x \Delta y \Delta z)^{1/3}$$

The SAMSM is then:

$$\nu_{SM}(\alpha) = C_M \left(|\mathcal{F}(\overline{u})|\right)^{\alpha} q^{\left(\frac{1-\alpha}{2}\right)} (\Delta_{iso} f(a_1, a_2))^{(1+\alpha)} \tag{11}$$

### 3.4. THE TENSORIAL ANISOTROPIC MIXED SCALE MODEL

The second proposed modification of the MSM to account for grid anisotropy is the Tensorial Anisotropic Mixed Scale Model (TAMSM). To describe the geometry of the mesh, we now use six different lengthscales. So the subgrid-scale viscosity is now defined as a $3 \times 3$ matrix, corresponding to the proposal of Zahrai (Zahrai *et al.*, 1995). For cartesian grids, the subgrid-scale viscosity is then defined as:

$$\nu_{SM}(\alpha) = C_M \left(|\mathcal{F}(\overline{u})|\right)^{\alpha} q^{\left(\frac{1-\alpha}{2}\right)} (\Delta_{iso})^{\alpha - 1/3} \begin{pmatrix} \Delta x^{4/3} & 0 & 0 \\ 0 & \Delta y^{4/3} & 0 \\ 0 & 0 & \Delta z^{4/3} \end{pmatrix} \tag{12}$$

## 4. Results

Two cases of transitional shear flow are considered: the flow past a backward facing step (BFS) at Re=11 200 and the round jet (RJ) at Re=23 000. The BFS case will be devoted to the analysis of the influence of the upstream boundary condition and the importance of the subgrid-scale model. The RJ case is dedicated to the comparison of the MSM, the SAMSM and the

TAMSM, and to the analysis of the influence of the subgrid-scale term formulation.

## 4.1. BACKWARD FACING STEP

Three simulations are presented here, associated to different choices of the subgrid scale model and upstream boundary conditions. The calculations are performed on a $201 \times 31 \times 51$ stretched cartesian grid with mesh refinement near solid boundaries and a time step $\Delta t = 0.005$ was used. The computational box is taken equal to $20H \times 4H \times 2.5H$, where H is the step height. The expansion ratio is equal to 5/3, corresponding to the experiments of Eaton and Johnston (Eaton *et al.*, 1980). The Smagorinsky constant $C_S$ and the MSM constant $C_M$ are respectively taken equal to 0.2 and 0.1, corresponding to the same theoretical energy transfer for homogeneous isotropic turbulence. The conservative form of the SGS terms is used for all the simulations. The cost of a run ranges from 15 to 18 CPU hours on a CRAY YMP computer.

The model dependency of both the time-averaged recirculation zone length $X_R$ and the vortex shedding Strouhal number $St$ were examined first. The results obtained with the two models for these two parameters are reported in Table 1.

TABLE 1. Backward Facing Step: Global Parameters

| | $X_R/H$ (Error) | St (Error) |
|---|---|---|
| Eaton & al. (exp.) | 6.97 (-) | 0.074 (-) |
| Smagorinsky (unsteady inflow) | 9.95 (42.7 % ) | 0.083 (12.1 % ) |
| MSM (unsteady inflow) | 7.46 (6.5 % ) | 0.083 (12.1 % ) |
| MSM (steady inflow) | 9.46 (35 % ) | 0.066 (10.8 % ) |

It can be clearly seen that the MSM gives very good results, while the Smagorinsky model leads to an error of 42 % in the recirculation length, because of its over-dissipative character. The MSM clearly alleviates that problem, leading to a good description of the dynamic of the transitioning separated shear layer. The fact that both models are able to predict the vortex shedding Strouhal with the same accuracy indicates that the Smagorinsky model only delays the Kelvin-Helmholtz instability of the separated shear layer, but does not destroy it.

The dependency on the upstream condition was investigated in a second step. Two simulations were performed with the MSM: the first one corresponds to a steady laminar velocity profile, while the second one includes an additional random noise of the same amplitude as the residual turbulence detected in the experiments (≈ 0.3 % turbulence intensity). Results are reported in Table 1. It can be seen that the absence of the residual noise leads to large error on the prediction of the recirculation length, but the Strouhal number is still accurately predicted. This can be explained by the fact that without any perturbation, the transition of the shear layer is delayed, leading to errors similar to the ones induced by an over-dissipation.

Time-averaged longitudinal velocity and longitudinal turbulence intensity profiles are compared to experimental data at stations $(X - X_R)/H = -4.97$ and $(X - X_R)/H = 1.03$ on Figures 1 and 2. The three simulations give very similar results, in very good agreement with experiments. The analysis of the turbulence profiles shows large discrepancies between the three cases. Before the reattachment point, the damping of the dynamics of the transitional shear layer due to the use of the Smagorinsky model or the imposition of a steady inflow velocity profile is clearly observable. The use of MSM with an unsteady inflow leads to a significant improvement of the results. After the reattachment point, all the calculations predict a too large value of the turbulence intensity, but it should be noticed that the MSM produce the better agreement. This fact indicates that too much energy is conserved by the two-dimensional modes, so that the kinetic energy transfer process towards the spanwise velocity component is underestimated. This can be due to a lack of resolution in the spanwise direction and/or to a high dissipation level.

## 4.2. ROUND JET

The calculations are performed on two cartesian grids: a "coarse" one ( $101 \times 101 \times 101$) and a fine one ( $101 \times 121 \times 121$). In the two cases, the jet is located in a regular grid (dimension $D/10 \times D/20 \times D/20$ for the coarse grid and $D/10 \times D/26 \times D/26$ for the fine one), where $D$ is the initial diameter of the jet. Outside the jet region, the mesh is stretched. The computational domain is $[0, 10D] \times [-5.5D, 5.5D] \times [-5.5D, 5.5D]$ for all the simulations. The Reynolds number $Re_D$, based on the diameter $D$ and the maximum of the axial velocity at the inlet $U_{ref}$, is equal to 23 000, corresponding to the experiments of Djeridane (1994). The value of the constant $C_M$ is 0.1 for all the simulations. The cost of a typical run is 40 CPU hours on a CRAY YMP computer on the coarse mesh, and 70 CPU hours on the fine mesh.

Four simulations are performed on each grid: MSM with both non-conservative (case C1) and Laplacian (case C2) formulations for the SGS terms, SAMSM (case C3) and TAMSM (case C4) with Laplacian formulation for the SGS terms.

The Strouhal number $St$ associated to the leading temporal frequency, computed on the fine grid in the transitional region of the jet is reported in Table 2. All the simulations are in good agreement with the experimental values, which range from 0.3 to 0.5. It should be noticed that the case C1 differs from the three others.

TABLE 2. Round Jet: Global Parameters

| Specifications | St | $dB/dx$ | $d\theta/dx$ |
|---|---|---|---|
| C1 (coarse mesh) | - | 0.226 | 0.012 |
| C1 (fine mesh) | 0.32 | 0.347 | 0.011 |
| C2 (coarse mesh) | - | 0.277 | 0.010 |
| C2 (fine mesh) | 0.25 | 0.363 | 0.015 |
| C3 (coarse mesh) | - | 0.225 | 0.012 |
| C3 (fine mesh) | 0.25 | 0.335 | 0.013 |
| C4 (coarse mesh) | - | 0.168 | 0.009 |
| C4 (fine mesh) | 0.25 | 0.296 | 0.010 |

The axial profiles of the thickness and the momentum thickness of the annular mixing layer which surrounds the potential cone are plotted on figure 3. It can be seen that on the coarse mesh all the models, except the MSM model under non-conservative form, lead to poor results, characterized by a spurious spreading of the mixing layer just after the inflow boundary. On the fine mesh, the spreading of the mixing layer is reduced by the introduction of anisotropy in the subgrid model, indicating a diminution of the induced dissipation. The axial evolutions of $L_u$ (the distance from the jet axis of the point where the velocity is equal to the half of the mean of the maximum velocity and the co-flow velocity) and the quantity $U_{ref}/(U_\infty - U_C(x))$ are compared to experimental data on Figure 4. A good agreement is observed on the fine grid from the inflow plane to the end of the potential core. It should be noticed that these two parameters seem to be insensitive to the subgrid scale model on the fine grid up to $X/D = 4$.

A finer analysis is done by looking at the time averaged flow characteristics. The dynamics of the axisymetric mixing layer which surrounds the potential core is investigated through the axial variation of its thickness $B$ and its momentum thickness $\theta$. The axial derivatives of these two parameters are reported in Table 2. Experimental values of $dB/dx$ range from 0.16 to 0.20 following Hussain (Hussain *et al.*, 1978), and from 0.015 to 0.04 for $d\theta/dx$ (Hussain *et al.*, 1978) (Browand *et al.*, 1979). Comparisons between the eight simulations demonstrate the sensitivity of the results to the formulation of the SGS terms and to the subgrid-scale model.

Axial profiles of the quantity $(U_{ref} - U_\infty)/(U_C(x) - U_\infty)$ are compared to experimental data on Figure 5, where $U_\infty$ is the velocity of the external flow and $U_C(x)$ the time-averaged streamwise velocity on the jet axis. On the coarse mesh, the cases C1, C3 and C4 exhibit a spurious behavior just near the inflow boundary, while the case C2, which corresponds to the use of the non-conservative form of the SGS terms, does not. That unphysical phenomenon is associated to a sudden growth of the annular mixing layer, which is induced by an over-dissipation. The use of a finer grid or of the non-conservative form of the SGS terms alleviates that problem. It should be noticed that the tensorial version of the MSM diminishes the initial expansion rate of the mixing layer on the coarse mesh. On the fine mesh, all the calculations are in good agreement with experimental data in the potential core, but some discrepancies appear at the end of that cone. The Laplacian form of the MSM predicts a too strong diminution of the mean velocity on the axis of the jet, induced by a too large expansion of the jet. This can be explained by a too high level of the diffusion. The use of the SAMSM and the TAMSM leads to a clear improvement of the results.

The evolution of the turbulent longitudinal intensity along the jet axis is shown on Figure 3. It can be seen that all the simulations predict the location of the maximum in the region near the end of the potential cone, in good qualitative agreement with the experiments. However, large discrepancies appear concerning the quantitative level: all the simulations exhibit a too large peak of the turbulence intensity, and this value seems to be very sensitive to the subgrid-scale model. Nevertheless, the computed values belong to the range found in other experiments (Browand *et al.*, 1979) (Hussain *et al.*, 1978). The discrepancy at the first station indicates that the upstream turbulence is not taken into account in an adequate way in the simulation: the same phenomenon was exhibited in the BFS case.

## 5. Conclusions

The sensitivity of large eddy simulation of transitional shear flows to various parameters has been investigated. The cases of backward facing step and the round jet were selected to perform the analysis. The strong influence of the upstream boundary condition was shown: a correct description of the upstream turbulence is of crucial importance to recover a good quantitative agreement with experimental data, but a good qualitative agreement relative to the topology of the flow, such as the existence of a recirculation zone or a potential core, is generally recovered. The quality of the results depends very much on the ability of the subgrid-scale model to capture the dynamics of the transition in an accurate way. The MSM was shown to greatly improve the results over the Smagorinsky model with the selected value for the constant $C_S$.

Two modifications of the MSM were proposed: the SAMSM, which is based on the Scotti formula, and the TAMSM, which incorporates the tensorial evaluation of the cut-off lengthscale by Zahrai. The results exhibit a significant dependency on the modification of the SGS model to account for grid anisotropy, which generally improves the results by lowering the dissipation, leading to a better capturing of the transition. Additional tests were performed to evaluate the influence of the formulation of the subgrid-scale term. The results obtained on the round jet case demonstrate that the Laplacian form is more diffusive than the non-conservative form, but further work is needed on that point.

Future works will deal with a more complete evaluation of the SAMSM and the TAMSM, the representation of turbulent upstream boundary condition, and the subgrid modelling, including buoyancy forces.

## 6. Acknowledgements

This work was supported by DGA/DRET, French Ministry of Defence.

## References

Bardina, J., Ferziger, J.H., Reynolds, W.C. (1983) Improved turbulence models based on large eddy simulation of homogeneous, incompressible, turbulent flows. *Technical Report TF-19, Stanford University, Dept. of Mechanical Engineering.*

Browand, F.K., Latigo, B.O. (1979) Growth of the two-dimensional mixing layer from a turbulent and a non-turbulent boundary layer, *Phys. Fluids*, **22(6)**, pp. 1011-1019.

Djeridane, T. (1994) Contribution à l'étude expérimentale de jets turbulents axisymétriques à densité variable, Thèse de Doctorat de l'Université d'Aix-Marseille II.

Eaton, J.K., Johnston, J.P. (1980) Turbulent flow reattachment: an experimental study of the flow and structure behind a backward-facing step. *Stanford University, Rep.* MD-39.

Husain, Z.D., Hussain, A.K.M.F. (1978) Axisymmetric mixing layer: influence of the initial and boundary conditions, *AIAA Journal*, vol. **17**, No. **1**, pp. 48-55.

Hussain, A.K.M.F., Zedan, M.F. (1978) Effects of the initial condition on the axisymmetric free shear layer: effect of the initial fluctuation level, *Phys. Fluids*, **21(9)**, pp. 1475-1481.

Sagaut, P., Troff, B., Lê, T.H., Ta, P.L. (1995) Large eddy simulation of turbulent flow past a backward facing step with a new Mixed Scales SGS model, IMACS-COST Conference Computational Fluid Dynamics Three-Dimensional Complex Flows, Lausanne, Sept. 13-15.

Sagaut, P., Lenormand, E., Klahr, D. (1996) Poursuite du développement du code PEGASE de simulation des grandes échelles des écoulements incompressibles, Technical Report RTS 18/4368, ONERA.

Scotti, A., Meneveau, C., Lilly, D.K. (1993) Generalized Smagorinsky model for anisotropic grids, *Phys. Fluids*, **5(9)**, pp. 2306-2308.

Zahrai, S., Bark, F.H., Karlsson, R.I. (1995) On anisotropic subgrid modelling, *Eur. J. Mech. B/Fluids*, **14(4)**, pp. 459-486.

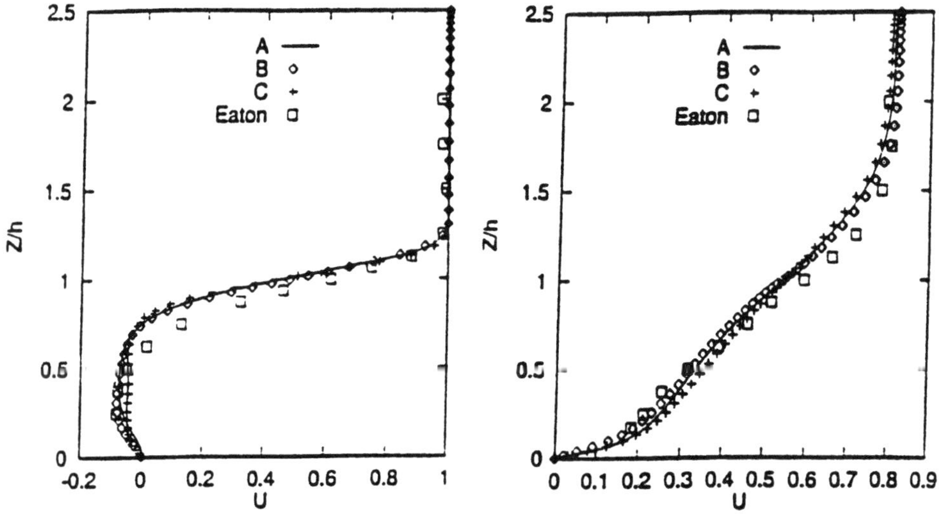


*Figure 1.* Backward Facing Step: Time-averaged longitudinal velocity profiles at stations $(X - X_R)/H = -4.97$ (left) and $(X - X_R)/H = 1.03$ (right) - A: MSM with unsteady inflow ; B: Smagorinsky with unsteady inflow ; C: MSM with steady inflow

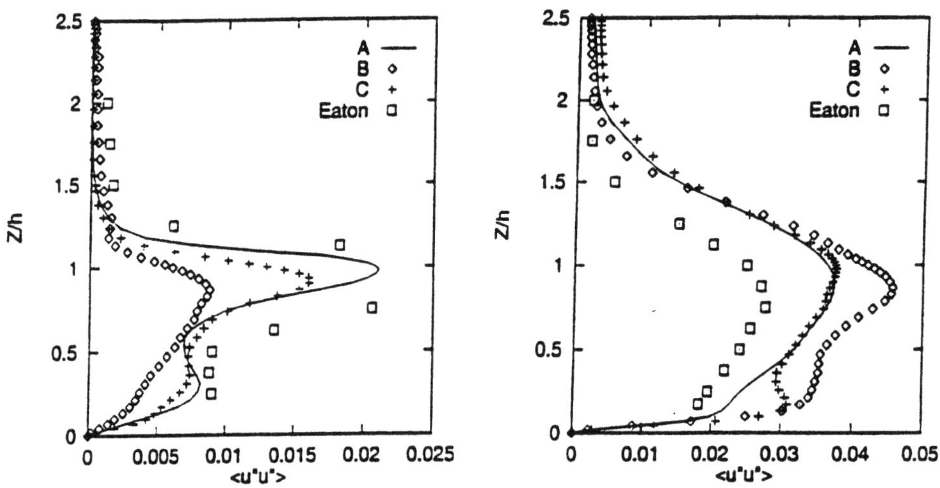


*Figure 2.* Backward Facing Step: longitudinal turbulence intensity profiles at stations $(X - X_R)/H = -4.97$ (left) and $(X - X_R)/H = 1.03$ (right) - A: MSM with unsteady inflow ; B: Smagorinsky with unsteady inflow ; C: MSM with steady inflow

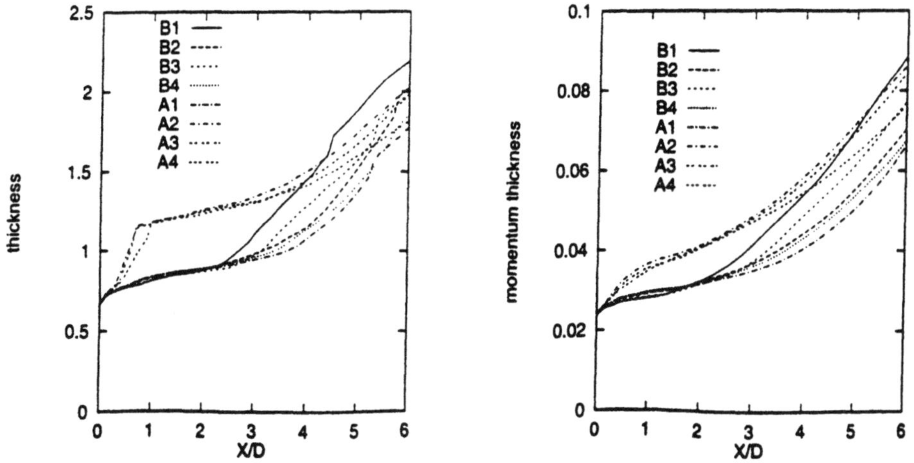


*Figure 3.* Round Jet: axial profiles of the thickness (left) and the momentum thickness (right) of the annular mixing layer ; A1: MSM, Laplacian form (coarse mesh) ; A2: MSM, non-conservative form (coarse mesh) ; A3: SAMSM, Laplacian form (coarse mesh) ; A4: TAMSM, Laplacian form (coarse mesh) ; B1: MSM, Laplacian form (fine mesh) ; B2: MSM, non-conservative form (fine mesh) ; B3: SAMSM, Laplacian form (fine mesh) ; B4: TAMSM, Laplacian form (fine mesh)

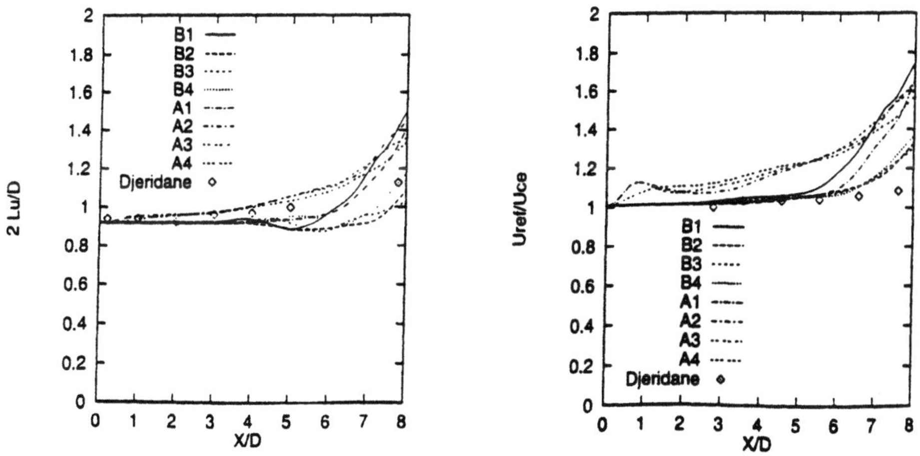


*Figure 4.* Round Jet: axial profiles of the quantity $2L_u/D$ (left) and $U_{ref}/(U_\infty - U_C(x))$ (right) ; A1: MSM, Laplacian form (coarse mesh) ; A2: MSM, non-conservative form (coarse mesh) ; A3: SAMSM, Laplacian form (coarse mesh) ; A4: TAMSM, Laplacian form (coarse mesh) ; B1: MSM, Laplacian form (fine mesh) ; B2: MSM, non-conservative form (fine mesh) ; B3: SAMSM, Laplacian form (fine mesh) ; B4: TAMSM, Laplacian form (fine mesh)

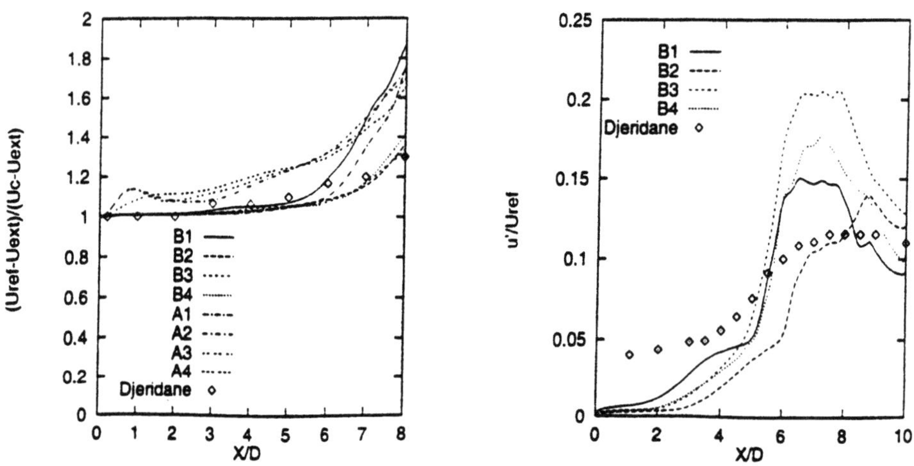


*Figure 5.* Round Jet: axial profiles of the quantity $(U_{ref} - U_\infty)/(U_C(x) - U_\infty)$ (left) and longitudinal turbulence intensity $u'/U_{ref}$ (right) ; A1: MSM, Laplacian form (coarse mesh) ; A2: MSM, non-conservative form (coarse mesh) ; A3: SAMSM, Laplacian form (coarse mesh) ; A4: TAMSM, Laplacian form (coarse mesh) ; B1: MSM, Laplacian form (fine mesh) ; B2: MSM, non-conservative form (fine mesh) ; B3: SAMSM, Laplacian form (fine mesh) ; B4: TAMSM, Laplacian form (fine mesh)

# A COMPARISON OF LOW-ORDER DNS, HIGH-ORDER DNS AND LES

R.W.C.P. VERSTAPPEN AND A.E.P. VELDMAN
*Department of Mathematics, University of Groningen,*
*P.O.Box 800, 9700 AV Groningen, The Netherlands*

**Abstract.** In this paper we compare a second-order accurate finite volume method and a fourth-order approach for a direct numerical simulation of the flow in a cubical driven cavity at $Re = 10,000$. Experimental results are available for comparison. The fourth-order method turns out to be the superior method. For a driven cavity with spanwise aspect ratio 0.5 at $Re = 10,000$, along with experimental results also the results of a LES (with a dynamic mixed subgrid-scale model) are available. We will demonstrate the challenge of turbulence modelling by comparing this LES on a $64 \times 64 \times 32$ grid, a fourth-order DNS on a $50^3$ grid and an experiment. Finally, using the fourth-order simulation method, a DNS of a turbulent flow in a cubical cavity at $Re = 50,000$ is performed using a $192^3$ grid. Mean velocities, turbulence intensities and power spectra are computed.

## 1. Introduction

In the early days of Direct Numerical Simulation (DNS) spectral and pseudo-spectral methods were the prevalent methods. In the search for methods that can handle geometrically complex configurations, second-order accurate finite-difference/volume methods became popular. The questioning of the accuracy, or better the cost-effectiveness, of these methods gave birth to higher-order finite-difference methods, often applied in combination with (pseudo-)spectral methods in homogeneous directions.

Rai and Moin (1989) introduced a fifth-order upwind-biased method for the convective terms and a sixth-order central method for the viscous terms; Fasel (1990) used a velocity/vorticity formulation with fourth-order finite differences; Joslin *et al.* (1992) have applied a fourth-order compact method for the momentum equations; Liu and Liu (1994) have developed



*J.-P. Chollet et al. (eds.), Direct and Large-Eddy Simulation II,* 93-102.

a transformation of a standard fourth-order difference scheme in computational space into physical space.

This brief enumeration of higher-order DNS methods shows that there are several ways of making them. Although Richardson extrapolation is a well-known way to increase the order of accuracy of numerical approximations, it has not yet been applied to obtain higher-order DNS methods (as far as we know). In this paper, we will focus on a fourth-order method that is constructed as the Richardson extrapolate of a classical second-order finite volume method. To evaluate this method we consider driven cavity flows (see Section 2). The main lines of the fourth-order finite volume method are outlined in Section 3; for a more detailed discussion the reader is referred to Verstappen and Veldman (1996).

The rest of this paper is organized as follows. In Section 4 the outcomes of the fourth-order accurate finite volume method and the underlying second-order method are compared for a direct numerical simulation of the flow in a cubical driven cavity at $Re = 10,000$. Experimental results are available for comparison. Next, we consider the flow in a cavity with a spanwise aspect ratio of 0.5. The reason for this is that for that flow both experimental results and the results of a LES are available. A comparison of LES, experiment, second-order DNS and fourth-order DNS is presented in Section 5. After our short excursion to non-cubical cavities we will revert to cubical cavities in Section 6. With the fourth-order simulation method a DNS of a turbulent flow in a lid-driven cavity at $Re = 50,000$ has been performed. Time-mean quantities and power spectra are presented.

## 2. Driven cavity flows

Prasad and Koseff (1989) have conducted a series of physical experiments in three-dimensional rectangular lid-driven cavities for Reynolds numbers varying between 3,200 and 10,000 and spanwise aspect ratios between one quarter and one. Figure 1 illustrates the locations of the measurements in the cavity: experimental results are available along the horizontal and the vertical centre line in the shaded, square cross-section, the central plane.

The velocity measurements have been performed using an optical laser-Doppler anemometer system. The overall error in an instantaneous measured velocity is about 0.6% of the speed at which the upper lid of the cavity is driven. The physical experiments reveal that the flow in a cubical lid-driven cavity is unsteady at $Re = 3,200$, and in an early stage of turbulence at $Re = 10,000$. That is, a part of the power spectrum at $Re = 10,000$ has a $-5/3$ slope.

Numerically the flow in the lid-driven cavity has been simulated by solving the unsteady, incompressible Navier-Stokes equation with no-slip

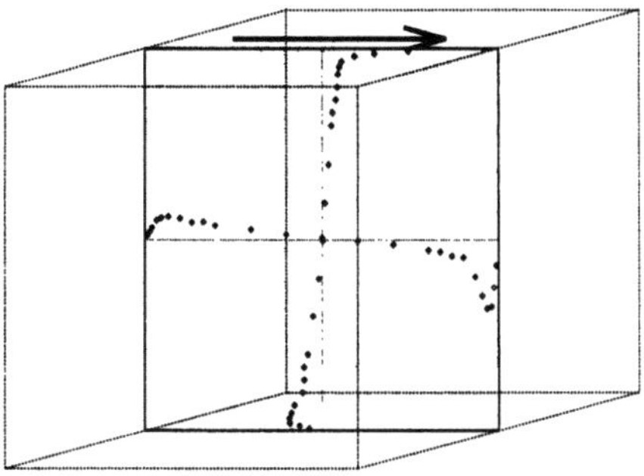

*Figure 1.* Locations of the measurements of Prasad and Koseff (1989). The orientation of the cavity is such that the upper-plane is driven from the left to the right. The points show the vertical component of the velocity along the horizontal centre line and the horizontal component of the velocity along the vertical centre line at $Re = 10,000$.

conditions imposed at all six the walls. The flow is sustained by the constant velocity that is prescribed at the upper lid of the cavity (as is indicated in Figure 1). The velocities at the other walls are zero. The Reynolds number is based upon the velocity of the lid and the height of the cavity.

## 3. A fourth-order finite volume method

The incompressible Navier-Stokes equations are discretized on an orthogonal, staggered grid as is proposed by Harlow and Welsh (1965). The pressure gradient and the incompressibility constraint are integrated implicitly in time; the convective and diffusive fluxes are treated explicitly (by a second-order Adams-Bashforth method). The time-discrete momentum equations are applied to control volumes. The left picture in Figure 2 shows a control volume for the second-order method in two spatial dimensions; the extension to 3D is straightforward. All spatial integrations (over the control volumes as well as over their surfaces) are approximated using the midpoint rule, with linear interpolations for the convective flux and central differences for the diffusive flux.

To eliminate the leading term of the truncation error of the second-order method the momentum equations are also integrated over three times larger control volumes. Figure 2 (right) illustrates the definition of a three times larger control volume for the horizontal component of the momentum equation. The definition of a control volume for the vertical component is obtained by rotating Figure 2 one quarter of a turn counter clock-wise.

The larger control volumes are chosen in such a way that their corners coincide with positions of the vertical velocity. Three times larger control volumes are the smallest possible larger control volumes with corners at these positions. The coinciding of corners with vertical velocities implies

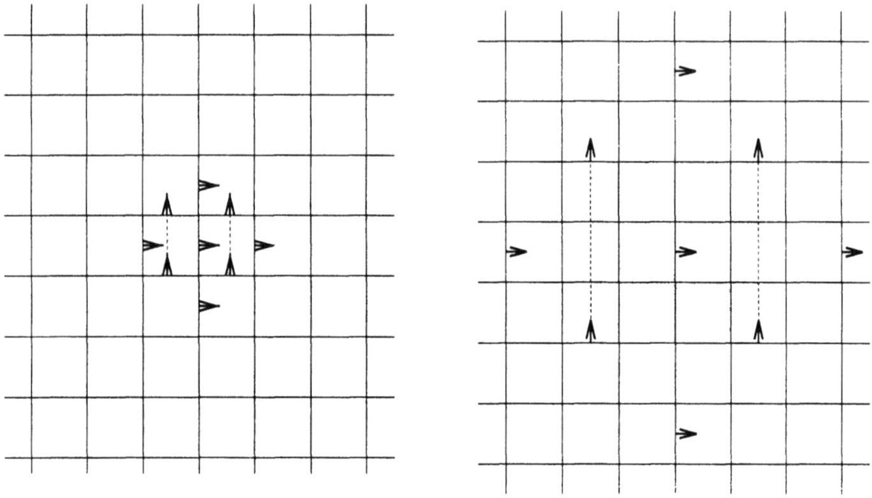

*Figure 2.* The left picture shows a control volume for the conservation of the horizontal component of the momentum field (in two spatial dimensions). The right picture shows a three-times larger control volume that is applied to eliminate the leading term of the truncation error. The arrows denote the components of the discrete velocity that are used to discretize the application of the conservation law to the control volume.

that one discretization method can be used for the spatial integration of the momentum equations over both the original and the larger control volume. Two times larger control volumes, for instance, have corners that do not coincide with positions of the vertical velocity, and thus do require additional high-order interpolations (for the cross terms in the convective flux, e.g.).

The discrete conservation law for the original control volume is expressed in terms of nearby velocities: for the horizontal component in terms of five nearby horizontal velocities and four nearby vertical velocities; see Figure 2 (left). The same interpolation, differentiation and integration rules are applied to the three times larger control volume. Hence, the truncation error in the integration of the momentum equations over the three times larger control volume is of the same form as the truncation error in the integration over the original control volume.

On a uniform grid, the leading term of the truncation error in the application of the conservation law to a three times larger control volume is precisely nine times as large as the leading term of the truncation error in the application to a corresponding original volume. Thus, subtracting 1/8 times the result of the integration of the conservation law over the large control volume from 9/8 times the result of the second-order integration eliminates the second-order terms of the truncation error and results in a fourth-order accurate spatial discretization of the momentum equations.

For a conservative discretization the flux through a surface has to be computed independently of the control volume in which it is considered: two adjacent control volumes see one flux through their common surface. This fundamental property is lost if the weights in the Richardson extrapolation depend on the control volumes. Therefore, we use the weights $9/8$ and $-1/8$ also for weakly stretched grids. Finally, it may be noted that for constant weights the matrix that represents the discretisation of a second-order derivative is symmetric; the discretisation of a first-order derivative is skew-symmetric.

In the vicinity of boundaries, some three times larger control volumes do not fit into the flow domain. A misfit is treated separately: here, the second-order finite volume discretization is used for misfitting parts. This approach is similar to the one that is followed in most of the higher-order methods that are mentioned in Section 1. The large high-order-accurate stencils used by Rai and Moin (1989), Fasel (1990), or by Liu and Liu (1994), e.g., can also not be applied near boundaries, since their is not enough space available to do so, and no attempt has been made to discretize with a high-order of accuracy in the neighborhood of boundaries.

The velocity is updated by integrating the momentum equations over one time-step and by adding a pressure correction such that the mass is conserved in control volumes centred on the pressures. Obviously, one can also apply the balance of mass to three times larger control volumes and use these control volumes to eliminate the leading term in the truncation error. Here, we do not follow that approach: we simply use the same control volumes as in a second-order finite volume method. The resulting discrete Poisson equation for the pressure correction is solved using a conjugate gradient method with modified incomplete Choleski preconditioning.

## 4. Second- versus fourth-order

The experimental data of the flow in a cubical lid-driven cavity at Re = 10,000 consists of mean velocities and turbulence intensities along two centre lines in the central plane. Figure 1 illustrates the locations of the measurements. In 1994 we have shown that the results of the second-order DNS on a $100^3$ grid agree well with the experimental data (Verstappen and Veldman, 1994). Recently, we have re-done this simulation with the fourth-order method using various grids. It turned out that one Richardson extrapolation improves the results significantly.

For example, on a stretched $50^3$ grid (the gridspacing is geometrically stretched away from the wall; the largest mesh width is approximately seven times the smallest), the agreement between the mean velocities and turbulence intensities obtained with the fourth-order method and the experiments is better than the agreement between the results of the second-order method on a $100^3$ grid and the experiments. As an illustration of this, mean velocities as obtained from both DNS's and from the experiment are compared in Figure 3.

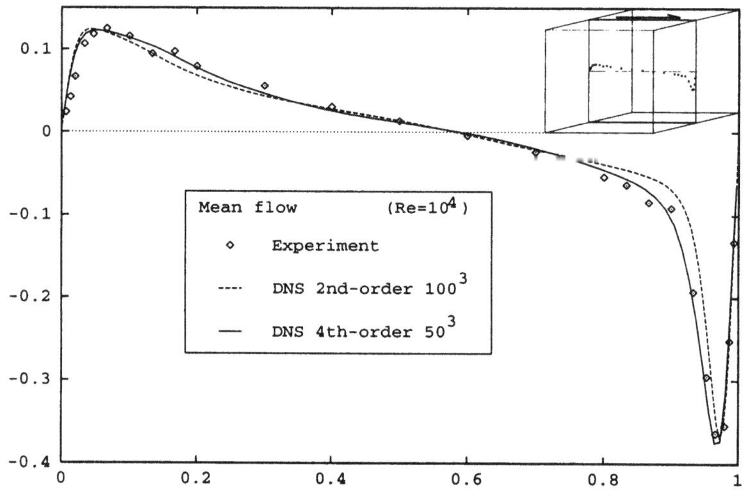


*Figure 3.* Second- and fourth-order DNS vs. experiment of turbulent flow in a cubical driven cavity at Re=10,000. Shown is the vertical mean velocity on the horizontal centre line in the central plane of the cavity.

In this example the fourth-order results are clearly superior to the second-order results, whereas their computational effort is about twenty times less. Indeed, the CPU-time per grid point and time-step of both methods is comparable, a $50^3$ grid has 8 times less grid-points than a $100^3$ grid, allows for a twice as large time-step (note that the time-step is restricted by the CFL-condition), and the number of iterations needed to solve the pressure correction from the Poisson equation is somewhat less for a $50^3$ grid than for a $100^3$ grid: $8 \times 2 \times 1.25 = 20$.

## 5. DNS versus LES

In this section we consider a cavity with a spanwise aspect ratio of 0.5. The reason for this is that Zang *et al.* (1993) have performed a large-eddy simulation of a driven cavity flow for this spanwise aspect ratio (at Reynolds numbers of 3,200, 7,500 and 10,000; we consider $Re = 10,000$ here). This LES uses a dynamic mixed subgrid-scale model. Prasad and Koseff (1989) have also measured the flow in the central plane of a cavity with aspect ratio

0.5. Hence, for a spanwise aspect ratio of 0.5, both experimental results and the results of a LES are available. The LES of Zang *et al.* applies a second-order method. In Figure 4 we demonstrate the challenge of LES by comparing the results of this LES, a second- and a fourth-order DNS and an experiment. The resolution of the LES and the DNS's are similar: the LES uses a stretched $64 \times 64 \times 32$ grid, the DNS's use a stretched $50^3$ grid; the two DNS's use the same grid.

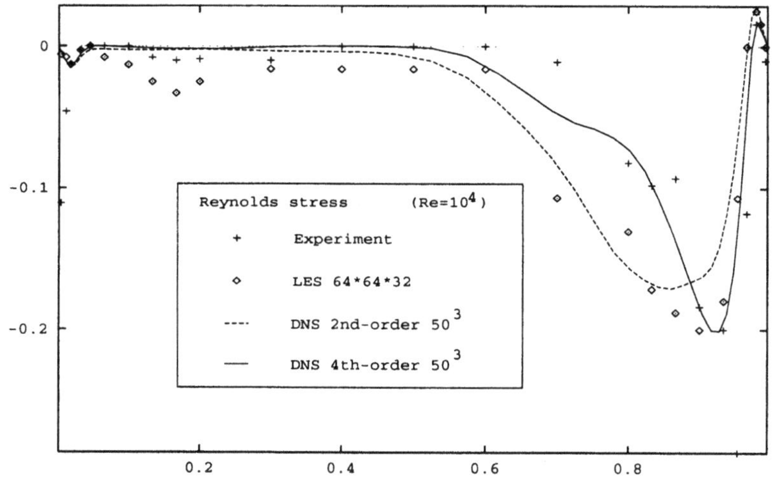


*Figure 4.* Comparison of experimental results, LES and DNS in a 3D driven cavity with aspect ratio 0.5 at Re=10,000. The LES uses approximately 131,000 grid-points; both the second- and fourth-order DNS use 125,000 grid-points. Shown is 500× the Reynolds stress $\overline{u'v'}$ along the vertical centre line in the central plane of the cavity. The LES-data is taken from Zang *et al.* (1993); the experimental data is taken from Prasad and Koseff (1989).

Again, this figure illustrates that the fourth-order DNS results are better than the second-order ones.

In addition, Figure 4 clearly shows that the dynamic mixed subgrid-scale model used in the LES is still open for improvement: the Reynolds stress is systematically larger than in the experiment, or stated otherwise, the subgrid model adds too much damping. Moreover, the second-order LES results do not lie nearer to the measurements than the second-order DNS: the subgrid modelling doesn't seem to improve the results significantly. Here the word 'seem' is to be emphasized since both the numerical method and the grid used for the LES and the second-order DNS are not exactly the same. The comparison between the LES and the fourth-order DNS turns out to favour the latter: the fourth-order DNS - not requiring any modelling and hence only containing numerical errors - does a better job.

## 6. Higher Reynolds numbers: a DNS at Re=50,000

As is shown in Section 4, the fourth-order simulation technique is more economical than the second-order method. Thus with the fourth-order method higher Reynolds numbers can be attacked. On a $192^3$ grid (the grid is stretched away from the walls; the smallest grid-size is $2.5\ 10^{-3}$; the largest is approximately five times larger) $Re = 50,000$ is feasible. In this Section some results of the fourth-order method at $Re = 50,000$ are presented. Mean velocities, turbulence intensities and power spectra are computed and compared to their equivalents at $Re = 10,000$.

Mean velocities along the centre lines in the central plane are compared in Figure 5. It may be noted that the velocities have been averaged over a

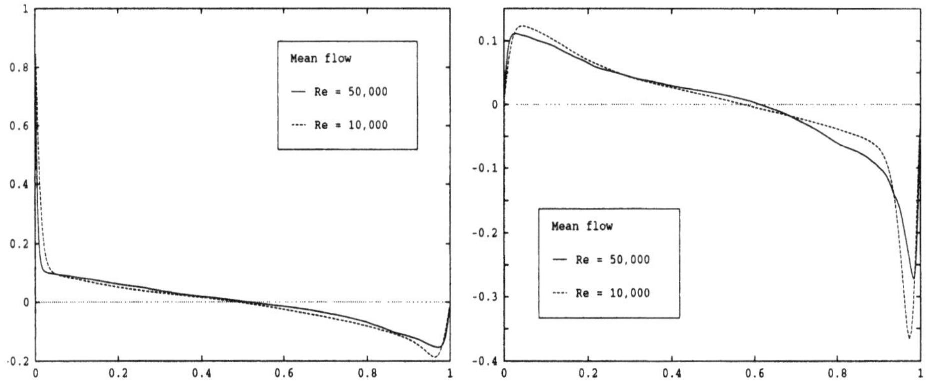


*Figure 5.* Comparison of mean velocities in the central plane of the cavity at Re = 10,000 and Re = 50,000. The continuous line corresponds to Re = 50,000; the dashed line corresponds to Re = 10,000. The left picture shows the horizontal velocity along the vertical centre line; the right picture shows the vertical velocity along the horizontal centre line.

relatively short period of time, namely over 150 characteristic time-units.

Table 1 shows the peak values (and their locations) of turbulence intensities along the horizontal centre line in the central plane at $Re = 10,000$ and $Re = 50,000$. As can be seen from Table 1, the maximum of $\overline{v'v'}$ lies approximately two and a half times closer to the wall at $Re = 50,000$ than at $Re = 10,000$. It may be remarked that at $Re = 50,000$ the maximum of $\overline{v'v'}$ is reached at the sixth grid point counted from the wall.

Unfiltered power spectra of a horizontal velocity near the bottom of the cavity are compared in Figure 6. In both spectra a $-5/3$ slope can be observed.

Figure 7 shows a snapshot of the vorticity in the central plane of the cavity at $Re = 50,000$. Compared to the flow at $Re = 10,000$ the number of small-scale vortical structures has increased drastically. To verify that

TABLE 1. Maximum values of $\overline{v'v'}$ and $\overline{u'v'}$ along the horizontal centre line in the central plane. Here $u'$ denotes the horizontal component of the fluctuating velocity; $v'$ is the vertical component.

| Reynolds number | $\overline{v'v'}$ | | $\overline{u'v'}$ | |
|---|---|---|---|---|
| | peak | location | peak | location |
| 10,000 | 0.104 | 0.957 | 1.11 $10^{-3}$ | 0.952 |
| 50,000 | 0.146 | 0.983 | 1.63 $10^{-3}$ | 0.956 |

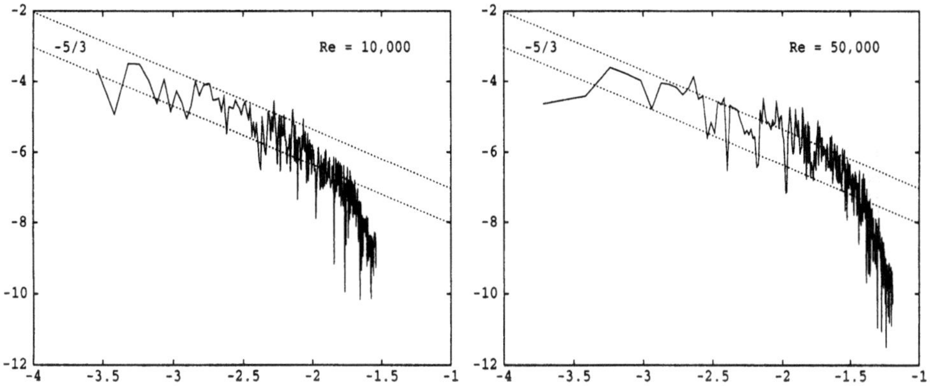


*Figure 6.* Comparison of power spectra at $Re = 10,000$ (left) and $Re = 50,000$ (right). Both pictures show the power spectrum of an u-velocity at a point at 2% of the height of the cavity along the vertical centre line in the central plane of the cavity (log-log plot).

all these small structures are indeed resolved, we have estimated the Taylor micro-scale from the flow data by approximating the distance between the location of the maximum and the location of the zero of the osculating parabola of a computed two-point velocity auto-correlation: the number of grid points per Taylor micro-scale is approximately equal to 8.

## Acknowledgements

Both the Stichting Nationale Computerfaciliteiten (National Computing Facilities Foundation, NCF) with financial support from the Nederlandse Organisatie voor Wetenschappelijk Onderzoek (Netherlands Organization for Scientific Research, NWO) and the National Aerospace Laboratory NLR are gratefully acknowledged for the use of supercomputer facilities. Moreover, the authors would like to thank Prof. A.K. Prasad for giving us the detailed data of the physical experiments.

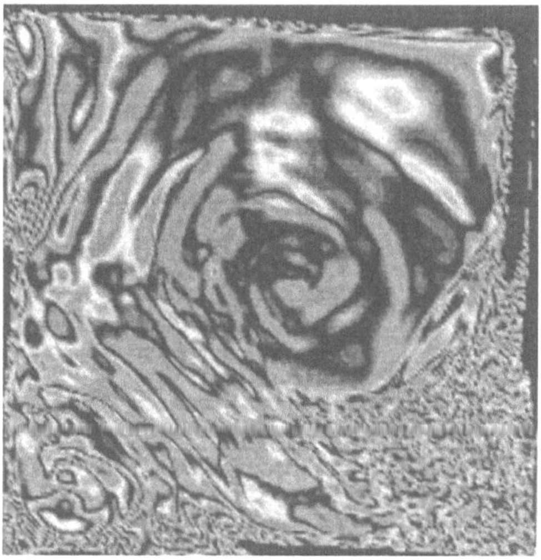

*Figure 7.* An illustration of the structures in the central plane of the cavity at Re=50,000. The orientation of the cavity is such that the upper lid is driven from the left to the right. Shown is the third component of the vorticity vector in the central plane of the cavity. Here, the vorticity is averaged over one large-eddy turn-around time.

## References

Fasel, H.F. (1990) Numerical simulation of instability and transition in boundary layer flows, *Laminar-Turbulent Transition*, D. Arnal & R. Michel (eds.). Springer-Verlag, Berlin.

Harlow, F.H. and Welsh, J.E. (1965) Numerical calculation of time-dependent viscous incompressible flow of fluid with free surface, *Physics of Fluids*, **8**, pp. 2182–2189.

Joslin, R.D., Streett, C.L. and Chang, C.L. (1992) Validation of three-dimensional incompressible spatial direct numerical simulation code - a comparison with linear theory and parabolic stability equation theories for boundary layer transition on a flat plate, NASA Technical Paper 3205.

Liu, Z. and Liu, C. (1994) Fourth order finite difference and multigrid methods for modeling instabilities in flat plate boundary layers - 2D and 3D approaches, *Computers & Fluids*, **7**, pp. 955–982.

Prasad, A.K. and Koseff, J.R. (1989) Reynolds number and end-wall effects on a lid-driven cavity flow, *Physics of Fluids A*, **1**, pp. 208–218.

Rai M. and Moin P. (1989) On direct simulations of turbulent flow using finite-difference schemes, AIAA-89-0369.

Verstappen, R.W.C.P. and Veldman, A.E.P. (1994) Direct numerical simulation of a 3D turbulent flow in a driven cavity at Re=10,000, *Computational Fluid Dynamics '94*, S. Wagner *et al.* (eds.), John Wiley & Sons, Chichester pp. 558–565.

Verstappen, R.W.C.P. and Veldman, A.E.P. (1996) A fourth-order finite volume method for direct numerical simulation of turbulence at higer Reynolds numbers, *Computational Fluid Dynamics '96*, J.A. Désidéri *et al.* (eds.), John Wiley & Sons, Chichester pp. 1073–1079.

Zang, Y.Z., Street, R.L. and Koseff, J.R. (1993) A dynamic mixed subgrid-scale model and its application to turbulent recirculating flow, *Physics of Fluids A*, **5**, pp. 3186–3196.

# SUBGRID-SCALE MODEL OF SCALAR MIXING FOR LARGE EDDY SIMULATIONS OF TURBULENT FLOWS

FABRICE MATHEY AND JEAN PIERRE CHOLLET
*Laboratoire des Ecoulements Géophysiques et Industriels*
*Univ. J. Fourier, INPG, CNRS, France*

**Abstract.** Results are presented for numerical simulations of passive scalar mixing in homogeneous and sheared turbulent flows. These results are generated with the Linear Eddy Model (LEM). The discussion is based on the subgrid-scale statistics and it is shown that the LEM model reproduces correctly the structure of the scalar field in the case of high Reynolds numbers when all the relevant length scales are represented. Thanks to the results obtained, a simpler closure model for the subgrid scalar fluctuations is presented.

## 1. Introduction

The understanding of turbulent mixing is crucial in the context of chemical reactions. For example the knowledge of quantities like mixing rates is determinant in chemical engineering. Large eddy simulations are very efficient for the calculation of mixing at large scales since these scales are dependent on initial and boundary conditions. For instance, in the mixing layers which are considered herein, the mixing between the two streams is quite well reproduced by computing explicitly the large scales. Because of the limited size of the computers, an explicit calculation of these large scales cannot be run simultaneously with the calculation at smaller scales. Concerning the velocity field, the contribution of these small length scales can be modeled by an eddy viscosity, but flows with chemical reactions are typical of problems where scalar mixing must also be described at all scales. The probability density function approach (p.d.f.) inspired from Reynolds average can provide this subgrid information, but needs other closures. We consider here the linear model of Kerstein (1988) used as a subgrid-



*J.-P. Chollet et al. (eds.), Direct and Large-Eddy Simulation II,* 103-114.

scale model. With this model, we investigate the entrainment and mixing processes, first in a homogeneous turbulence and then in a spatial three-dimensional turbulent mixing layer. The results are used to derive a simple model of subgrid scale fluctuations as a closure for the presumed p.d.f. approach.

## 2. Large eddy simulation

The large eddy field $\overline{f}$, which is explicitly calculated in the LES, can be obtained from its full field by using a filter $G$ of specified width $\Delta$ (typically the grid mesh size) :

$$\overline{f}(x,t) = \int_{-\infty}^{\infty} f(x',t)G(x'-x)dx' \tag{1}$$

By applying this filter to the compressible Navier-Stokes equations, we obtain a set of equations which describes the evolution of the resolved part of the flow. The unresolved quantities which appear in these equations are the subgrid stress tensor and the unresolved enthalpy flux term. These terms can be modeled with an eddy-viscosity and an eddy-diffusivity. In the present simulations, we have used the dynamic modeling procedure of Germano *et al.* (1991), extended to compressible flows by Moin *et al.* (1991). In this model, the coefficients are dynamically estimated by introducing a test filter whose size is wider than the grid filter $G$. If the grid filter is now applied to the scalar $\phi$ evolution equation, we get:

$$\frac{\partial \overline{\rho}\widetilde{\phi}}{\partial t} + \frac{\partial}{\partial x_k}(\overline{\rho}\tilde{u}_k\widetilde{\phi}) = \frac{\partial}{\partial x_k}(\overline{\rho}D_m\frac{\partial \widetilde{\phi}}{\partial x_k}) - \frac{\partial}{\partial x_k}(q_k) + \widetilde{\omega(\phi)} \tag{2}$$

where $D_m$ is the molecular scalar diffusivity. The equation is written in terms of the Favre-filtered quantities defined as $\widetilde{f} = \overline{\rho f}/\overline{\rho}$.

The last two terms of the right-hand side of the equation, the unresolved scalar fluxes $q_k$ and the filtered reaction rate $\widetilde{\omega(\phi)}$, need closure hypotheses. Eddy-diffusivity models similar to eddy viscosity ones can be developed for the unresolved flux term :

$$q_k = (\overline{\rho u_k\phi} - (1/\overline{\rho})\overline{\rho u_k}\,\overline{\rho\phi}) = -\overline{\rho}D_t\,\frac{\partial\widetilde{\phi}}{\partial x_k} \tag{3}$$

The eddy-diffusivity $D_t$ can be obtained with the dynamical procedure. The second term, the filtered reaction rate, involves strong correlations between concentrations of different species, and between concentrations and temperature in case of heat release. An accurate modeling of this term requires the knowledge of the subgrid field structure. Inspired from the

Reynolds average approach for combustion, we can model $\widetilde{\omega(\phi)}$ according to Gao and O'Brien (1993) :

$$\widetilde{\omega(\phi; \vec{x}, t)} = \int_{\psi=-\infty}^{\psi=+\infty} \omega(\psi) p(\psi; \vec{x}, t)) d\psi \tag{4}$$

where $\psi$ represents the probability space variable corresponding to the scalar $\phi$ and $p()$ the subgrid p.d.f. Once $p()$ is known, we can express the correct value of the filtered reaction rate term. The subgrid p.d.f. can be determined by solving a transport equation but this equation needs other closure hypotheses and can be very expensive to solve. A less expensive solution is to presume this p.d.f. according to its two first moments ($\overline{\phi}$ and $\overline{\phi'^2}$). The beta function is selected as evolving from a gaussian distribution centered around a mean value corresponding to perfectly mixed species to a two-peak distribution corresponding to perfectly segregated species. The scalar variance $\overline{\phi'^2}$ is calculated as $\overline{\phi^2} - \overline{\phi}^2$. Cook and Riley (1994) have suggested to evaluate $\overline{\phi'^2}$ according to a scale similarity model:

$$\overline{\phi^2} - \overline{\phi}^2 = c_f \left( \widehat{\overline{\phi}^2} - \widehat{\overline{\phi}}^2 \right) \tag{5}$$

where the symbol $\widehat{\ }$ corresponds to the same kind of test filter as used in the dynamical model of Germano with a size wider than the size of the mesh. Comparing with low Reynolds number DNS results, they found $c_f \approx 1$. With our simulation of high Reynolds number homogeneous turbulence, we will see that the constant $c_f$ depends on the size of the grid mesh. In order to correct this effect, we suggest a new model in the following form:

$$\overline{\phi^2} - \overline{\phi}^2 = \alpha\, \overline{\Delta}^2 \left( \frac{\partial \mathrm{L}(\overline{\phi})}{\partial x_i} \frac{\partial \mathrm{L}(\overline{\phi})}{\partial x_i} \right) \tag{6}$$

where L denotes a high pass filter applied on the scalar field, and $\overline{\Delta}$ the typical size of the mesh. The basis of this model is the same as in the LEM (figure 1) where fluctuations are first generated from gradients at large scale. As shown in section 3 this formulation leads to a constant value of $\alpha$ and then to a subgrid model independent of the size of the mesh.

In our study, the Linear Eddy Model of Kerstein will be used in order to compute high Reynolds number flows. Two configurations will be considered : the LEM in its original form to compute stationary homogeneous turbulence and the LEM coupled with a Large Eddy Simulation of a turbulent shear layer. The aim of these simulations is to compare the scalar subgrid fluctuations generated with the LEM with the subgrid fluctuations computed with the simpler models presented in this section.

## 3. Linear Eddy Model

The subgrid mixing model is based on the linear-eddy model, developed by Kerstein (1988), and extended to Large Eddy Simulation of turbulent combustion by Menon *et al.* (1993). Because it solves all relevant length scales of the flow, from the integral length-scale to the Kolmogorov scale, the LEM is able to separate turbulent stirring from molecular diffusion mechanisms which both govern the evolution of a scalar field. This is done through the consideration of a one dimensional domain, the linear domain, along which the 1D diffusion equation is solved :

$$\frac{\partial \phi}{\partial t} = D_m \frac{\partial^2 \phi}{\partial x} + \omega(\phi) \tag{7}$$

The turbulent stirring mechanism is modeled by a rearrangement process applied to the 1D scalar field of the linear domain. This process consists of a series of events, called triplet map, which rearrange the distribution of the scalar $c(x)$ along a given segment of size $l$, according to figure 1. Each rearrangement can be seen as the effect of a single turbulent structure on a scalar distribution. Rearrangement events are governed in space by the probability density function $f(l)$ and in time by the frequency of stirring $\lambda$. $f(l)$ is the p.d.f. of the eddy size $l$ of the triplet map. Based on scaling arguments, it can be shown that these parameters satisfy :

$$f(l) = \frac{5}{3}\frac{l^{8/3}}{\eta^{-5/3} - L^{-5/3}} \qquad \lambda = \frac{54}{5}\frac{\nu R_{eL}}{L^3}\left(\frac{L}{\eta}\right)^{5/3} \tag{8}$$

where $L$ is the integral scale and $\eta$ the Kolmogorov scale ($\eta < l < L$).

The Reynolds number $R_{eL}$ is operationally defined as $(L/\eta)^{4/3}$. In the case of a Large Eddy simulation, the linear domain is a statistical representation of the subgrid scalar fluctuations. The splicing process of Menon *et al.* (1993) is used to convect the mean and fluctuating scalar fields across each cell according to the resolved velocity field. This is done by taking into account LES resolved convective fluxes and an additional transport due to turbulent fluctuations through each cell of the LES computation. A segment of each linear domain is copied in the adjacent cells according to these fluxes. We have slightly improved this process by using the PPM advection scheme (piecewise parabolic method, Colella & Woodward, 1984) in order to ensure mass and species conservation for the mean scalar fields.

## 4. Homogeneous turbulence

In this part the Linear Eddy Model is used to model the evolution of a scalar field in a steady homogeneous turbulent flow field. The velocity and scalar

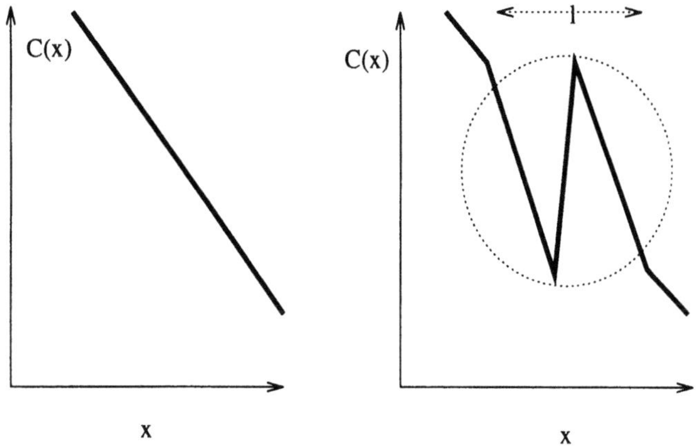


*Figure 1.* The basic mechanism of the LEM model : the triplet map

fields are not calculated with a LES. The velocity field has the properties of a homogeneous turbulence and the scalar is computed with the rearrangement events described in the previous section. The reason for using the LEM in this configuration is its relatively low computational cost, and the fact that it provides a way of simulating high Reynolds number flows. Mac Murtry et al. (1993) have shown that in this configuration the LEM is able to reproduce the results of the DNS of Eswaran and Pope (1988). In a recent paper, Jaberi *et al.* (1996) use the LEM and a DNS to study the nature of mixing and the scalar p.d.f.. Following this approach, we have used the LEM to compute high Reynolds number homogeneous turbulence in order to make "a priori" tests of the models presented in section 2. DNS will be carried out in a future work to validate the present results.

The LEM calculations deal with the transport of a passive and conserved scalar variable. The resolution requirements depend on the Reynolds number, $Re_L = (L/\eta)^{4/3}$, where $L$ is defined as the largest eddy size (approximately the size of the domain). For $Re_L = 600$ and with six computational elements to resolve the eddies at the Kolmogorov scale, 700 elements are needed to resolve the complete flow field. The statistics are computed by averaging several independent realizations. For the simulations presented below, the scalar field statistics were averaged over 1000 realizations.

The scalar field which can be interpreted as a mixture fraction $f$, is initially permitted to take the values of $+1$ or $-1$ only, with equal probability. The transition through the interface is smoothed in order to avoid strong gradients. The specification of this initial scalar field is described by Eswaran and Pope (1988).

The figure 2 shows the evolution of the scalar p.d.f. during the LEM calculations. At the beginning, the p.d.f. exhibits two peaks at $+1$ and $-1$

corresponding to the two unmixed species. Then, when mixing occurs, the distribution evolves towards a one peak configuration corresponding to the value 0.5 of perfect mixing.

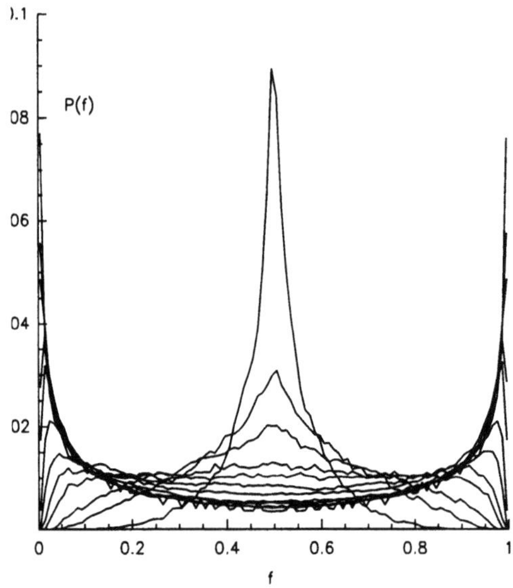


*Figure 2.* Evolution with time of the p.d.f. of the mixture fraction $f$ in a homogeneous turbulence, from LEM calculations

The scalar field computed by the LEM is filtered with different box filter sizes $fi$ (corresponding to the notation $f1 = 8\Delta$ to $f4 = 32\Delta$). The subgrid scalar fluctuations are computed from *(i)* the filtered exact scalar field, *(ii)* the subgrid model of Cook and Riley and *(iii)* the gradient model suggested in section 2, with (GM model) or without (FGM model) contributions from the largest scales. The results are given in figure 3. The exact subgrid fluctuations (corresponding to the LEM curve) first increase due to the growing contribution of the small scales and then decrease as energy decays. In fact both the model of Cook (CM) and our first formulation without the Laplacian filter (GM model) over-predict the subgrid fluctuation at the beginning.

The figures 4.a) and 4.b) display the values of the coefficients $c_f$ and $\alpha$ of equations (5) and (6) for the different models. The different curves correspond to the different filters $fi$ in use. Both coefficients depend on the filter size. But concerning the GM model, we can explain the disagreement with the LEM at the beginning of the simulation. Srong gradients initially develop at larges scales without significant fluctuations. In order to take into account this effect the model has to separate the large scale components of the resolved field from the smaller ones. This problem is similar to the problem of evaluation of the filtered structure function model solved by Ducros *et al.* (1996). To get rid of the large scales, the filtered structure function model consists in filtering the field with a Laplacian filter. In the same way the scalar field $\phi$ is filtered before computing the gradient in equation (6). The filtered model (FGM) (figure 3) captures correctly the initial behavior of the filtered scalar fluctuations. Moreover, the new constant $\alpha$ in figure 4.c is observed to be independent of the size of the mesh. $\overline{\phi'^2}$

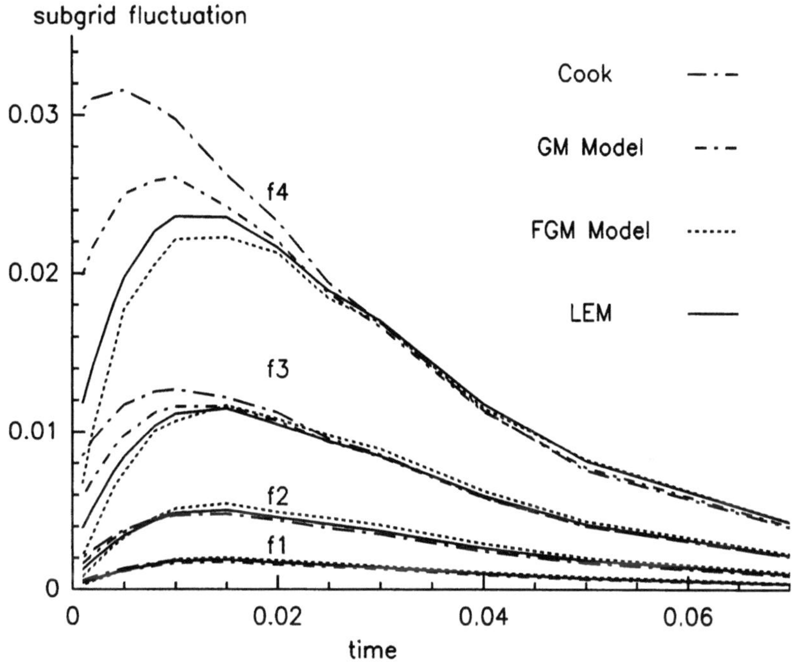


*Figure 3.* Subgrid fluctuation $\overline{\phi'^2}$ computed with different models and different sizes of the filter

## 5. Weakly compressible turbulent shear layer

The turbulent shear layer investigated in this section corresponds to the experimental investigations of Mungal *et al.* (1985) and the recent numerical large eddy 2D-simulations of Calhoon *et al.* (1996). The filtered compressible Navier-Stokes equations are numerically solved with our LES code (developped by Gathmann, 1993) based on the piecewise parabolic method. The Reynolds subgrid terms are evaluated with the dynamical model. Like in the previous section the subgrid scalar field is computed with the LEM. Each cell of the computational domain is coupled with a subgrid domain, the linear domain. The numerical resolution is $200 \times 64 \times 48 \times 150$. The last

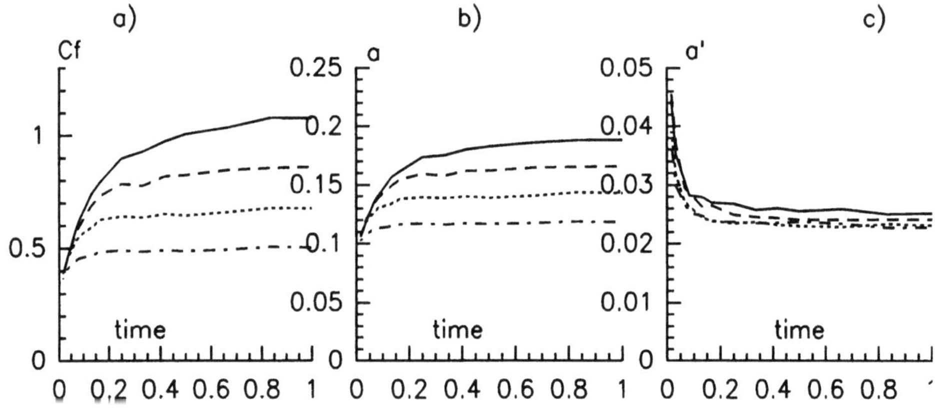


*Figure 4.* Coefficients $c_f$ and $\alpha$ versus time for different sizes of the filter ; from left to right : a) Cook model, b) GM gradient model, c) FGM : filtered gradient model

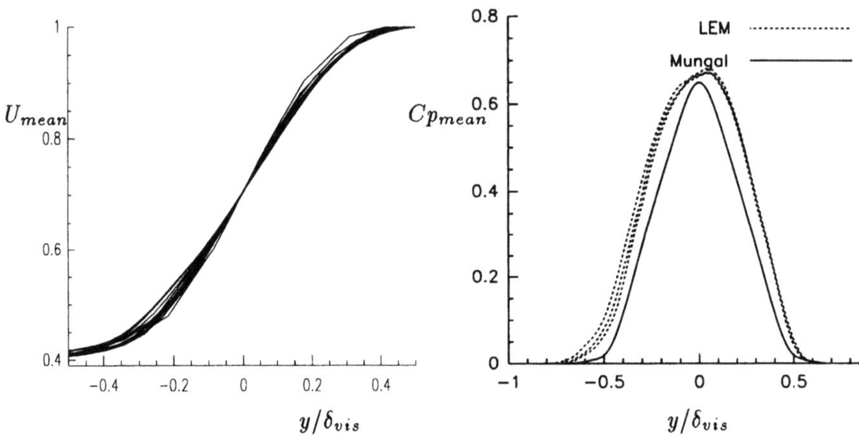


*Figure 5.* a) mean velocity profile b) mean product concentration profile at different streamwise locations

dimension corresponds to the linear domain for the subgrid scalar mixing, each node represents a step in the splicing process. The subgrid statistics are computed by filtering the scalar field with a filter of the size of the mesh.

The species which are transported by each of the two streams of the mixing layer are supposed to be subjected to an ireversible chemical reaction $F + O -> P$, with the assumption of infinite rate chemistry. Equal molecular diffusivities $D_m$ are assumed for the species and the passive scalar $\phi$ transported under this hypothesis is the Shvab-Zel'dovich mixture fraction $f$. The product concentration $C_p$ can be evaluated from the mixture fraction by:

$$C_p = f \quad if \quad f < \frac{C_a}{C_A + C_B} \qquad C_p = 1 - f \quad if \quad f > \frac{C_a}{C_A + C_B} \tag{9}$$

## 5.1. SUBGRID-SCALE WITH THE LEM MODEL

The LEM is used here as the subgrid-scale scalar mixing model. The results are normalized by the visual thickness of the layer ($\delta_{vis}$), defined as the width of the domain where the streamwise velocity ranges from 99 to 101% of its value at the center of the layer. The mean velocity profiles obtained from the simulation are found to be self-similar (figure 5.a). The growth rate of the layer is about 0.18. The mean concentration profiles across the layer are given in figure 5.b, for different positions along the stream-wise axis, and are in relatively good agreement with the experimental results. The maximum concentration value slightly over-predicts the product formation. This small discrepancy is due to the estimation of the local turbulent Reynolds as a parameter of the LEM, calculated with the eddy viscosity derived from the Germano's model.

The Broadwell and Breidenthal model (BBM) of mixing (1982) suggests that the total amount of product formed in a shear layer is a decreasing function of the Reynolds number. The experiences of Mungal *et al.* (1985) and the two dimensional temporal simulations of Calhoon and Menon (1996) confirm this trend. The BBM model suggests that the amount of product can be divided into two parts : a Reynolds independent part, determined by the global entrainment ratio, which consists in the homogeneous mixture in the core of the rolls, and a Reynolds dependent part, the flame sheet region, where new pure unmixed fluid from one side diffuses across a thin layer into the homogeneous part or into the unmixed fluid from the other side. The figure 6 displays the product thickness $\delta_p$ obtained from the simulation, (defined from $\delta_p = \int_{-\infty}^{+\infty} C_p(y)dy$ and normalized by $\delta_{vis}$). On the same figure, we have plotted the experimental results of Mungal. After the mixing transition, the results are similar and confirm that the total amount of product formed in a shear layer is a decreasing function of the Reynolds number. This agreement shows that the LEM is able to reproduce correctly the small-scale effects.

## 5.2. SUBGRID-SCALE WITH THE GRADIENT MODEL

The subgrid scalar field obtained in the previous section with the LEM is now compared with the corresponding field calculated with the simpler subgrid-scale model, the GM model presented in section 2. The figure 7

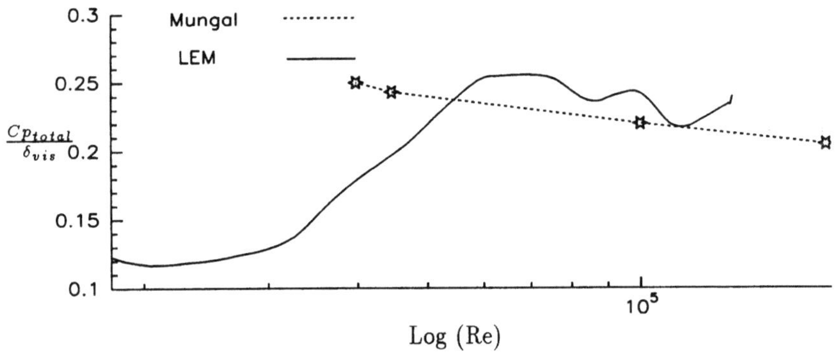


*Figure 6.* Product thickness versus the Reynolds number

shows the instantaneous evolution of the subgrid fluctuations along the streamwise axis of the mixing layer. The results obtained with the GM model (without the Laplacian filter) are observed to be in good agreement with the results given by the LEM, past the mixing transition. But at the beginning of the computational domain the GM model overpredicts the subgrid fluctuations (not shown on the figure). This effect is probably due to large scales of the scalar fields present in this region as mentioned in the case of homogeneous turbulence. We have not yet computed the simulation with the filtered version of the G model (FGM) which is expected to improve the results in this region of the flow where large scales dominate. The correlations between the subgrid scalar fluctuations $\overline{\phi'^2}$ calculated first with the LEM and the model of Cook, then with the LEM and the GM gradient model are shown figure 8. The values of $\overline{\phi'^2}$ at each space location are averaged along the spanwise direction. In this case the results are similar. But we recall that with the formulation (6) we can dynamically compute the value of the coefficient $\alpha$.

## 6. Conclusion

In this paper we have shown that the subgrid scalar fluctuations can be efficiently modeled with the (GM) gradient model in the case of high Reynolds number homogeneous turbulence. The constant of the model is found to be independent of the size of the mesh. The combination of a low pass filter, here a Laplacian filter, with the usual high pass filter of the LES significantly improves the behaviour of the subgrid-scale model. In the case of a high Reynolds number turbulent mixing layer, the subgrid fluctuations computed by the gradient model are in good agreement with the results from the Linear Eddy Model. Going further, the formulation can be used in conjunction with the dynamic procedure to get a subgrid-scale model

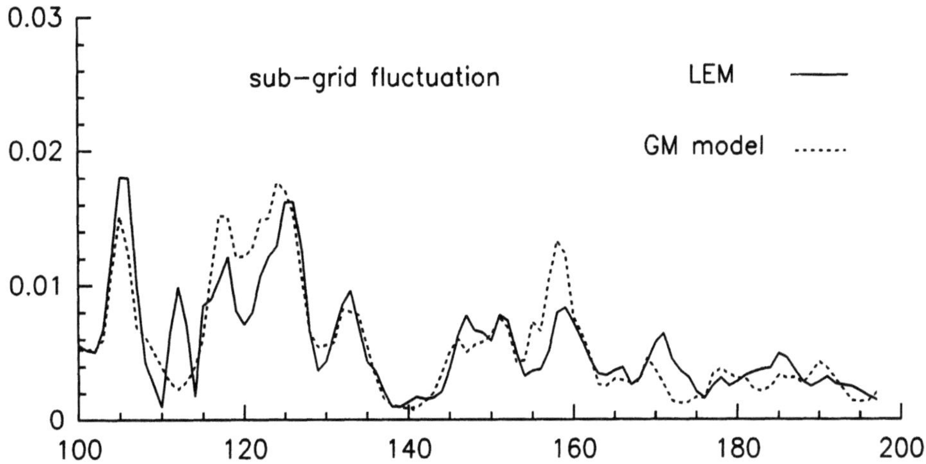


*Figure 7.* Subgrid fluctuations $\overline{\phi'^2}$ of the scalar

without any adjustable parameters for subgrid modeling. It would make possible the extension of the present work to reacting flows with finite rate chemistry .

## Acknowledgements

This work was supported in part by the SNECMA. The numerical calculations were carried out on the CRAY C98 of the IDRIS (CNRS) and on the CRAY C94 of the CENG computing center.

## References

Broadwell, J.E., Breidenthal, R.E. 1982 A simple model of mixing and chemical reaction in a turbulent shear layer, *J.Fluid Mech.* **125**, 1982, 397-410.

Calhoon, H., Menon, S. 1996 Subgrid Modeling for Reacting Large Eddy Simulations, *AIAA paper* 96-0516, 34th Aerospace Sciences Meeting, Reno.

Colella, P., Woodward, P.R. 1984 The piecewise parabolic method for gas-dynamical simulations, *J. Comp. Phys.* **54**, 174-201.

Cook, A.W., Riley, J.J. 1994 A subgrid model for equilibrium chemistry in turbulent flows, *Phys. Fluids* **6** (8), 2868-2870.

Ducros, F., Comte, P., Lesieur, M. 1996 Large-eddy simulation of transition to turbulence in a boundary layer developing spatially over a flat plate, *J. Fluid Mech.* **326**, 1-36.

Eswaran, V., Pope, S.B. 1988 Direct numerical simulations of the turbulent mixing of a passive scalar, *Phys. Fluids* **31** (3), 506-520.

Gao, F., O'Brien, E.E. 1993 A large-eddy simulation scheme for turbulent reacting flows, *Phys. Fluids* A **5**, 1282-1284.

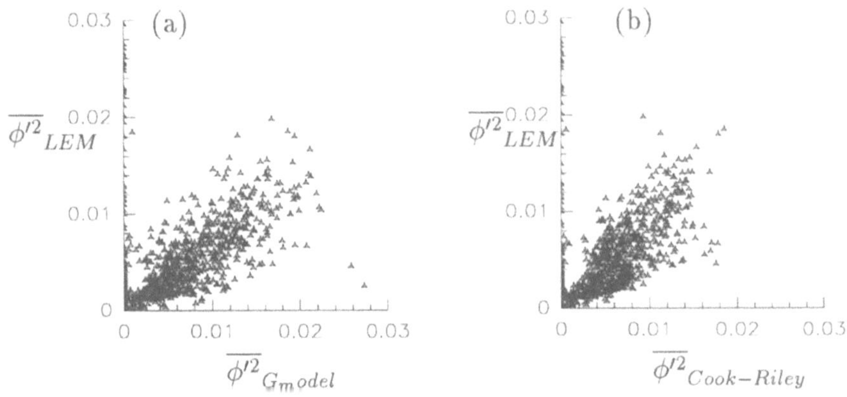


*Figure 8.* Subgrid scalar fluctuations $\overline{\phi'^2}$ : a) LEM versus GM model ; b) LEM versus Cook model

Gathmann, R.J. 1993 Analyse d'écoulements supersoniques et réactifs par simulation numérique instationnaire tridimensionnelle, Thèse de l'Universite Joseph-Fourier - Laboratoire des Ecoulements Géophysiques et Industriels.

Germano, M., Piomelli, U., Moin, P., Cabot, W.H. 1991 A dynamic subgrid-scale eddy viscosity model , *Phys. Fluids* A3, 1760-1765.

Jaberi, F.A., Miller, R.S., Madnia, C.K., Givi, P. 1996 Non-Gaussian scalar statistics in homogeneous turbulence, *J. Fluid Mech.* **313**, 241-282.

Kerstein, A.R. 1998 Linear-Eddy Model of Turbulent Scalar Transport and Mixing, *Comb.Sci Tech.* **60**, 391-421.

Menon, S., Mac Murtry, A., Kerstein, A.R. 1993 A Linear Eddy Mixing Model for Large Eddy Simulation of Turbulent Combustion, in *Large Eddy Simulations of Complex Engineering and Geophysical Flows*, pp.287-314, B. Galperin, S. Orsag, eds. Cambridge University Press.

Moin, P., Squires, K., Cabot, W., Lee, S. 1991 A dynamic subgrid-scale model for compressible turbulence and scalar transport, *Phys. Fluids* A **3**, 2746-2757.

Mungal, M.G., Hermanson, J.C., Dimotakis, P.E. 1985 Reynolds Number effects on Mixing and Combustion in a Reacting Shear Layer, *AIAA journal* **23** (9), 1418-1423.

Mac Murtry, P.A., Gansauge, T.C., Kerstein, A.R. Krueger, S.K. 1993 Linear eddy simulation of mixing in a homogeneous turbulent flow, *Phys. Fluids* A **5** (4), 1023-1034.

# TURBULENT SIMULATION OF A FLAT PLATE BOUNDARY LAYER AND NEAR WAKE

SHIAN GAO
*Department of Engineering, The University of Leicester*
*Leicester LE1 7RH, UK*

AND

PETER R. VOKE AND TIM GOUGH
*Department of Mechanical Engineering*
*The University of Surrey, Guildford GU2 5XH, UK*

**Abstract.** Extensive LES studies have been made of the low-Reynolds-number turbulent boundary layer following a trip, and the near wake behind a thin flat plate. The numerical results are compared with those from parallel experiments of identical flow conditions conducted in conjunction with the simulations. Apart from the integral parameters and profiles of mean quantities and fluctuations, the balance analyses of Reynolds stress, kinetic energy and dissipation have also been derived. These linked studies allow us to understand in greater depth the complex dynamics that operates in both wall-bounded and spatially developing turbulent flows.

## 1. Introduction

The near-wake region behind a thin body has always presented difficulties to closure modellers. It is known that the flow in this region is far from an equilibrium state, as the sudden removal of the constraint of the solid wall results in a flow which is dynamically out of balance, and two boundary layers previously separated by the solid body suddenly come into contact. Understanding and modelling of this region is very important to engineering, since trailing-edge flows are critical in determining the behaviour of lifting surfaces in both turbomachinery and aircraft design.

Studying the dynamics of this situation experimentally or by numerical simulation is also challenging. A great deal of critical activity takes



*J.-P. Chollet et al. (eds.), Direct and Large-Eddy Simulation II,* 115-124.

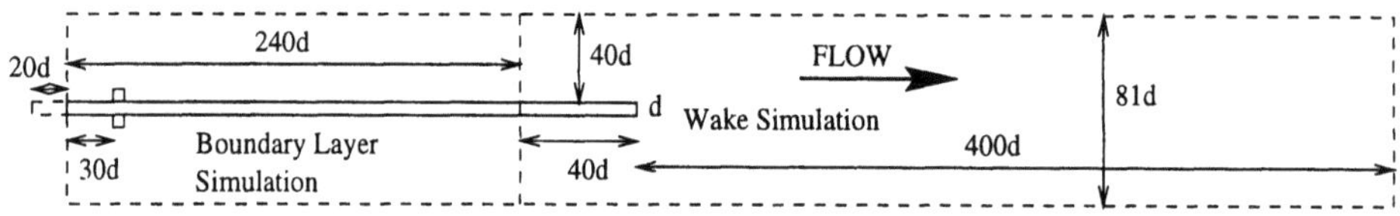


*Figure 1.* Geometry of the simulations (thickness and trip not to scale).

place just behind the trailing edge, and adequate resolution (experimental or computational) of the flow there is not always practical to achieve. In the present study, we also took the view that the nature of the boundary layers as they approach the trailing edge might also be very important, and that a crude representation of a boundary layer, for instance using pseudo-random disturbances at an inflow plane, would not be convincing. The critical nature of the flow up to and just after the trailing edge tends to confirm this view that the approaching boundary layers do need to be simulated with some veracity.

The organisation of our simulations is illustrated in Figure 1. The boundary layer is simulated separately, and data from this simulation at widely separated times is transferred to the inflow plane of the wake simulation to represent two statistically independent and uncorrelated boundary layers on either side of the plate of uniform thickness $d$. These layers develop in the wake simulation for a short distance and then interact beyond the trailing edge. Upstream effects of the trailing edge on the boundary layers are thus substantially accounted for.

We have analysed in considerable detail the statistical and structural properties of both the boundary layer and wake simulations, in comparison with coupled experiments by Gough *et al.* (1996a and b). The results presented here are selected to highlight key conclusions.

## 2. Methods

### 2.1. NUMERICAL TECHNIQUES

The second-order finite volume code employed in the study solves the incompressible Navier-Stokes equations on a staggered mesh. The pressure field is processed efficiently using Fourier decomposition in the $z$ (span) direction and a highly vectorised cyclic reduction algorithm in the other two directions. This allows the $x$ and $y$ dimensions to accommodate spatial stretching of the mesh and spatial development of the flow, solid walls and strong shear.

Each complete simulation is carried out in two separate parts. The boundary layer simulation is run first and the velocities on a predetermined plane (known as a slicing plane) are stored at every time step. These data are then used to define the inflow boundary conditions for the subsequent wake simulation. The inflow condition of the boundary layer simulation is a Blasius profile at $Re_x = 13241$. The transition is triggered by inserting a force in the streamwise direction, varying in such a way as to oppose the flow, in a small region that corresponds to the position of the solid trip in the experiment, $x/d = 50$ or $Re_x = 33100$. The slicing plane is at $x/d = 260$ ($Re_x = 172120$) measured from the leading edge. Note that the leading edge of the plate itself cannot be simulated.

The trailing edge is blunt with thickness $d$ and the length of the plate is $x_t = 300d$. In the wake simulation a length of the trailing edge is included (from $x/d = 260$ to $x/d = 300$), allowing interactive effects between the flow downstream of the trailing edge and the boundary layer near the edge. The intruding length of trailing edge is treated by the methods described by Voke and Potamitis (1994), though in other respects the present simulations are superior to that work. The inflow boundary conditions for the wake are defined by imposing the velocities from the slicing plane, recorded at two widely separated times.

## 2.2. RESOLUTION

The computational domains and meshing for the simulations are given in Table 1. Dimensions are in terms of the thickness of the plate $d$, which was uniform. The Reynolds number based on $d$ and the upstream velocity $U_0$ is 662.

TABLE 1. Domain and Mesh

| | Layer | Wake | | Layer | Wake |
|---|---|---|---|---|---|
| $L_x/d$ | 320 | 440 | $\Delta_x/d$ (in) | 3 | 2.8 |
| $L_y/d$ | 40 | 81 | $\Delta_x/d$ (min) | 1 | 0.1 |
| $L_z/d$ | 40 | 40 | $\Delta_x/d$ (out) | 2.8 | 10.6 |
| $N_x$ | 127 | 127 | $\Delta_y/d$ (min) | 0.03 | 0.1 |
| $N_y$ | 64 | 92 | $\Delta_y/d$ (max) | 2.8 | 4.0 |
| $N_z$ | 72 | 72 | $\Delta_z/d$ | 0.556 | 0.556 |

Note that non-uniform grids have been used in both $x$ and $y$ directions. The finest $x$ resolution is at the trip plane in the boundary layer and at the trailing edge for the wake simulation. The mesh is compressed from the inflow plane of the wake simulation, $40d$ upstream of the trailing edge, and the stretching of the $x$ mesh beyond the trailing edge is gradual.

The $y$ meshes are stretched, the small value of $\Delta_y$ at the solid surface being determined by our desire to avoid using a wall-function boundary condition. The cell nearest the wall has dimension approximately $\Delta_y^+ = 1$ in terms of wall units for the turbulent boundary layer towards the end of the plate. (To convert any dimension in terms of $d$ to these wall units, multiply by about 30.) The trip was $0.4d$ high and $1d$ long. The mesh in the wake simulation must be coarser in $y$ that for the boundary layer simulation, otherwise vertical ($v$) motions across the highly refined mesh result in an excessively strict Courant limit to the time step. Note also that the $y$ meshes are not matched between the simulations from the slicing to the inflow plane. Instead, the velocities are interpolated between the meshes in a simple manner that strictly conserves the mass flux.

### 2.3. SUBGRID-SCALE MODEL

The simulations were carried out using a subgrid-scale model similar to that of Smagorinsky,

$$\nu_s = (\Delta C_s)^2 \sqrt{2 s_{ij} s_{ij}} \,, \tag{1}$$

modified to allow for the cutoff falling in the dissipation portion of the spectrum as described by Voke (1996), with $\beta = 0.22$:

$$\nu_e = \nu_s - \beta\nu \left[1 - \exp(-\nu_s/\beta\nu)\right] . \tag{2}$$

The length scale $\Delta$ was related to the mesh and further reduced near the wall by a damping factor of standard form:

$$\Delta = (\Delta x \Delta y \Delta z)^{1/3} \left[1 - \exp(-y^+/A^+)\right] \tag{3}$$

where $A^+ = A u_\tau / \nu = 25$.

The constant of the subgrid model has to be altered systematically both in the region near the numerical trip and in the region after the plate trailing edge in order to obtain reasonable comparisons with the experimental data. In both cases it is believed that inadequate resolution of the actual physical behaviour of the fluid in the small separated regions is to blame. The need to alter $C_s$ suggests a dynamic subgrid scale formulation should be used to handle naturally the variation in the strength of the subgrid terms in the near-wake region; this possibility was not part of the present project.

The value of $C_s$ was fixed at 0.1 in the boundary-layer simulation except just downstream of the trip where a value 0.4 was imposed locally; this was needed to remove excessive fluctuations. In the wake, the value of $C_s = 0.1$ was maintained up to the trailing edge, and then rose in the near wake gradually to a value of 0.35 far downstream.

### 2.4. BOUNDARY CONDITIONS

The inflow condition for the boundary layer was a Blasius profile, with free-stream velocity $U_0$. For the wake simulations the inflows were taken from the boundary layer simulation as described above.

The solid surfaces were all sufficiently well resolved to allow the use of natural no-slip boundary conditions, though the near-wall resolution of the boundary layer in the wake simulation ($y_1^+ \approx 3$) was inferior to that in the boundary layer simulation ($y_1^+ \approx 1$).

The upper boundaries (and lower in the wake simulation) were free-slip with the mass flux through the boundaries carefully adjusted to produce a very low or zero pressure gradient in the free stream. This resulted in the requirement for mass to be removed (at a constant rate) through the upper surface from the boundary layer simulation to accommodate the growth of the displacement thickness in the layer, and a similar small inflow of mass through the lateral boundaries of the wake simulation to balance the reduction of displacement thickness as the wake fills in.

For all simulations the outflow boundaries were treated by the well-known advective boundary condition in the same manner as Voke and Potamitis (1994) with the mass flux through the outflow boundary balancing precisely the net mass inflow through the other boundaries.

### 2.5. SIMULATIONS

The time step was $0.0768d/U_0$ in the boundary layer simulation. In total the layer was run for $61250\Delta t$ on a single processor of a Cray Y-MP 8, at a cost of about 65 cpu hours. 18% of this time was required for conditioning, and statistics and sliced data were gathered over the remainder. The initial condition in the boundary layer simulation is a Blasius solution which rapidly breaks down from the trip position.

The wake simulation is started from zero flow but establishes a wake-like organisation as soon as the inflow and outflow boundary conditions are imposed. The flow is taken to be statistically stationary after the flow has passed entirely though the volume three times. The wake required a time step a quarter of that used for the boundary layer, owing to the Courant restriction arising from lateral ($v$) fluctuations through the refined $y$ mesh in the central region. Each inflow data plane was therefore reused four times. 120000 time steps were carried out for the wake, costing about 145 cpu hours, with averages taken over the last 100000 steps.

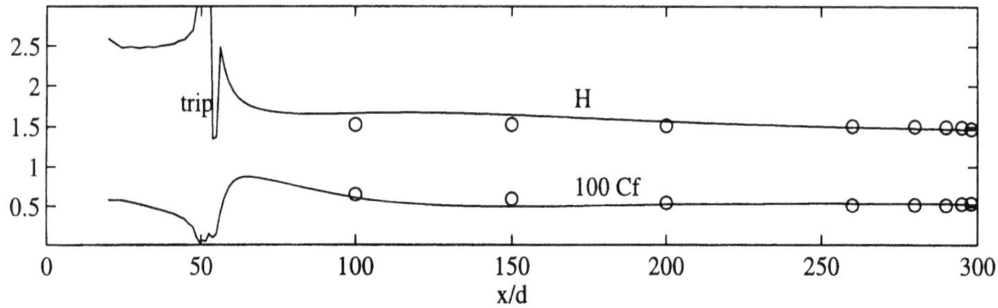


*Figure 2.* Streamwise distributions of shape factor $H$ and friction factor $c_f$ for the boundary-layer. Lines, simulation; symbols, experiment.

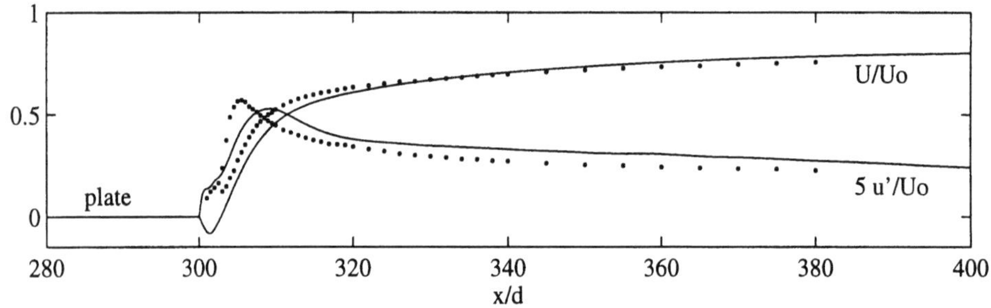


*Figure 3.* Mean $U$ and r.m.s. fluctuation $u'$ along the centreline in the wake. Lines, simulation; points, experiment.

## 3. Results

The streamwise distribution of shape and friction factors in Figure 2 agree well, and indicate that the boundary layer is typical of low Reynolds number layers. The real flow may settle a little faster than the simulation to the standard turbulent shape factor of about 1.4.

To compare the development of the wake, we look at the distribution along the wake centreline of the streamwise mean velocity and r.m.s. fluctuation, Figure 3. It is immediately obvious that the experimental measurements indicate a faster initial response. The experimental points very close to the trailing edge, in the region where the simulation shows a small mean recirculation, are omitted since the use of a single hot wire does not allow the determination of negative velocities. Nevertheless it is apparent that the physical recirculation is shorter and that the initial reaction of the flow is faster than the LES predicts. Detailed inspection of the development of a number of quantities indicates that in the near wake region many phenomena occur 25% closer to the trailing edge than in the LES. Aside from this, the qualitative agreement is generally good.

To study in further detail the differences between the LES and experi-

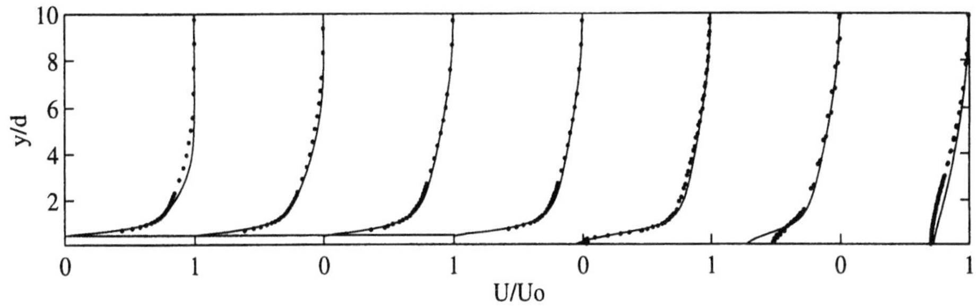


*Figure 4.* Velocity profiles $U/U_0$ at (left to right) $x/d = 100, 200, 280, 298, 302, 308, 340$.

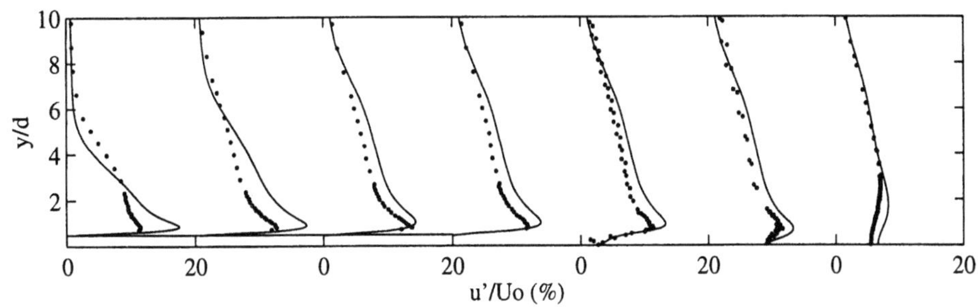


*Figure 5.* Profiles of r.m.s. fluctuation $u'$ as % of $U_0$; as Fig. 4.

mental measurements, profiles will be discussed at seven stations, four in the boundary layer including one just before the trailing edge, and three in the wake including one just after the trailing edge. Figure 4 shows the profiles at these stations of $U$, compared with experimental data. $y/d$ is measured from the centreline of the plate of thickness $d$. The agreement is good, although the physical boundary layer appears to be somewhat more developed at the earliest station shown than the LES. In the wake, the LES data is averaged over both halves of the simulation, while individual experimental measurements taken both above and below the centreline ($y = 0$) are shown.

Figure 5 shows the $u'$ profiles at the same stations. The LES, using its crude representation of the trip, has a higher peak value of $u'$ at the earliest station and this tendency persists downstream right into the wake. The peak in the boundary layer and in the very near wake is further from the wall or the centreline than the experiment indicates, and slightly higher fluctuation levels persist in the outer layers until the wake is established.

Figure 6 shows the $v'$ profiles, and Figure 7 those for $w'$ at the same stations. Experimental measurements of these quantities are not available

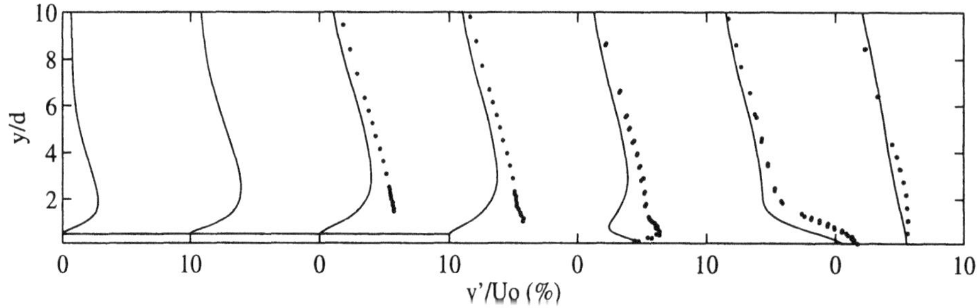


*Figure 6.* Profiles of r.m.s. $v'$ as % of $U_0$; as Fig. 4.

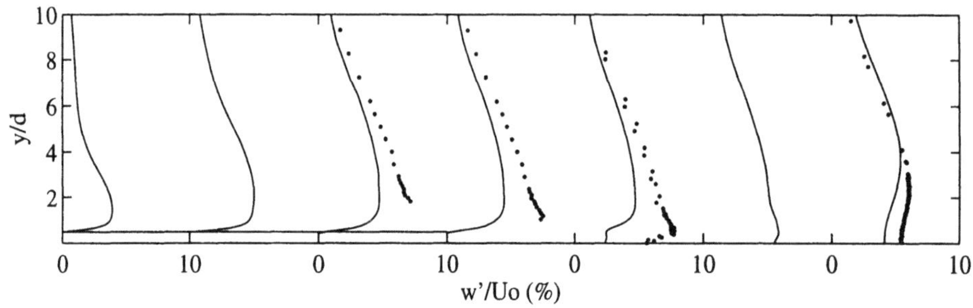


*Figure 7.* Profiles of r.m.s. $w'$ as % of $U_0$; as Fig. 4.

at the first two stations, nor is $w'$ at $x/d = 308$. The LES predicts lower values in the boundary layer and near wake, and a longer persistence beyond the trailing edge of the central dip in both these fluctuation quantities. We do agree qualitatively with the finding that $v'$ acquires a sharp central spike (see Figure 6, sixth profile at $x/d = 308$) in the near wake. This is may be partly due to the presence of semi-coherent vortex shedding from the trailing edge, although it is also suggested that the persistence of boundary-layer turbulent stress distributions in the near wake may also cause such overshoots. It is also known that the appearance of a strong streamwise strain $dU/dx$ may greatly enhance production.

Figure 8 shows the profiles of the off-diagonal stress $\overline{-u'v'}$ at the same seven stations. Agreement again is adequate or excellent at the earlier boundary layer station and the last wake station shown, but there is disagreement at the station just beyond the trailing edge at $x/d = 302$, again perhaps owing to poor resolution by the LES of the detailed flow in the very near wake This station is only $2d$ downstream of the blunt trailing

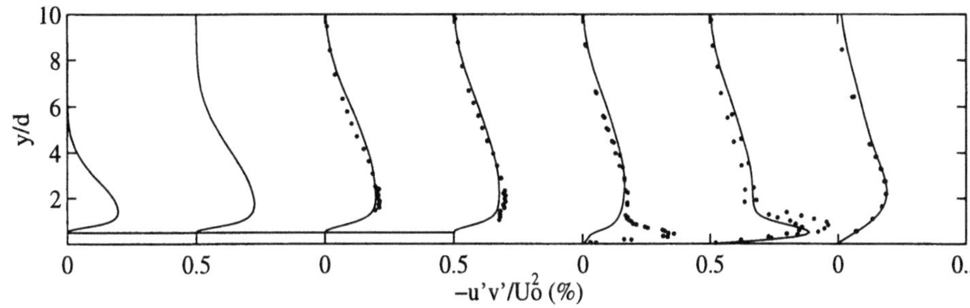


*Figure 8.* Profiles of $\overline{-u'v'}$ as % of $U_0^2$; as Fig. 4

edge of thickness $d$ and in the region of the small mean recirculation in the simulation of length about $2.5d$; the cross-wire data showing a central spike developing may therefore not be reliable here. However the LES does indicate the development of such a central spike in $\overline{-u'v'}$ slightly later than the experiment: see the next profile at $x/d = 308$. By this stage the experimental profile is already beginning to develop a central dip in $\overline{-u'v'}$ as the maximum of the Reynolds stress starts to move outward into the region of highest shear. (The experimental results at the $x/d = 308$ station are not symmetrical on the two sides of the plate owing to hot-wire drift.) Further downstream, where the development is taking place at a less rapid pace, the agreement between experiment and LES is good.

It is clear that most of the differences found between the simulation and experimental data can be attributed either to the crude representation of the trip which results in a persistent high level of $u'$, or to the slightly slow rate of streamwise development of the simulation in the very early wake just beyond the trailing edge. Many of the interesting phenomena found experimentally in the very near wake are also present in the LES but occur a little further downstream, suggesting that the subgrid diffusion of momentum towards the centreline of the wake may be low.

The critical nature of the near wake flow in this study suggests that the use of genuine turbulent boundary layer data obtained from the separate (but relatively cheap) boundary layer simulation is important. It appears unlikely that better agreement could be obtained using pseudo-random or analytically defined disturbances at the inflow plane of the wake simulation, unless instabilities in a transitional wake are the subject of study.

Budgets of Reynolds stress, kinetic energy and dissipation have also been derived from the LES at four of the stations and some comparable budgets from the experiment. These are in broad agreement and give further insight into the mechanisms at work in the near wake which support our conclusions. The details are discussed by Gough *et al.* (1996a and 1996b).

## 4. Conclusions

Large-eddy simulation has been used for the study of a tripped boundary layer and near wake behind a plate, in a comparative study with experiment. Velocity data from the boundary layer simulation are used to define the inflow for the simulation of the near wake.

It is found that the main sources of disagreement can be traced to poor resolution of the physically small elements of the geometry: the trip and the blunt trailing edge. In particular, the simulated near wake undergoes spatial development slightly slower than the physical near wake, though many significant dynamical features are present in both the real and simulated flows. There is evidence of semi-coherent vortex shedding and other complex features affecting the turbulent stress levels in the central very-near-wake region. Spectra indicating the presence of vortex shedding at specific frequencies will be presented elsewhere (Gough and Hancock 1996). The observed features may also arise from the persistence of boundary-layer turbulent transport processes for a short distance beyond the trailing edge or enhanced production due to the appearance of high streamwise gradients.

## Acknowledgements

This work was supported under grant GR/H 22439 from the U.K. EPSRC and Ministry of Defence, and the computer simulations were carried out at the Atlas Laboratory. We are very grateful to Philip Hancock for his supervision of the experimental work that yielded the data used here for comparison.

## References

Gough, T., Gao, S., Hancock, P.E. and Voke, P.R. (1996a) Experiment and Simulation of the Tripped Boundary Layer on a Flat Plate: Comparative Study. *Report ME-FD/96.40*, Dept. of Mechanical Engineering, The University of Surrey, Guildford GU2 5XH U.K.

Gough, T., Gao, S., Hancock, P.E. and Voke, P.R. (1996b) Experiment and Simulation of the Near Wake behind a Flat Plate: Comparative Study. *Report ME-FD/96.41*, Dept. of Mechanical Engineering, The University of Surrey, Guildford GU2 5XH U.K.

Gough, T.D. and Hancock. P.E. (1996) Low Reynolds Number Turbulent Near Wakes, *Advances in Turbulence VI*, (ed. S. Gavrilakis, L. Machiels, P.A. Monkewitz). Kluwer Academic, Dordrecht. 445–448.

Voke, P.R. (1996) Subgrid-scale Modelling at Low Mesh Reynolds Number, *Theoret. Comput. Fluid Dynamics*, **8**, pp 131– 143.

Voke, P.R. and Potamitis, S.G. (1994) Numerical Simulation of a Low-Reynolds-Number Turbulent Wake behind a Flat Plate, *Int. J. Num. Methods Fluids*, **19**, pp 377–393.

# SIMULATION OF LAMINAR SEPARATION BUBBLE INSTABILITIES

M. ALAM AND N. D. SANDHAM
*Department of Engineering,*
*Queen Mary and Westfield College,*
*Mile End Road, London E1 4NS, U.K.*

**Abstract.** Direct numerical simulations of the three-dimensional (3-D) time-dependent incompressible Navier-Stokes equations are performed to investigate the growth and stability of separation bubbles, generated by suction of fluid through the upper boundary of the simulation. Temporal stability of a family of matching analytic profiles for steady laminar separation bubbles is investigated by solving the Orr-Sommerfeld equation. A wave packet analysis of the growing modes suggests that absolute instability can exist in the strong reverse flow region of a laminar separation bubble. A comparative study is made of three cases. In the first a high suction strength is used to generate a self-excited irregular vortex shedding. The second and third cases use a lower suction strength and compare two and three-dimensional forced bubbles. Small scale structures in the 3-D forced bubble were found to have little effect on the time-averaged bubble length.

## 1. Introduction

When a laminar boundary layer encounters a strong enough adverse pressure gradient it separates to form a laminar free shear layer. The separated shear layer undergoes transition to turbulent flow and the turbulent shear layer may reattach to form a closed laminar separation bubble. Gaster (1963) and more recently Dovgal *et al.* (1994) have shown both by theoretical and experimental means that downstream of separation the laminar part of the bubble is a spatial filter-amplifier of upstream disturbances. It is generally accepted that transition takes place due to non-linear breakdown of spatially-growing travelling waves in the separated shear layer. Such flows where disturbances grow in space, convecting away from the



*J.-P. Chollet et al. (eds.), Direct and Large-Eddy Simulation II,* 125-136.

source, are termed *convectively unstable* (see figure 1a). The flow is termed *absolutely unstable* if the disturbances grow in time (see figure 1b) and gradually contaminate the flow everywhere. Gaster (1963) raised the idea of the possible existence of regions of absolute instability in separation bubbles and expected this to be manifested by some kind of dynamic response from the shear layer. Niew (1993) carried out a linear wave packet analysis to show that velocity profiles with strong reverse flow and high enough Reynolds number exhibit characteristics of absolute instability. If this occurs in real bubbles then the transition mechanism of the free shear layer and the reattachment process is likely to be affected.

Recent advances in computer power have increased the use of direct numerical simulations (DNS) as a tool to investigate complex flow problems. There have been several investigations of two-dimensional (2-D) separation bubbles on flat plates using DNS. Pauley *et al.* (1990) used a method of suction of fluid from the upper boundary opposite the flat plate, whereas Bestek *et al.* (1993) and Maucher *et al.* (1984) used a decelerating freestream velocity to generate the adverse pressure gradient. With a suitable adverse pressure gradient the flow separated from the surface and reattached further downstream to form a closed separation bubble. For strong pressure gradient self-excited oscillations developed where the upstream part of the bubble remained steady and the reattachment region shed vortices. Niew (1993) reviewed the work of Pauley *et al.* (1990) and Bestek *et al.* (1993) and suggested that these self-excited oscillations may be due to regions of absolute instability in the flow field.

Pauley *et al.* (1990) carried out further analysis of 2-D separation bubbles by comparing the time-averaged bubble characteristics with the steady separation experiments of Gaster (1963). Since this work there have been number of articles (such as Ripley & Pauley (1993), Lin & Pauley (1993 a & b), Wilson & Pauley (1995)) which seem to have accurately reproduced numerous experimental results of separation bubbles even though the simulations were 2-D. The general mechanism put forward is that the large-scale 2-D structures control bubble reattachment and small-scale turbulence contributes a secondary role. This mechanism is somewhat different from the classical view of reattachment in which the fully 3-D turbulent mixing process plays an important role in energising the dividing streamline sufficiently for reattachment to be possible. The semi-empirical models of Horton (1968) and Robert (1980) (a modified version of Horton's model) make use of the classical view of the reattachment process. So it is clear that the reattachment region needs to be investigated in order to gain a clearer understanding of the mechanisms involved.

The objective of this paper is firstly to check for the existence of local regions of absolute instability in separation bubbles. This is done by the use

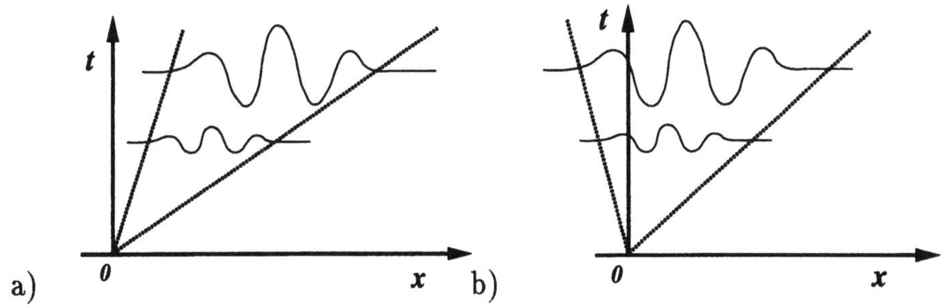


*Figure 1.* Stability of a typical impulse response a) convectively unstable, b) absolutely unstable.

of a wave packet analysis of Niew (1993) with a family of analytic velocity profiles which match the separated boundary layer profiles, and by running numerical simulations with strong adverse pressure gradient to try to create regions of absolute instability. The structure of the reattachment region is then investigated by analysing velocity fluctuation statistics. The statistics of a 2-D self-excited vortex-shedding bubble and a forced 2-D bubble are compared to a bubble with fully 3-D turbulent reattachment in order to verify whether 2-D vortex shedding captures the main features of separation bubbles with turbulent reattachment.

## 2. Direct Numerical Simulation

The spectral numerical method (due to Kleiser & Schumann (1980)) solves the three-dimensional time-dependent incompressible Navier-Stokes equations. Fourier discretisation is used for the periodic spanwise and streamwise directions and Chebyshev discretisation is used in the wall normal direction. The time discretisation is achieved with a compact third-order Runge-Kutta method for the convective terms and the Crank-Nicolson method for the viscous and pressure terms. This numerical method has been set up to run on the Cray T3D massively parallel computer at Edinburgh Parallel Computing Centre (for further details of numerical methods and parallel implementation see Sandham & Howard (1995)). The periodic boundary conditions mentioned above are suitable for flows that are temporal in nature. The separation bubble problem, on the other hand, is of a spatial nature and non periodic in the streamwise as well as the wall normal direction. Changing the streamwise discretisation would have meant using a less efficient numerical method. An alternative approach has been adopted which keeps the periodic boundary condition and also allows spatial development of the flow. This is done by introducing a zone where the flow is artificially damped back to initial laminar flow as prescribed at the in-

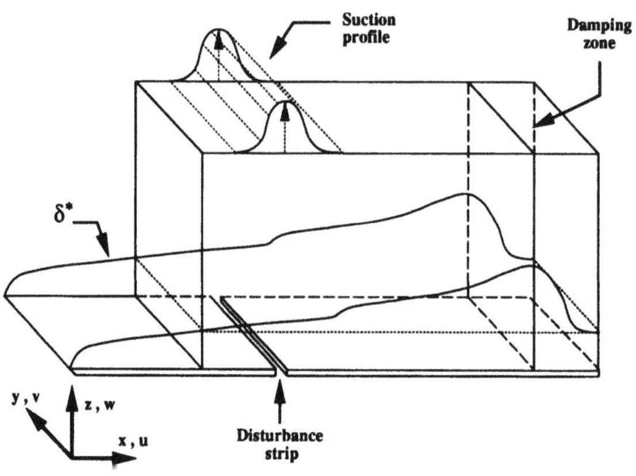


*Figure 2.* The geometry of the computational domain.

flow boundary. The implementation of the damping zone is made by the introduction of a term (denoted as $M$) in the right hand side of the time advancement,

$$M = \Delta u_i \times f(x) \tag{1}$$

where $\Delta u_i$ is the difference in the velocity components in the damping zone compared to the velocities at the inflow and $f(x)$ is zero everywhere except in the damping zone where it increases according to a modified Gaussian profile which reaches a maximum value of one and then returns to zero at the end of the damping zone. Various parameters, including the length of this damping zone, have been tuned so that there is minimal numerical propagation of disturbances through the damping zone. The cost of the damping zone is 25 streamwise grid points at the end of the computational domain. The basic geometry of the flow is shown on figure 2. The reference quantities are the freestream velocity and the displacement thickness of the incoming boundary layer. A Blasius boundary layer profile is prescribed at the inflow boundary and a no-slip boundary condition is used at the flat plate. A Gaussian profile is prescribed for the wall-normal velocity on the upper boundary in order to suck fluid out of the computational domain to obtain separation from the lower boundary. The strength of the pressure gradient is controlled by varying the height and the width of the Gaussian profile. The basic 2-D computations were run on a Silicon Graphics workstation (R8000) with $96 \times 4 \times 120$ grid points, while the 3-D computations were run on a Cray T3D (64 processors) with $96 \times 64 \times 120$ grid points. Although the number of grid points for 3-D computations is sufficient to capture the important features of transition, the grid does not fully resolve the smallest turbulence scales downstream of reattachment. Both 2-D and 3-D simulations were carried out for same physical box lengths

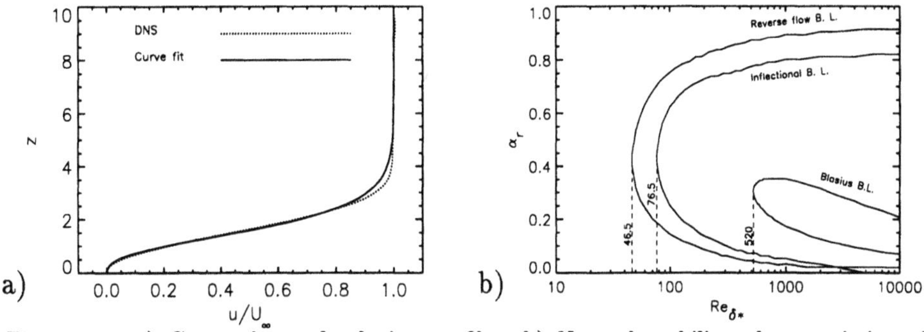


*Figure 3.* a) Comparison of velocity profiles, b) Neutral stability characteristics of laminar boundary layer ($A$=0.65, $B$=1.5 for inflectional boundary layer, and $A$=0.88, $B$=1.52 for reverse flow boundary layer)

for comparative purposes. The streamwise domain length was 200 displacement thicknesses($\delta^*$) based on the inflow boundary layer. The wall normal height was $10\delta^*$ and the spanwise width for the 3-D computation was $20\delta^*$.

## 3. Stability analysis of the separated shear layer

The stability of separation bubbles can be investigated by analysing the local separated velocity profiles. This is done by devising a flexible family of velocity profiles which can be matched to the simulation results,

$$\frac{u}{U_\infty} = \tanh(z) - 2A\frac{\tanh(z/B)}{\cosh^2(z/B)} \tag{2}$$

where the constant $A$ controls the amount of reverse flow and the constant $B$ controls the distance of the inflection point from the wall. Figure 3a shows a typical result, comparing a matched profile to a DNS profile. The devised profile is thought to be adequate for our investigation.

We assume a disturbance of the form

$$u'(z,t) = \hat{u}(z)e^{i(\alpha_r x - \omega_r t)}e^{\omega_i t} \tag{3}$$

where $u'$, $\hat{u}$, $\alpha_r$, $\omega_r$ and $\omega_i$ denote a velocity component of the wave, eigenfunction, spatial wave number, frequency and temporal growth rate respectively. Positive $\omega_i$ means temporal amplification of the disturbance. Substitution of this disturbance profile into the linearised Navier-Stokes equations yields the Orr-Sommerfeld equation. The Orr-Sommerfeld equation is numerically solved by a boundary value approach where it is re-arranged to isolate $\omega$ (in complex form) and discretised in wall normal direction using Chebyshev polynomials to get a generalised matrix eigenvalue problem. An eigenvalue solver is used to solve for $\omega$ for a given spatial wave number, Reynolds number and velocity profile (together with its second derivative).

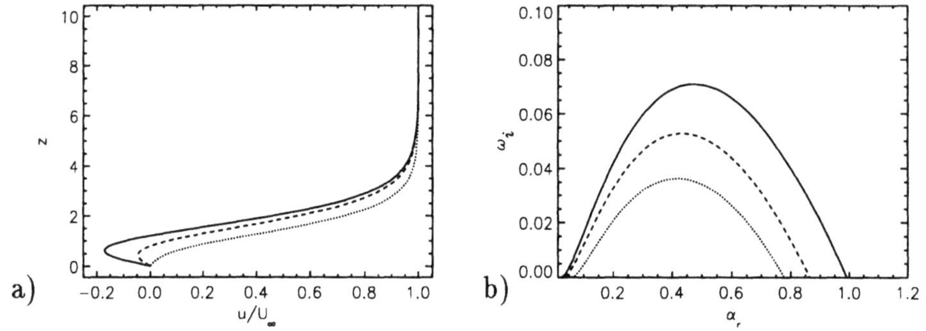


*Figure 4.* Profile shapes (a) and stability characteristics (b). The various cases are 17% reverse flow (solid line, $A = 1.1, B = 1.55$), 5% reverse flow (dashed line, $A = 0.9, B = 1.55$) and inflectional profile (dotted line, $A = 0.65, B = 1.45$).

The highly unstable nature of the velocity profiles that occur in separated flow due to the presence of an inflection point is demonstrated by figure 3b. The critical Reynolds number of the separated profiles decreases significantly and the wave number range for amplification increases. The unstable nature of these profiles can be further analysed to see the effect of increasing reverse flow. Referring to figure 4, which shows stability calculations at $Re_{\delta}* = 500$, the temporal growth rate and the range of amplified wavenumbers increases with the increase of reverse flow. It was also found that profiles with the inflection point closer to the wall were more unstable.

A cheap and simple method of Niew (1993) can be used to get an estimate of the local velocity profiles required for absolute instability. He simplified Gaster's (1981) wave packet integral for an impulse response and reformulated it as a complex summation of the temporal dispersion relation of a particular profile without any branching behaviour. This complex summation (denoted as $G$) is evaluated and the resulting modulus ($\mid G \mid$) together with the real part ($G_r$) is used to illustrate graphically the absolute/convective nature of a given velocity profile. Figure 5 shows a wave packet growth both in space and time for 5 % reverse flow. Along the line $x = 0$, the wave packet decays in time, so this case is clearly convectively unstable. When the amount of reverse flow is increased to 20 % (see figure 6) the wave packet becomes much wider and its boundary crosses the line $x = 0$, implying that this flow is absolutely unstable. This analysis was repeated for various amounts of reverse flow profiles to find a critical amount of reverse flow below which the profiles are convectively unstable. The critical amount of reverse flow was found to be 15% for the current family of velocity profiles, increasing as Reynolds number was reduced.

Two-dimensional simulations have been run for various strengths of pressure gradient. Weak pressure gradient produced closed steady laminar separation bubbles with weak reverse flow in the separated region. Small

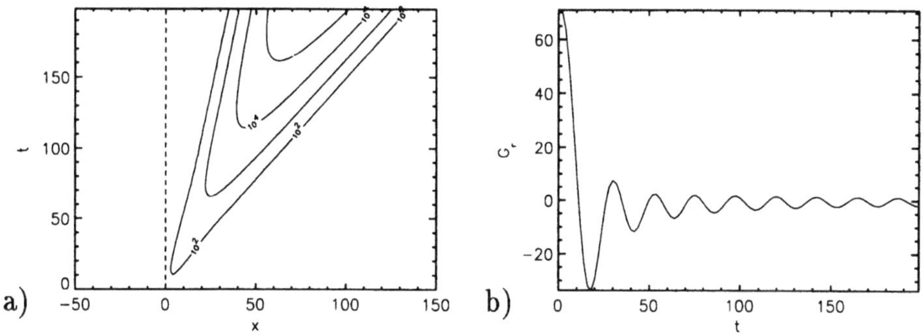


*Figure 5.* Convectively unstable 5 % reverse flow profile at $Re_{\delta}* = 500$ a) contour of $\mid G \mid$ and b) $G_r$ along the line $x = 0$ .

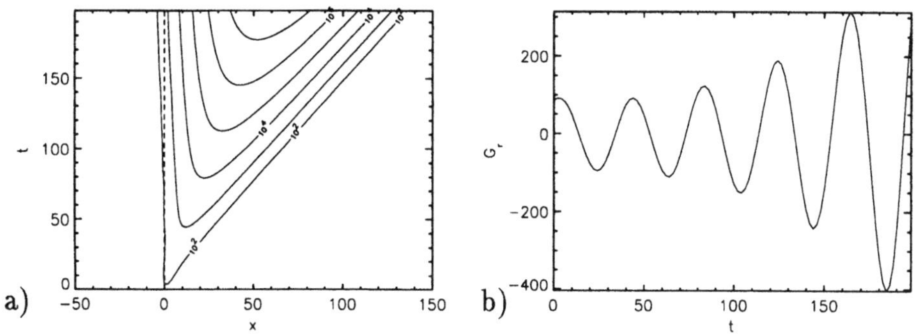


*Figure 6.* Absolutely unstable 20 % reverse flow profile at $Re_{\delta}* = 500$ a) contour of $\mid G \mid$ and b) $G_r$ along the line $x = 0$.

amplitude forcing was introduced upstream of this bubble to see the response of the free shear layer. The free shear layer amplified the upstream disturbances and shed vortices periodically. The amplified frequency range was within the range predicted by linear stability theory. These steady bubbles can be classed as convectively unstable, since they act as a spatial filter-amplifier of upstream disturbances. However, results from a series of simulations with strong adverse pressure gradient suggest that strong reverse flows of 15 % or more can exist and, furthermore, self-excited irregular oscillations developed. Figure 7 shows how the reattachment region sheds vortices in a similar manner to that reported by Pauley *et al.* (1990). The minimum recorded reverse flow in the recirculation region for this self-excited vortex shedding case was 14 % and the maximum was 28 %. Reduction of the suction strength resulted in a rapid drop of the minimum reverse flow velocities (well below the 15 % needed for absolute instability) and at the same time the strength of the vortex shedding reduced considerably. It can be shown that laminar separation bubbles may contain regions of absolute instability, but it is difficult to reach definite conclusions on the

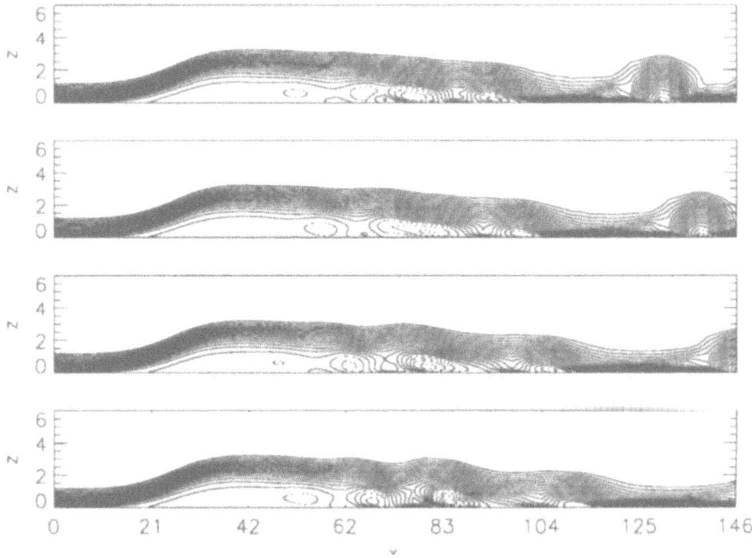

*Figure 7.* Streamwise velocity contours of a self-excited vortex shedding process evolving in time.

role of absolute instability in separation bubbles due to numerical effects (e.g. damping zone).

## 4. Comparison of the fluctuation statistics

For comparison purposes three types of separation bubbles were simulated. In the first case a 2-D self-excited vortex shedding was generated by strong suction strength (29% of the incoming mass was removed through the upper boundary), and high Reynolds number ($Re_{\delta *} = 700$). In the other two cases a steady bubble was generated by moderate suction strength (19% of the incoming mass was removed through the upper boundary) and moderate Reynolds number ($Re_{\delta *} = 500$). In a 2-D case sinusoidal forcing was placed upstream of the steady bubble. The forcing amplitude was 3 percent of the freestream velocity and the non-dimensional frequency was 0.15. These disturbances were readily amplified to large vortical structures near separation which then convected downstream (see figure 8, top figure). A 3-D bubble with turbulent reattachment was generated by the addition of oblique waves upstream of the steady bubble. The oblique waves consisted of two components, one along the span and the other in time. The amplitude was 3 percent of the freestream velocity. The spanwise wavelength was half the spanwise domain length and the non-dimensional forcing frequency was 0.15. These oblique waves travelled downstream along the laminar free shear layer and caused a rapid breakdown of the free shear layer just be-

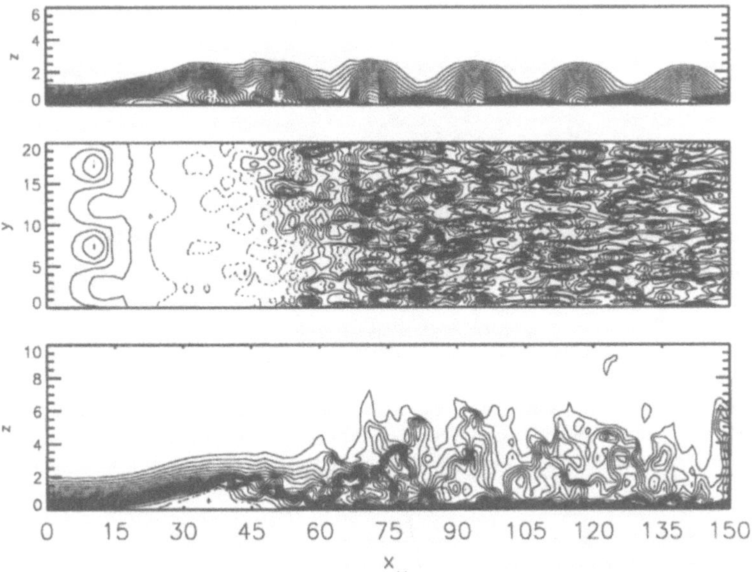


*Figure 8.* The top figure shows the streamwise velocity contours of forced 2-D vortex shedding. The lower two figures show the 3-D bubble with turbulent reattachment. The middle figure shows a x-y plane just above the flat plate and the bottom figure shows a x-z plane.

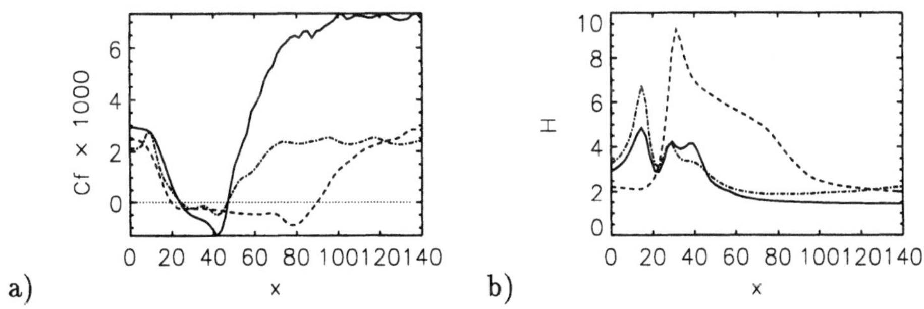


*Figure 9.* Skin-friction (a) and shape factor (b) along the flat plate (dashed line: self-excited vortex shedding, dash dot line: 2-D forcing, solid line: 3-D turbulent reattachment).

fore reattachment (see figure 8). Downstream of the reattachment the shear layer continued to spread further into the freestream, the skin-friction increased significantly and the shape factor settled down to 1.5 (see figure 9b) indicating that the transition to a turbulent boundary layer was complete.

Bubble lengths were obtained from the time-averaged skin-friction plot (figure 9a). The bubble with self-excited vortex shedding is much longer than the other two cases because it required much higher suction strength and Reynolds number compared to the two forced cases. The bubble lengths

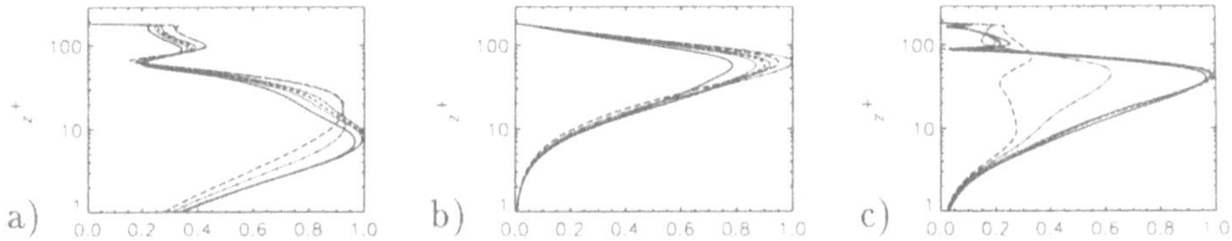

*Figure 10.* Velocity fluctuation statistics of self-excited vortex shedding, a) $\sqrt{(\overline{u'u'})}/0.2$, b) $\sqrt{(\overline{w'w'})}/0.1$, c) $\sqrt{(\overline{u'w'})}/0.06$. Line styles; solid: $x = 91$ (reattachment), dots: $x = 96$, short dash: $x = 98$, dash dot: $x = 100$, dash triple dot: $x = 116$, long dash: $x = 129$.

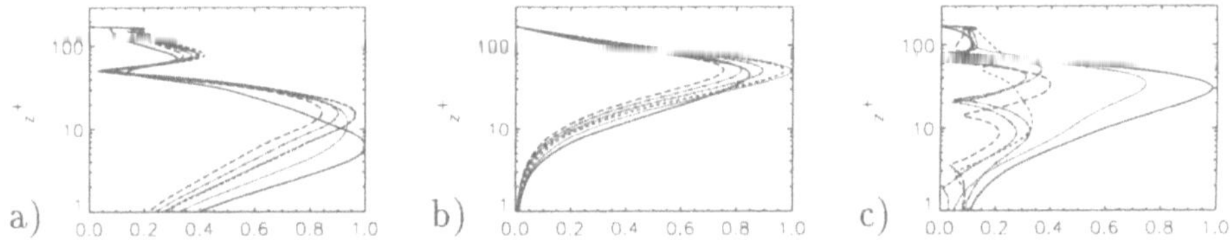

*Figure 11.* Velocity fluctuation statistics of forced vortex shedding, a) $\sqrt{(\overline{u'u'})}/0.21$, b) $\sqrt{(\overline{w'w'})}/0.13$, c) $\sqrt{(\overline{u'w'})}/0.07$. Line styles; solid: $x = 50$ (reattachment), dots: $x = 67$, short dash: $x = 83$, dash dot: $x = 100$, dash triple dot: $x = 116$, long dash: $x = 129$.

are equal for the two forced cases even though they are physically very different. The 3-D case corresponds to the traditional view of separation bubble structure as a well defined bubble with superimposed turbulence, whereas in the 2-D case the large structures dominate and no clear bubble is visible in instantaneous views. Although the two averaged bubble lengths are similar the minimum skin friction is much lower in the 3-D than in the 2-D case.

The reattachment regions of the three cases are further investigated by analysing the velocity fluctuation statistics $\sqrt{(\overline{u'u'})}$, $\sqrt{(\overline{w'w'})}$, $\sqrt{(\overline{u'w'})}$. Figures 10, 11 and 12 show these statistics against the wall co-ordinates ($z^+$) for the three cases of self-excited vortex shedding, 2-D, and 3-D forced bubbles respectively. At first glance the velocity fluctuation profiles at reattachment and further downstream look similar. However, close inspection of the fig-

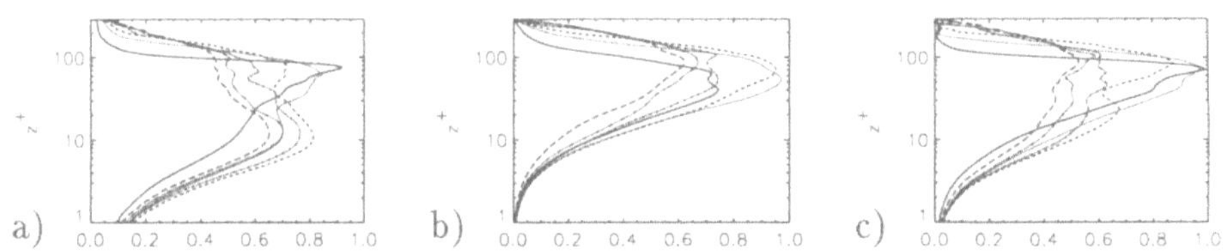

*Figure 12.* Velocity fluctuation statistics of turbulent reattachment, a) $\sqrt{(\overline{u'u'})}/0.17$, b) $\sqrt{(\overline{w'w'})}/0.11$, c) $\sqrt{(\overline{u'w'})}/0.01$. Line styles same as figure 11.

ures reveals that there are several differences between the 2-D and the 3-D cases. At reattachment the peaks of $\surd(\overline{u'u'})$, $\surd(\overline{w'w'})$ and $\surd(\overline{u'w'})$ are at higher values of $z^+$ for the 3-D case compared to the 2-D cases. Downstream of reattachment the fluctuation profiles of 3-D case looks similar to that of fully developed turbulent boundary layer, but a significant dip in $\surd(\overline{u'u'})$ is present in the 2-D cases at $z^+ \approx 50$ which is absent in the 3-D case. The second peak in the vicinity $z^+ = 100$ is present in all three cases. The peaks of $\surd(\overline{w'w'})$ and $\surd(\overline{u'w'})$ downstream of reattachment are at lower heights for the two 2-D cases compared to the 3-D case. A significant drop of the magnitudes of $\surd(\overline{u'w'})$ is seen in all the three cases but with differing amplitude. Although the 2-D case has the same time-averaged bubble length and some statistical similarities to the 3-D case, these do not appear to be sufficient to use results from the 2-D simulations to validate empirical models for separation bubbles.

## 5. Conclusions

A wave packet analysis has been used to show that velocity profiles with weak reverse flow exhibit characteristics of convective instability, whereas velocity profiles with strong reverse flow (more than 15%) exhibit characteristics of absolute instability. Results of 2-D numerical simulation show that weak pressure gradients give rise to steady separation bubbles which act to filter upstream disturbances and can be classed as convectively unstable. Strong pressure gradients give rise to separation bubbles which shed vortices from the reattachment part. A direct comparison of fluctuation statistics of 2-D bubbles (a self-excited vortex shedding bubble and a forced steady bubble) with a 3-D bubble with turbulent reattachment show that the general flow characteristics downstream of reattachment are significantly different. However the length of the bubble is not affected.

## 6. Acknowledgements

This research is supported by British Aerospace (Operations) Limited, Sowerby Research centre. Time on the Cray T3D is provided by EPSRC under the grant GR/K 43957.

## References

Bestek, H.,Gruber, K. and Fasel, H., 1993: Self-excited unsteadiness of laminar separation bubbles caused by natural transition, *Proceedings of RAeS Conference on 'The prediction and exploitation of separated flow'*, 14.1-14.16.

Dovgal, A.V., Kozlov, V.V. and Michalke, A., 1994: Laminar boundary layer separation: instability and associated phenomena, *Prog. Aerospace Sci.* **30**, 61-94.

Gaster, M., 1963: On stability of parallel flows and the behaviour of separation bubbles, *PhD Dissertation*, University of London.

Gaster, M., 1981: Propagation of linear wave packet in laminar boundary layers, *AIAA J.* **19**, (No. 4), 419-423.

Horton, H.P., 1968: Laminar separation in two and three dimensional incompressible flow, *PhD Dissertation*, University of London.

Kleiser, L. and Schumann, U., 1980: Treatment of incompressibility and boundary conditions in 3-D numerical spectral simulation of plane channel flows, *Proc. 3rd GAMM Conf. on Numerical Methods in Fluid Mechanics (Ed. E. H. Hirschel)*, 165-173.

Lin, J.C.M., and Pauley, L.L., 1993 a: Unsteady laminar separation on low-Reynolds number airfoils,*AIAA paper 93-0209.*

Lin, J.C.M., and Pauley, L.L., 1993 b: Unsteady view of transitional separation bubbles, *Separated flow symposium of the ASME Fluids Engineering Conference.*

Maucher, U., and Rist, U. and Wagner S., 1994: Direct numerical simulation of airfoil separation bubbles, *ECCOMAS.*

Niew, T.R., 1993: The stability of the flow in a laminar separation bubble, *PhD Dissertation*, University of Cambridge.

Pauley, L.L., Moin, P. and Reynolds, W.C., 1990: The structure of two dimensional separation, *J. Fluid Mech.* **220**, 397-411.

Ripley, M.D. and Pauley, L.L., 1993: The unsteady structure of two-dimensional separation, *Phys. Fluids a*,**5**,(12), 3099-3106.

Robert, W.B., 1980: Calculation of laminar separation bubbles and their effect on airfoil performance, *AIAA J.* **18**, 25-31.

Sandham, N.D. and Howard, R.J.A., 1995: Statistics database from direct numerical simulation of fully-developed turbulent channel flow, *Queen Mary and Westfield College, Department of Engineering*, QMW-EP-1106.

Wilson, P.G. and Pauley, L.L., 1995: Two-dimensional large eddy simulation of a transitional separation bubble, *Symposium on separated and complex flows ASME/JSME Fluid Engineering Conference.*

# LARGE-EDDY SIMULATION OF SEPARATED BOUNDARY LAYER TRANSITION

ZHIYIN YANG
*Department of Mechanical Engineering, University of Surrey, Guildford GU2 5XH, U.K.*

**Abstract.** Large-eddy simulation has been employed to investigate the mechanisms of separated boundary layer transition under zero and 2.5% free stream turbulence on a flat plate. The geometry is a flat plate with a semi-circular leading edge, which induces a separation bubble at the blend point where the curvature of the surface changes. The simulated results are compared with the available experimental data. The separation bubble at the leading edge is unstable and plays an important role in triggering transition. Data from the simulation have been gathered which allow us to extract all terms in the derived equations for the Reynolds stresses. Aspects of these balances are presented which allow new insights into the physical mechanisms at work.

## 1. Introduction

Boundary layer transition from the laminar state to turbulence is of great engineering importance but is still only partly understood. It is well established that boundary-layer transition on a flat plate without separation can be classified into two categories, i.e., natural transition where the transition process consists of several stages starting with 2D TS wave, nonlinear amplification, 3D vortices, turbulent spots, eventually breaking dwon to turbulence, and bypass transition where transition occurs very rapidly and those stages in the natural transition are not observed when some certain conditios, such as high free stream turbulence, are satisfied. Transition is quite sensitive to many factors and conditions and the mechanisms at work are different for the above two categories.

Large-Eddy Simulation (LES) and Direct Numerical Simulation (DNS) have become powerful tools for investigating boundary layer transition due



*J.-P. Chollet et al. (eds.), Direct and Large-Eddy Simulation II*, 137-146.

to rapid advances of supercomputer power and numerical techniques. Yang and Voke (1992,1993,1994,1995) have applied LES successfully to bypass transition of the boundary layer on a flat plate with zero pressure gradient, and DNS have been performed of natural boundary layer transition by Spalart and Yang (1987), Laurien and Kleiser (1989), and Fasel et al. (1990). However, little simulation work has been done on boundary layer transition with leading edge shapes which cause a separation bubble. The purpose of this paper is to study the separated boundary layer transition involved in such a case with zero and 2.5% free stream turbulence.

## 2. Methods

The numerical methods employed for the present LES are direct descendants of well-known finite-volume techniques successfully used for many high-Reynolds-number LES studies, and recently for flat plate bypass transition by Voke and Yang (1995). The 2D curvilinear geometry requires the use of general coordinates in the $x$ and $y$ dimensions, though the simplicity of the geometry in the present case means that the coordinate system in the $(x, y)$ plane can be orthogonal. The fully covariant general Navier-Stokes equations are discretised on a staggered mesh, using forms of the equations that employ the Jacobian of the transformation so that the advection term is relatively simple and involves no connection coefficients (Voke and Collins 1984). The numerical discretisation conserves momentum to high accuracy volume by volume and globally, and conserves mass to a specified accuracy through a pressure correction algorithm. The discretisation also conserves kinetic energy (to the same accuracy as the mass) in the advection term, ensuring that the removal of energy from the simulation takes place through the discrete approximation to the viscous term. The governing equations are advanced in time explicitly by the second-order Adams-Bashforth scheme. The pressure is found by solving a discrete Poisson equation.

The coordinate system metric and connection coefficients enter the pressure and viscous terms but since the coordinates are orthogonal for the present case the metric tensor is diagonal. Therefore the Poisson equation for pressure does not have cross-derivatives and so is formally similar to that in Cartesian coordinates. Because the metric tensor and hence the coefficients of the discrete Laplace operator vary between the cylindrical region of the mesh round the leading edge and the Cartesian region above the flat plate, it is impossible to use direct methods such as cyclic reduction for the solution of the pressure equation. An iterative pressure solver has been developed which is as fast as a vectorised cyclic reduction solver (Gavrilakis 1992) used in previous Cartesian LES studies. It makes use of the homogeneity of the spanwise dimension $(z)$ to Fourier transform the pressure equation, decomposing the 3D Poisson equation into a sequence of Klein-Gordon-Fokker equations which increase in diagonal dominance as the spanwise wavenumber increases. These decoupled equations are then solved by striped line relaxation in the $(x, y)$ plane with multigrid acceleration applied to speed the convergence for lower wavenumber of the discrete Fourier transform. It is found that the highest wavenumber matrices are so diagonally dominant that multigrid acceleration is not beneficial. This hybrid Fourier-multigrid pressure solver (Voke and Yang 1995) has proven to be efficient and robust.

The computational box is $16d \times 7.5d \times 2d$, curving around the bottom, front and top of the plate; $d$ is the leading edge diameter, equivalent to 10 mm in the experiment. The meshing used in the simulation was $264 \times 72 \times 64$ stretched in the $y$ direction away from the wall and also in the $x$ direction to ensure sufficient resolution in the separation bubble; $\Delta x^+$ varies from 5.6 to 86, $\Delta y^+$ varies from 0.7 to 137 and $\Delta z^+ = 13.3$. (The wall units are based on the friction velocity at $x = 7d$ from the blend point where the cylindrical leading edge meets the flat plate.) The Reynolds number is about 3333 based on the uniform inflow velocity of 5 m/s and the leading edge diameter of 10 mm.

The subgrid-scale model used is based on a Smagorinsky model modified by Voke (1996) to allow for low-Reynolds-number viscous effects. We find that using a single value for the Smagorinsky constant $C_s$ in the whole computational domain does not give satisfactory results since there are two different flow regions. Several tests have been conducted for this geometry:

a) Using a single value of $C_s = 0.1$.
b) Using two values, $C_s = 0.23$ up to reattachment and $C_s = 0.1$ beyond.
c) Using a variable $C_s$ changing linearly from 0.27 at separation to 0.1 beyond reattachment.
d) Using a dynamic subgrid-scale model with the Smagorisky model as a base.

It has been found that the third of these methods gives satisfactory behaviour of the separation bubble, reattachment and transition, and the results presented in this paper are extracted from this simulation. Further work using the dynamic approach is required to verify its utility for this flow.

## 3. Results

Figure 1 shows the comparison between the simulated mean streamwise velocity profiles and the experimental data (Coupland 1994) for the zero free stream turbulence case (no disturbances was imposed at inflow boundary) at three streamwise stations, two in the bubble, one after the separation

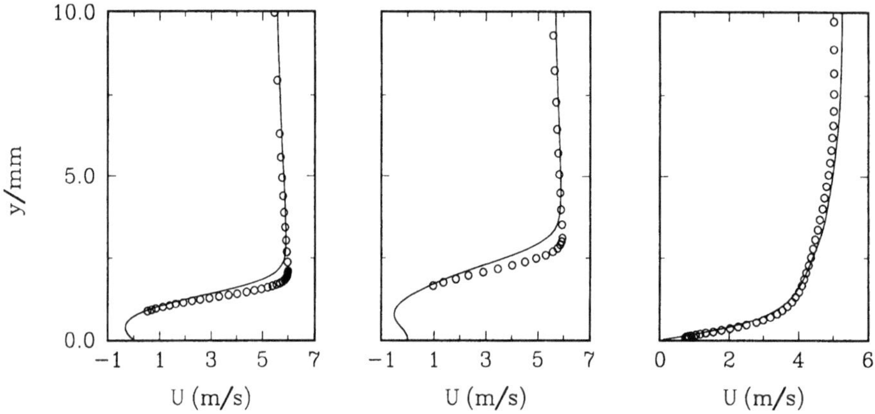


*Figure 1.* Mean $U$ profiles for the zero FST case at $x/x_r = 0.22$ (left), 0.67 (middle), 2.55 (right). Solid line, LES; symbols, experimental data.

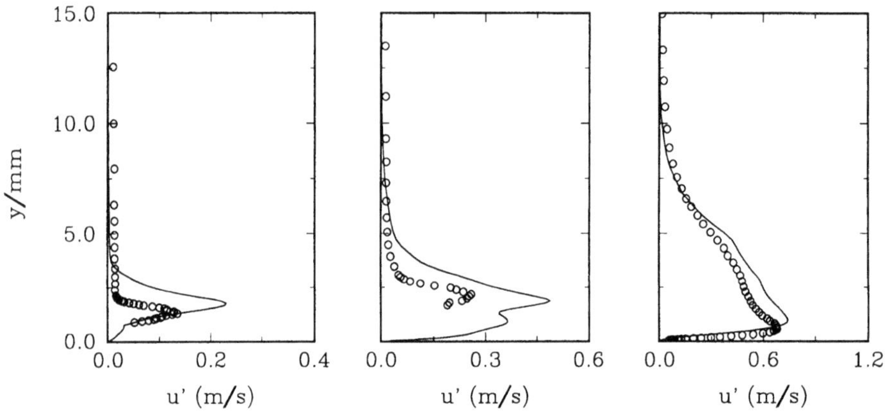


*Figure 2.* R.m.s fluctuation of u profiles for the zero FST case. Other details as in Figure 1.

at further downstream station. Separation occurs immediately at the blend point and the streamwise distance is normalised by the bubble length. As can be seen from the figure, good agreement has been obtained at all stations and by $x/x_r = 2.55$ the numerical results and the experimental data both show a well developed turbulent boundary layer velocity profile.

Figure 2 shows the simulated streamwise rms intensity and the corresponding experimental data at the same stations for the zero free stream turbulence case. The numerical results compare well with the experimental data at the third stations but overpredicted the turbulence intensity at the first and second stations. The double peaks predicted by the simulation within the separation bubble are genuine, not due to lack of statistical

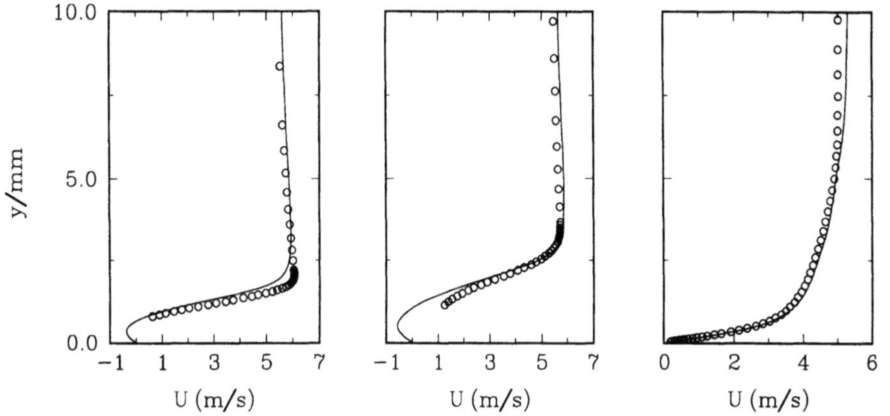


*Figure 3.* Mean $U$ profiles for the 2.5% FST case at $x/x_r = 0.37$ (left), 0.75 (middle), 3.2 (right). Solid line, LES; symbols, experimental data.

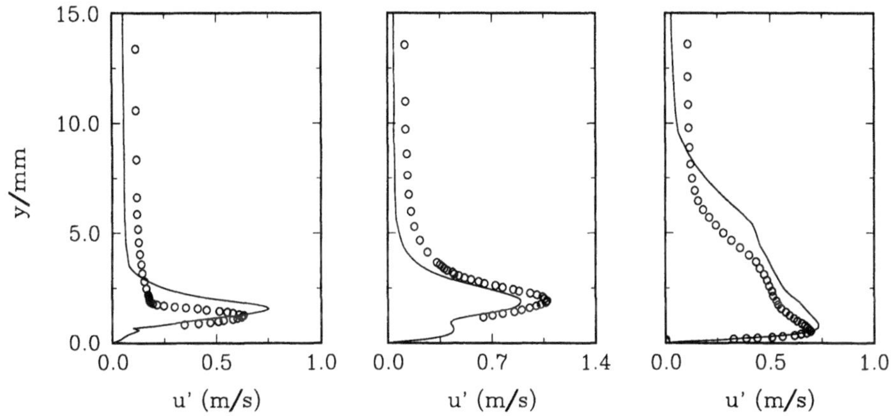


*Figure 4.* R.m.s fluctuation $u'$ profiles for the 2.5% FST case. As Figure 3.

samples, but arise from two different events, one happening in the free shear layer and another in the wall layer. This can be confirmed from the Reynolds stress balances shown in figures 9 and 10.

Figure 3 and 4 show the mean streamwise velocity and the rms intensity for the 2.5% isotropic free stream turbulence case at three streamwise stations. It can be seen from the figures that reasonable good agreement has been obtained at all stations and the profiles in both cases are quite similar apart from the fact that the peak rms intensities at the first two stations are higher for the 2.5% case. The peak levels of rms intensity for both cases are more or less the same at the last station where the turbulent boundary layer has already been developed.

In both cases the predicted mean bubble lengths are longer than those

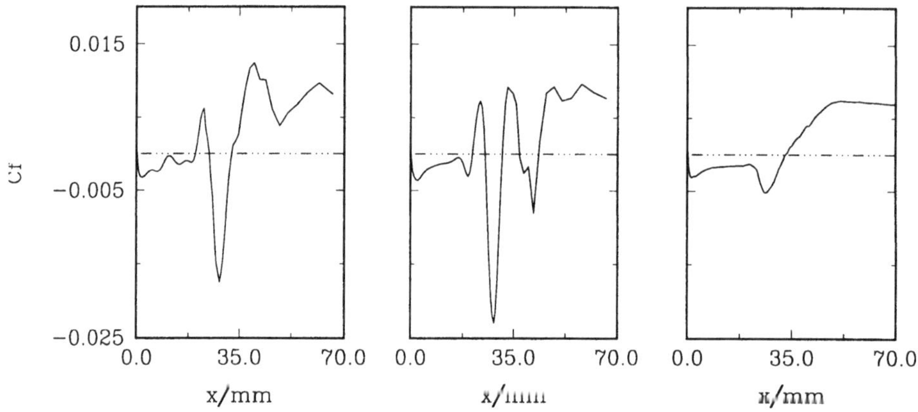


*Figure 5.* Skin friction coefficient, left and middle, instantaneous; right, time averaged.

of experiment. The simulated mean bubble length is about $3.3d$ while the experimental one is about $2.8d$ for the zero free stream turbulence case. For the 2.5% case the predicted mean bubble length is about $2.2d$ while the experimental one is about 16mm. The free stream turbulence mainly accelerates the transition process and makes the bubble shorter.

Figure 5 shows the plot of instantaneous spanwise-averaged skin friction coefficient against streamwise distance from the blend point for zero free stream turbulence. From the time-averaged profile shown in the right figure it can be seen that there is only one separation bubble formed immediately after the blend point. However, the instantaneous profiles show some quite different structures and it is difficult to define 'a reattachment point'. The flow shows a two bubble structure at a certain time, left, and three bubble structure at another time, middle. These structures can be confirmed by looking at the instantaneous velocity vectors in the $(x, y)$ plane as shown in figure 6.

Figures 7 and 8 show the cross-correlation of $u'$ in the $z$ direction at three streamwise stations for both the zero and 2.5% free stream turbulence cases. The high correlations at the first station for both cases mean that the flow is still mainly two-dimensional within the first half of the separation bubble. This confirms that the transition starts with a 2D instability for both cases and breaks down into three-dimensional motion further downstream as shown in the figure at the last streamwise station for the zero FST case where the correlation value drops rapidly to around zero. For the 2.5% FST case the correlation value drops down to a much lower value even at the second streamwise station which means that the breakdown from 2D to 3D takes place earlier with the shorter bubble as mentioned above. The fact that the cross-correlation value drops to about zero further

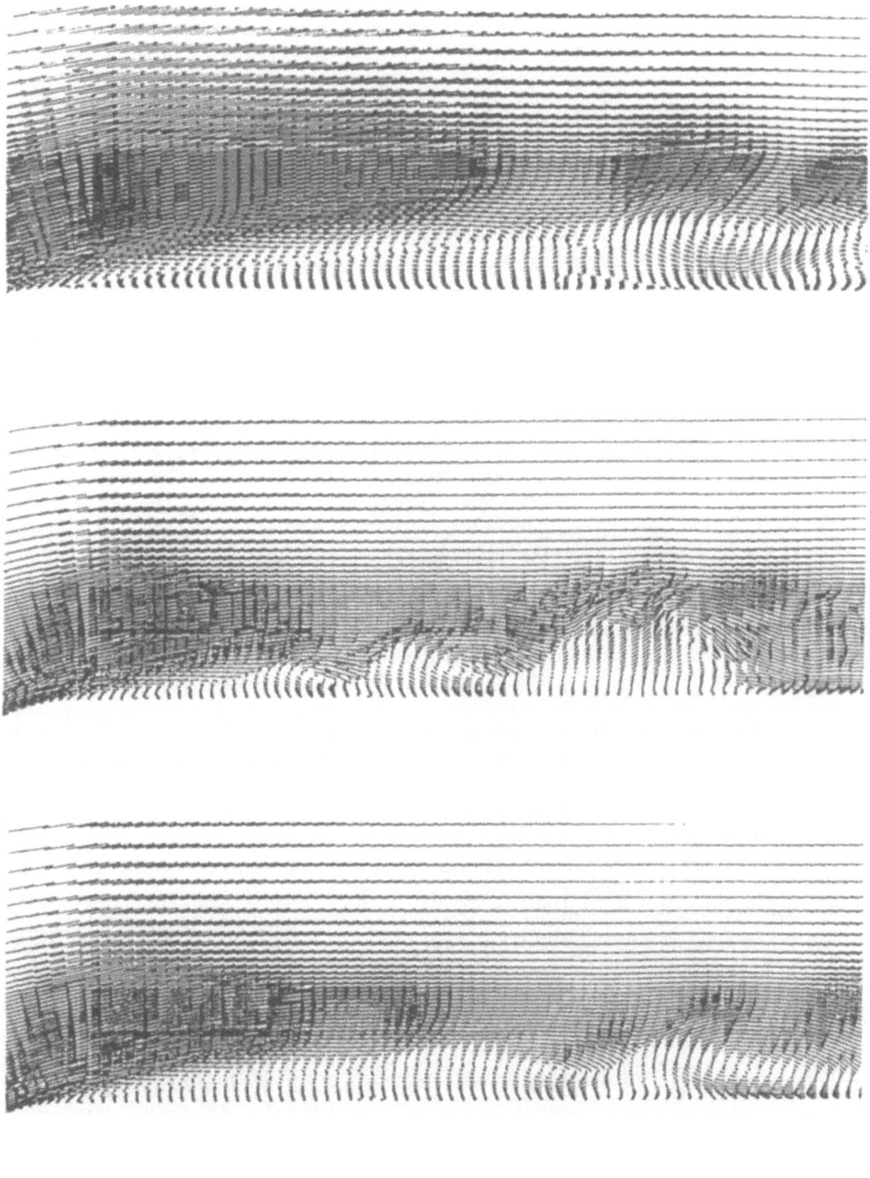

*Figure 6.* Instantaneous velocity vectors in the $(x, y)$ plane.

downstream also justifies that the computational box dimension in the $z$ direction is sufficient.

Figures 9 and 10 show the Reynolds stress balance terms for $u'^2$, $v'^2$ and $w'^2$ at $x/x_r = 0.55$ and $x/x_r = 0.76$ for the zero FST case. Subgrid contribution is quite samll and not presented in those figures. As shown in

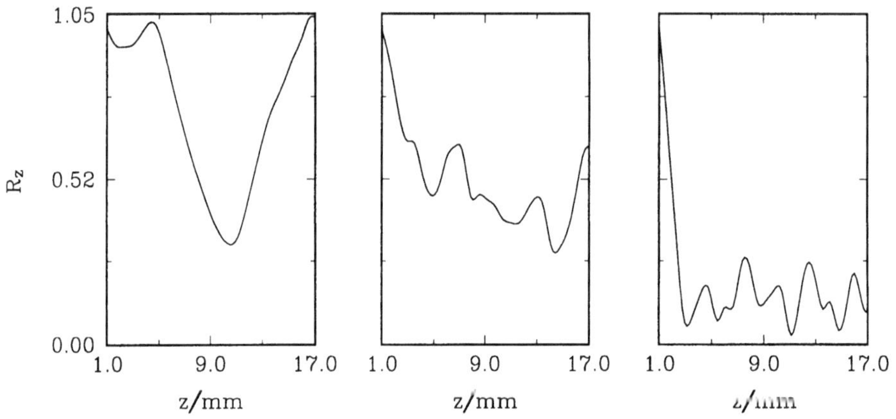


*Figure 7.* Cross-correlation of $u'$ in $z$ direction for the zero FST case at $(x,y,z)$ = (15mm,0.5mm,1mm) (left); (35mm,0.5mm,1mm) (middle); (70mm,0.5mm,1mm) (right).

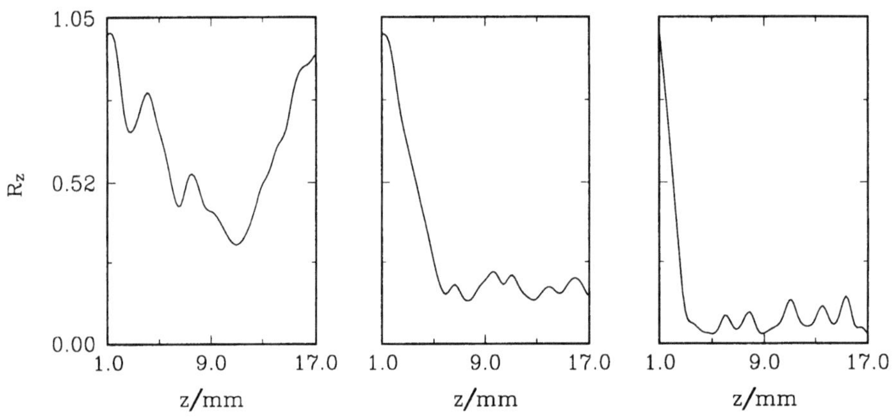


*Figure 8.* Cross-correlation of $u'$ in $z$ direction for the 2.5% FST case at $(x,y,z)$ = (15mm,0.5mm,1mm) (left); (35mm,0.5mm,1mm) (middle); (70mm,0.5mm,1mm) (right).

the figures, the production of $u'^2$ is balanced mainly by the pressure term and redistributed into $v'^2$ and $w'^2$ even at early upstream stations within the bubble. This is similar to the bypass transition in many ways as described by Voke and Yang (1993,1995) although the initial $v'$ in the present case comes from the unstable free shear layer within the bubble, which provides a rich source of stimulus for the production of $u'v'$ and $u'^2$ further downstream. Transition occurs immediately around the reattachment point.

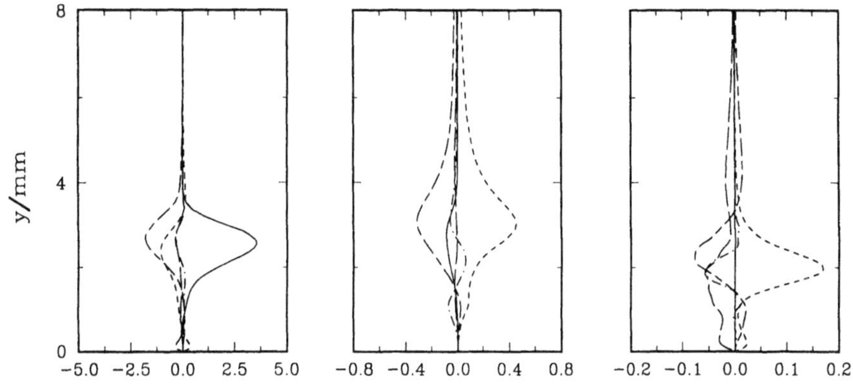


*Figure 9.* Terms in the balance of $u'^2$ (left), $v'^2$ (middle) and $w'^2$ (right) at $x/x_r = 0.55$. Solid line, production; alt dash, convection; dot dash, turbulent transport; short dash, pressure; long dash, viscous.

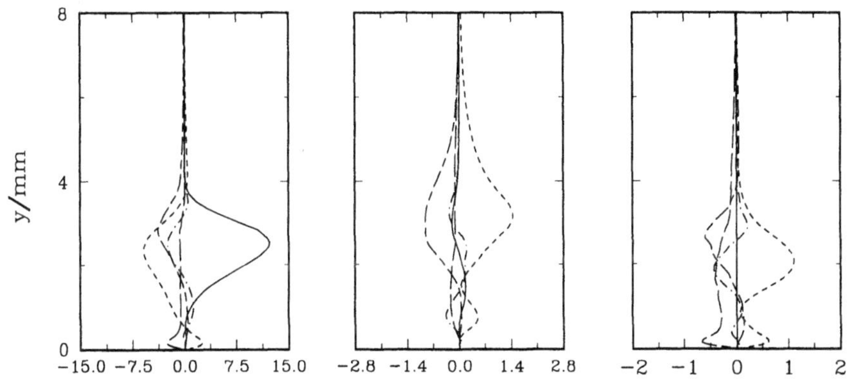


*Figure 10.* Terms in the balance of $u'^2$ (left), $v'^2$ (middle) and $w'^2$ (right) at $x/x_r = 0.76$. Solid line, production; alt dash, convection; dot dash, turbulent transport; short dash, pressure; long dash, viscous.

## 4. Conclusions

It is evident from the simulation that the transition process begins with a 2D instability of the large-scale unstable free shear layer in the separation bubble, which provides initial $v'$ for the further production of $u'v'$ and $u'$. There are not major differences between the zero and 2.5% free stream turbulence cases. The presence of free stream turbulence speeds up the transition process and hence produces a shorter bubble. Three-dimensional motions follow rapidly, in which pressure redistribution of streamwise disturbances into wall-normal and spanwise disturbances plays a key role. Full transition occurs rapidly around the reattachment point in a way similar in certain aspects to bypass transition under free-stream turbulence though in the zero free stream turbulence case the external disturbances originate in the large-scale instability of the separated shear layer. The boundary layer develops into a fully turbulent state a short way downstream.

## 5. Acknowledgements

The author gratefully acknowledges the support of Rolls-Royce plc and U.K. EPSRC for this research under grant number GR/J 07334. The supercomputer time was provided on the Y-MP8 by DRAL. My special thanks to Dr. P. R. Voke for supervising the work, and to Drs. A.M. Savill and J. Coupland for their useful advice and suggestions.

## References

Coupland, J. 1994 Private Communication.

Fasel, H., Rist, U. and Konzelmann, U. 1990 Numerical investigation of the three-dimensional development in boundary-layer transition, *AIAA J.* **28**, 29-37.

Gavrilakis, S. 1992 Numerical simulation of low-Reynolds-number flow through a straight square duct, *J. Fluid Mech.* **224**, 101-129.

Laurien, E. and Kleiser, L. 1989 Numerical simulation of boundary-layer transition and transition control, *J. Fluid Mech.* **199**, 403-440.

Spalart, P.R. and Yang, K.S. 1987 Numerical study of ribbon-induced transition in Blasius flow, *J. Fluid Mech.* **178**, 345-365.

Voke, P.R. and Collins, M.W. 1984 Forms of the generalised Navier-Stokes equations, *J. Eng. Math.* **18**, 219-233.

Voke, P.R. and Yang, Z.Y. 1993 Numerical studies of mechanisms of bypass transition in the flat plate boundary layer, *Ninth Symp. Turbulent Shear Flows* **2**, 21.2-6.

Voke, P.R. and Yang, Z.Y. 1995 Numerical study of bypass transition, *Phys. Fluids* **7**, 2256-2264.

Voke, P.R. and Yang, Z.Y. 1995 Hybrid Fourier-multigrid pressure solution for Navier-Stokes Simulations, in K.W. Morton and M.J. Baines (eds.), *Numerical Methods for Fluid Dynamics V*, Clarendon, Oxford, 615-621.

Voke, P.R. 1996 Subgrid-scale modelling at low mesh Reynolds number, *Theoretical and Computational Fluid Dynamics.* **8**, 131-143.

Yang, Z.Y. and Voke, P.R. 1992 Numerical simulation of boundary layer transition in the presence of free stream turbulence, in O. Pironneau et al. (eds.), *Numerical Simulation of Unsteady Flows and Transition to Turbulence*, C.U.P, New York, 398-402.

Yang, Z.Y. and Voke, P.R. 1993 Large-eddy simulation studies of bypass transition, in W. Rodi et al. (eds.), *Engineering Turbulence Modelling and Experiments* **2**, Elsevier science, Amsterdam, 603-611.

Yang, Z.Y., Voke, P.R. and Savill, A.M. 1994 Mechanisms and models of boundary layer receptivity deduced from large-eddy simulation of bypass transition, in P.R. Voke et al (eds.), *Direct and Large-Eddy Simulation* **I**, Kluwer Academic, Dordrecht, 225-236.

# LES OF TURBULENT THERMAL CONVECTION

ANTONELLA ABBÀ, CARLO CERCIGNANI, LORENZO VALDETTARO AND PAOLO ZANINI
*Department of Mathematics, Politecnico di Milano*
*Piazza L. da Vinci 32, 20133 Milano (Italy)*

**Abstract.** We describe results obtained from Large Eddy Simulations (LES) of turbulent natural convection flow in an infinite horizontal fluid layer heated from below. We use a dynamic subgrid scale model in scaling formulation, and discuss an anisotropic modification of it. The largest Rayleigh number successfully simulated is $10^8$ with an aspect ratio of 7. Average properties of the resulting flow are computed and compared with DNS data in the literature. An *a priori* test is done, that allows us to compare the models. The probability distribution functions of temperature fluctuations at different Rayleigh numbers are compared. Finally, flow structures at largest simulated Rayleigh number are investigated.

## 1. LES of turbulent natural convection flow

In order to solve the Navier-Stokes equations in Boussinesq approximation using LES, the subgrid-scale (SGS) stress tensor divided by $\rho$ (taken equal to 1, as usual when dealing with incompressible, constant density flows)

$$\tau_{ij} = \overline{u_i u_j} - \overline{u}_i \overline{u}_j$$

and SGS heat flux

$$q_i = \overline{\vartheta u_i} - \overline{\vartheta}\overline{u_i}$$

must be modelled; **u** is the velocity field and $\vartheta$ the temperature. The overbar represents the filtering on the grid scale.

We model $\tau_{ij}$ and $q_i$ using an eddy kinetic viscosity $\nu_t$ and an eddy thermal diffusivity $\nu_\vartheta$

$$\tau_{ij} - \frac{\delta_{ij}}{3}\tau_{kk} = \tau_{ij}^a = -2\nu_t \overline{S}_{ij} \tag{1}$$



*J.-P. Chollet et al. (eds.), Direct and Large-Eddy Simulation II*, 147-156.

$$q_i = -\nu_\vartheta \frac{\partial \overline{\vartheta}}{\partial x_i} \tag{2}$$

where $\overline{S}_{ij} = \frac{1}{2}\left(\frac{\partial \overline{u}_i}{\partial x_j} + \frac{\partial \overline{u}_j}{\partial x_i}\right)$.

In the scaling formulation proposed by Wong & Lilly (1994) the eddy coefficients are expressed in terms of the grid-filter width $\overline{\Delta}$

$$\nu_t = C_\varepsilon \overline{\Delta}^{4/3}, \qquad \nu_\vartheta = \frac{C_\varepsilon}{Pr_t}\overline{\Delta}^{4/3}; \tag{3}$$

the model coefficients $C_\varepsilon \propto \varepsilon^{1/3}$ and $Pr_t$ have to be determined.

In the dynamic model (Germano *et al.* 1991, Wong & Lilly 1994) the test-scale stress tensor and the test-scale heat flux

$$T_{ij} = \widehat{\overline{u_i u_j}} - \widehat{\overline{u}}_i \widehat{\overline{u}}_j, \qquad Q_i = \widehat{\overline{\vartheta u_i}} - \widehat{\overline{\vartheta}}\widehat{\overline{u}}_i$$

are introduced; the 'hat' represents the test-filtering, with a larger filter width than the grid filter, and $\widehat{\overline{\ }}$ denotes the composed filtering. The fluxes $\tau_{ij}$, $q_i$, $T_{ij}$ and $Q_i$ are linked to the resolved turbulent stress tensor and to the resolved turbulent heat flux

$$\mathcal{L}_{ij} = \widehat{\overline{u}_i \overline{u}_j} - \widehat{\overline{u}}_i \widehat{\overline{u}}_j, \qquad \mathcal{S}_i = \widehat{\overline{\vartheta}\overline{u}_i} - \widehat{\overline{\vartheta}}\widehat{\overline{u}}_i$$

by the identities

$$\mathcal{L}_{ij} = T_{ij} - \widehat{\tau}_{ij} \tag{4}$$

$$\mathcal{S}_i = Q_i - \widehat{q}_i. \tag{5}$$

If $T_{ij}$ and $Q_i$ are modelled as in (1)-(3)

$$T_{ij}^a = -2\nu_t \widehat{\overline{S}}_{ij} = -2C_\varepsilon \widehat{\overline{\Delta}}^{4/3} \widehat{\overline{S}}_{ij}, \quad Q_i = -\nu_\varepsilon \frac{\partial \widehat{\overline{\theta}}}{\partial x_i} = \frac{C_\varepsilon}{Pr_t}\widehat{\overline{\Delta}}^{4/3} \frac{\partial \widehat{\overline{\theta}}}{\partial x_i},$$

the identities (4) and (5) can be used to determine $C_\varepsilon$ and $Pr_t$ (Lilly 1992; Wong & Lilly 1994)

$$2\overline{\Delta}^{4/3} C_\varepsilon = \frac{\langle \left(\mathcal{L}_{ij} - \frac{\delta_{ij}}{3}\mathcal{L}_{kk}\right) \widehat{\overline{S}}_{ij} \rangle}{\left(1 - \left(\widehat{\overline{\Delta}}/\overline{\Delta}\right)^{4/3}\right) \langle \widehat{\overline{S}}_{lm} \widehat{\overline{S}}_{lm} \rangle} \tag{6}$$

$$\overline{\Delta}^{4/3} \frac{C_\varepsilon}{Pr_t} = \frac{\langle \mathcal{S}_i \left(\frac{\partial \widehat{\overline{\vartheta}}}{\partial x_i}\right) \rangle}{\left(1 - \left(\widehat{\overline{\Delta}}/\overline{\Delta}\right)^{4/3}\right) \langle \frac{\partial \widehat{\overline{\vartheta}}}{\partial x_j} \frac{\partial \widehat{\overline{\vartheta}}}{\partial x_j} \rangle}, \tag{7}$$

where <> indicates average over homogeneous direction, that are horizontal planes in our case.

## 2. An anisotropic LES model

In (1) and (2) the SGS tensor and heat flux are assumed to be strictly proportional to the grid-scale strain rate tensor and heat flux, respectively. Under this hypothesis, the anisotropy of the small scale structures, particularly important near the walls, cannot be captured.

Following Abbà *et al.* (1995a, 1995b) we propose here an anisotropic modification of (1) using an eddy viscosity tensor model instead of a scalar one

$$\tau_{ij}^a = -2\sum_{r,s} \nu_{ijrs}^t \overline{S}_{rs} + \frac{2}{3}\delta_{ij} \sum_{k,r,s} \nu_{kkrs}^t \overline{S}_{rs}. \tag{8}$$

Here and in the sequel the usual sum convention on repeated indices has been abolished for the sake of clearness.

Even if we impose all the possible relations among the coefficients of $\nu_{ijrs}^t$ that arise from symmetry considerations and lack of trace, we cannot get the number of components down to 5, which is the number of conditions provided by identity (4). To avoid this difficulty, we reduce the number of unknown coefficients by using, at every grid point, a cartesian local frame $a_\alpha$ ($\alpha = 1, 2, 3$), to be suitably chosen (see later), and write $\nu_{ijrs}^t$ in terms of $a_{i\alpha}$

$$\nu_{ijrs}^t = \overline{\Delta}^{4/3} \sum_{\alpha,\beta} C_{\alpha\beta}^\varepsilon a_{i\alpha} a_{j\beta} a_{r\alpha} a_{s\beta}; \tag{9}$$

the coefficients $C_{\alpha\beta}^\varepsilon$ ($= C_{\beta\alpha}^\varepsilon$) vary in space and time, like $C^\varepsilon$ in the original dynamic model, and are completely determined by imposing identity (4).

The $3 \times 3$ matrix $a_{i\alpha}$ is unitary; in particular $a^{-1} = a^T$, where $a^T$ is the transposed matrix.

Inserting the explicit expression (9) for $\nu_{ijrs}^t$ into (8) and into a similar expression for $T_{ij}$, we obtain

$$\tau_{ij}^a = -2\overline{\Delta}^{4/3} \sum_{\alpha,\beta,r,s} C_{\alpha\beta}^\varepsilon a_{i\alpha} a_{j\beta} a_{r\alpha} a_{s\beta} \overline{S}_{rs} + \frac{2}{3}\delta_{ij} \overline{\Delta}^{4/3} \sum_{\alpha,r,s} C_{\alpha\alpha}^\varepsilon a_{r\alpha} a_{s\alpha} \overline{S}_{rs}$$

$$T_{ij}^a = -2\hat{\overline{\Delta}}^{4/3} \sum_{\alpha,\beta,r,s} C_{\alpha\beta}^\varepsilon a_{i\alpha} a_{j\beta} a_{r\alpha} a_{s\beta} \hat{\overline{S}}_{rs} + \frac{2}{3}\delta_{ij} \hat{\overline{\Delta}}^{4/3} \sum_{\alpha,r,s} C_{\alpha\alpha}^\varepsilon a_{r\alpha} a_{s\alpha} \hat{\overline{S}}_{rs}.$$

The identity (4) gives

$$\mathcal{L}_{ij} - \frac{\delta_{ij}}{3}\sum_k \mathcal{L}_{kk} = -2\left(\hat{\overline{\Delta}}^{4/3} - \overline{\Delta}^{4/3}\right)\sum_{\alpha,\beta,r,s} C^{\varepsilon}_{\alpha\beta} a_{i\alpha}a_{j\beta}a_{r\alpha}a_{s\beta}\hat{\overline{S}}_{rs} + \frac{2}{3}\delta_{ij}\left(\hat{\overline{\Delta}}^{4/3} - \overline{\Delta}^{4/3}\right)\sum_{\alpha,r,s} C^{\varepsilon}_{\alpha\alpha} a_{r\alpha}a_{s\alpha}\hat{\overline{S}}_{rs}. \quad (10)$$

We define

$$\lambda_{\alpha\beta} \equiv -2\left(\hat{\overline{\Delta}}^{4/3} - \overline{\Delta}^{4/3}\right)\sum_{r,s} a_{r\alpha}a_{s\beta}\hat{\overline{S}}_{rs}, \qquad (C^{\varepsilon})'_{\alpha\beta} \equiv C^{\varepsilon}_{\alpha\beta}\lambda_{\alpha\beta}.$$

In matrix notation (10) reads

$$\mathcal{L} - \frac{1}{3}I\,\mathrm{Tr}(\mathcal{L}) = a(C^{\varepsilon})'a^T - \frac{1}{3}I\,\mathrm{Tr}(C^{\varepsilon})', \quad (11)$$

where $I$ is the identity matrix, $\mathrm{Tr}(C^{\varepsilon})'$ denotes the trace of $(C^{\varepsilon})'$ and we have taken into account the invariance of the trace by unitary transformations. $\mathrm{Tr}(C^{\varepsilon})'$ in (11) is arbitrary; we can thus normalize our unknowns so that $\mathrm{Tr}(C^{\varepsilon})' = 0$. Then (11) has the solution

$$(C^{\varepsilon})' = a^{-1}\left(\mathcal{L} - \frac{1}{3}I\,\mathrm{Tr}(\mathcal{L})\right)\left(a^T\right)^{-1} = a^T\left(\mathcal{L} - \frac{1}{3}I\,\mathrm{Tr}(\mathcal{L})\right)a$$

and finally, if we define $\mathcal{L}' = a^T\left(\mathcal{L} - \frac{1}{3}I\,\mathrm{Tr}(\mathcal{L})\right)a$,

$$C^{\varepsilon}_{\alpha\beta} = \frac{\mathcal{L}'_{\alpha\beta}}{\lambda_{\alpha\beta}}. \quad (12)$$

This anisotropic modification of the dynamic SGS closure can be extended straigthforwardly to model the unresolved heat flux. Using an eddy diffusivity tensor $\nu^{\vartheta}_{ir}$, instead of (2) we have

$$q_i = -\sum_r \nu^{\vartheta}_{ir}\frac{\partial\overline{\vartheta}}{\partial x_r}. \quad (13)$$

Starting from the decomposition

$$\nu^{\vartheta}_{ir} = \overline{\Delta}^{4/3}\sum_{\alpha}\left(\frac{C^{\varepsilon}}{Pr_t}\right)_{\alpha} a_{i\alpha}a_{r\alpha} \quad (14)$$

we get:

$$\left(\frac{C^{\varepsilon}}{Pr_t}\right)_{\alpha} = -\frac{a^T\mathcal{S}\Big|_{\alpha}}{\left(\hat{\overline{\Delta}}^{4/3} - \overline{\Delta}^{4/3}\right)\sum_r a_{r\alpha}\frac{\partial\hat{\overline{\theta}}}{\partial x_r}} \quad (15)$$

## 3. Results

The LES equations are discretized over a staggered grid using a centered second order finite differences scheme. The time integration is performed using a Projection Method (Le & Moin 1991) combined with an explicit third order, three steps Runge-Kutta scheme. The Poisson equation for the pressure is solved using a Fast Poisson Solver in the horizontal directions of periodicity.

We do not use an explicit filter as grid-filter; indeed the discretization of the equations over the computational grid is equivalent to using a box filter with the same width of the mesh. For the simulations done using the original dynamic model (1)-(3), (6) and (7), a two-dimensional box filter in homogeneous horizontal directions (with $\widehat{\overline{\Delta}} = 2\overline{\Delta}$) is used as test-filter, while for the simulations employing the anisotropic modification (8), (9), (12)-(15) a three-dimensional box filter is employed.

Cyclic boundary conditions are taken in horizontal directions. At the walls, no-slip conditions for the velocity and constant Dirichlet conditions for temperature are taken. To solve the Poisson problem for pressure we also impose homogeneous Neumann conditions for pressure. At the initial time the velocity field is set to zero while the pressure field is the one corresponding to hydrostatic distribution. Random temperature variations with maximum amplitudes of $\pm$ 2% have been superimposed on the conductive temperature profile to trigger instability.

Simulations of turbulent flows have been done at Rayleigh numbers up to $Ra = 10^8$ and with Prandtl number of 0.71 which is suitable for air at normal conditions.

A first series of runs at $Ra = 2 \times 10^7$ and with an aspect ratio $A$ (periodicity length over height) of 6 has been done to compare the performances of the original dynamic model with those of the anisotropic one. A grid of $24 \times 24 \times 31$ cells has been used. Because of the strong vertical gradients, we have taken a higher resolution in $z$ direction. In order to resolve the flow near the walls, the vertical grid has been stretched. With the resolution adopted in our simulations we get 4-5 points in the thermal boundary layer.

Two choices of the local frame $\mathbf{a}_\alpha$ have been tested. The first one consists, following Abbà *et al.* (1995b), in using

$$\mathbf{a}_1 = \frac{\mathbf{u}}{|\mathbf{u}|}, \ \mathbf{a}_3 = \frac{\nabla(|\mathbf{u}|)^2 \times \mathbf{u}}{|\nabla(|\mathbf{u}|)^2 \times \mathbf{u}|}, \ \mathbf{a}_2 = \mathbf{a}_3 \times \mathbf{a}_1$$

for the SGS stress tensor model and

$$\mathbf{a}_1' = \frac{\mathbf{u}}{|\mathbf{u}|}, \ \mathbf{a}_3' = \frac{\nabla\vartheta \times \mathbf{u}}{|\nabla\vartheta \times \mathbf{u}|}, \ \mathbf{a}_2' = \mathbf{a}_3' \times \mathbf{a}_1'$$

for the SGS heat flux model. The second choice consists in taking the orthogonal frame of the normalized eigenvectors of the grid-scale strain rate tensor $\overline{S}_{ij}$ at the last time-step, both for SGS stress tensor and heat flux model. This choice has the advantage of not depending on the particular flow considered and being automatically Galilei invariant.

Results obtained using the two proposed sets of local frames show insignificant differences and so we will refer in the sequel to the anisotropic model without specifying the local frame used.

In figures 1-3 we present the averages in time and over horizontal planes of the fluctuations of temperature and velocity components as a function of the height $z$. The time averages are done only after a statistically stationary state is reached and on an interval of some large eddy turnover times. We have plotted our LES results together with those of a direct numerical simulation (DNS) of Kerr (1996). The anisotropic model appears to be everywhere less dissipative than the original dynamic one and in the region close to the thermal boundary layer it agrees better with DNS data.

To get a deeper insight in the performances of the models an *a priori* test has been done using the fields obtained filtering those of a separate DNS that we performed at $Ra = 4 \times 10^6$ with a grid of $80 \times 80 \times 50$ cells. This grid is not very fine and could be improved, but it is sufficient to resolve the thermal boundary layers and neighbouring regions, that are the regions where anisotropy occurs. The comparison is done between unresolved terms computed from data of DNS and those obtained using the SGS models on filtered distributions. Figures 4 and 5 show correlations of dissipation and of kinetic energy nonlinear transfers

$$\varepsilon_{SGS} = \tau_{ij}\overline{S}_{ij} \qquad \Phi_{SGS} = \frac{\partial}{\partial x_j}\left(\overline{u}_i\tau_{ij}\right) - \varepsilon_{SGS}$$

for Smagorinsky, original dynamic model and anisotropic model. Correlation between quantities $a$ and $b$ is computed as

$$\mathcal{C}(a,b) = \frac{< ab > - < a >< b >}{\sqrt{(< a^2 > - < a >^2)(< b^2 > - < b >^2)}}.$$

We clearly see that the models work about in the same way in the interior of the flow. Important differences instead are visible near the walls where dynamic models, and in particular the one with anisotropic modification, are better.

A large aspect ratio is desired to minimize the effect of periodic boundary conditions in the horizontal directions. A minimum of $A \simeq 6$ is necessary to include the largest natural wavelength as suggested by the work of Fitzjarrald (1976) for Rayleigh numbers greater than $10^6$. The effects of the value of the aspect ratio are discussed in Grötzbach (1983).

The use of a subgrid model allows for larger values of Rayleigh number and of aspect ratio compared to DNS. We have successfully simulated flows at $Ra = 10^8$ and with an aspect ratio $A = 7$ using $80 \times 80 \times 64$ cells. According to Grötzbach's estimates a grid of about $450 \times 450 \times 64$ points would be necessary for a direct simulation in this case.

Experimental studies by Heslot, Castaing & Libchaber (1987) and by Castaing *et al.* (1989) (the Chicago group) at large Rayleigh numbers have shown that there is a transition in the turbulent flow from a regime of so called soft turbulence in the range $2 \times 10^5 < Ra < 4 \times 10^7$ to a regime of so called hard turbulence in the range $4 \times 10^7 < Ra < 6 \times 10^{12}$. In particular the probability distribution function of temperature fluctuations in the interior of the flow shows a switch from a Gaussian to an approximately exponential distribution. Christie & Domaradzki (1993, 1994) have found numerical evidence for nonuniversality of the soft/hard turbulence classification by making DNS for moderate Rayleigh numbers ($Ra \leq 6.3 \times 10^5$). They confirm, as was stated by the Chicago group, that, in addition to Rayleigh number, possible important dependencies on Prandtl number and aspect ratio should be considered.

In figure 6 we present the probability distribution function of temperature fluctuations obtained in two LES simulations at $Ra = 6.3 \times 10^5$ and $Ra = 10^8$. The curve corresponding to the larger Rayleigh number clearly displays an exponential behaviour at all the scales, which might be the signature of the hard turbulence regime. For the smaller Rayleigh number we obtain a mixed behaviour, close to Gaussian for the smallest temperature fluctuations and close to exponential for the tails. This mixed behaviour is similar to the one found by Christie & Domaradzki (1993).

Finally we have investigated the patterns of turbulent convection. Figures 7 and 8 show the contour plots of the temperature fluctuation in horizontal planes for $Ra = 10^8$, respectively near the lower wall and in the middle of the layer. Figures 9 and 10 show isosurfaces of constant temperature fluctuations. We clearly see hot buoyant sheets leaving the lower surface (figure 7), and cold descending plumes expanding near the bottom. This picture is in agreement with the results of a DNS at $Ra = 2 \times 10^7$ by Kerr (1996). Looking in more detail at the structure of the buoyant sheets we recognize that they are composed of a network of plumes (figure 9), the most intense being those at the intersection of the sheets. We also note that the rising plumes are not axisymmetric; rather they appear to be distorted by the wind blowing at the base of the convection zone (figure 10). This local wind appears to be produced by the large downwelling plumes. It might be interpreted as the wind responsible for the instability of the thermal boundary layer (Castaing *et al.* 1989). We note finally that in the middle of the layer the fine structure observed in the boundary layer disappears, and there remain only the most intense plumes (figure 8). In this region there is good symmetry between downwelling and upwelling structures.

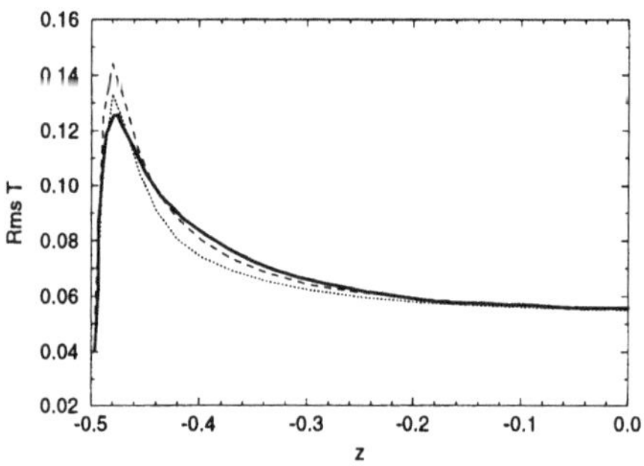


*Figure 1.* Temperature fluctuations (RMS values) at $Ra = 2 \times 10^7$ and $A = 6$. —— : DNS (Kerr 1996), resolution $288 \times 288 \times 96$ cells; ··· : original dynamic model, resolution $24 \times 24 \times 31$ cells ; - - - : anisotropic model, resolution $24 \times 24 \times 31$ cells. $\Delta T_{\mathrm{wall}} = 1$, $z_{\mathrm{bottom}} = -0.5$, $z_{\mathrm{top}} = +0.5$.

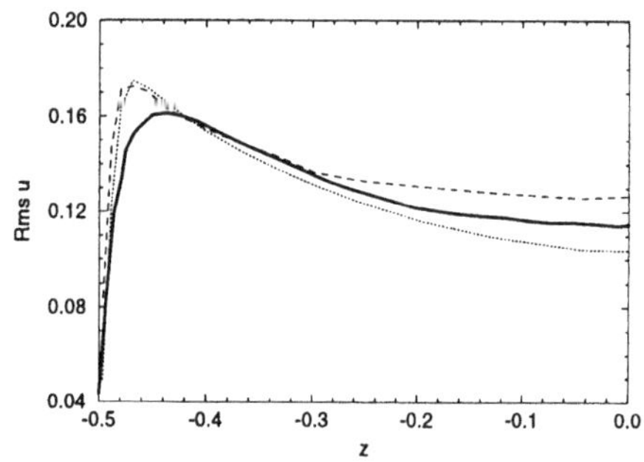


*Figure 2.* Horizontal fluctuating velocity component (RMS values) at $Ra = 2 \times 10^7$ and $A = 6$. —— : DNS (Kerr 1996), resolution $288 \times 288 \times 96$ cells; ··· : original dynamic model, resolution $24 \times 24 \times 31$ cells; - - - : anisotropic model, resolution $24 \times 24 \times 31$ cells. Velocity is in units of $Ra\, \nu\, k/L^2$ ($L$ is the height of the layer).

## References

Abbà, A., Bucci, R., Cercignani, C., & Valdettaro, L., 1995a: New variants to the dynamic subgrid scale model and new non reflecting outlet boundary conditions in channels and annuli. In *Proceedings of the Workshop on small-scale structures in three-dimensional hydro and magnetohydrodynamic turbulence*, Nice. Springer-Verlag.

Abbà, A., Bucci, R., Cercignani, C., & Valdettaro, L., 1995b: A new approach to the dynamic subgrid scale model. *Submitted for publication.*

Castaing. B., Gunaratne, G., Heslot, F., Kadanoff, L., Libchaber, A., Thomae, S., Wu, X.-Z., Zaleski, S., & Zanetti, G., 1989: Scaling of hard thermal turbulence in Rayleigh-Bénard convection. *J. Fluid Mech.* **204**, 1-30.

Christie, S.L., & Domaradzki, J.A., 1993: Numerical evidence for nonuniversality of the soft/hard turbulence classification for thermal convection. *Phys. Fluids A* **5**(2), 412-421.

Christie, S.L., & Domaradzki, J.A., 1994: Scale dependence of the statistical character of turbulent fluctuations in thermal convection. *Phys. Fluids* **6**(5), 1848-1855.

Fitzjarrald, D.E., 1976: An experimental study of turbulent convection in air. *J. Fluid Mech.* **73**, 693-719.

Grötzbach, G., 1983: Spatial resolution requirements for direct numerical simulation of the Rayleigh-Bénard convection. *Journal of computational physics* **49**, 241-264.

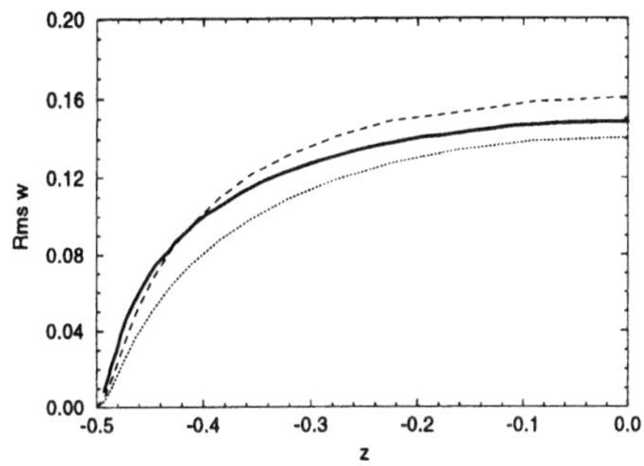


*Figure 3.* Normal fluctuating velocity component (RMS values) at $Ra = 2 \times 10^7$ and $A = 6$. —— : DNS (Kerr 1996), resolution $288 \times 288 \times 96$ cells; ⋯ : original dynamic model, resolution $24 \times 24 \times 31$ cells; - - - : anisotropic model, resolution $24 \times 24 \times 31$ cells. Velocity is in units of $Ra\, \nu\, k/L^2$ ($L$ is the height of the layer).

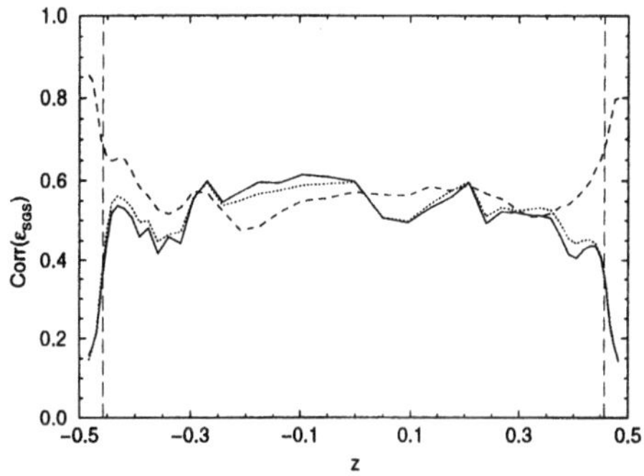


*Figure 4.* Correlation of $\varepsilon_{SGS}$ as a function of height $z$ at $Ra = 4 \times 10^6$ and $A = 5.6$. —— : Smagorinsky model; ⋯ : original dynamic model; - - - : anisotropic model. The long-dashed vertical lines delimit the thermal boundary layer, computed as one half of the Nusselt number.

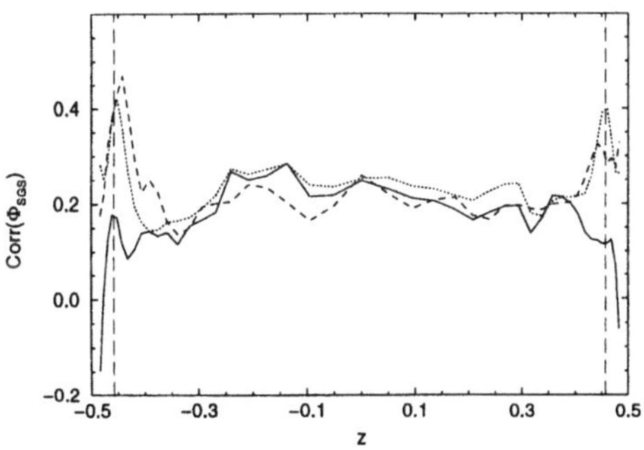


*Figure 5.* Correlation of $\Phi_{SGS}$ as a function of height $z$ at $Ra = 4 \times 10^6$ and $A = 5.6$. —— : Smagorinsky model; ⋯ : original dynamic model; - - - : anisotropic model. The long-dashed vertical lines delimit the thermal boundary layer, computed as one half of the Nusselt number.

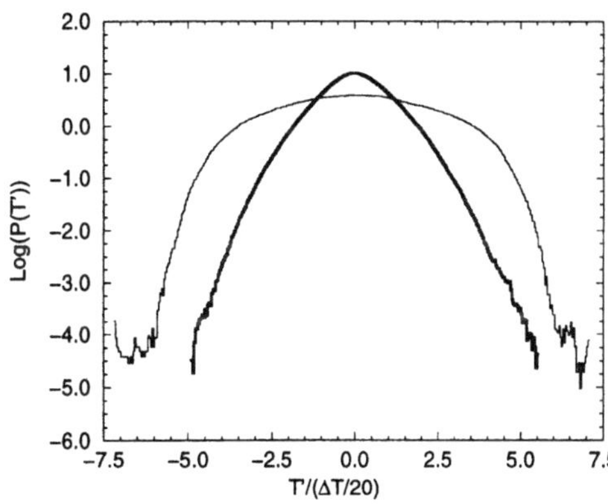


*Figure 6.* Probability distribution function of temperature fluctuations. —— : $Ra = 10^8$. —— : $Ra = 6.3 \times 10^5$.

Germano, M., Piomelli, U., Moin, P., & Cabot, H., 1991: A dynamic subgrid-scale eddy viscosity model. *Phys.Fluids A* **3**(7), 1760-1765.

Heslot, F., Castaing, B., & Libchaber, A., 1987: Transition to turbulence in helium gas. *Phys. Rev. A* **36**, 5870-5873.

Kerr, R.M., 1996: Rayleigh number scaling in numerical convection. *J. Fluid Mech.* **310**, 139-179.

Le, H., & Moin, P., 1991: An improvement of fractional step methods for the incompressible Navier-Stokes equations. *Journal of Computational Physics* **92**, 369-379.

Lilly, D.K., 1992: A proposed modification of the Germano subgrid-scale closure method.

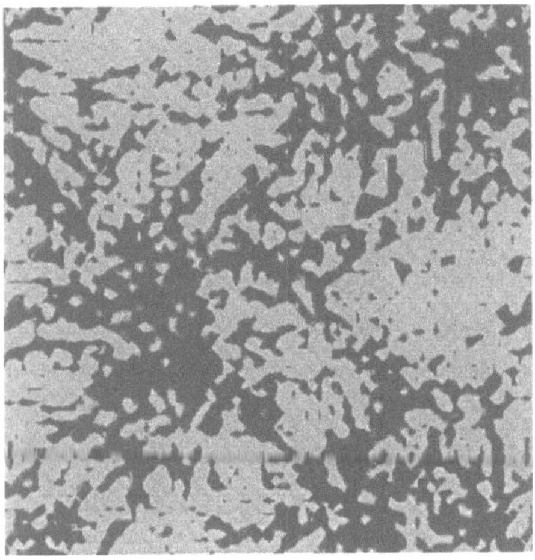

*Figure 7.* Contour plot of the temperature fluctuations in horizontal plane for $Ra = 10^8$. The surface is at $z = -0.488$ (the bottom wall is at $z = -0.5$ and the limit of the thermal boundary layer is at $z \sim -0.48$). Dark color corresponds to hot fluid, light color to cold fluid.

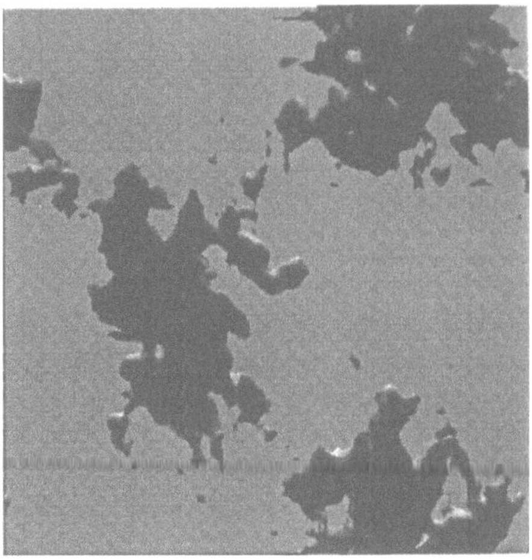

*Figure 8.* Contour plot of the temperature fluctuations in horizontal plane for $Ra = 10^8$. The surface is at $z = 0$ (the middle of the layer). Dark color corresponds to hot fluid, light color to cold fluid.

*Figure 9.* Isosurfaces of temperature fluctuations at $Ra = 10^8$. The chosen isosurface value is 0.23 scaled with the maximum value.

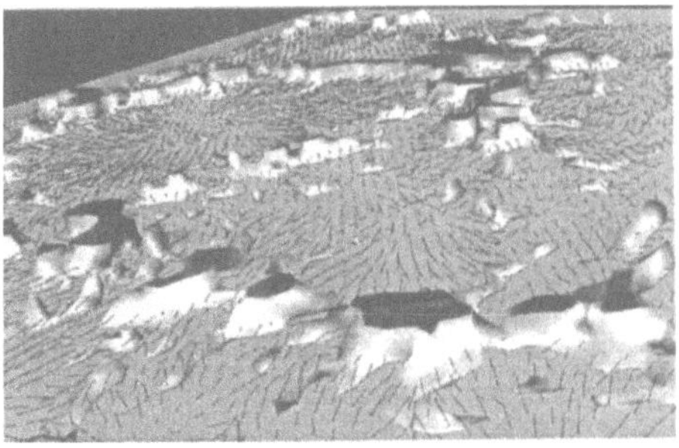

*Figure 10.* Isosurfaces of temperature fluctuations at $Ra = 10^8$. The isosurface value is 0.23 scaled with the maximum value. The isosurfaces are cut at $z = -0.35$ for an easier interpretation. Also shown are the velocity vectors at the horizontal plane $z = -0.47$.

*Phys.Fluids A* **4**(3), 633-635.

Wong, V.C., & Lilly, D.K., 1994: A comparison of two dynamic subgrid closure methods for turbulent thermal convection. *Phys. Fluids* **6**(2), 1016-1023.

# LARGE EDDY SIMULATIONS OF STRATIFIED ATMOSPHERIC FLOWS WITHIN A DEEP VALLEY

C. GUILBAUD, J.P. CHOLLET AND S. ANQUETIN
*Laboratoire des Ecoulements Géophysiques et Industriels*
*UJF-INPG-CNRS, BP 53, 38041 GRENOBLE cedex 9, France*

**Abstract.** Large Eddy Simulations are used to compute atmospheric thermal circulations in complex terrain. The subgrid models are based either on the Smagorinsky formulation or on a subgrid kinetic energy equation. They are tested in a convective boundary layer and then used in a comparison with available measurements. The formulation with the subgrid kinetic energy gives as good results as the Smagorinsky formulation, but with more information available from the subgrid energy. The large eddy simulation of the atmosphere in a deep valley allows reproduction of the diurnal cycle of thermal winds and the progression of the inversion layer.

## 1. Introduction

Large eddy simulations are commonly used to model atmospheric flows because of very large length scales and their dependence on time. The non stationarity is due to : *(i)* the synoptic wind which evolves with meteorological conditions and *(ii)* the thermal flux at the ground which changes with solar radiation on a 24 hour time scale. Whatever the computing power will be in the future, there is no hope of simulating explicitly the whole range from planetary scales (hundreds of kilometers) to dissipative ones (few centimeters). So atmosphere modelling usually combines LES, i.e. explicit computation of large scales, with statistical modelling of small scales.

Turbulent closure models (e.g. $k - \epsilon$, $k - l$) are not considered here because of the strong coupling between large and small scales. These models should be restricted to problems either at shorter length scale or quasi stationary states which can be viewed as short time scale processes.



*J.-P. Chollet et al. (eds.), Direct and Large-Eddy Simulation II,* 157-166.

LES of atmospheric flows relies on the usual set of equations for balance of mass, momentum and energy. This set of equations has to be completed with boundary conditions and a subgrid scale model whose formulation may depend on the problem under consideration.

The present work is restricted to the so called meso $\gamma$ scales which are of particular interest for urban atmosphere and related pollutant studies. At these scales, especially in the case of mountainous area, flows are strongly three dimensional. Down- or up- slope movements are driven by buoyancy forces. Inversion layers may develop and break up. The boundary condition at the surface plays an important part in the dynamics through both the drag force due to the surface roughness and the heat exchange with the air situated above. These boundary conditions at the ground evolve with time according to diurnal changes in solar radiation. This paper focuses on flows driven by thermal process in low or zero mean wind conditions.

First, the basic features of the model are recalled, including the set of flow equations and the soil model. Then, two subgrid models are considered, the Smagorinsky model and a model based on an equation for the evolution of subgrid kinetic energy. The third part is devoted to a comparison of these models in simulations of a steady atmospheric convective boundary layer and the diurnal varying boundary layer measured in Wangara in August 1967 (Yamada & Mellor, 1975). Finally, the large eddy simulation is used to model the effect of the diurnal cycle on thermal flows in a deep valley.

## 2. Large eddy simulation and subgrid scale models

### 2.1. LARGE EDDY SIMULATIONS

The governing set of equations is based on balance of mass, momentum and energy. Because of strong vertical accelerations which may develop in the flows under consideration, no hydrostatic hypothesis is used. The equations are written for a compressible gas while neglecting some terms because of low values of the Mach number. Therefore, the usual Boussinesq assumptions are not needed, and the velocity field is not required to be divergent free. The formulation is fully three dimensional in space in order to let turbulent interactions develop under vorticity stretching with the associated mean drain of energy from large to small scales.

The boundary conditions can be split into two classes : (1) the conditions needed by the limited extent of the computational domain ($\mathcal{D}$) which acts as a supergrid scale model : an absorption layer at the top and lateral free outflow boundaries, (2) conditions at the ground which force the flow through heat exchanges. A parameterization is needed at the ground to take account of the mean drag on the large scale flow, the flux of small scale energy produced by surface roughness and the thermal flux. In the case under consideration here, this thermal flux plays the major part in sustaining the large scale flow. This parametrization is obtained by using a soil model coupled with a wall model. The soil model $SM2 - Isba$ (Guilbaud 1996) is based on the one developed by Noilhan & Planton (1989). This is a two layer soil model which uses the force-restore method. The determination of the surface fluxes (momentum and heat) are given by an explicit method developed by Louis *et al.* (1981) for the stable case and by Byun (1990) for the unstable case.

The code, Submeso, which is used here has been developed on the basis of ARPS (Xue *et al.*, 1995). The equations are solved on a staggered C-Arakawa grid under finite difference approximations. Terrain following coordinates (Gal-Chen & Sommerville, 1975) are used.

## 2.2. SUBGRID MODELS

Most subgrid models assume that the balance between large and small scale energy must be achieved at each time step and in each space location, which tends to become less realistic as $\Delta x$ gets larger. A significant amount of small scale energy is produced near the ground by either the interacting wakes induced by roughness elements or heat flux fluctuations. Much of this energy is not dissipated there, but transported and dissipated at higher levels. This is the reason why one of the two subgrid models considered hereafter takes into account the evolution of the subgrid energy $e$. Temperature stratification may also develop in the atmosphere and reduce the subgrid transfer in the vertical direction compared to its horizontal counterpart. As atmospheric flows are so sensitive to thermal circulations, accurate subgrid models have to be used not only for kinetic energy but also for heat fluxes. Subgrid transfers are modelled with an eddy viscosity $\nu_t$, where $\nu_t \propto \mathcal{L}u$, and a thermal diffusivity $\kappa_{t\theta}$ and the associated turbulent Prandtl number $Pr_t = \nu_t/\kappa_{t\theta}$. Usually $\mathcal{L}$ and $u$ are typical scales of length and velocity respectively.

The Smagorinsky model (Smagorinsky 1963), noted later as $SMAGO$, assumes, at each time step, a local balance between production and dissipation of turbulent kinetic energy. The eddy viscosity $\nu_t$ is related to the strain rate $S_{ij}$ which is computed from the large scale velocity field. The length scale $\mathcal{L}$ is assumed to be equal to the equivalent mesh size $\Delta_{eq} = (\Delta x \Delta y \Delta z)^{\frac{1}{3}}$. As $S_{ij}$ is usually at a maximum near the ground, $\nu_t$ can be expected to be maximum there, leading to an overestimation of the energy dissipation.

The extension by Lilly (Lilly 1967) takes account of the stratification which may significantly reduce $\nu_t$ by making it dependent on the Brunt

Väisälä frequency $N$. The constant $C_{SL}$ is chosen to be equal to 0.21 according to Deardorff (1972).

$$\nu_t = C_{SL}\Delta_{eq}^2\sqrt{2\left(S_{ij}\right)^2 - \frac{N^2}{Pr_t}} \tag{1}$$

The second model ($TKE$) is based on the calculation of the subgrid kinetic energy $e$. The kinetic energy of the subgrid scales, $e$, is produced, transported by the large scale velocity $U_j$ and diffused according to the equation :

$$\frac{\partial e}{\partial t} + U_j\frac{\partial e}{\partial x_j} = -\tau_{ij}S_{ij} + \frac{g}{\Theta_{init}}\tau_{\theta w} - C_\varepsilon\frac{e^{\frac{3}{2}}}{\ell} + \frac{\partial}{\partial x_i}\left(2\nu_t\frac{\partial e}{\partial x_i}\right) \tag{2}$$

with, on the right hand side terms representing : production through large scale shear, production/destruction by buoyancy, dissipation through viscosity and diffusion. This last includes pressure-velocity interactions. The eddy viscosity $\nu_t \propto \mathcal{L}u$ is built on the velocity $u \propto \sqrt{e}$ times the length $\mathcal{L} \propto \ell$. The subgrid mixing length $\ell$ is made dependent on the stratification by switching from $\ell = \Delta_{eq}$ in unstable atmosphere to $\ell = 0.76\sqrt{e}/N$ for stable stratification (Deardorff 1980). Thermal diffusion is also made stratification dependent with $Pr_t\left(\ell\right)$

$$\nu_t = C_k\,\ell\sqrt{e} \quad ; \quad Pr_t = \frac{1}{1 + \frac{2\ell}{\Delta_{eq}}} \tag{3}$$

The constants $C_k$ and $C_\varepsilon$ are given by Moeng & Wyngaard (1988). If we assume equilibrium in the $e$ equation, this choice of constants makes the two models equivalent.

## 3. Subgrid models in atmospheric boundary layer simulations

The two subgrid models are compared when used in the simulation of a convective boundary layer in order to point out the additional information given, of course at a higher computing cost, by the method based on the $e$ equation. The influence of a non uniform vertical distribution of the mesh size is also considered. Then the simulation is run in conditions which make comparisons with experimental data possible in order to assess the capability of subgrid models to represent real flow features.

### 3.1. CONVECTIVE BOUNDARY LAYER

The case under consideration here is a pure convective boundary layer with steady thermal flux from the ground. It is based on the configuration studied by Nieuwstadt *et al.*, (1993) with three simulations performed : (1) $TKE_A$ : TKE model with cartesian grid, (2) $TKE_B$ : TKE model with non uniform grid, (3) $SMAGO$ : Smagorinsky model with cartesian grid. Conditions for the numerical test are described in table 1.

TABLE 1. Simulations of convective boundary layer

| SGS Model | $SMAGO$ | $TKE_A$ | $TKE_B$ |
|---|---|---|---|
| resolution | 43x43x53 | 43x43x53 | 43x43x53 |
| $dx = dy,\ dz\ (m)$ | 160.0, 48.0 | 160.0, 48.0 | 160.0, 24.0→72.0 |
| $\Delta t\ (s)$ | 2.0 | 2.0 | 1.0 |
| $T_{CPU},\ CRAY\ (s)$ | 5400.0 | 6200.0 | 13000.0 |

Boundary conditions are periodic in horizontal $x$, $y$ directions. Because of zero mean external wind, the flows develop under thermal effects only. The surface is characterized by a roughness $z_0 = 0.16m$. The steady thermal flux is $Q_s = 0.06Kms^{-1}$. The scaling length is the final height of the CBL : $z_{i0} = 1400\ m$. The extension of the domain is $4.6z_{i0}$ in the horizontal directions and $1.7z_{i0}$ in the vertical direction. The results are scaled by the velocity $w_{*0} = 1.40\ ms^{-1}$, the potential temperature $\Theta_{*0} = 4.28\ 10^{-2}K$, and time $t_* = 1000\ s$.

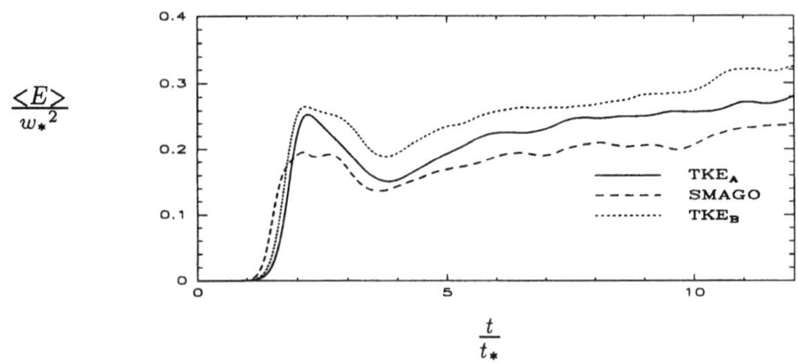


*Figure 1.* Mean total kinetic energy on a 1200 $m$ high layer, versus time $t/t_*$.

Figure (1) shows that the total amount of kinetic energy in the CBL tends to reach a quasi steady value. The differences between the results from the simulations are small but remain significant as time proceeds. The subgrid contribution $e$ to $< E >$ is explicitly calculated from the $TKE$ method when it is only estimated with the $SMAGO$ method.

Figure (2a) shows profiles of the vertical heat flux averaged over the horizontal computational domain. The profiles remain linear in the range

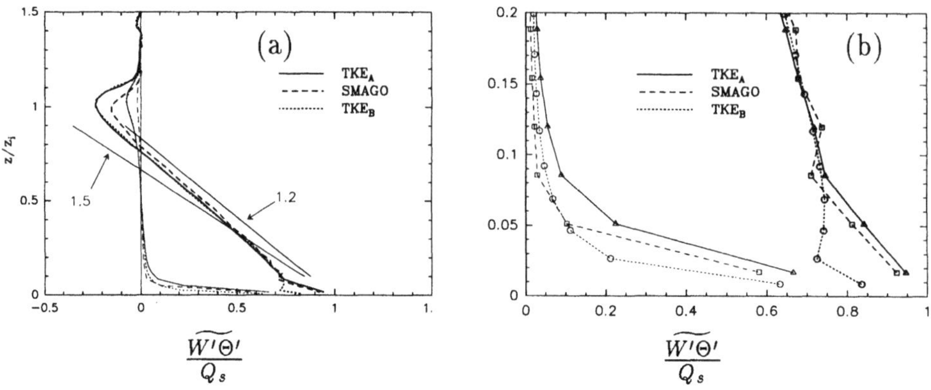


*Figure 2.* heat flux $\frac{\widetilde{W'\Theta'}}{Q_s}$ versus altitude : a) across the CBL, b) near the ground ; *bold lines* : total heat flux, *thin lines* : subgrid contribution

$z/z_i = [0.1, 0.8]$ in agreement with the results of a similarity theory by Stull (1988). Negative fluxes above $z/z_i = 0.8$ correspond to the *entrainment zone* and vary from $-0.15$ (SMAGO) to $-0.23$, to be compared to $-0.17$ obtained by Schmidt & Schumann (1989) and $-0.25$ in experiments or atmospheric measurements collected by Young (1988). Figure (2b) gives a closer look at the distribution of the heat flux near the ground ($z/z_i < 0.2$) where the results are more sensitive to the model in use. The solution is also observed to depend on the grid mesh since $TKE_A$ and $TKE_B$ differ only by $\Delta z$. Subgrid models always depend on the numerical scheme because of its filtering properties. This dependency is also explicit since $\Delta_{eq}$ is often used as a characteristic mixing length for subgrid scale modelling.

The differences between the two models remain small without visible effect on large scale convective structures, at least as far as mean quantities are concerned.

## 3.2. WANGARA BOUNDARY LAYER

A full size experiment is used here to test the realism of the simulation especially as regards the predicted height of the planetary boundary layer and the time scale which characterizes its growth. Day 33 of the Wangara experiment was selected, as being already used as a reference by other authors among them Deardorff (1974) with a 3D simulation. Our simulation was run on $23 \times 23 \times 63$ grid for 10 hours with a 2 *sec* time step and a resolution $\Delta x = \Delta y = 150$ $m$ and $\Delta z = 40$ $m$. Contrary to the previous case, the boundary condition at the bottom is made representative of the behaviour of a real soil (loam) under the effect of solar radiation with the soil model $SM2 - Isba$ (Guilbaud 1996, Noilhan & Planton, 1989). The flow is forced by the observed geostrophic wind.

Figure (3) shows the growth of the boundary layer under the effect of convective activity due to heat flux at the ground. The results from the simulation are in good agreement with the measurements, and are better than the results from the Deardorff (1974)'s simulation referred to later as DR74, in particular during the convective boundary layer growth. Therefore, large eddy simulations of atmospheres prove to accurately predict not only the height of the convective boundary layer but also the time scale which characterizes its growth.

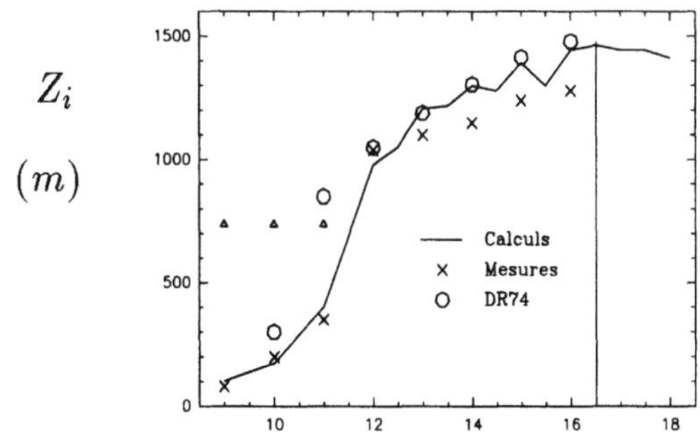


*Figure 3.* Altitude of the CBL ($z_i$) versus time (in hours) : *solidline :* simulation ; (×) measurements ; o Deardorff (1974) ; △ CBL on the previous day

The large scale flow is not very sensitive to the subgrid model in use, which makes the Smagorinsky-Lilly model still relevant in many cases. Nieuwstadt *et al.* (1993) have found the same properties for different subgrid models in the convective boundary layer. The $TKE$ formulation costs the same time as the $SMAGO$ model (table 1). However, as it gives more subgrid information, we will use it in future studies. It should be noted that the evaluation of the subgrid kinetic energy is useful not only for subgrid models but also for the calculation of mixing properties of various scalar quantities of interest in atmospheric processes (e.g. moisture, pollutant concentration).

## 4. Simulation of atmospheric flows in a deep valley

The objective is to produce quantitative information on the evolution of the atmosphere within a deep valley under zero mean wind conditions considering : (i) the winds which develop along the slopes and (ii) the growth of the inversion layer during the day and the collapse of the mixed layer at night. A large eddy simulation with the $TKE$ subgrid model was performed on the computational domain 12 $km \times 3$ $km$ horizontally and 5 $km$ in the vertical direction. The valley is 3 $km$ wide and 1.26 $km$ high. Although the valley is 2 dimensional, the computation is fully three dimensionsal with 83 × 18 nodes horizontally and 53 nodes vertically. The simulation was run for 48 hours in order to use 24 hours as the time required to make the solution less dependent on initial conditions and the results from the second day are the only ones to be analysed. The results are detailed in Guilbaud (1996), with a discussion of seasonal effects. We will mention here summer situations only.

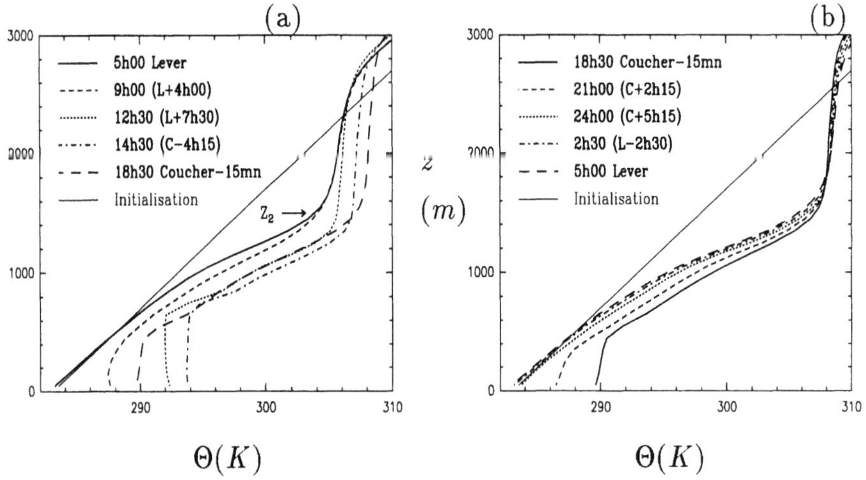


*Figure 4.* Vertical profiles of potential temperature on a summer day (a) from sunrise (5:00) to sunset (18:00), (b) from sunset (19:00) to sunrise (5:00)

Figure (4a) shows the vertical potential temperature profile, averaged over all the valley bottom, at different times from sunrise to sunset. At 5 : 00, the atmosphere is split into three layers : a stable layer from 0 to $h_{st} = 1600m$, a quasi neutral layer from 1600 to $2400m$ and a stable layer above. The convective mixed layer grows from the ground level, but in the case under consideration here, there is always stable air which persists above. The inversion is not totally broken up. Figure (4b) shows the same quantity from sunset to sunrise with cooling and hence stabilization which proceeds from the ground to higher levels.

Together with the evolution of the stratification, winds blow up-slope in the morning and blow down-slope at the beginning of night. A complex pattern of wind develops, as shown in figure (5), with velocity streamlines and isovalue surfaces of vertical velocity. At 9:00 a.m., ascending currents develop on the slope exposed to the rising sun (left side of the picture). At 11:00 a.m., most of the convective activity takes place above the centre of the valley. The limited width of the valley clearly makes the phenomena more intricate by inducing recirculations of slope winds under the stable layers which form the inversion.

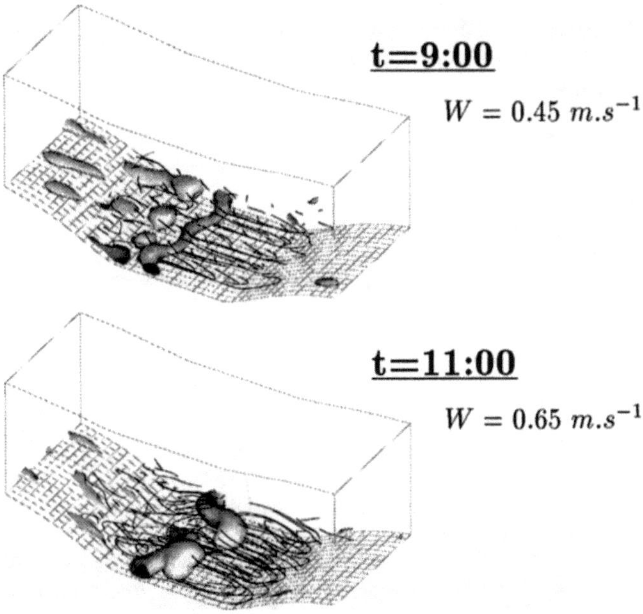


*Figure 5.* Isocontour surface of vertical velocity and streamlines in the valley, in the morning of a summer's day. The velocity written in the picture is the level of the isocontour.

## 5. Conclusions and prospects

Diurnal cycles of atmospheric flows within deep valleys are a good test of the capabilities of LES since they are strongly unsteady and driven by intricate phenomena including thermal effects. There are numerous applications, such as pollutant dispersion and mixing in urban sites. The zero geostrophic wind condition is often the most sensitive case for pollution.

Large scale flows do not seem to be very sensitive to the subgrid model in use. Nevertheless the $TKE$ model based on the subgrid kinetic energy equation gives more information than the Smagorinsky-Lilly model for the same computational cost. And we expect that it gives a good representation near the ground, a region which needs to be well represented because of the dominant effect of surface heat flux and humidity flux, which depends on the temperature and the humidity from the soil model, in driving the flow. The effect of vertical grid size is emphasized.

For future development of these models to study pollution, the inclusion of moisture and reacting chemical species will be considered. An estimation of small scale mixing and hence of energy at subgrid scale is required. The TKE formulation allows evaluation of this energy and its variability with a good accuracy. A challenge in the future could be caused by the use of nonuniform (and presumably unstructured) meshes in order to well-represent complex terrain. New problems then arise with varying filter width $\Delta x(\vec{x})$ and anisotropy.

## Acknowledgements

The computations were carried out on the CRAY C98 of the IDRIS (CNRS) and the CRAY C94 of the Grenoble CEA computing center. This work has been partly supported by the GDR CNRS 1102, Atmosphère urbaine à submeso échelle.

## References

Byun, D. W., 1990: On the analytical solutions of flux-profile relationships for the atmospheric surface layer. *J. Appl. Met.* **25**, 1205-1212.

Deardorff, J. W., 1972: Numerical investigation of neutral and unstable planetary boundary layer. *J. Atmos. Sci.* **29**, 91-115.

Deardorff, J. W., 1974: Three-dimensional numerical study of the height and mean structure of a heated planetary boundary layer. *Boundary-layer Met.* **7**, 81-106.

Deardorff, J. W., 1980: Stratocumulus-capped mixed layers derived from a three-dimensional model. *Phys. Fluids.* **A7**, 1760-1771.

Gal-Chen, T., & R. C. J. Sommerville, 1975: On the use of a coordinate transformation for the solution of the Navier-Stokes equations. *J. Comput. Phys.* **17**, 209-228.

Guilbaud, C., 1996, Etude des inversions thermiques : application aux écoulements atmosphériques dans des vallées encaissées, *thèse de l'Université Joseph Fourier*, Grenoble

Lilly, D. K., 1967: The representation of small-scales turbulence in numerical simulation experiments. In *Proc. IBM Sci. Comput. Symp. on Enivironmental Sci., Nov. 14-16, 1966, Thomas J. Watson Res. Center, Yorktown Heights, N.Y., IBM Form* 320-1951, pp. 195-210.

Louis, J. F., M. Tiedke, & J. F. Geleyn, 1981: A short history of the operational PBL parametrization of the ECMWF. *Workshop on planetary boundary layer parametrization, November 1981, ECMWF, Reading, England.* 59-79.

Moeng, C.-H., & J. C. Wyngaard, 1988: Spectral analysis of large-eddy-simulation of the convective boundary layer. *J. Atmos. Sci.* **45**, 3573-3587.

Nieuwstadt, F. T. M., P. J. Mason, C.-H. Moeng, & U. Schumann 1993: Large-eddy-simulation of the convective boundary layer: A comparison of four computer codes. *Turbulent Shear Flows 8*, Durst *et al*, Eds. Springer-Verlag. 431 pp.

Noilhan, J., & S. Planton, 1989: A simple parametrization of the land surface processes for meteorological models. *Mon. Wea. Rev.* **117**, 536-549.

Schmidt, H., & U. Schumann, 1989: Coherent structure of the convective boundary layer derived from large-eddy-simulations. *J. Fluid. Mech.* **200**, 511-562.

Smagorinsky, J., 1963: General simulation experiments with the primitive equations. *Mon. Wea. Rev.* **91**, 99-164.

Stull, R. B., 1988: *An introduction to Boundary layer Meteorology.* Kluwer Academic Publishers. 666pp.

Xue, M., K. K. Droegemeier, V. Wong, A. Shapiro, & K. Brewster, 1995: *Advanced regional prediction system* (A.R.P.S.), *version 4.0, user's guide.* Center for analysis and prediction of storms. 374 pp.

Yamada, T., & G. Mellor, 1975: A simulation of the Wangara atmospheric boundary layer data. *J. Atmos. Sci.* **32**, 2309-2329.

Young, E.L., 1988 : Turbulence structure of the convective boundary layer. Part *I* : Variability of normalized turbulenc statistics. *J. Atmos. Sci.* **45**, 719-726.

# PRESSURE AND DILATATION EFFECTS IN HIGH-SPEED TURBULENCE

K. H. LUO
*Department of Engineering*
*Queen Mary and Westfield College*
*University of London*

**Abstract.**
This paper examines the pressure and dilatation effects in high-speed turbulence on the basis of direct numerical simulation (DNS) databases of three-dimensional compressible mixing layers at convective Mach numbers $M_c = 0.8$, 1.1 and 1.2. Fluctuating pressure effects are found to be significant for local turbulence energy balance and scalar transport. However, due to its acoustic nature, fluctuating pressure is mainly responsible for energy redistribution rather than energy creation or destruction. Models for velocity- and scalar-pressure gradients are developed and tested, which give satisfactory agreement with DNS data. Eddy shocklets are observed at $M_c = 1.1$ and above. Their effects are spatially and temporally localized but have been substantially over-predicted by all existing models.

## 1. Introduction

Interest in high-speed turbulence is closely associated with high-speed combustion, which finds many industrial applications, such as in hypersonic propulsion systems. Fundamental features that set high-speed turbulence and combustion apart from their low-speed counterparts include the presence of a non-zero dilatation and a pressure wave travelling with a finite characteristic speed. The effects of dilatation and pressure, which increase with the Mach number (compressibility) and chemical heat release, have been subject to intense study but many key issues remain unsolved (Bray *et al.* 1994). For example, the mean pressure term in the turbulence energy equation was found to be important to turbulence generation and chemical reaction rate (Bray *et al.* 1981) but the fluctuating pressure term



*J.-P. Chollet et al. (eds.), Direct and Large-Eddy Simulation II*, 167-178.

was negligible (Bilger 1976; Bray *et al.* 1981). A recent study in premixed flames by Zhang and Rutland (1995) showed that fluctuating and mean pressure terms were equally important. Even more controversies surround the effects of dilatation. Zeman (1990) contended that dilatation-related dissipation played a crucial role in high-speed turbulence. He believed that eddy shocklets existed in high-speed flow, which can lead to large enough dilatation terms to cause the well-known phenomenon of the mixing layer growth rate reduction with increasing Mach number. However, such claims have not been supported by experiments or independent numerical calculations. Vreman *et al.* (1995) argued that it was the reduced pressure-strain term that was responsible for the growth rate reduction. In any case, to determine the minimal Mach number at which shocklets will appear in *three-dimensional* turbulence is of considerable interest.

The study of pressure and dilatation effects is made particularly difficult by a lack of experimental techniques to measure flow fluctuations accurately, especially the pressure fluctuations. Numerically, only DNS databases contain full information about these fluctuations. However, DNS has been limited to isotropic and homogeneous flows (e.g. Sarkar 1992; Sarkar *et al.* 1991; Zeman 1991) and low Mach numbers ( e.g. Zhang and Rutland 1995). The present paper extends the study to inhomogeneous flow on the basis of three DNS databases of turbulent mixing layers of $M_c = 0.8$, 1.1 and 1.2. The convective Mach number $M_c$, first defined by Lin (1953), is written as $M_c = (U_1 - U_2)/(c_1 + c_2)$, where $U_1$ and $U_2$ are the upper and lower free-stream velocities respectively, and $c_1$ and $c_2$ are the corresponding sound speeds. It is widely used as a measure of compressibility. Turbulence modelling in compressible flow has received insufficient attention and remains relatively unexplored. This paper looks at some of the key remaining issues with the aid of DNS data.

## 2. Turbulence Models

The turbulence kinetic energy equation for compressible flow is written as

$$\frac{\partial}{\partial t}\bar{\rho}\tilde{k} + \frac{\partial}{\partial x_l}(\bar{\rho}\tilde{k}\tilde{u}_l) = P_{kk} + T_{kk} + \Pi_{kk} + D_{kk} + \varepsilon_{kk} \tag{1}$$

where $P_{kk} = -\overline{\rho u_i'' u_l''}\partial\tilde{u}_i/\partial x_l$ is the production, $T_{kk} = -\partial(\overline{\rho u_i'' u_i'' u_l''/2})/\partial x_l$ the triple correlation, $\Pi_{kk} = -\overline{u_i''\partial p/\partial x_i}$ the pressure work, $D_{kk} = \partial\overline{u_i''\sigma_{li}}/\partial x_l$ the viscous diffusion and $\varepsilon_{kk} = -\overline{\sigma_{li}\partial u_i''/\partial x_l}$ the dissipation. Here an overbar and a tilde indicate the Reynolds and Favre averages respectively. A prime and double primes denote the corresponding fluctuations. The usual symbols are used for time ($t$), coordinates ($x_i$, $i = 1,2,3$), velocity components ($u_i$), density ($\rho$), pressure ($p$), turbulence kinetic energy

($\tilde{k} \equiv \widetilde{u_i''u_i''}/2$), the dissipation rate ($\varepsilon$) and viscous stress tensor ($\sigma_{ij}$). Among the terms on the r.h.s. of Eq. (1), $\Pi_{kk}$ and $\varepsilon_{kk}$ explicitly contain the pressure and/or dilatation fluctuations. They are the focus of this study since they have not been subject to sufficient scrutiny in inhomogeneous compressible turbulence before. The other terms have been the subject of numerous modelling work for incompressible flow, a review of which has been written by Launder (1989).

The pressure work $\Pi_{kk}$ can be split into the following terms

$$\Pi_{kk} = -\overline{u_i''}\frac{\partial \bar{p}}{\partial x_i} - \overline{u_i''\frac{\partial p'}{\partial x_i}} \tag{2}$$

$$= -\overline{u_i''}\frac{\partial \bar{p}}{\partial x_i} - \frac{\partial}{\partial x_i}\overline{p'u_i''} + \overline{p'\frac{\partial u_i''}{\partial x_i}} \tag{3}$$

The first term on the r.h.s. of Eq. (2) is the mean pressure work. It is zero in a constant density flow since in that case $\overline{u_i''} = \overline{u_i'} = 0$. It represents a coupling between density inhomogeneity and the mean pressure gradient, responsible for non-gradient and counter-gradient diffusion and turbulence production in premixed turbulent flames (Bray *et al.* 1981). The second term on the r.h.s. of Eq. (3) is the pressure-velocity correlation, whose physical meaning is still unclear. Conventionally, it is treated as a diffusion term and absorbed into the triple correlation $T_{kk}$ in turbulence modelling. An alternative view was offered by Strahle (1983), which considered the term not to be diffusive in nature but related to acoustics. Indeed, the correlation $\overline{p'u_i''}$ is the acoustic intensity in linear acoustics. The third term on the r.h.s. of Eq. (3) is the pressure-dilatation, which is nonzero only in compressible flow and often neglected (Bilger 1976; Bray *et al.* 1981). Bilger (1976) argued that it is negligible because it is acoustic in nature and the interaction of a pressure fluctuation with an undirected volume expansion cannot affect the vorticity of a low Mach number flow. However, Zeman (1991) and Sarkar (1992) considered it to be significant in the turbulence energy balance. Zeman (1991) proposed to relate the pressure-dilatation to the temporal change of pressure variance

$$\overline{p'\frac{\partial u_i''}{\partial x_i}} = -\frac{1}{2\bar{\rho}\tilde{c}^2}\frac{\partial}{\partial t}\overline{p'^2} = \frac{1}{2\bar{\rho}\tilde{c}^2}\frac{\overline{p'^2} - p_e^2}{\tau_a} \tag{4}$$

where $\tau_a$ is an acoustic timescale and the equilibrium pressure variance $p_e$ is determined by $p_e^2/\bar{\rho}\tilde{q}^2\tilde{c}^2 = c_z(\alpha M_t^2 + \beta M_t^4)/(1 + \alpha M_t^2 + \beta M_t^4)$, where the model constants $(\alpha, \beta, c_z) = (1, 2, 1)$. $M_t = \tilde{q}/\tilde{c}$ is the turbulent Mach number, where $\tilde{q} = \sqrt{2\tilde{k}}$ and $\tilde{c}$ is the sound speed based on Favre mean temperature. Sarkar (1992) also modelled pressure-dilatation but used closure approaches for incompressible flow

$$\overline{p'\frac{\partial u_i''}{\partial x_i}} = -c_{s1}M_t P_{kk} + c_{s2}M_t^2\bar{\rho}\varepsilon_s \tag{5}$$

where constants $(c_{s1}, c_{s2}) = (0.15, 0.2)$ and $\varepsilon_s$ is the solenoidal dissipation.

Strahle (1983) questioned the wisdom of modelling the pressure terms separately and recommended modelling the term $\Pi_{kl} = -\overline{u_k''\partial p/\partial x_l}$ as a whole. In a radical departure from tradition, he constructed a Poisson equation for a scalar variable $S$ defined as

$$\frac{\partial S}{\partial x_l} = \frac{1}{\rho}\frac{\partial p}{\partial x_l} \tag{6}$$

instead of pressure itself. As a result, the pressure strain is determined as $\Pi_{kl} = -\overline{\rho u_k''\partial S'/\partial x_l}$ since $\overline{\rho u_k''} \equiv 0$. Therefore, $\Pi_{kl}$ is dependent on the fluctuating field of S only, irrespective of the mean field. Neglecting the viscous terms, the Poisson equation for $S'$ is derived from the momentum equation as:

$$\Delta^2 S' = -2\frac{\partial u_j'}{\partial x_i}\frac{\partial \bar{u}_i}{\partial x_j} - u_j'\frac{\partial}{\partial x_j}\left(\frac{\partial \bar{u}_k}{\partial x_k}\right) - \frac{\partial}{\partial x_i}\left[u_j'\frac{\partial u_i'}{\partial x_j} - \overline{u_j'\frac{\partial u_i'}{\partial x_j}}\right] \tag{7}$$

In this equation, a term representing the time derivative of the potential velocity field has been ignored. For free shear flow away from solid walls, the free space Green function can be used to solve for $S'$ from Eq. (7). With a further near-homogeneity assumption, $\Pi_{kl}$ is modelled as:

$$-\overline{u_k''\frac{\partial p}{\partial x_l}} = c_1'\,\overline{\rho u_k'' u_j''}\frac{\partial \bar{u}_l}{\partial x_j} + c_2'\,\overline{\rho u_k'' u_l''}\frac{\partial \bar{u}_i}{\partial x_i} \tag{8}$$

To construct the above model, the last term in Eq. (7) has been neglected, mainly because no good approximation can be found for the term. Limited testing of the model has been performed in combusting flows with high exothermity and mean strain rates (Strahle 1983). In the present study, the focus is on pressure-generated turbulence and the model for pressure work is easily obtained by contraction in Eq. (8). However, unlike Strahle's case, the flow of concern has large fluctuations compared with the mean. In particular the fluctuating dilatation is a dominant term. With this consideration in mind, the model for pressure work is written as:

$$-\overline{u_i''\frac{\partial p}{\partial x_i}} = c_1\,\overline{\rho u_i'' u_j''}\frac{\partial \bar{u}_i}{\partial x_j} + c_2\,\bar{\rho}\tilde{q}^2\frac{\partial \tilde{u}_i}{\partial x_i} + c_3\,\bar{\rho}\tilde{q}^2\frac{\partial \overline{u_i''}}{\partial x_i} \tag{9}$$

The split of the dilatation into a mean and a fluctuation with different model constants partly offsets the effect of neglecting the last term in

Eq. (7), which contains fluctuating dilatation. This becomes clearer with $\partial[u'_j(\partial u'_i/\partial x_j)]/\partial x_i = (\partial u'_j/\partial x_i)(\partial u'_i/\partial x_j) + u'_j\partial(\partial u'_i/\partial x_i)/\partial x_j$. The first term on the r.h.s. should be small while the second term bears strong resemblance to the mean dilatation term in Eq. (7). To determine $\overline{u''_i}$, Jones (1994) suggested the approximation $\overline{u''_i} = \overline{\xi''}\ \overline{\rho u''_i \xi''}/\overline{\rho \xi''^2}$, where the quantities related to the scalar ($\xi$) are readily available in the conserved scalar, presumed pdf approach. This expression has been verified with reference to DNS data in the present study.

Similar pressure-related terms exist in the conserved scalar flux equation, which is written for compressible flow in symbolic form as

$$\frac{\partial}{\partial t}\overline{\rho\xi'' u''_i} + \frac{\partial}{\partial x_l}(\overline{\rho\xi'' u''_i}\tilde{u}_l) = P_{\xi i} + T_{\xi i} + \Pi_{\xi i} + D_{\xi i} + \varepsilon_{\xi i} \tag{10}$$

where terms on the r.h.s. have similar forms and meanings as those in Eq. (1). At the second-moment closure level, the major task is to model the scalar-pressure gradient correlation $\Pi_{\xi i} = -\overline{\xi''\partial p/\partial x_i}$. By analogy with the decomposition of the pressure work in Eqs. (2) and (3), $\Pi_{\xi i}$ can be split up into similar terms and each modelled individually. However, the modelling of these terms is subject to greater uncertainty than of their counterparts in the Reynolds stress equation (Jones 1994; Launder 1989). In fact, none of the existing models have been derived directly from a compressible formulation. The usual practice is to recast incompressible models into compressible ones. An elaborate model for $\Pi_{\xi i}$ as a whole resulting from this practice is based on that of Jones and Musonge (1988)

$$\begin{aligned} -\overline{\xi''\frac{\partial p}{\partial x_i}} &= -c'_{\xi 1}\frac{\overline{\rho\xi'' u''_i}}{\tau} + c'_{\xi 2}\bar{\rho}\tilde{k}a_{il}\frac{\partial\tilde{\xi}}{\partial x_l} \\ &\quad + c'_{\xi 3}\overline{\rho\xi'' u''_l}\frac{\partial\tilde{u}_i}{\partial x_l} + c'_{\xi 4}\overline{\rho\xi'' u''_l}\frac{\partial\tilde{u}_l}{\partial x_i} + c'_{\xi 5}\overline{\rho\xi'' u''_i}\frac{\partial\tilde{u}_l}{\partial x_l} \end{aligned} \tag{11}$$

where $a_{il} = \widetilde{u''_i u''_l}/\tilde{k} - \frac{2}{3}\delta_{il}$ is the anisotropy tensor and $\tau$ is a time scale to be specified. Jones and Musonge (1988) used $\tau = (1 + c'_{\xi 0}\sqrt{a^2})\tilde{k}/\varepsilon$, where $a^2 = a_{il}a_{il}/4$. The recommended model constants are $(c'_{\xi 0}, c'_{\xi 1}, c'_{\xi 2}, c'_{\xi 3}, c'_{\xi 4}, c'_{\xi 5}) = (3.0, 3.0, 0.12, 1.09, 0.51, 0.0)$. The basic model described by Launder (1989) has only the first and third terms in Eq. (11) with constants $(c'_{\xi 1}, c'_{\xi 3}) = (2.9, 0.4)$ and $\tau$ is specified as the turbulence energy turnover time $\tilde{k}/\varepsilon$. Other forms of $\tau$ have also been used in the literature, for example by setting $\tau$ to be a geometric mean of the time scales of the turbulence and the scalar field.

| $M_c$ | $\mathrm{Re}_s$ | $\mathrm{Re}_f$ | $L_1$ | $L_2$ | $L_3$ | grid | symmetry in $x_3$ |
|---|---|---|---|---|---|---|---|
| 0.8 | 280 | 3360 | 13.4 | 30.0 | 13.4 | $144 \times 221 \times 160$ | yes |
| 1.1 | 200 | 1800 | 20.0 | 45.0 | 11.5 | $144 \times 384 \times 128$ | no |
| 1.2 | 200 | 1800 | 20.0 | 30.0 | 11.5 | $96 \times 301 \times 96$ | yes |

TABLE 1. DNS database parameters. $\mathrm{Re}_s$ and $\mathrm{Re}_f$ are initial and final Reynolds numbers, $L_i$ is the computational box length in each direction.

By analogy with the models in Eqs. (8) and (9), a compressible model for the scalar-pressure gradient can be constructed as

$$-\overline{\xi'' \frac{\partial p}{\partial x_i}} = c_{\xi 1}\, \overline{\rho \xi'' u_l''} \frac{\partial \overline{u}_i}{\partial x_l} + c_{\xi 2}\, \overline{\rho u_i'' u_l''} \frac{\partial \tilde{\xi}}{\partial x_l} + c_{\xi 3}\, \overline{\rho \xi'' u_i''} \frac{\partial \tilde{u}_l}{\partial x_l} + c_{\xi 4}\, \overline{\rho \xi'' u_i''} \frac{\partial \overline{u_l''}}{\partial x_l} \qquad (12)$$

where the four model constants are to be determined with reference to the DNS data. An apparent advantage of this model is that the uncertainty of specifying a time scale $\tau$, as in the conventional modelling, is removed.

## 3. DNS Databases

Three DNS databases of temporally-developing compressible mixing layers are used for the present analysis. The simulation conditions are shown in Table 1. In all cases, length was normalized by the initial vorticity thickness and other quantities by their values in the upper free-stream. The normalized mean pressure was $1/[\gamma(M_c)^2]$, where $\gamma$ is the ratio of specific heats. The Lewis number and Prandtl number were both unity. The full three-dimensional compressible Navier-Stokes equations were solved directly along with a transport equation for the conserved scalar $\xi$. Fourier spectral methods were used for homogeneous streamwise ($x_1$) and spanwise ($x_3$) directions while a modified sixth-order compact finite difference scheme was adopted in the inhomogeneous lateral ($x_2$) direction, in which free open boundaries were specified and characteristic non-reflecting boundary conditions imposed. A third-order compact-storage Runge-Kutta method was employed for time advancement. Resolution was monitored by checking the two-dimensional energy spectrum $E(k_1, k_3)$ to ensure that *all* the scales were adequately resolved. All three mixing layers began as laminar flow, went through transition and became turbulent at later stages. The transition processes for the mixing layer at $M_c = 0.8$ were described in detail in Luo and Sandham (1994). The turbulence thus obtained is more "natural" than that "prescribed" by some energy spectra often seen in the literature.

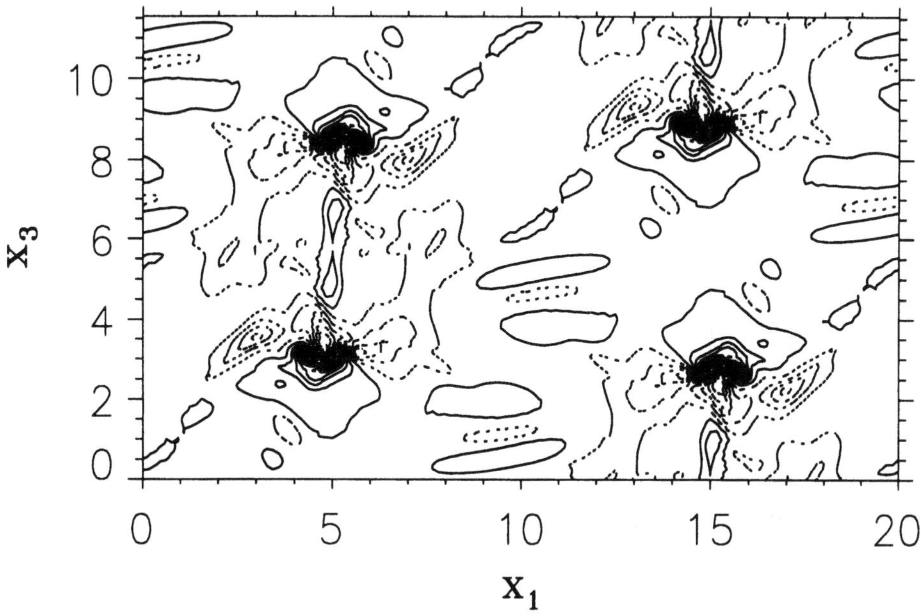


*Figure 1.* Plan view of the dilatation field in the central plane ($x_2 \approx 0$) at $t = 60.5$, showing the formation of shocklets in the mixing layer at $M_c = 1.1$. ($d_{min} = -2.25$, $d_{max} = 0.44$; 35 contour lines, solid — negative, dotted — positive.)

## 4. Results and Discussions

### 4.1. SHOCKLETS AND DILATATION

Previously, Sandham and Reynolds (1991) did not see any shocklets with $M_c$ up to 1.05 but their DNS did not go beyond mixing transition. Vreman *et al.* (1995) observed shocklets in *confined* mixing layers at $M_c = 1.2$ while the present author and colleagues observed shocklets in *free* mixing layers at the same $M_c$ (Vreman *et al.* 1996). However, if the dilatation dissipation theory of Zeman (1990) were correct, then strong shocklets would exist at $M_c$ much less than unity. The present study shows that shocklets do not exist at $M_c = 0.8$ but appear at $M_c = 1.1$, as shown in Figure 1. This Mach number is the smallest so far at which shocklets have ever been observed in *three-dimensional* free turbulence. These shocklets do not last long (about 5 time units), though, before they disappear. They may reappear again at later times, as observed by Vreman *et al.* (1995).

In compressible flow, the total dissipation in Eq. (1) can be decomposed into a solenoidal dissipation and a dilatation dissipation: $\varepsilon_{kk} \approx -\bar{\rho}(\varepsilon_s + \varepsilon_d)$, where $\varepsilon_s \equiv (\tilde{\nu}/Re)\overline{\omega_i''\omega_i''}$ and $\varepsilon_d \equiv (4\tilde{\nu}/3Re)\overline{(\partial u_i''/\partial x_i)^2}$. Here, $\nu$ is the kinematic viscosity and $\omega_i$ the vorticity components. Sarkar *et al.* (1991) proposed $\varepsilon_d = c_{ds}\varepsilon_s M_t^2$, where $c_{ds} = 1$, based on theories of linear acoustics. Zeman (1990) developed a similar model assuming existence of shocklets

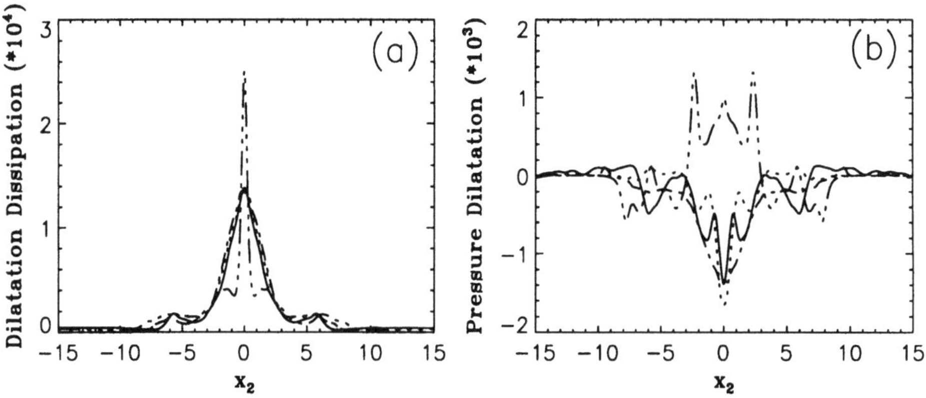


*Figure 2.* Mixing layer at $M_c = 1.2$. (a) Dilatation dissipation. For $t = 100$, solid — DNS, dotted — Sarkar model (1991) with $c_{ds} = 0.042$; dash dot — Zeman model (1991) with $c_{dz} = 0.032$. Dash triple dots — DNS at $t = 76$. (b) Pressure-dilatation. For $t = 100$, solid — DNS, dotted — Sarkar model (1992) with $(c_{s1}, c_{s2} = 0.15, 0.2)$, dash dot — Zeman model (1991) with $(\alpha, \beta, c_z = 1, 2, 0.1)$. Dash triple dots — DNS at $t = 76$.

in the flow: $\varepsilon_d = c_{dz}\varepsilon_s F(M_t, Ku)$, where $Ku = 6 \sim 8$ is the kurtosis of the streamwise velocity fluctuation and $c_{dz} = 0.75$.

To see the effects of shocklets, dilatation dissipation for the $M_c = 1.2$ simulation at $t = 76$ and 100 is plotted in Fig. 2(a). For this case, shocklets appear around $t = 76$ and result in a huge peak in the dilatation dissipation profile. This peak value is about twice as much as that at $t = 100$ where shocklets are absent. By adjusting constants $c_{ds}$ and $c_{dz}$, the models of Sarkar *et al.* (1991) and Zeman (1990) give similar profiles that fit the DNS data at $t = 100$. However, the models cannot reproduce the isolated peak generated by shocklets. In any case, dilatation dissipation is just a very small fraction (up to 3%) of the total dissipation, even though the turbulent Mach number $M_t$ is as high as 0.65 at $t = 100$.

## 4.2. PREDICTIONS OF PRESSURE TERMS

The major terms in turbulence kinetic energy and scalar flux budgets for the mixing layer of $M_c = 1.2$ at $t = 100$ are shown in Fig. 3. Apart from $\Pi_{kk}$ and $\Pi_{\xi 1}$, redistributive terms in Eqs. (1) and (10) are small and oscillating, and thus not shown. It is seen that $\Pi_{kk}$ and $\Pi_{\xi 1}$ are not insignificant in relation to the production and dissipation terms. In the absence of a mean pressure gradient, only pressure-dilatation $\overline{p'\partial u_i''/\partial x_i}$ is able to make net contributions (after integration over $x_2$) to the energy budget. This term is plotted in Fig. 2(b), along with its profile at an earlier time $t = 76$. This term, like $\Pi_{kk}$, is wavelike and can be negative (an energy sink) or positive (an energy source). Also shown are model predictions for the term at $t = 100$. Sarkar's (1992) model predicts more or less the right order of mag-

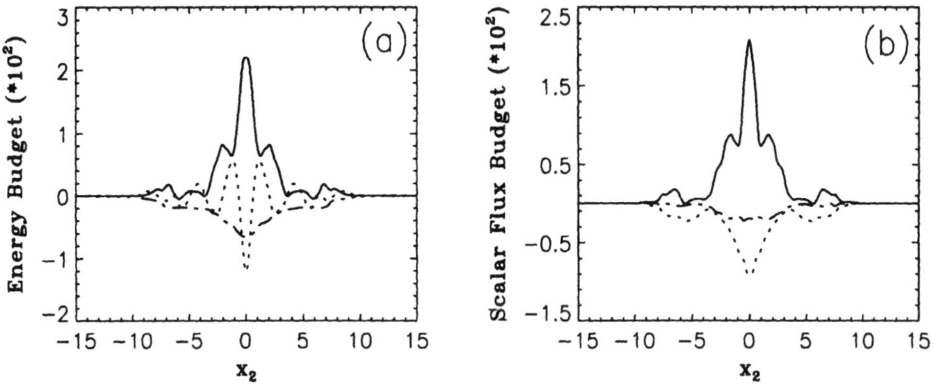


*Figure 3.* Turbulence kinetic energy budget (a) and conserved scalar flux budget (b) of the mixing layer of $M_c = 1.2$ at $t = 100$. Solid — production ($P_{kk}$, $P_{\xi 1}$), dotted — pressure term ($\Pi_{kk}$, $\Pi_{\xi 1}$), dash dot — dissipation ($\varepsilon_{kk}$, $\varepsilon_{\xi 1}$).

nitude without any adjustment. In contrast, Zeman's (1991) model would give a peak value ten times too high, without changing the constant $c_z$ from 1 to 0.1. Both models could not even predict the correct sign for $\overline{p'\partial u_i''/\partial x_i}$ at $t = 76$ when there is a large positive region around the shear layer center. The phenomenon of pressure-dilatation oscillating and changing sign from one time to another is also observed at other Mach numbers and is not entirely due to presence of shocklets. This fact makes its physical nature elusive and its value hard to predict. It should be noted that the magnitude of this term is much smaller than $\Pi_{kk}$ as a whole, suggesting that at least for non-reacting flows the pressure terms are mainly responsible for energy redistribution rather than energy generation or destruction.

The model for $\Pi_{kk}$ in Eq. (9) is now assessed with reference to DNS data as shown in Fig. 4. The notable feature of this term is the wild oscillation at all Mach numbers. There doesn't seem to be any trend with increasing $M_c$ in terms of the number and the magnitude of oscillations. With constants fixed at $(c_1, c_2, c_3) = (0.2, 0, 2.5)$, the model gives good overall predictions for all Mach numbers at different times, which is remarkable. (Note that the three cases have different initial disturbance levels too.) The predictions are not so good at the outer edges of the shear layer, where intermittency and large-scale structures dominate. Modelling these effects is particularly difficult with statistical models based on Reynolds averages.

Finally, model predictions for scalar-pressure gradient $\Pi_{\xi i}$ are compared with DNS data from the simulation of $M_c = 1.2$. Only two components of $\Pi_{\xi i}$ are compared because of symmetry in $x_3$ for the cases $M_c = 0.8$ and 1.2 and near symmetry for the case with $M_c = 1.1$. Figure 5(a) indicates that both the basic model (Launder 1989) and the Jones and Musonge model (Eq. (11)) overpredict $\Pi_{\xi 1}$ by up to 100%. In the meantime, $\Pi_{\xi 2}$ is substantially underpredicted by the former but overpredicted by the latter. The

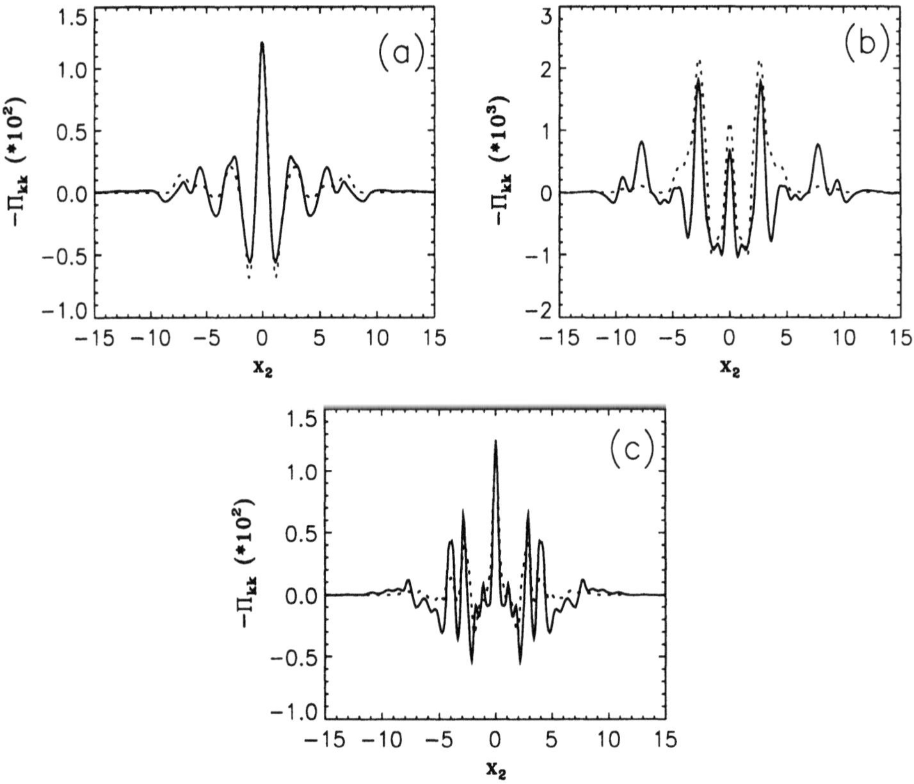


*Figure 4.* Comparison of pressure work $(\overline{u_i''\partial p/\partial x_i})$ from DNS and model predictions with $(c_1, c_2, c_3) = (0.2, 0, 2.5)$. (a) $M_c = 1.2$, $t = 100$; (b) $M_c = 1.1$, $t = 90$; (c) $M_c = 0.8$, $t = 80$. Solid — DNS, dotted — model.

underprediction by the basic model is attributed to the absence of a mean scalar gradient term in the model. With the model constants changed to $(c'_{\xi 0}, c'_{\xi 1}, c'_{\xi 2}, c'_{\xi 3}, c'_{\xi 4}, c'_{\xi 5}) = (3.0, 2.0, 0.09, 0.5, 0.3, 0.5)$, the Jones and Musonge model gives reasonable agreement with the DNS data as shown in Figure 5(b). The prediction by the model in Eq. (12) is also compared with DNS data in the same figure. The agreement is very encouraging. The interesting feature is that the humps at the two outer edges of the mixing layer are partly predicted by the model. The model constants used are $(c_{\xi 1}, c_{\xi 2}, c_{\xi 3}, c_{\xi 4}) = (0.1, 1.0, 1.0, 0.1)$. These values are chosen to give good overall predictions at all three Mach numbers. In principle, realizability constraints such as good model behaviour in the limit of incompressible and/or two-dimensional flow can be used to determine some or all of the constants. However, there is some doubt about whether or not second-moment closure models should or are able to satisfy all realizability constraints. The fact is that turbulence models use only limited time and length scale information to predict the effects of full spectra of scales in real turbulence. Thus any attempt to construct "universal" models is likely to produce exceedingly

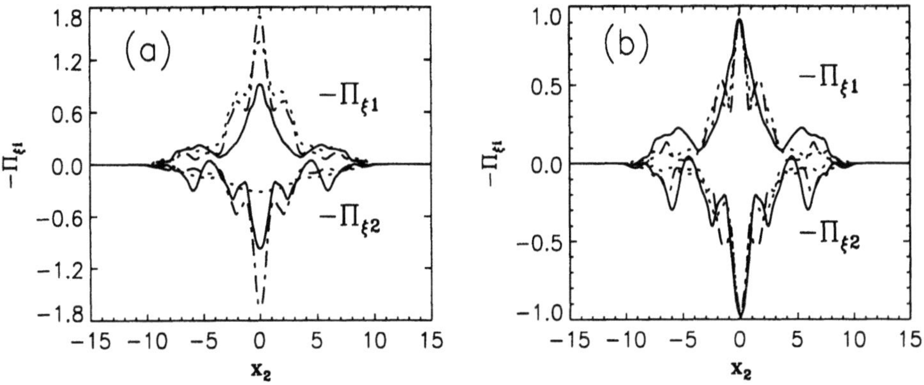


*Figure 5.* Scalar-pressure gradient ($\Pi_{\xi i} = -\overline{\xi'' \partial p / \partial x_i}$) for the mixing layer of $M_c = 1.2$ at $t = 100$. (a) Solid — DNS; dotted — the basic model (Launder, 1989); dash dot — the standard Jones & Musonge model (1988). (b) Solid — DNS; dotted — Jones & Musonge model (1988) with $(c'_{\xi 0}, c'_{\xi 1}, c'_{\xi 2}, c'_{\xi 3}, c'_{\xi 4}, c'_{\xi 5}) = (2.0, 3.0, 0.09, 0.5, 0.3, 0.5)$; dash dot — new model with constants $(c_{\xi 1}, c_{\xi 2}, c_{\xi 3}, c_{\xi 4}) = (0.1, 1.0, 1.0, 0.1)$.

bulky expressions that no one except the developer can implement. The advantage of the models in Eqs. (8), (9) and (12) is their simplicity. No complicated tensors are involved, nor any arbitrary time or length scales.

## 5. Concluding Remarks

Fluctuating pressure and dilatation effects have been studied at convective Mach numbers $M_c = 0.8$, 1.1 and 1.2 on the basis of DNS databases of compressible turbulent mixing layers. Fluctuating pressure effects are found to be significant for local turbulence energy balance and scalar transport in high-speed turbulence. The pressure terms in the turbulence energy and the conserved scalar flux equations are associated with strong oscillations, indicating their acoustic nature. As a result, the net contribution of these terms to turbulence energy, which can be a source or a sink, is relatively small, suggesting that fluctuating pressure is mainly responsible for energy redistribution rather than energy creation or destruction. Velocity-pressure gradient and scalar-pressure gradient correlations are modelled and tested at three Mach numbers. The agreement between model predictions and DNS data is remarkably good.

Eddy shocklets are found to exist in turbulent mixing layers at $M_c = 1.1$ and above but not at $M_c = 0.8$. The $M_c = 1.1$ is the lowest Mach number in which eddy shocklets have ever been found in three-dimensional free turbulence. Dilatation dissipation increases dramatically in the regions with shocklets but is never greater than 3% of the solenoidal dissipation even at a turbulent Mach number $M_t$ up to 0.65. Existing models for dilatation dissipation cannot predict the effects of eddy shocklets, which are spatially and temporally localized. These models substantially over-predict dilatation dissipation, regardless of the existence of eddy shocklets in the flow.

## 6. Acknowledgements

Numerical simulations were performed under the grant GR/H40518 from the U.K. EPSRC and with the Cray T3D time from the DNS Consortium under the grant GR/K43957. The author is grateful to Dr. N. D. Sandham for his generous support for this work.

## References

Bilger, R. W. (1976) Turbulent jet diffusion flames, *Prog. Energy Combust. Sci.*, **1**, pp. 87-109.

Bray, K. N. C., Libby, P. A. and Williams, F. A. (1994) High-speed turbulent combustion, in *Turbulent Reactive Flows* (P. A. Libby and F. A. Williams, Ed.), Academic Press, pp. 609-638.

Bray, K. N. C., Libby, P. A., Masuya, G. and Mass, J.B. (1981) Turbulence production in premixed turbulent flames, *Combust. Sci. Technol.*, **25**, pp. 127-140.

Jones, W. P. (1994) Turbulence modelling and numerical solution methods for variable density and combusting flows, in *Turbulent Reactive Flows* (P. A. Libby and F. A. Williams, Ed.), Academic Press, pp. 309-374.

Jones, W. P. and Musonge, P. (1988) Closure of the Reynolds stress and scalar flux equations, *Phys. Fluids*, **31** (12), pp. 3589-3604.

Launder, B. E. (1989) Second-moment closure: present and future? *Int. J. Heat and Fluid Flow*, **10**, No. 4, pp. 282-300.

Lin, C. C. (1953) On the stability of the laminar mixing region between two parallel streams in a gas, *NACA Technical Note*, **No. 2887**, pp. 1-50.

Luo, K. H. and Sandham, N. D. (1994) On the formation of small scales in a compressible mixing layer, in *Fluid Mechanics and its Applications: Direct and Large-Eddy Simulation I* (P. R. Voke, L. Kleiser and J.-P. Chollet, Ed.), Kluwer Academic Publishers, **26**, pp. 335-346.

Sandham, N. D. and Reynolds, W. C. (1991) Three-dimensional simulations of large eddies in the compressible mixing layers, *J. Fluid Mech.*, **224**, pp. 133-158.

Sarkar, S. (1992) The pressure-dilatation correlation in compressible flows, *Phys. Fluids A*, **4** (12), pp. 2674-2682.

Sarkar, S., Erlebacher, G., Hussaini, M. Y. and Kreiss, H. O. (1991) The analysis and modelling of dilatational terms in compressible turbulence, *J. Fluid Mech.*, **227**, pp. 473-493.

Strahle, W. C. (1983) Velocity-pressure gradient correlation in reactive turbulent flows, *Combust. Sci. Technol.*, **32**, pp. 289-305.

Vreman, A. W., Kuerten, H. and Geurts, B. (1995) Shocks in direct numerical simulation of the confined three-dimensional mixing layer, *Phys. Fluids*, **7** (9), pp. 2105-2107.

Vreman, A. W., Sandham, N. D. and Luo, K. H. (1996) Compressible mixing layer growth rate and turbulence characteristics, *J. Fluid Mech.*, **320**, 235-258.

Zeman, O. (1990) Dilatation dissipation: the concept and application in modelling compressible mixing layers, *Phys. Fluids A*, **2** (2), pp. 178-188.

Zeman, O. (1991) On the decay of compressible isotropic turbulence, *Phys. Fluids A*, **3** (5), pp. 951-955.

Zhang, S. W. and Rutland, C. J. (1995) Premixed flame effects on turbulence and pressure-related terms, *Combust. Flame*, **102**, pp. 447-461.

# BREAKDOWN OF A CROSSFLOW VORTEX IN A THREE-DIMENSIONAL BOUNDARY LAYER

TORSTEN WINTERGERSTE AND LEONHARD KLEISER
*Institute of Fluid Dynamics*
*ETH Zürich*
*ETH-Zentrum, CH-8092 Zürich, Switzerland*

**Abstract.**
Transition to turbulence in a three-dimensional boundary layer is investigated by a highly resolved direct numerical simulation which has been adapted to a swept flat plate transition experiment performed at DLR Göttingen. As shown earlier, our temporal simulation results can be related to the spatial disturbance development in the experiment and show a good agreement with measurements. The crossflow vortices, which develop as a primary instability of the laminar flow, break down at the late stages of the transition process. At the end of transition, the shape factor and the local skin friction coefficient of the computed mean velocity profile reach their turbulent levels. Particular attention is given to the development of flow structures in the breakdown stage. A new three-dimensional vortical structure is found to emerge which propagates with approximately 70% of the free-stream velocity in the streamwise direction. Vortical structures are identified by different criteria based on local pressure minima or the velocity-gradient tensor $\nabla u$.

## 1. Introduction

In three-dimensional boundary layers on swept wings of modern transport aircraft different primary instability mechanisms may be responsible for the onset of laminar-turbulent transition in different sections of the wing. At the leading edge disturbances travelling along the attachment line towards the wing tip can be amplified. In the middle part of the wing the well-known Tollmien-Schlichting instability may occur. The crossflow instability dominates transition in regions of strong negative pressure gradients. Transition



*J.-P. Chollet et al. (eds.), Direct and Large-Eddy Simulation II,* 179-190.

due to crossflow instability is the subject of the present investigations. According to classical (primary) linear instability theory of the laminar base flow, a range of instability waves may be amplified. In particular, corotating stationary vortices – the so-called crossflow vortices – develop which are approximately aligned with the streamline of the inviscid outer flow. This disturbance type is also well known from rotating-disk flow as described e.g. in the review by Reed & Saric (1989).

Experiments and theoretical investigations of the crossflow instability have shown that nonlinear interactions play an important role already at early stages of the transition process. These stages are characterized by interactions between the crossflow vortices and the travelling waves. At the highly nonlinear stages of transition, a secondary instability with a frequency an order of magnitude larger than the travelling primary disturbances was observed in the experiments by Kohama et al. (1991) and Bippes (1996). This instability was also observed in theoretical investigations by Malik et al. (1994) based on the parabolized stability equations (PSE) in a swept Hiemenz flow. They found that the high-frequency instability is associated with a shear layer on the upper side of the crossflow vortex. The processes at the late stages which lead to the breakdown of the crossflow vortex and to the onset of turbulence are not yet fully understood.

Our investigations are adapted to the transition experiment by Bippes et al. (1991). Fischer & Dallmann (1992) interpret travelling disturbances in the three-dimensional boundary layer of this experiment as secondary instabilities of the base flow deformed by the stationary vortices. Meyer & Kleiser (1989, 1990) and Wagner (1992) carried out temporal simulations of the 3D boundary-layer transition. With growing disturbance amplitudes they observed a strong deformation of the mean flow in good agreement with the experiment. These simulations already reached highly developed stages of transition shortly before breakdown of the crossflow vortices, but were not fully resolved in the late stages of transition to turbulence.

In a continuation of this work, we present results of a simulation which includes the complete breakdown process and finally reaches the turbulent regime. Thus our interest will be focused on the flow phenomena occurring in the late stages of transition where the crossflow vortices break down as the boundary layer becomes turbulent.

## 2. Base flow

In the DLR transition experiment (Bippes et al. 1991) a 3-D boundary-layer develops on a swept flat plate of chord length $\tilde{c} = 0.5\ m$. A displacement body mounted above the plate produces an essentially constant negative pressure gradient in the $x_c$-direction normal to the leading edge. Figure 1

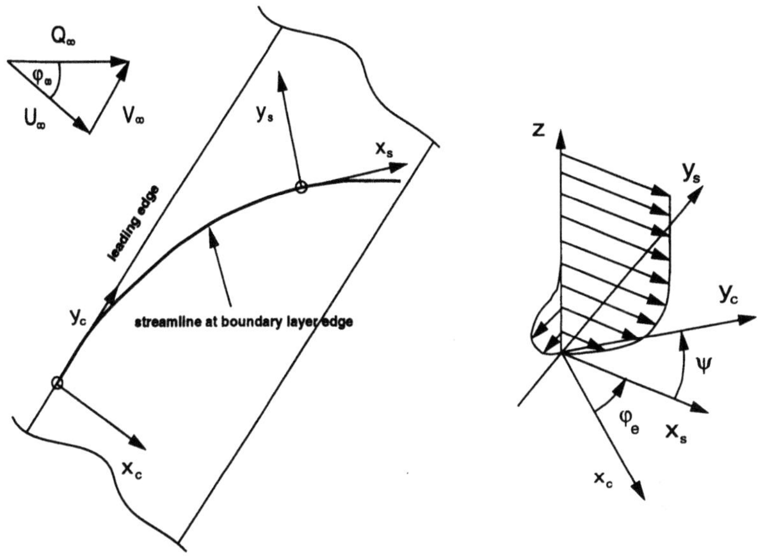


*Figure 1.* Sketch of the DLR swept plate transition experiment (Bippes et al. 1991).

includes a sketch of the various coordinate systems used. The index $c$ denotes a system oriented perpendicular ($x_c$) and parallel ($y_c$) to the leading edge of the plate.

Besides the streamwise velocity component also a crossflow component appears, which is already inviscidly unstable due to its inflectional shape. $U_{0c}$ and $V_{0c}$ are the streamwise and spanwise velocity components of the laminar flow in the body-oriented coordinate system. Due to the infinite-swept wing assumption the component $V_{0c,e}$ at the boundary-layer edge is constant, whereas $U_{0c,e}$ increases continuously with $x_c$. This results in a curved streamline of the potential flow. The local coordinate system aligned with this streamline is denoted by $(x_s, y_s)$. In the vortex-oriented system the coordinate $x_v$ points into the direction of the crossflow vortex axis and includes a (small) angle $\epsilon$ with the $x_s$ direction. The angle $\epsilon$ is determined by linear stability analysis as the angle between the local streamline and the selected stationary primary disturbance, i.e. the crossflow vortex. The simulations have been carried out in this vortex-oriented coordinate system $(x_v, y_v)$. The wall-normal direction is denoted by $z$ in all coordinate systems. All quantities are non-dimensionalized by the reference length $\tilde{d} = [(\tilde{\nu}\tilde{x}_c)/\tilde{U}_{0c,e}]^{1/2}$ (where $\tilde{\nu}$ is the kinematic viscosity) and the magnitude of the local velocity at the boundary layer edge, $\tilde{Q}_{0,e}$. The local Reynolds number is defined by $Re = (\tilde{Q}_{0,e}\tilde{d})/\tilde{\nu}$.

It has been demonstrated previously by Meyer & Kleiser (1989, 1990) that the laminar base flow can be approximated very well locally by the Falkner-Skan-Cooke (FSC) similarity solution for the boundary-layer flow on a yawed wedge (Cooke 1950). In the region of small disturbance amplitudes, up to about 60% of the chord length of the plate. The FSC profiles agree well with the experimentally observed mean flow (see Meyer & Kleiser 1989). The FSC profiles depend on two parameters, the local sweep angle $\varphi_e$ and the pressure-gradient (Hartree) parameter $\beta_H$. With given sweep angle $\varphi_\infty$ of the oncoming flow and pressure distribution these parameters can be determined as a function of the chordwise position $x_c$. The free-stream parameters are $\tilde{Q}_\infty = 19\ m/s$ and $\varphi_\infty = 45^o$. The variation of the boundary-layer quantities is weak in the region $x_c > 0.5$ due to the fact that the Hartree parameter remains nearly constant there and the shape of the FSC-profiles varies only slightly when $\varphi_\infty \approx 45^o$. Thus it appears acceptable to perform temporal simulations of transition at a fixed reference position in the rear part of the plate assuming a constant laminar flow which is supposed to be representative for the transition region. We select $\tilde{x}_{c,ref} = 0.8\ \tilde{c}$ as the reference position for the simulations presented herein. There we obtain the local Reynolds number $Re = 826$. The local sweep angle is chosen as $\varphi_e = 46.9^o$, the Hartree parameter as $\beta_H = 0.63$.

## 3. Numerical method

In our numerical simulations we compute the nonlinear development of streamwise and spanwise periodic disturbances by solving the 3D incompressible time-dependent Navier-Stokes equations. While the basic flow and the Reynolds number remain fixed during the simulation, the formulation allows for a deformation of the wall-parallel mean flow (averaged in horizontal planes) via the Reynolds stress. The results obtained from this temporal simulation model (Meyer & Kleiser 1989, 1990 and Wagner 1992) show a good agreement with the spatial disturbance development as observed in the experiments.

Except for a modified time integration the same numerical method has been used as described in Meyer & Kleiser (1989). For the time differencing we treat the nonlinear terms explicitly by a third-order four-stage Runge-Kutta scheme and the viscous terms implicitly by the Crank-Nicolson scheme. For the spatial discretization we use a Fourier/Chebychev spectral method. The nonlinear terms are computed, alias-free, in $x$ and $y$ by a pseudo-spectral approximation.

## 4. Results

In the present investigations we are mainly interested in the processes occurring in the highly nonlinear stages of transition, when the stationary crossflow vortex breaks down and the flow becomes turbulent. The spatial discretization of the computational domain has been increased up to $N_x \cdot N_y \cdot N_z = 240 \cdot 192 \cdot 160$ grid points. This number is increased by a factor of 3/2 in the wall-parallel directions for the alias-free calculation of the nonlinear terms.

The characteristic parameters of the flow are chosen following Meyer & Kleiser (1989). The computational domain is aligned with the axis of the stationary crossflow vortex which includes an angle of $\epsilon = 4.3^o$ with the potential flow streamline. The streamwise and spanwise wave numbers are chosen as $\alpha_v = 0.08$ and $\beta_v = 0.48$. The computational domain comprises one crossflow vortex in the spanwise direction. The selected spanwise wave number gives the maximum secondary amplification rate as predicted by the secondary instability theory (Fischer & Dallmann 1992) for the selected mode $(\alpha_v, \beta_v)$. The crossflow vortex was ed with an amplitude $A_{CF} =$ 0.25% of the free-stream velocity $U_v$. We computed two cases with different amplitudes of the additionally imposed travelling disturbances.

### 4.1. CASE 1: RANDOM NOISE

At first the crossflow vortex was superimposed with low-amplitude random noise. Figure 2 shows the temporal development of the maximum in the wall-normal direction of the Fourier-amplitudes $|\hat{u}|(k_x, k_y)$. The amplitude of the fundamental crossflow vortex mode $(0,1)$ grows linearly at early times. Higher harmonics of the fundamental mode are generated due to the nonlinearity of the basic equations. The travelling modes $|\hat{u}|(k_x, 1)$ are growing also with a growth rate predicted by linear instability theory, after an initial transient, but with lower amplitudes. The crossflow vortex modes $|\hat{u}|(0, k_y)$ reach a saturated state at non-dimensional time $t \approx 850$. The amplitude of the fundamental mode $(0,1)$ attains a value of approximately 29%. The mean flow deformation mode $(0,0)$ keeps growing. At time $t \approx 800$ the modes $(6,1)$ and $(7,1)$ – the latter one becomes dominant at later stages – begin to grow with a much higher growth rate than the primary instability disturbances. The symmetric modes $(6,-1)$ and $(7,-1)$ have the same growth rates. These three-dimensional disturbances appear as an instability of the nonlinear crossflow vortex. It has not yet been determined if the growth rates agree with the secondary instability theory. They initiate the breakdown of the crossflow vortex and the onset of turbulence. Our results shown that the instability results depend strongly on the amplitude and the higher harmonics of the fundamental crossflow vortex mode. Approximately 80 Fourier-modes had to be used to resolve the nonlinear saturation of the crossflow vortex sufficiently.

### 4.2. CASE 2: FINITE AMPLITUDE TRAVELLING DISTURBANCE

In the second computation the crossflow vortex was superimposed with a travelling three-dimensional linear instability wave with an initial am-

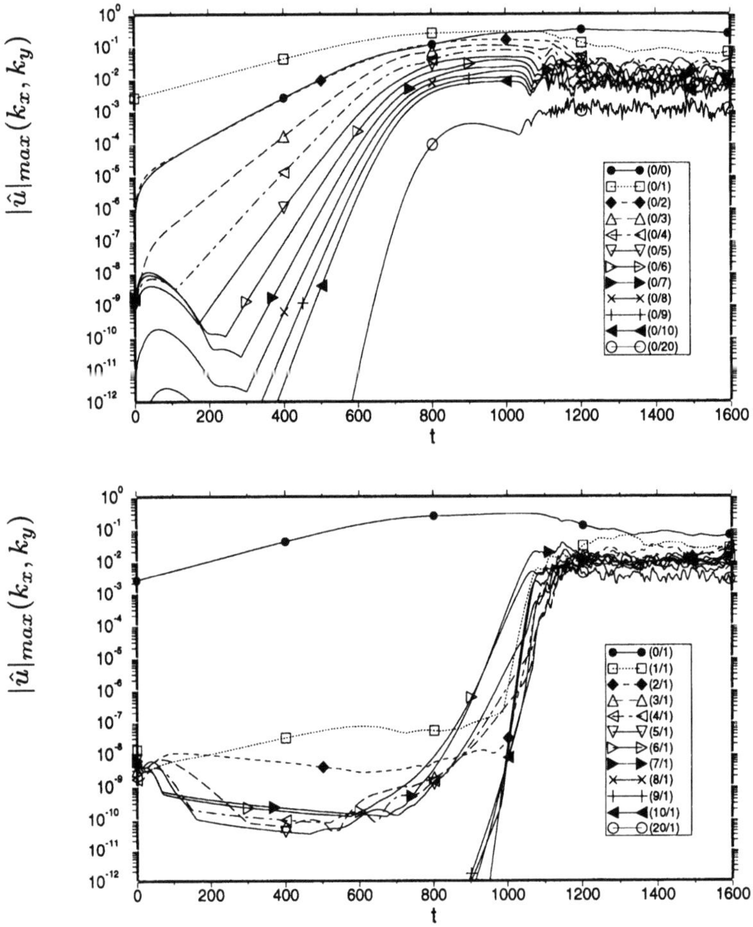


*Figure 2.* Development of the maximum in the wall-normal direction of the Fourier amplitudes $|\hat{u}|(k_x, k_y)$ for case 1.

plitude of $A_{tr} = 0.1\%$ based on the streamwise velocity component. The temporal development of selected modes is shown in figure 3. Both initially excited modes grow exponentially up to the time $t \approx 600$ with the rate predicted by the linear instability theory. Between time $t \approx 700$ and $t \approx 920$ the crossflow vortex modes $|\hat{u}|(0, k_y)$ are approximately saturated, while the travelling wave modes $|\hat{u}|(k_x, 0)$ are still growing (see Wintergerste & Kleiser 1996). The maximum amplitude of the crossflow vortex reaches a value of $A_{CF} \approx 23\%$ and the amplitude of the travelling wave $(1, 1)$ a value of $A_{tr} \approx 3\%$. Beyond time $t \approx 920$ the amplitudes are highly fluctuating, and the boundary layer rapidly becomes turbulent. A "secondary" instability, similar to case 1, cannot be clearly identified in this simulation.

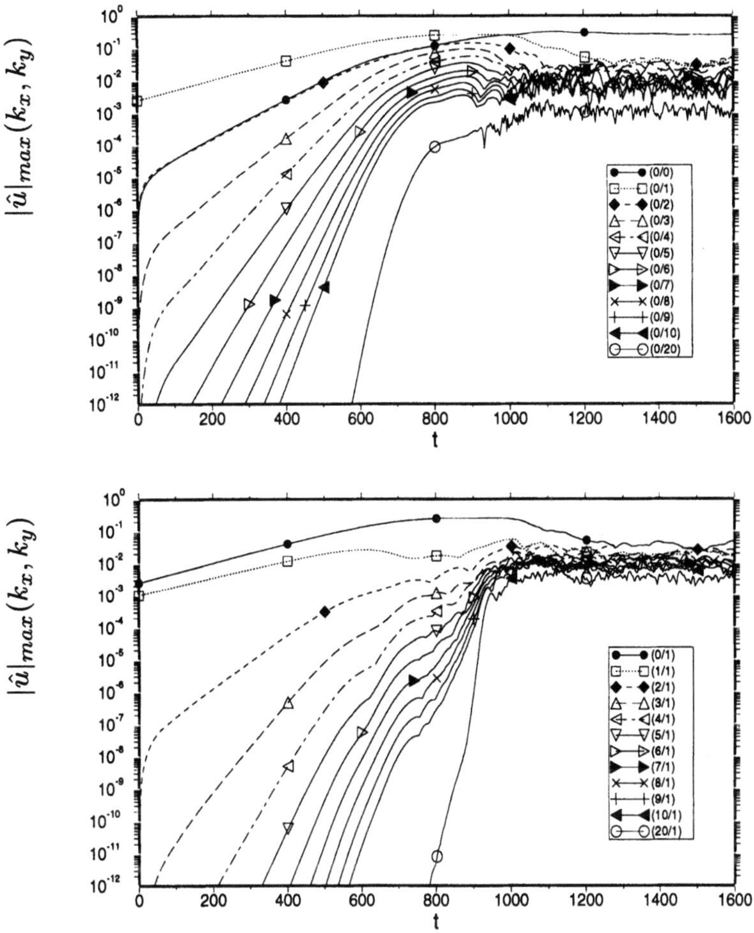


*Figure 3.* Development of the maximum in the wall-normal direction of the Fourier amplitudes $|\hat{u}|(k_x, k_y)$ for case 2.

The development of the shape factor $H_{12}$ is shown in figure 4. $H_{12} = \delta_1/\delta_2$ is defined as the ratio between the displacement thickness $\delta_1$ and the momentum thickness $\delta_2$ of the horizontally averaged velocity profile and characterizes the development of the mean flow. It changes its value from the laminar level $H_{12} \approx 2.38$ to the turbulent level $H_{12} \approx 1.41$. The nearly constant turbulent value is reached for times $t > 1200$. In figure 5 the minimal and maximal pressure fluctuations are shown. The pressure fluctuations grow linearly for the early stages. During the breakdown of the crossflow vortex the pressure fluctuations increase dramatically. The highest minimum of the pressure fluctuations is reached at $t \approx 1120$. At this stage the highest spatial discretization of the computational domain is

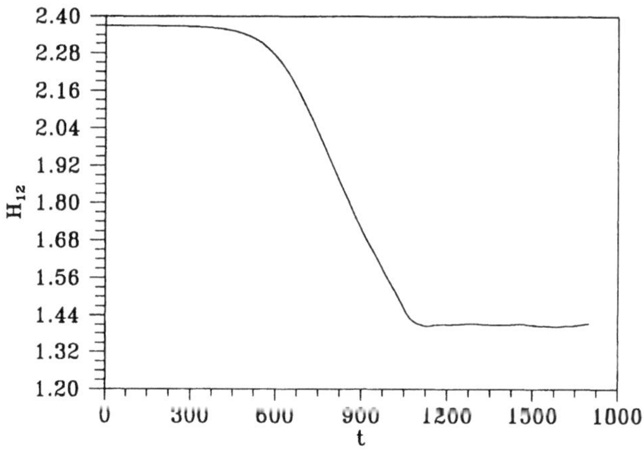


*Figure 4.* Development of the shape factor $H_{12}$ in the streamline oriented system during the transition process.

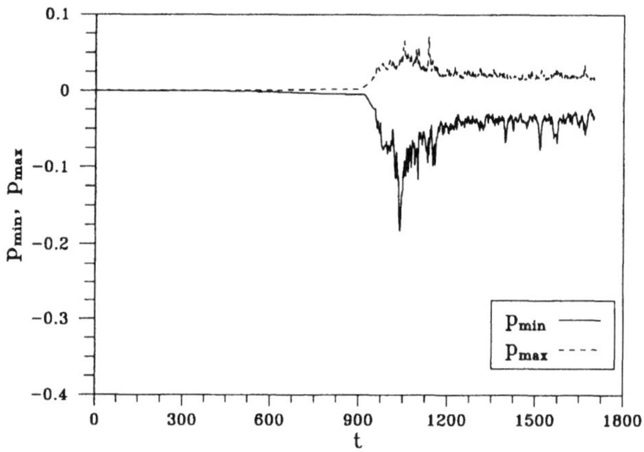


*Figure 5.* Development of the minimal and maximal pressure fluctuations during the simulation.

necessary. The pressure fluctuations decrease with the onset of turbulence and reach a quasi-constant fluctuating level.

## 4.3. VISUALIZATION OF CASE 2

Viewed in the streamwise direction, the crossflow vortex rotates in a counterclockwise direction. Thus, the vortex moves slow fluid from near the wall to the outer part of the boundary layer and generates rather narrow layers of high shear $\partial u/\partial y$ and $\partial u/\partial z$. Conversely, faster fluid from the outer part of the boundary layer moves closer to the wall. Velocity profiles with two inflection points can be obtained due to the upward motion of the slow

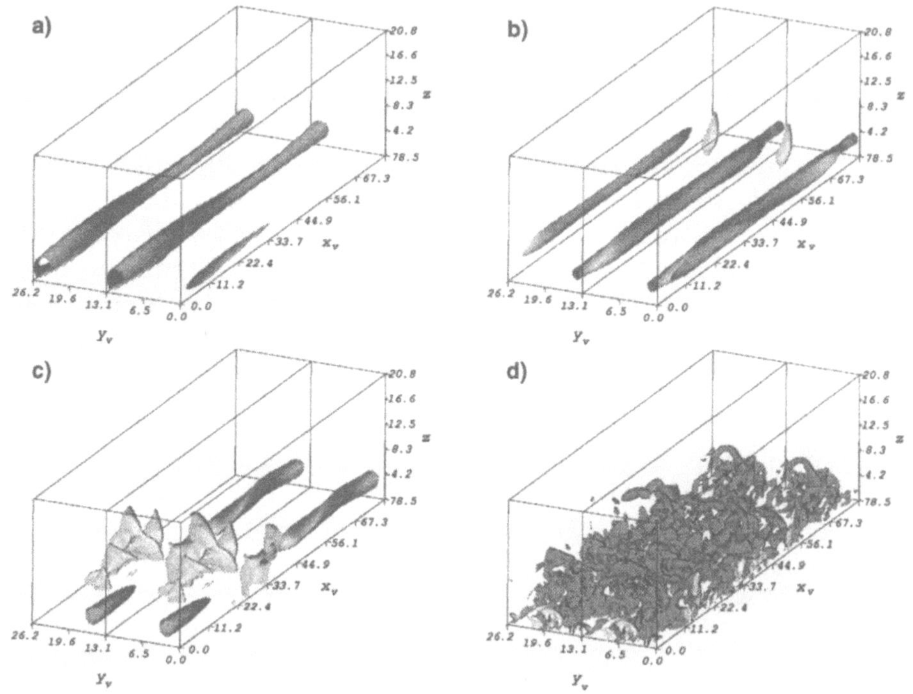


*Figure 6.* Visualization of vortical structures by isosurfaces of low static pressure. For visualization twice the computational domain in the spanwise direction $y$ is shown. a): $t = 800$, $p_{iso} = -0.0022$; b): $t = 920$, $p_{iso} = -0.0031$; c): $t = 970$, $p_{iso} = -0.004$; d): $t = 1300$, $p_{iso} = -0.005$.

fluid.

Vortical structures are important for the energy transfer from the large scales to the smaller ones. Different criteria for their identification can be applied to the three-dimensional flow field (e.g. Wintergerste et al. 1995).

In figure 6a the crossflow vortex is visualized by isosurfaces of low static pressure at time $t = 800$ where a near-saturation state of the crossflow vortex is obtained. The modulation in the $x_v$−direction is caused by the imposed travelling wave. Figure 6b shows the vortical structures identified at a later time $t = 920$. In addition to the crossflow vortex, a new three-dimensional structure is formed in the downstream part of the domain. The foot-print of this new structure can be found in the wall-pressure distribution (not shown here), where it marks a strong minimum. Downstream of the three-dimensional vortex a region of high pressure fluctuations is observed. The identification of this flow structure as a vortex can be confirmed by calculating streamlines in a moving frame of reference. This vortex, which generates new shear layers, moves downstream rapidly and initiates the local breakdown of the crossflow vortex. The visualization of the flow fields at later times shows that the breakdown spreads from this new vortex while in the remaining part of the flow the saturated crossflow vortex persists (figure 6c). When the spreading has covered the full domain the entire flow field becomes turbulent (figure 6d). The propagation velocity of the new three-dimensional structure at time $t = 920$ can be estimated as approximately 70% of the free-stream velocity by correlations of the wall-pressure distribution at different times (Wintergerste & Kleiser 1996). The propagation velocity, which is directly associated with the frequency, is one magnitude larger than the velocity of the primary travelling instability wave.

The identification of vortical structures by isosurfaces of low static pressure can only capture vortices with the largest pressure minimum. This criterion fails if vortical structures with different strength appear in the flow. We have shown earlier (Wintergerste & Kleiser 1996) that the emergence of the new three-dimensional vortical structure is associated with the interaction of the stationary crossflow vortex and a secondary near-wall streamwise vortex. The new vortex is induced by the co-rotating crossflow vortices in the presence of the wall. The secondary flow, i.e. the flow in the spanwise direction near the wall, separates due to the high pressure gradient between the crossflow vortices. The identification of this new vortex has been done by a "*second eigenvalue criterion*" based on the velocity gradient tensor $\nabla u$, which was first used by Jeong & Hussain (1995). A still open question is the beginning of the interaction of these two vortices.

A similar near-wall streamwise vortex was also found by Malik et al. (1994) in their simulation of swept Hiemenz flow. They found that the high-frequency instability is associated with a shear layer above the crossflow vortex. Their analysis was based on a secondary stability analysis of the nonlinear deformed crossflow vortex. In figure 7 the crossflow vortex and the near-wall streamwise vortex are visualized by the "*second eigenvalue criterion*" at time $t = 800$. A deformation of the isosurface at the upper side of the crossflow vortex is observed at $x \approx 33$. This deformation suggests a local instability at the upper side of the crossflow vortex near the edge of the boundary layer. Extended investigations have shown that the interaction of the crossflow vortex and the near-wall streamwise vortex starts at this location. The interaction results in an upward movement of the near-wall vortex away from the wall and finally to the emergence of the new three-dimensional vortical structure mentioned above. The high-shear layers $\partial U/\partial x$ and $\partial U/\partial z$ at $x = 33.134$, where the local instability can be observed, are shown in figure 8. These observations agree well with the results of the secondary stability analysis of Malik et al. (1994).

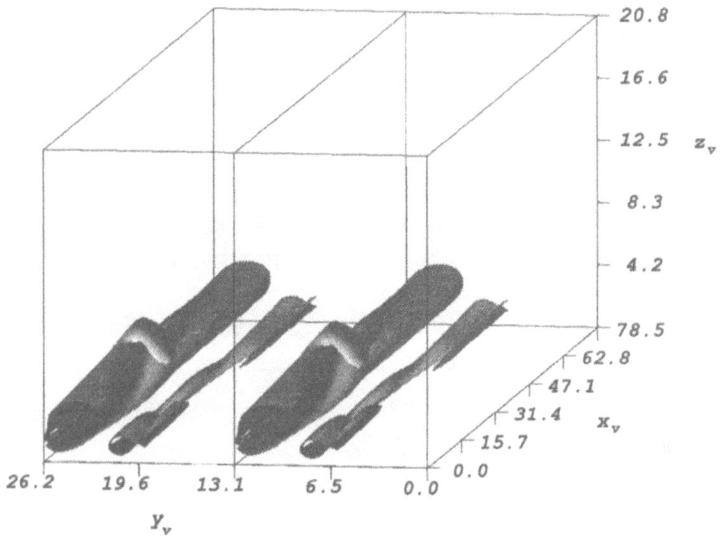


*Figure 7.* Visualization of the vortical structures at time $t = 800$ by the second eigenvalue criterion. The isovalues are set to -0.002 for the crossflow vortex and -0.001 for the near-wall streamwise vortex.

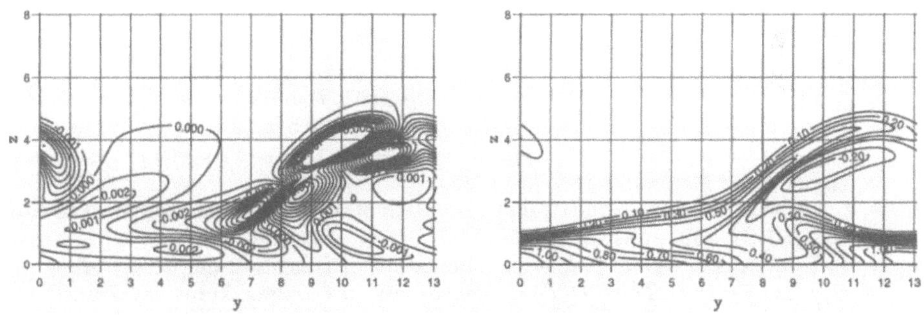


*Figure 8.* Variation of $\partial U/\partial x$ (left) and $\partial U/\partial z$ (right) in a $(y,z)$–plane at time $t = 800$. The streamwise position is $x = 33.143$. The incrementals are 0.001 (left) and 0.1 (right).

## 5. Conclusions

The transition process in a three-dimensional boundary layer has been simulated using the temporal model. The temporal development of the various spectral modes depend strongly on their initial amplitudes. An instability with much higher growth rates than the primary disturbances is observed. This local instability cannot be identified if the initial amplitudes of the stationary crossflow vortex and the travelling waves are of the same order.

Small secondary streamwise vortices appear close to the wall next to each crossflow vortex. In the case of comparable initial amplitudes of stationary crossflow vortex and travelling waves, the interaction of these vortices and the roll-up on the upper side of the crossflow vortex (Wintergerste & Kleiser 1996) starts with a local shear-layer instability at the upper side of the crossflow vortex. From this interaction and roll-up new three-dimensional vortices emerge which locally initiate the final breakdown of the crossflow vortices. The three-dimensional vortices move downstream with a propagation velocity of approximately 70% of the free-stream velocity, which is an order of magnitude larger than the phase velocity of the initially excited travelling wave. This local breakdown finally spreads over the entire domain and the flow becomes turbulent.

## References

Bippes, H., B. Müller & M. Wagner, 1991: Measurements and stability calculations of the disturbance growth in an unstable three-dimensional boundary layer. *Physics of Fluids A* **3**, pp. 2371–2377.

Bippes, H., 1996: Private communication.

Cooke, J. C., 1950: The boundary layer of a class of infinite yawed cylinders. *Proc. Cambridge Phil. Soc.* **46**, pp. 645–648.

Fischer, T. M. & U. Dallmann, 1992: Primary and secondary stability analysis of a three-dimensional boundary-layer flow. *Physics of Fluids A* **3**, pp. 2378–2391.

Jeong, J. & F. Hussain, 1995: On the identification of a vortex. *Journal of Fluid Mechanics* **285**, pp. 69–94.

Kohama, Y., W.S. Saric & J.A. Hoos, 1991: A high frequency secondary instability of crossflow vortices that leads to transition. In: *Proc. of Boundary Layer Transition and Control Conference*, The Royal Aeronautical Society, Cambridge UK, pp. 4.1–4.13.

Malik, M. R., F. Li & C.-L. Chang, 1994: Crossflow disturbances in three-dimensional boundary layers: non-linear development, wave interaction and secondary instability. *Journal of Fluid Mechanics* **268**, pp. 1–36.

Meyer, F. & L. Kleiser, 1989: Numerical investigation of transition in 3D boundary layers. In: *AGARD-CP-438*, pp. 16-1–16-17.

Meyer, F. & L. Kleiser, 1990: Numerical simulation of transition due to crossflow instability. In: *Proc. of Third IUTAM Symposium on Laminar-Turbulent Transition*, Editors: D. Arnal and R. Michel, Springer Verlag, pp. 609–619.

Reed, H. L. & W. S. Saric, 1989: Stability of three-dimensional boundary layers. *Annual Review of Fluid Mechanics* **21**, pp. 235–284.

Wagner, M., 1992: Numerische Untersuchungen zum laminar-turbulenten Übergang in zwei- und dreidimensionalen Grenzschichten, *Doct. Dissertation, Universität Karlsruhe, Report DLR-FB 92-36.*

Wintergerste, T., D. Wilhelm & L. Kleiser, 1995: Visualization of time-dependent three-dimensional transitional and turbulent flow fields. In: *Visualization Methods in High Performance Computing and Flow Simulation. Proceedings of the International Workshop on Visualization, Paderborn Jan. 18-21, 1994*, Editors: W. Borchers, G. Domick, D. Kröner, R. Rautmann and D. Saupe, International Science Publishers.

Wintergerste, T. & L. Kleiser, 1996: Direct numerical simulation of transition in a three-dimensional boundary layer. To appear in: *Proceedings of the Colloquium on Transitional Boundary Layers in Aeronautics*, Royal Netherlands Academy of Arts and Sciences, Amsterdam, Dec. 6–8, 1995.

# INTENSE VISCOUS DISSIPATION EVENTS AND THE VORTICITY FIELD IN NEAR-WALL TURBULENCE

S. GAVRILAKIS AND M. O. DEVILLE
*LMF-DGM, Swiss Federal Institute of Technology*
*1015 Lausanne, Switzerland*

## 1. Introduction

In high Reynolds number turbulent flows there exists a close statistical relationship between the mean viscous dissipation rate $\epsilon$ and the mean enstrophy field (Tennekes & Lumley, 1972). For near-wall shear-driven turbulent flows the intensities of the turbulent velocity and vorticity fields attain their maxima close to the boundary in an area of high viscous dissipation. This suggests that the processes underlying turbulent energy, enstrophy and dissipation generation may be (and are) related, but the constraints imposed by the wall leads to the expectation that the same flow structures may be involved in all three processes. Conditional sampling of the turbulent flow field around ejection and sweep events shows that most of the dissipation and enstrophy generation occurs near the energy producing events – with some of these events involving greater dissipation than production of turbulent energy (Gavrilakis, 1996). The present report looks at the relation between the generation of the turbulent energy dissipation and enstrophy fields in the vicinity of a smooth wall.

## 2. The Database

The near-wall statistics in the present analysis have been obtained from the database generated by direct numerical simulation (DNS) of the turbulent flow through a straight duct of rectangular cross section and aspect ratio of 2 (Figure 1). The mean flow is in the homogeneous (x) direction. The direction parallel to the z-axis will be referred to as spanwise whereas the direction parallel to the y-axis normal. Some of the flow and simulation parameters are shown in table 1.



*J.-P. Chollet et al. (eds.), Direct and Large-Eddy Simulation II,* 191-200.

TABLE 1. Physical and numerical parameters of the simulation : $U_b$=Bulk velocity; $D$=Hydraulic diameter of the duct; $u_\tau$=Average wall friction velocity.

| | |
|---|---|
| $Re = U_b D/\nu$ | 6780 |
| $Re^+ = u_\tau D/\nu$ | 436 |
| Domain size (x,y,z) | $4\pi h(streamwise) \times 2h \times 4h$ |
| Number of collocation points | 144(x) × 113(y) × 211(z) |
| Streamwise grid ($\Delta x$) | 14.5 $\nu/u_\tau$ |
| Normal grid ($\Delta y$) | 0.6-5.1 $\nu/u_\tau$ |
| Spanwise grid ($\Delta z$) | 0.6-5.1 $\nu/u_\tau$ |
| Time span of database | 1.5 $h/u_\tau$ |

The method of simulation is based on the discretised incompressible Navier-Stokes equations. Fourier expansions are used in the homogeneous direction by assuming periodicity in x. Discretisation in the two inhomogeneous directions is based on the Legendre spectral element technique (Maday & Patera, 1989; Rønquist, 1991). Time advancement is by a fourth-order explicit Runge-Kutta scheme which is applied to all (convective and viscous) terms. The flow is driven by adjusting the mean pressure gradient at every time step so that a constant flow rate is maintained throughout the simulation (Gavrilakis *et al.* 1996).

Here we are not concerned with the effects of the corners and all statistics have been extracted from the part of the flow enclosed by the segments of the two longer walls that are at least 115 wall units away for the nearest short wall (see Figure 1). In effect, the data away from the short walls is treated like that of a plane channel. Typical velocity spectra in the buffer region are shown in figure 2a. No accumulation of energy at high wavenumbers is evident even though the method has no numerical viscosity. Figure 2b shows the variation of the terms of the equation for the mean turbulent kinetic energy along a line normal to the longer walls at their mid-point. Note that the superscript (+) is used to denote quantities scaled with the kinematic molecular viscosity $\nu$ and the four-wall average friction velocity $u_\tau$. For comparison the corresponding data from the plane channel simulation of Gilbert & Kleiser (1991), (see also Gilbert 1993), is shown. Noting that the statistical sample of the present simulation is somewhat short, the agreement between the two datasets is good. The duct data shows higher dissipation levels around $y^+ = 11$ and with slightly reduced production rates but the turbulent and viscous diffusion out of this region is also less than that of the plane channel. Thus, a greater proportion of the turbulent energy which is produced at that region is dissipated locally.

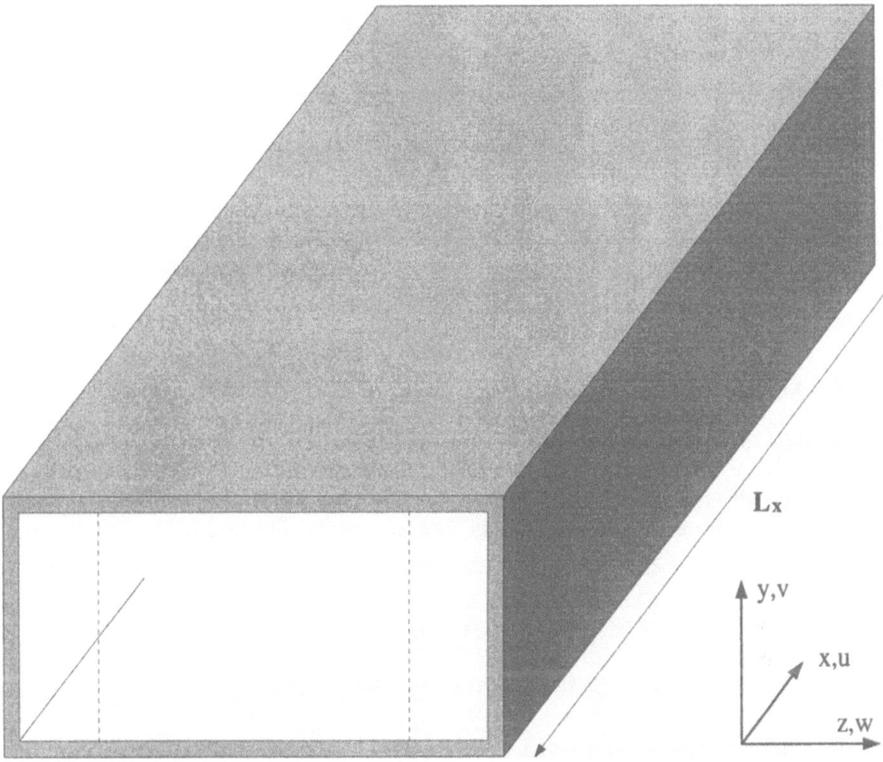


*Figure 1.* Flow geometry and axes system. Only the flow domain enclosed between the vertical lines is used herein.

## 3. Enstrophy and Dissipation Production Terms

The contributions of the various flow processes to the generation of dissipation and enstrophy are charted using the leading contributions to production terms of the dissipation rate and enstrophy equations. Here we consider the homogeneous part of the energy dissipation rate

$$\epsilon' = 2\nu \frac{\partial u_i}{\partial x_k} \frac{\partial u_i}{\partial x_k} \tag{1}$$

where the summation over repeated subscripts has been assumed. The mean value of $\epsilon'$ is $\epsilon$. Note that the above definition gives the total dissipation

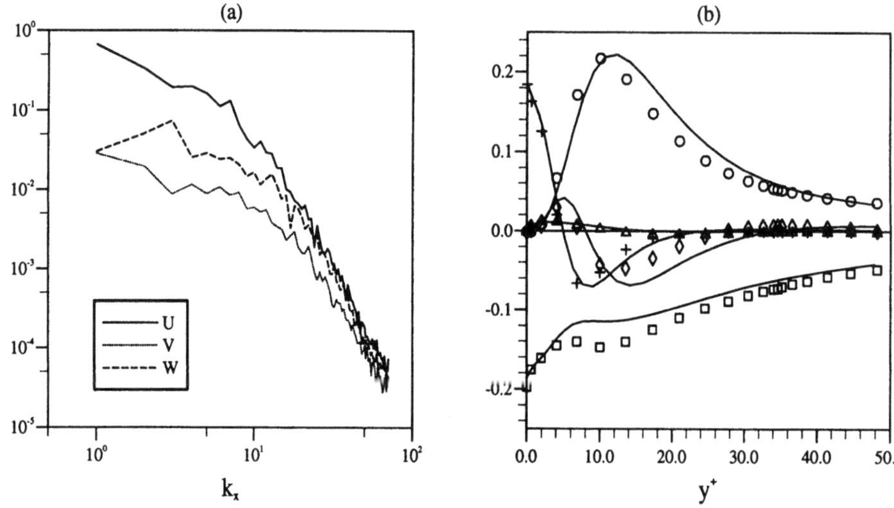


*Figure 2.* a) One-dimensional velocity spectra in the near-wall buffer region. b) Terms of the equation for the mean turbulent kinetic energy along the mid-plane ($z = 2h$) of the rectangular duct : o, production; □, viscous dissipation; △, pressure diffusion; ◇, turbulent diffusion; +, viscous diffusion. Solid lines show the corresponding data from the direct simulation of a turbulent flow in a plane channel (Gilbert & Kleiser, 1991; Gilbert, 1993). All terms have been scaled using wall units. The wall is at $y^+ = 0$.

rate in homogeneous flows only. An additional contribution to the viscous dissipation in vicinity of smooth walls due to the mean inhomogeneity is not considered here but has been estimated by Bradshaw and Perot (1993) to be at most 6 percent of the homogeneous value. Lower case letters are used for the fluctuating quantities. Letting $\omega_i$ denote the i-th component of the fluctuating vorticity, the instantaneous value of turbulent enstrophy is defined as

$$e' = \omega_i \omega_i / 2. \tag{2}$$

$e$ is the mean value of $e'$. The main production terms in the equation for the homogeneous dissipation rate are (Mansour *et al.* 1988)

$$\begin{aligned} P_\epsilon^1 &= -2\frac{\partial u_i}{\partial x_j}\frac{\partial u_k}{\partial x_j}S_{ik}, \\ P_\epsilon^2 &= -2\frac{\partial u_i}{\partial x_k}\frac{\partial u_i}{\partial x_m}S_{km}, \\ P_\epsilon^4 &= -2\frac{\partial u_i}{\partial x_k}\frac{\partial u_i}{\partial x_m}\frac{\partial u_k}{\partial x_m}, \end{aligned}$$

where capital letters are used to denote mean quantities. $S_{ij} = 0.5(U_{i,j} + U_{j,i})$ and $s_{ij} = 0.5(u_{i,j} + u_{j,i})$ are used to denote the mean and fluctuating strain rates respectively. The significant production terms in the equation for the enstrophy are (Tennekes & Lumley, 1972)

$$\begin{aligned} P_e^1 &= 2\omega_i s_{ij} \Omega_j, \\ P_e^2 &= 2\omega_i \omega_j S_{ij}, \\ P_e^4 &= 2\omega_i \omega_j s_{ij}, \end{aligned}$$

where $\Omega$ is the mean vorticity field.

In the rectangular duct configuration all terms but those involving derivatives of mean quantities in the homogeneous direction, are in principle non-zero. Away from the side walls the terms involving the mean field can be simplified, to a good approximation, to the form when only one component of the mean velocity gradient is present. Thus, for the dissipation rate

$$\begin{aligned} P_\epsilon^1 &= -\left(\frac{\partial u}{\partial x}\frac{\partial v}{\partial x} + \frac{\partial u}{\partial y}\frac{\partial v}{\partial y} + \frac{\partial u}{\partial z}\frac{\partial v}{\partial z}\right)\frac{\partial U}{\partial y} \\ P_\epsilon^2 &= -\left(\frac{\partial u}{\partial x}\frac{\partial u}{\partial y} + \frac{\partial v}{\partial x}\frac{\partial v}{\partial y} + \frac{\partial w}{\partial x}\frac{\partial w}{\partial y}\right)\frac{\partial U}{\partial y} \end{aligned}$$

and for the enstrophy

$$\begin{aligned} P_e^1 &= 2\left(\omega_1 s_{13} + \omega_2 s_{23} + \omega_3 s_{33}\right)\Omega_3, \\ P_e^2 &= (\omega_1 \omega_2)\frac{\partial U}{\partial y}. \end{aligned}$$

The near-wall average values of the terms are shown in figure 3. For both $\epsilon'$ and $e'$ the terms involving the mean field are significant for $y^+ < 30$. It appears from this figure that $P_\epsilon^1 + P_\epsilon^2$ would be of the same order as $P_e^1 + P_e^2$. The turbulent vortex-line stretching term ($P_e^4$) and the dissipation production $P_\epsilon^4$ also have approximately the same values. The peak of $P_\epsilon^2$ at $y^+ \simeq 11$ is associated with the lifting of turbulent fluid away from the boundary (not the ejections themselves – ejections and the convection of intense vorticity are generally found to be quite separate processes), a situation which would occur ahead of a separation point in the streamwise direction – necessarily a near wall phenomenon which requires the presence of the mean velocity gradient for the dissipation to arise. A second process contributing to this term is the skewing of the high speed streaks in the spanwise direction.

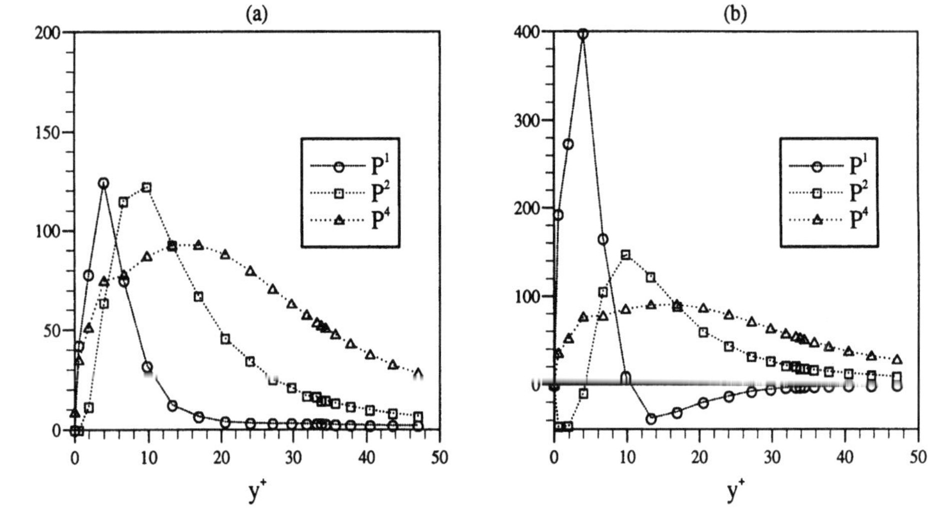


*Figure 3.* Near-wall variation of the averaged production terms for (a) the viscous dissipation rate, and (b) enstrophy. All terms are scaled with the friction velocity and $h$.

This is quite common occurrence in the near wall turbulence. This part of the dissipation contains motions that correspond to $\omega_1$ (due to the spanwise motion over the fixed wall) and $\omega_2$ (from the spanwise gradient of the $u$ component and the streamwise gradient of $w$) and therefore is closely related to the $P_e^2$ term. This term, $P_e^2$, turns negative very close to the wall ($y^+ < 3$); this phenomenon is associated with the change of sign in the instantaneous streamwise vorticity due to the zero-slip velocity boundary condition required at the smooth boundary.

Term $P_\epsilon^1$ has a peak very close to the wall at $y^+ \simeq 2$. The terms contributing to it (not shown) suggest that this term accounts for the dissipation of the sweep flow events in the presence of the mean gradient. The corresponding peak in $P_e^1$ is associated with the same events but dominates over the other production terms. In detail, the correspondence between these two production terms can be traced in the effects of turbulent convection towards the wall. Sweep motions that produce high dissipation carry at the same time fluid from areas of different mean vorticity. In addition, the normal motion is converted to (i.e. fluid has to be accelerated to) motions parallel to the wall giving rise to high instantaneous gradients in the spanwise direction – high values of $s_{33}$. Thus $\omega_3 s_{33}$ is one of the two important terms in the $P_e^1$ production and is essentially the same as the dissipative event contributing to $P_\epsilon^1$. The second important contribution is that of $\omega_1 s_{13}$ where the streamwise vorticity is generated by spanwise motion due to the sweep together with the no-slip boundary condition at

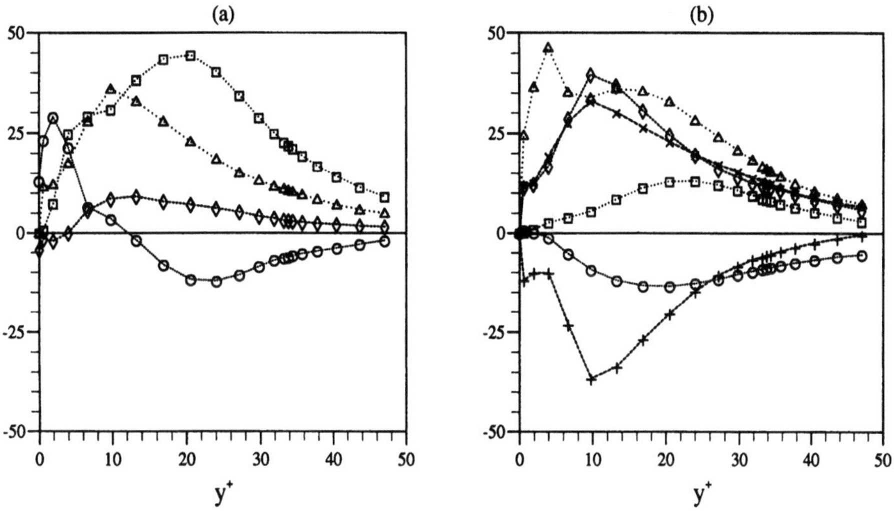


*Figure 4.* Near-wall variation of the averaged contributions to $P_\epsilon^4$ and $P_e^4$. (a) $P_\epsilon^4$ terms : ○, $-(\partial v/\partial y)(\partial u/\partial y)^2$; □, $-(\partial u/\partial x)(\partial u/\partial y)^2$; △, $-(\partial u/\partial y)(\partial u/\partial z)(\partial w/\partial y)$; ◇, $-(\partial v/\partial y)(\partial w/\partial y)^2$. (b) $P_e^4$ terms : ○, $\omega_1^2 s_{11}$; □, $\omega_2^2 s_{22}$; △, $\omega_3^2 s_{33}$; ◇, $\omega_1\omega_3 s_{13}$; +, $\omega_1\omega_2 s_{12}$; ×, $\omega_2\omega_3 s_{23}$. $s_{ij}$ is the fluctuating strain rate.

the wall. The importance of $s_{13}$ during sweeps can be deduced from the localized nature of the sweep-wall interaction.

The $P^4$ terms contain many contributions of which several have been found to be important. These are shown in figure 4. In $P_\epsilon^4$ four terms are dominant and are shown in figure 4a. The three largest of these all involve gradients of the streamwise fluctuating velocity $u$. The terms involving the normal gradient of $u$ are high very close to the boundary. The term $-(\partial v/\partial y)(\partial u/\partial y)^2$ will be high at sweep events since the high $\partial u/\partial y$ will associated with negative $\partial v/\partial y$ – decelerating flow towards the wall. High $\partial u/\partial y$ values are also found to contribute to the $-(\partial u/\partial x)(\partial u/\partial y)^2$ which is significant outside the viscous sublayer up to the buffer layer. The peak value is found at $y^+ \simeq 22$. At this distance this term would attain large values in high-speed streaks that are strongly deflected from the streamwise direction. The form $-(\partial u/\partial y)(\partial u/\partial z)(\partial w/\partial y)$ becomes significant in the interaction between low- and high-speed streaks which attains its maximum at $y^+ \simeq 10$. The weakest of the four terms is $-(\partial v/\partial y)(\partial w/\partial y)^2$ and attains most of its high instantaneous values when a streamwise vortex comes into contact with the walls.

In the production term $P_e^4$ for the enstrophy all terms appear to be significant (Figure 4b). The 'stretching' terms $\omega_1^2 s_{11}$ and $\omega_2^2 s_{22}$ attain their peak values well outside the viscous sublayer, the first being a sink of en-

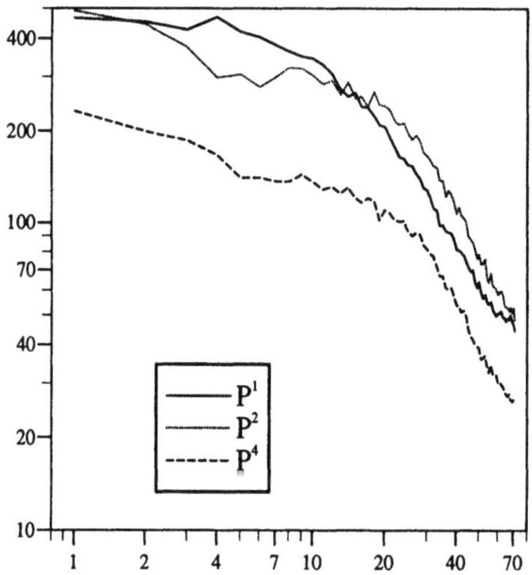


*Figure 5.* One dimensional spectra of the dissipation production terms at $y^+ = 11$.

strophy, the latter a generation term. Their sum is small compared with the other terms. The $\omega_3^2 s_{33}$ term has a maximum at a distance where the production of turbulent kinetic energy $k$ has its maximum. This term is closely correlated to the $-(\partial u/\partial y)(\partial u/\partial z)(\partial w/\partial y)$ of the dissipation equation. The $\omega_1\omega_2 s_{12}$ appear to have a marked negative peak just outside the viscous sublayer $y^+ \simeq 10$ and is also related to the $-(\partial u/\partial y)(\partial u/\partial z)(\partial w/\partial y)$ term in the dissipation equation but with a change in sign. The sum $\omega_1\omega_3 s_{13} + \omega_1\omega_2 s_{12}$ is thus quite small relative to other terms close to the boundary. This is due to the presence of the $(\partial u/\partial y)(\partial u/\partial z)(\partial w/\partial y)$ term in each expression but with opposite sign. High instantaneous values of this term tend to have negative sign. This corresponds to spanwise convection of high-speed $(u > 0)$ over adjacent low-speed streaks $(u < 0)$. In these events $\partial u/\partial z$ and $\partial w/\partial y$ have opposite signs whereas $\partial u/\partial y$ is always positive in these regions. In these events one observes a high speed streak 'crossing over' a low-speed one and this is one of the mechanisms whereby the observed intense shear layers are generated – this also accounts for the spanwise dimensions of these shear layers. The remaining dominant terms are $\omega_3^2 s_{33}$ and $\omega_2\omega_3 s_{23}$ both related to the presence of 'oblique' shear layers. Their origin may be due to the generation and tilting of wall-generated vorticity as well as the contact of high- and low-speed streaks. It appears therefore that the enstrophy production by vortex-line stretching is dominated by the cross stream fluctuating vorticity and strain rates.

In figure 5 the one-dimensional spectra of the dissipation-generation terms are shown. It is somewhat surprising that no peak in the spectra is found. Contributions are now made by a broader range of wavenumbers (implying a short correlation length) but the low wavenumbers still dominate the production of dissipation. However, it should be noted that the dissipation terms are effectively determined by cross-flow derivatives and thus the structures are Kolmogoroff-thin in the cross-stream directions only. This is consistent with the presence of high- and low-speed streaks. Essentially, the statistics suggest that the energy carried by one velocity component, in this case $u$, is dissipated in long thin shear layers. Visual observation of the instantaneous distribution of $\epsilon$ (not shown here) supports this view.

## 4. Summary

The statistics of the dissipation and enstrophy generation terms in their respective budget equations have been calculated from a database generated by turbulent DNS in a rectangular duct. This study looks at the effects of the smooth walls away from the corners. Correspondence between the enstrophy generation and dissipation can be found within and outside the viscous sublayer.

At about 2 wall units from the wall the production of dissipation and enstrophy are due to intense sweep motions. The convection towards the wall produces high dissipation in the convected fluid as it is pressed against the boundary. This affects the enstrophy generation in two ways. First the spanwise gradient of the spanwise velocity stretches the existing mean spanwise vorticity ('splatting'). Second, because the opposing fluid streams are in contact with the wall it results in the generation of the streamwise vorticity which is 'tilted' and stretched by flow gradients parallel to the bounding wall.

Just outside the viscous sublayer another set of phenomena lead to the generation of dissipation and enstrophy. These are the separation from the wall of high-speed streaks, and the sideways deflection of such streaks. The latter phenomenon is directly linked to the generation of enstrophy.

In the region $50 < y^+$ the homogeneous parts of the production rates for enstrophy and dissipation are determined by the same flow structures. This is specific to the near-wall turbulence and both quantities are determined by the strong cross-stream gradients in the fluctuating streamwise velocity component.

## References

Bradshaw, P., & J. B. Perot, 1993: A note on turbulent energy dissipation in viscous wall region. *Phys. Fluids*, **5**, 3305-3306.

Gavrilakis, S., 1996: Dissipative flow structures within near-wall turbulence-producing events. *Advances in Turbulence VI*, Gavrilakis, Machiels & Monkewitz (Eds.), Kluwer Academic.

Gavrilakis, S., L. Machiels, & M. O. Deville, 1996: A spectral element method for direct simulation of turbulent flows with two inhomogeneous directions. *Proceeding of the ECCOMAS 96 Conference*, Paris, September 9-13.

Gilbert, N., 1993: Turbulence model data derived from direct numerical simulations. *DLR Internal Report IB 221-93 A 14*.

Gilbert, N., & L. Kleiser, 1991: Turbulence model testing with the aid of direct numerical simulation results. *Proc. of 8th Symposium on Turbulent Shear Flows*, Paper 26-1, Munich, Sept. 9-11.

Maday, Y., & A. T. Patera, 1989: Spectral element methods for the Navier-Stokes equations. *State-of-the art surveys in computational mechanics*, ASME p. 71, A.K. Noor and J.T. Oden, Editors, New York.

Mansour, N. N., J. Kim, & P. Moin, 1988: Reynolds-stress and dissipation-rate budgets in a turbulent channel flow. *J. Fluid Mech.*, **Vol. 194**, pp. 15-44.

Rønquist, E. M., 1991: Spectral element methods for the unsteady Navier-Stokes equations. *Von Karman Institute Lecture Series.*

Tennekes, H. & J. L. Lumley, 1972: *A First Course in Turbulence*, MIT Press.

# ON THE MECHANISM OF ATTENUATION OF TURBULENCE BY SPANWISE WALL OSCILLATION

Y. MIYAKE, K. TSUJIMOTO
*Dept. of Mech. Eng., Osaka University,*
*2-1, Yamda-oka, Suita, Osaka, 565, Japan*

AND

M. TAKAHASHI
*Toshiba Electronic Co.,*
*2-4, Suehiro-cho, Tsurumi-ku, Yokohama, 230, Japan*

**Abstract.**

In this work, the mechanism of reduction of wall shear stress caused by spanwise oscillation of a wall is discussed by means of a DNS of a channel flow. For the case of wall oscillation where about 35% reduction in shear stress is attained, the detailed attenuation process of turbulence is discussed. It is demonstrated that the fall of production rate of quasi-streamwise vorticity due to stretching effect is substantial for the phenomenon. The periodical fluctuation of the attitude angle, *i.e.*, the inclination angle to the wall and the slant angle of a vortex to the mean flow, enhances the production. This mechanism is a straightforward extension of the one commonly found in the usual near wall turbulence.

## 1. Introduction

Many direct numerical simulations including those of the present authors revealed that quasi-streamwise vortices which have their axis nearly in the mean flow direction play a key role for turbulence in the near wall region. The present authors investigated carefully the behavior of the quasi-streamwise vortices and found that they constitute a particular zig-zag pattern by repeated regeneration (Miyake *et al.*, 1995). They also found the mechanism of self-sustenance of quasi-streamwise vortices (Miyake & Tsu-



*J.-P. Chollet et al. (eds.), Direct and Large-Eddy Simulation II,* 201-212.

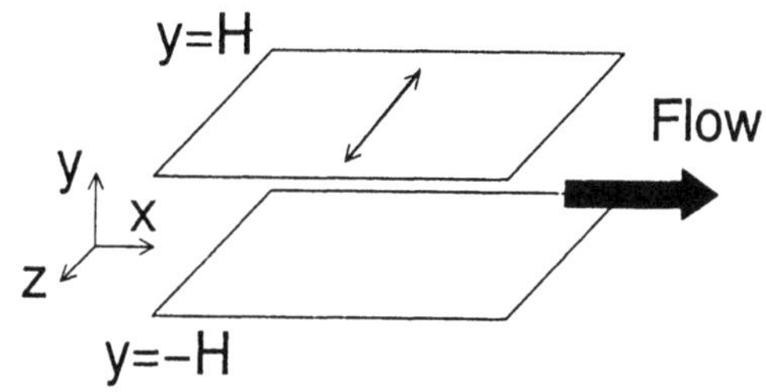


*Figure 1.* Two dimensional straight channel, top wall oscillating spanwise and bottom one stationary.

jimoto, 1996). These findings are the basis of the discussion of modification of near-wall turbulence described in the following.

Control of turbulence, especially active control modifying the quasi-streamwise vortices has come up as one of the most promising targets of turbulence research, for which computer power allowing large scale direct numerical simulations of turbulent flows is crucially important.

The reduction of wall shear stress is the most primitive measure of estimating the effect of turbulence control.

Several reports describing attenuation of turbulence in near-wall region are found in the literature. Bradshow *et al.* (1985) demonstrated experimentally that turbulence is reduced for some time interval after sudden transverse flow is applied in a boundary layer flow. This attenuation was confirmed numerically in 2-D channel flow by DNS conducted by Moin *et al.* (1990). On the basis of these findings, Jung *et al.* (1992) arrived at an idea to apply continuous spanwise wall motion to make the reduction last long. Their idea is to oscillate the wall normal to the main flow, in the spanwise direction. The drag reduction is confirmed numerically (Jung *et al.* 1992) and experimentally (Laadhari *et al.* 1994). But the mechanism of this reduction of friction drag is not well understood up to now, despite that it is expected that it includes the key mechanism of controlling wall shear stress.

The present work has been carried out to resolve this problem using direct numerical simulation. It turned out that the reduction mechanism exists on the extension of the findings of the present authors concerning the sustenance of quasi-streamwise vortices.

## 2. Methods

A direct numerical simulation is carried out for a turbulent flow in a channel of half width $H$ whose one wall is stationary but the other oscillates in

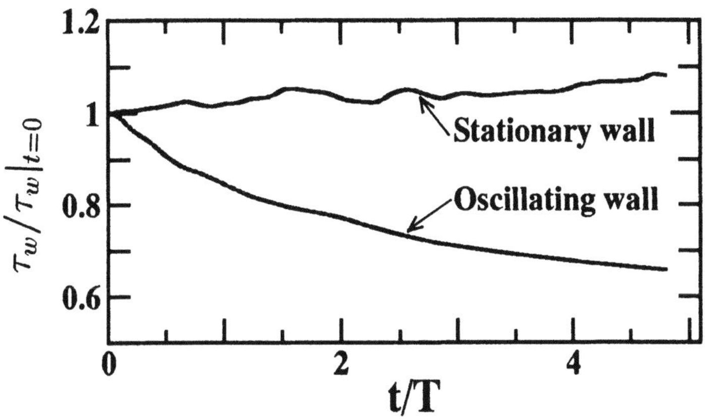


*Figure 2.* Time evolution of wall shear stress $\tau_w$ after the start of wall oscillation in a channel whose one wall is stationary and the other, in spanwise sinusoidal oscillation with period $T^+ = T\bar{u}_{\tau 0}^2/\nu = 100$.

the spanwise direction with speed $w_w = \overline{W}_0 \cos(2\pi t/T)$ where $\overline{W}_0$ means the amplitude and $T$ the period, as shown in Fig.1. Numerical procedures are given in Miyake & Tsujimoto(1995) in which periodic boundary conditions are assumed both in streamwise and spanwise directions and a no-slip condition is applied on the walls. The Reynolds number based on the mean friction velocity $\bar{u}_{\tau 0}$ which corresponds to the mean shear stress of two walls and $H$ is $Re_\tau = 150$. A spectral method is used with grid number $128(x) \times 129(y) \times 64(z)$. The wall oscillation is applied after a long enough calculation of non-oscillating channel flow and the time $t$ is measured from the instant of the start of oscillation.

## 3. Results

### 3.1. MEAN PROPERTY

Figure 2 shows the time evolution of wall friction $\tau_w/\tau_w|_{t=0}$ where $\tau_w|_{t=0}$ is the mean shear stress at $t = 0$. The period of oscillation is $T^+ = T\bar{u}_{\tau 0}^2/\nu = 100$ which is the case giving the largest shear stress reduction (Jung *et al.*, 1992) and the amplitude is $\bar{W}_0 = 0.4Q/H = 12$ where $Q$ is the mean flow rate at the initial time. The wall shear stress on the oscillating wall falls as time marches and that of the stationary wall rises, the mean pressure gradient being kept unchanged throughout the simulation.

Accordingly, the streamwise mean velocity $\bar{u}$ which is given in the form of wall units $\bar{u}^+ = \bar{u}/\bar{u}_{\tau 0}$ grows in the region near the oscillating wall, keeping that in the stationary wall side unchanged, as shown in Fig.3. This modi-

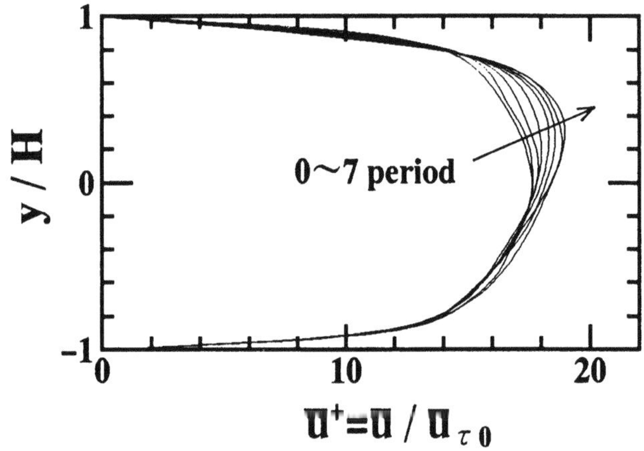


*Figure 3.* Evolution of mean streamwise velocity distribution across the channel. Near top wall $y/H = 1$, acceleration occurs while near the stationary wall $y/H = -1$, it stays unchanged.

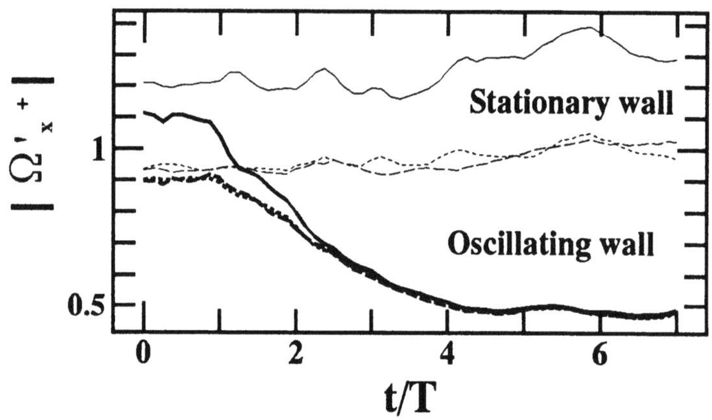


*Figure 4.* Time evolution of total streamwise vorticity $|\Omega_x'^+|$ integrated in the upper- and lower-half computational box. Former (thick lines) is $|\Omega_x'^+|$ in oscillating wall side half and the latter (thin lines), stationary wall side. Solid lines : total vorticity ($\omega_x > 0$), broken lines : vortices of $-0.2 < \omega_x'^+ < 0$, dotted lines : vortices of $0 < \omega_x'^+ < 0.2$.

fication of the flow on the oscillating wall side suggests relaminarization-like behavior

Figure 4 shows how streamwise vortices attenuate after the start of wall oscillation. The curves mean total vorticity $|\Omega_x'^+|$ $(=\int |\omega_x'| dV/(H^3 \bar{u}_{\tau 0}^2/\nu))$ integrated in upper (oscillating wall side)- and lower (stationary wall side) half computational volume. The thick curves are for the oscillating wall side and the thin ones, for the stationary side. To demonstrate the rapid attenuation of strong vortices, the attenuation process of weak vortices ($|\omega_x'^+| < 0.2$) are shown by dotted curves ($\omega_x' > 0$) and broken curves ($\omega_x' < 0$), as well as the total vorticity (solid curves).

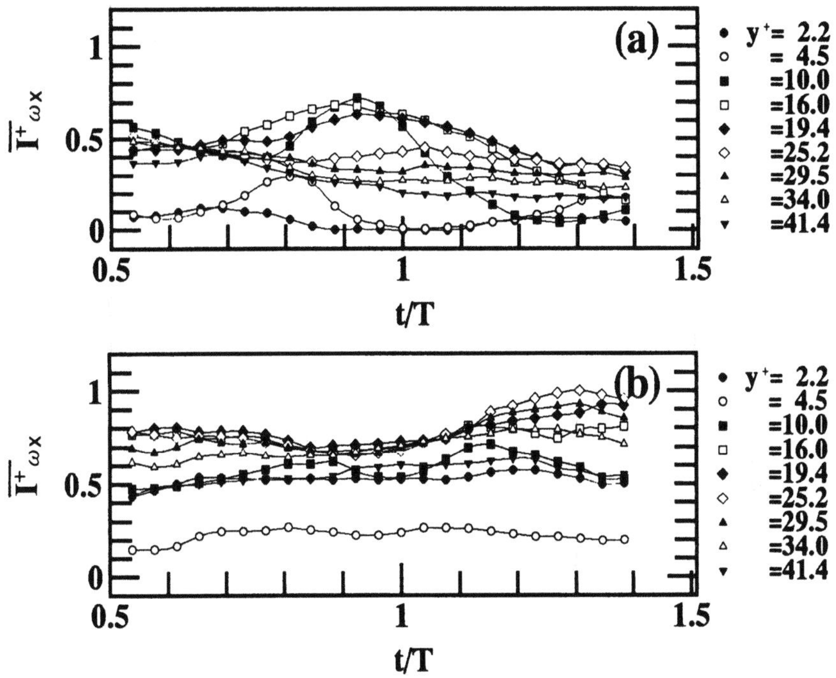


*Figure 5.* Time evolution of streamwise vorticity of strong vortices of $\omega_x'^+ \geq 0.2$ integrated in planes parallel to wall. (a) : Oscillating wall side, (b) : stationary wall side.

Quasi-streamwise vortices have been commonly identified by using $|\omega_x^+|$ >0.2 as the most simple and useful indicator(Brooke *et al.* 1993 & Miyake *et al.* 1995). The quasi-streamwise vortices thus identified are elongated long in the streamwise direction and have a distinct coherent structure in the near wall region.

The reduction process of streamwise vorticity is synchronized with that of $\tau_w$ and consequently the attenuation of streamwise vortices causes the reduction of shear stress. Corresponding to the increase of total flow rate, the total vorticity in the stationary wall side volume is gradually strengthened. It is notable that strong streamwise vortices disappear within the early stages of vortex attenuation as found by the difference between solid lines and broken or dotted lines, indicating that wall oscillation affects most strongly the strongest vortices firstly and then gradually the weaker ones.

But the attenuation is not straightforward. Figure 5 show an example of the time evolution of strong streamwise vortices in the period $t/T$ =0.5 to 1.5 when the shear stress on the oscillating wall falls fastest. In the fig.5, $\bar{I}^+_{\omega_x}(=\int \omega_x' dxdz/(H^2\bar{u}^2_{\tau 0}/\nu))$ is the integration of strong vorticity of $\omega_x'^+ \geq$ 0.2 within each plane parallel to the wall. Compared with the fluctuation in the stationary wall side, the total vorticity $\bar{I}^+_{\omega_x}$ in the buffer layer *i.e.*,

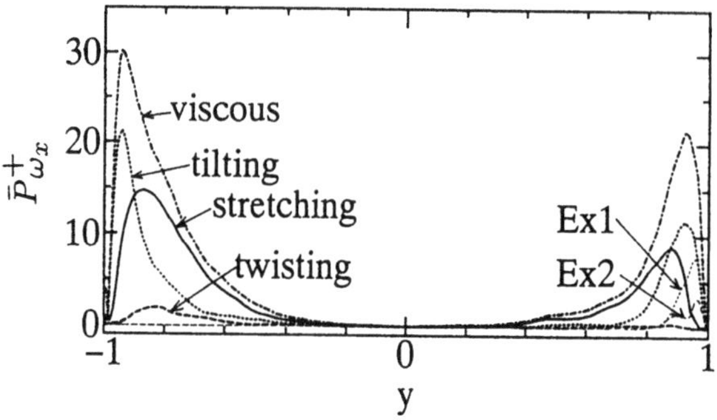


*Figure 6.* The wall normal distribution of mean absolute value of production rate which are averaged in plane parallel to the wall in t/T=0.5 to 1.5.

$y^+ = 10 \sim 20$ is enhanced strongly in some time interval. The time instant when the vorticity takes its peak value depends on the distance from the wall, suggesting that the effect of the wall oscillation propagates with a time lag away from the wall.

The transport equation of turbulent streamwise vorticity $\omega'_x$ is given by

$$\frac{D\omega'_x}{Dt} = \omega'_x \frac{\partial u}{\partial x} - \frac{\partial w}{\partial x}\frac{\partial u}{\partial y} + \frac{\partial v}{\partial x}\frac{\partial u}{\partial z} + \frac{1}{Re_\tau}\nabla^2 \omega'_x - v\frac{\partial \bar{\omega}_x}{\partial y} + \bar{\omega}_x \frac{\partial u}{\partial x} \quad (1)$$

where $u, v, w$ are velocity components in the $x, y, z$ direction, respectively and $\bar{\omega}_x$ is the streamwise vorticity of the flow induced by wall oscillation averaged in the plane parallel to the wall. The right-hand side terms of the above equation are the production-rate terms representing the effects of stretching, tilting, twisting and viscous dissipation in order from left to right. The last two terms EX1 and EX2 are extra terms added to the non-oscillatory wall cases. EX1 means the augmentation of $\omega'_x$ due to the entrainment of $\bar{\omega}_x$ by the wall-normal velocity fluctuation induced by streamwise vortices and EX2 means that due to stretching of vorticity $\bar{\omega}_x$.

Figure 6 show the wall normal distribution of mean absolute value of production rate terms. These terms are averaged in plane parallel to the wall in a time span of unique oscillating period between $0.5t/T$ and $1.5t/T$. Except for Ex1 and Ex2 terms, all the production rate terms near oscillating wall side are reduced, compared with those of stationary side. But qualitatively, their distribution are similar. The stretching and tilting term are dominant as in stationary wall side. The Ex1 term will be commented later.

Of the terms of production rate of $\omega'_x$, the time evolution of the stretching term most closely corresponds to that of fluctuating vorticity $\omega'_x$ itself. Figures 7 show the time evolution of the production rate $\bar{P}^+_{\omega_x} (= \int P_{\omega_x} dx dz \, / \, (H\bar{u}^2_{\tau 0}/\nu)^2)$ due to stretching of a strong positive streamwise vortex having $|\omega'^+_x| > 0.2$, integrated in several planes parallel to the wall. The bottom

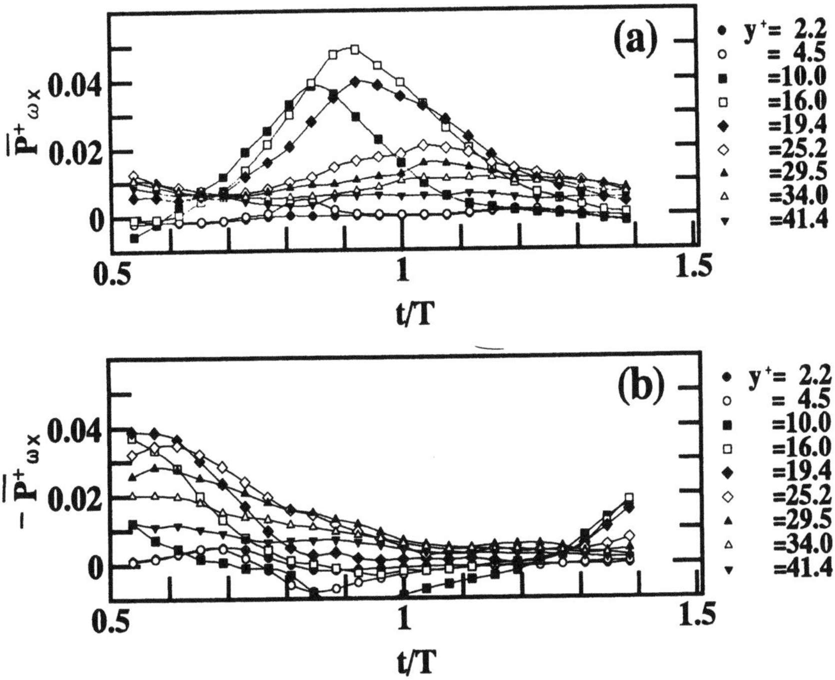


*Figure 7.* Time evolution of production rate $\bar{P}^+_{\omega_x}$ of $\omega'^+_x$ integrated in a plane parallel to the wall for strong vortices $|\omega'^+_x| > 0.2$, in $t/T = 0.5 \sim 1.5$. (a) : $\omega'^+_x > 0.2$, (b) : $\omega'^+_x < -0.2$.

figure is for the stationary wall and the top, the oscillating wall. The curves do not differ substantially from those for all positive vortices including weak ones. The layer where the production due to stretching is modified seriously is up to $y^+ \sim 20$ which is farther away from the wall than the distance the wall oscillation reaches, the thickness of the Stokes layer in this case being $l^+_{stokes}(= \bar{u}_{\tau 0}\sqrt{T/\pi\nu}) = 5.8$. This suggests that the modification of turbulence due to wall oscillation is not by direct modulation of the Stokes flow oscillation. In the time interval where the curves have humps in the figure 7(a), the wall oscillation enhances the production of streamwise vorticity due to stretching and otherwise suppresses. The time mean of $I_{\omega_x}$ in the time interval of $t/T = 0.5 \sim 1.5$, is larger at the stationary wall side than at the oscillating wall side. However, in the layer near the oscillating wall, the production rate becomes enhanced in a limited interval which depends on the distance from the wall. That is, in the layer at $y^+ = 10.0$, the wall oscillation begins to enhance the production the earliest and then the influence propagates farther away from the wall up to $y^+ = 25.2$ with a time lag.

The tilting term which is important next to the stretching term shows

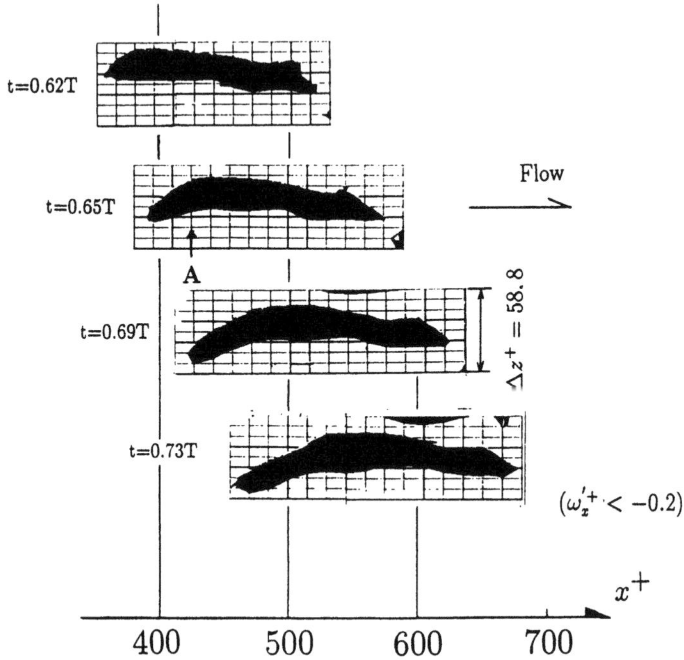


*Figure 8.* Top view of the evolution of typical quasi-streamwise vortex ($\omega_x'^+ \leq -0.2$) found near the oscillating wall.

similar time evolution. But careful investigation (Miyake & Tsujimoto 1996) revealed that although its contribution is as large as that of stretching term so far as its integrated amount is concerned, it is not substantial in sustaining coherent structure of near-wall turbulence, as is usual in any steady flat channel.

Therefore, the mechanism of rapid reduction of stretching effect on production after its peak found in Fig.7 is the key. In the authors' previous work (Miyake & Tsujimoto, 1996), it was found that the fundamental mechanism of the stretching effect is to wind the spanwise vorticity within a streamwise vortex and to stretch it by steep wall-normal gradients of streamwise velocity.

### 3.2. INDIVIDUAL QUASI-STREAMWISE VORTEX

To investigate the modification of the stretching effect for a streamwise vortex, the behavior of a particular representative streamwise vortex found in the layer near the oscillating wall is discussed in the following. Figure 8 shows the top view of a streamwise vortex, chosen as an example. Even in a globally attenuating period, some individual vortices are in a growing stage since not all the individual vortices synchronously attenuate. The

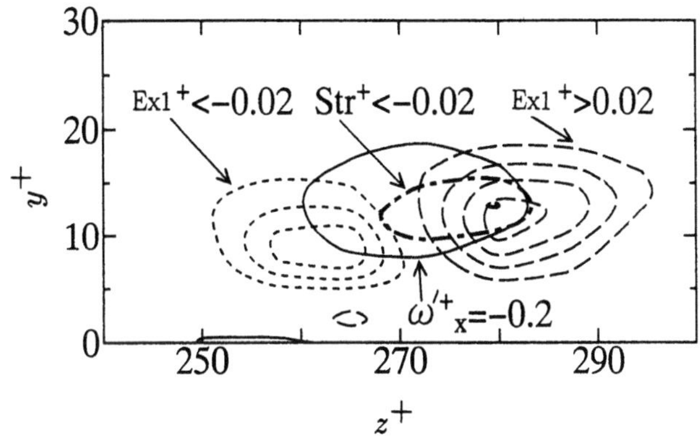


*Figure 9.* Cross-sectional distribution of production rate of a quasi-streamwise vortex at the position A in Fig. 8. (Solid lines:$\omega_x'^+$=-0.2, chained lines:$P^+_{\omega_x,str}$ < -0.02, broken lines:$|P^+_{\omega_x,Ex1}|$ >0.02)

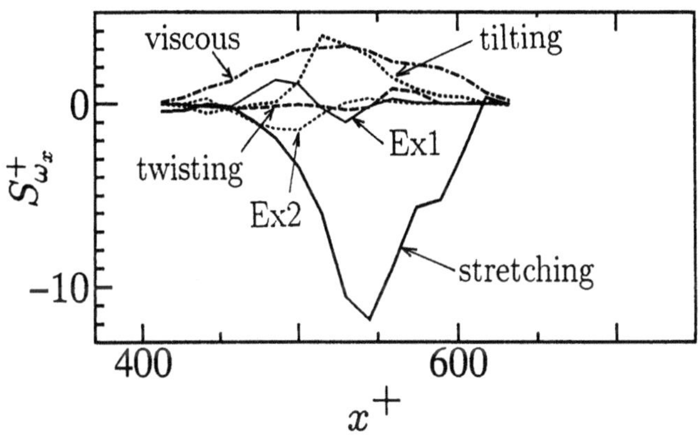


*Figure 10.* Streamwise distribution of cross sectional production rate of streamwise vorticity $S^+_{\omega_x}$ integrated in the cross sectional area of $\omega_x'^+ < -0.2$,of a typical quasi-streamwise vortex.$t/T = 0.69$.

upstream-end part of the vortex which is located close to the wall follows the wall movement, the rest keeping its attitude unchanged. This movement is caused by the term EX1, *i.e.*, by implementing strong streamwise vorticity included in the Stokes layer flow by vertical flow of the vortex itself and not by viscous entrainment in Fig. 9.

Streamwise distribution of the production rate terms $S^+_{\omega_x}(=\int P^+_{\omega_x}dy^+ dz^+)$ of a typical quasi-streamwise vortex at an early stage of an attenuating period (t/T= 0.69) is shown in Fig. 10 where each term on the righthand side of eq.(1) is integrated in the cross section of the vortex within the area $\omega_x'^+ < -0.2$ and the correspondence of each curve to the terms in eq.(1) is denoted in the figure.

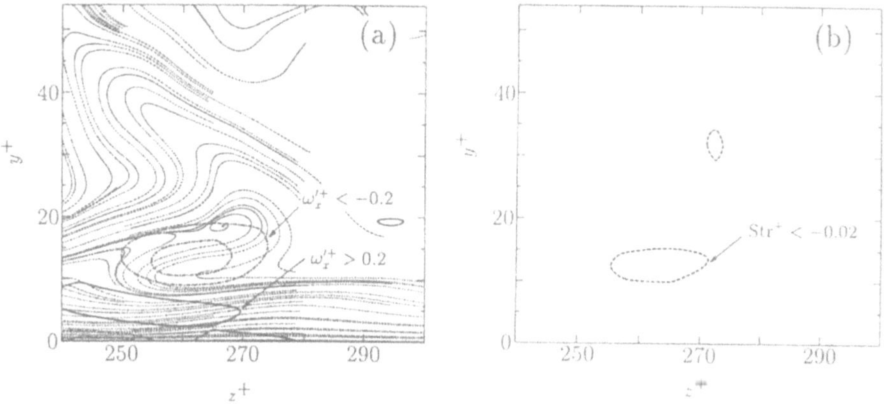


*Figure 11.* 2-D vortex lines (dotted lines) , the production rate due to stretching and the vorticity area ($\omega_x'^+ < -0.2$, chained lines) at $x^+$=486, t/T=0.69. (a) : vortex lines, (b) : production rate due to stretching.

The figure 10 indicates that stretching by turbulent flow occupies most of the production of $\omega_x'^+$ even in this case, as in the usual steady flow. The rest of the terms in eq.(1) generate opposite vorticity but since sum of these opposite terms is less than the stretching term, the vortex is still in a growing stage. Notable is that in individual strong vortices, the tilting effect functions to suppress the production despite that as a whole, the tilting effect contributes favorably to sustain the vorticity at this time instant, as mentioned earlier in conjunction with Fig.7. This tendency of the tilting effect manifests itself in a similar manner in any other wall turbulence (Miyake & Tsujimoto, 1996). Tilting production takes place more commonly than stretching production but it is mostly in weak, chaotic vortices.

In non-oscillating wall turbulence, a quasi-streamwise vortex sustains itself by supplying vorticity mainly in its upstream-end 1/3 portion. But in this particular vortex shown in Fig. 8, the upstream-end contributes quite negligibly and the production in the more downstream portion, compared to the case of stationary wall, covers the loss in the upstream-end portion. This modification demonstrates the fundamental mechanism of modulation of turbulence by wall oscillation.

Figures 11 (a)(b) show the two-dimensional vortex lines and the production rate due to stretching in a cross section of $x^+ = 486$ which is obtained by integrating $dy/\omega_y' = dz/\omega_z'$ in each section. It has been confirmed that the vortex lines thus obtained are nearly identical with the projection of three-dimensional vortex lines to a plane normal to the mean flow.

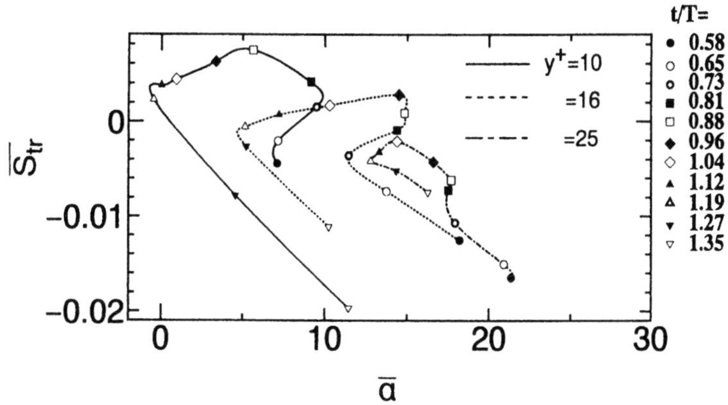


*Figure 12.* Time evolution of mean angle $\alpha$ versus mean production rate due to stretching in the plane $y^+$=10,16 and 25, for the vortices ($\omega_x'^+ < -0.2$)

It is observed in the figure that at the area where the $\omega_x'^+ < -0.2$, most of the production rate due to stretching takes place. At $x^+ = 486$, the vortex lines passing through the body of the streamwise vortex is too much inclined to the horizontal plane which results in the weakening of the stretching effect by the wall-normal gradient of mean flow. The upstream-end portion of a quasi-streamwise vortex found in the usual non-oscillating wall region is located close to the wall but its center is at $y^+ \sim 25$ where enough velocity gradient to stretch vortex is realized. But wall oscillation entrains this vortex to the spanwise direction and brings it closer to the wall; the vortex section becomes flattened and approaches more to the wall.

In the time period from $t/T = 0.5$ to $1.5$, Figure 12 shows a variation of the mean inclination angle $\bar{\alpha}$(=-$\int \tan^{-1}[\omega_x'/\sqrt{\omega_x'^2+\omega_z^2}]$dxdz / ($\int$ dxdz)) to the wall, of strong negative quasi-streamwise vortices and mean strength of production rate due to stretching $\overline{S_{tr}}$(=$\int P^+_{\omega_x,str}$dxdz / ($\int$ dxdz)) of those strong vortices. Namely, the integration of production rate is done within the area where vorticity $\omega_x'$ is stronger than -0.2. A strong coherency which is similar but different in amplitude is found between the angle $\alpha$ and the production rate due to stretching.

The above observation suggests that the effect of wall oscillation is to modify the strength and the manner of winding of the spanwise vortex layer. That is, the inclination of the vortex lines passing through the vortex cross section is modified by the wall oscillation and as a result, the efficiency of conversion of the spanwise vorticity to streamwise due to stretching of the entrained vortex filament is modified. The inclination angle varies continuously from the most efficient one to the least efficient in one oscillating period and on average, reduction is far larger than enhancement.

The suppression of the production rate causes the disappearance of concentrated streamwise vortices due to rapid diffusion by viscosity. So, in the next enhancement period, most of the quasi-streamwise vortices have to grow up newly. If the oscillation time is long enough to allow sufficient growth the reduction does not appear. In the case where the period is short enough, the effect of the wall does not appear too, since some lead time is necessary to allow the wall to affect the flow. So, an optimum oscillating period which is around $T^+ \sim 100$ exists.

## 4. Conclusions

In this work, the mechanism of the reduction of wall shear stress by spanwise wall oscillation is discussed by using direct numerical simulation.

The finding of this work is that the modification in strength and manner of winding of spanwise vortex layer of mean flow (which is commonly found as the key process of self-sustenance of streamwise vortices in the near wall region), is crucial for the reduction of wall shear stress. This modification enhances/suppresses the stretching effect by wall oscillation .

In one period of oscillation, the oscillation has both effects of enhancement and of suppression of the production of streamwise vorticity but it depends strongly on the phase of the oscillation and even at the most efficient condition of wall oscillation, the enhancement effect is far smaller than the suppression effect. In the time mean, the latter causes the reduction in the streamwise vorticity.

## References

Bradshaw,P. & Pontikos 1985 : Measurements in the turbulent boundary layer on an 'infinite' swept wing. *J. Fluid Mech.* **159**, 105-130.

Brooke,J.W. & Hanratty,T.J. 1993 : Origin of turbulence-producing eddies in a channel flow. *Phys. Fluids*, **A5**-4 1011-1022.

Jung,W.J., Mangiavacchi,N. & Akhavan,R. 1992 : Suppression of turbulence in wall-boundes flows by high-frequency spanwise oscillations. *Phys. Fluids*, **A4**-8 1605-1608.

Laadhari,F., Skandaji,L. & Morel,R. 1994 : Turbulence reduction in a boundary layer by a local spanwise oscillating surface. *Phys. Fluids*, **A6**-10 3218-3220.

Miyake,Y., Ushiro,R. & Morikawa,T. 1995 : On regeneration of quasi-streamwise eddies. *Proc. 10th Symp. Turb. Shear Flows, Penn.State, Aug.14-16, 1995*, pp.15.7-15.12

Miyake,Y. & Tsujimoto,K. 1996 : Mechanism of Self-Sustenance and Regeneration of Quasi-streamwise Vortices of Near-Wall Turbulence. *Trans. JSME, ser.B,* **62**-598 2172-2179.

Miyake,Y. & Tsujimoto,K. 1995 : Direct Numerical Simulation of Turbulent Channel Flow with periodic Pressure Gradient. *Trans. JSME, ser.B,* **61**-587, 2401-2408. (*in Japanese*)

Moin,P., Shih,T.-H. Driver,D. & Mansour,N.N. 1990 : Direct numerical simulation of a three-dimensional turbulent boundary layer. *Phys. Fluids*, **A2**-10 1846-1853.

# SIMULATION AND MODELLING OF THE SKEW RESPONSE OF TURBULENT CHANNEL FLOW TO SPANWISE FLOW DEFORMATION.

R.J.A.HOWARD AND N.D.SANDHAM
*London University, Queen Mary and Westfield College,*
*Mile End Road , London, E1 4NS, U.K.*

**Abstract.** Skewed turbulent channel flow is investigated by observing the response of a steady equilibrium turbulent channel flow to pressure driven and wall shear driven spanwise deformation. The results produced by direct numerical simulation (DNS) of the incompressible Navier-Stokes equations are compared with those generated by several low Reynolds number two-equation turbulence models. The DNS confirms earlier studies, showing that the initial response of the flow to spanwise deformations is a drop in the turbulence intensity, caused by a partial breakdown of the near-wall streak structures. None of the two-equation models adequately capture this phenomenon. The DNS also shows a lag between the shear stress and strain rate angles of the flow. Standard two-equation models are unable to show this as it is assumed that the eddy viscosity is independent of flow direction. Possible improvements of the models are discussed.

## 1. Introduction

Steady two dimensional (2-D) boundary layers undergo skewed deformation in flows such as those near a wing body junction, on swept wings and past a rotating body. Bradshaw & Pontikos (1985) carried out experiments on an 'infinite' swept wing, where a spanwise pressure gradient is applied to the boundary layer, and observed that the turbulent activity decreased as the flow became more three dimensional (3-D). A suggestion for the cause of the decrease in the turbulent activity was that the spanwise pressure gradient tilted the large eddies over sideways, confused their behaviour and reduced their efficiency as producers of shear stress. The 'amount' of



*J.-P. Chollet et al. (eds.), Direct and Large-Eddy Simulation II,* 213-224.

three dimensionality of the flow was dependent on downstream distance and magnitude of spanwise pressure gradient.

In DNS it is rather difficult and expensive to set up this spatial problem. However it is possible to set an analogous temporal problem in which a spanwise pressure gradient is imposed on an initially 2-D turbulent flow. This has been carried out and analysed by Moin *et al.* (1990) and Sendstad & Moin (1991). The initial 2-D flow involved was turbulent channel flow. The principal features observed were initial drops in the peak of the turbulence intensity, Reynolds shear stress and wall shear stress. Near-wall streaks were observed to be broken as quasi-streamwise vortices, convected by the spanwise pressure gradient, crossed over them.

In a similar way turbulent flow over a rotating or moving surface can be simulated by impulsively moving the walls in a turbulent channel flow and observing the initial response of the flow. Coleman *et al.* (1995) have investigated the response to the movement of one wall. They also observed drops in the turbulence intensity and shear stress and found that the near wall streaks were weakened and 'torn' into smaller structures before becoming stronger and more elongated as the flow re-aligned to the new direction.

In this paper we use DNS results to check two-equation turbulence models, which remain the standard method of calculating 3-D flows in practice. Test cases involve the response of steady 2-D turbulent channel flow to a spanwise pressure gradient (pressure driven disturbance) and the response to impulsive movement of both walls in the spanwise direction (wall shear driven disturbance). The results produced by the DNS are compared with those produced by the turbulence models of Chien (1982), Launder & Sharma (1975), Kawamura & Kawashima (1995) and Kalitzin *et al.* (1996) (hereafter referred to as CH, LS, KK and $k-g$ respectively).

## 2. Direct numerical simulations

The complete incompressible Navier-Stokes equations are solved using a spectral (Fourier - Chebyshev) method see Sandham & Howard (1995), Canuto *et al.* (1988) and Kleiser & Schumann (1980). Time discretisation is carried out using a compact 3rd order Runge-Kutta method for the convective terms and the Crank-Nicolson method for the viscous and pressure terms. The code has been written for use on massively parallel machines as discussed in Sandham & Howard (1995). The box has dimensions of $13d \times 6d \times 2d$ in the $x$ (streamwise)$\times z$ (spanwise) $\times y$(wall normal) directions, where $d$ is the channel half width. These are large enough to capture the initial effects of the spanwise disturbance. At the latest stage in the wall shear driven flow the grid size perpendicular to the streaks in wall units increases from 8.4 to 14.0. The computational grid is $128 \times 128 \times 80$, resolving all the essential scales of turbulent motion. The initial Reynolds number of the flow is 180 based on the friction velocity $u_\tau = \sqrt{\nu \partial \bar{u}/\partial y|_{wall}}$ and $d$. To improve statistics we use ensemble averages from eight independent simulations that were run in parallel, each using 64 processors.

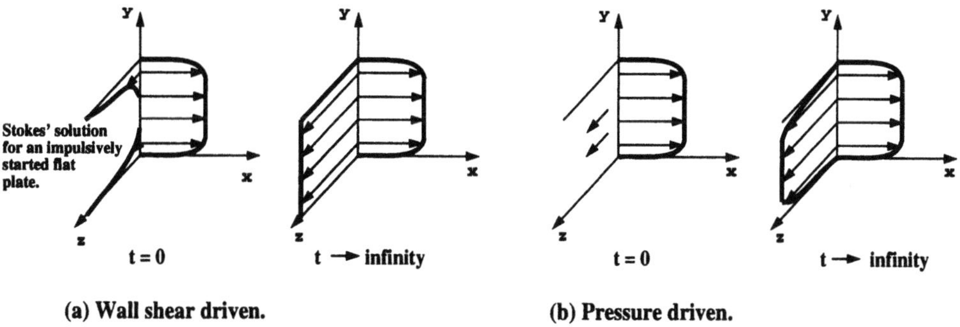


*Figure 1.* Two types of skewed turbulent channel flow.

The deformations to turbulent channel flow are shown on figure 1. For the pressure driven case an instantaneous spanwise pressure gradient $-\partial p/\partial z = 10{u_{\tau_0}}^2\rho/d$ was applied. For the wall shear driven case, in order to avoid a discontinuity in $y$, Stokes solution for the impulsively started flat plate was used to specify the initial distribution of $w(y)$. The Stokes solution corresponded to $t = 0.0045$ as in Coleman *et al.* (1995). The wall velocities were $w_{lower} = w_{upper} = 10u_{\tau_0}$. In each case $u_{\tau_0}$ represents the initial 2-D value for the friction velocity.

Profiles of the turbulent kinetic energy, $k = \overline{(u'u' + v'v' + w'w')}/2$, are shown in Figure 2 at different times in the skewing process. These plots clearly show the initial drop in the turbulence in the region $10 < y^+ < 30$, due to the skew disturbance, as observed by other researchers. This drop

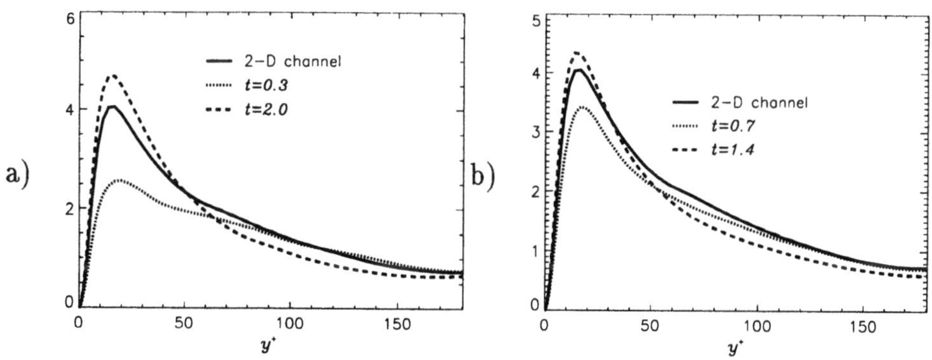


*Figure 2.* Turbulence kinetic energy profiles at different times for (a) wall shear driven and (b) pressure driven cases.

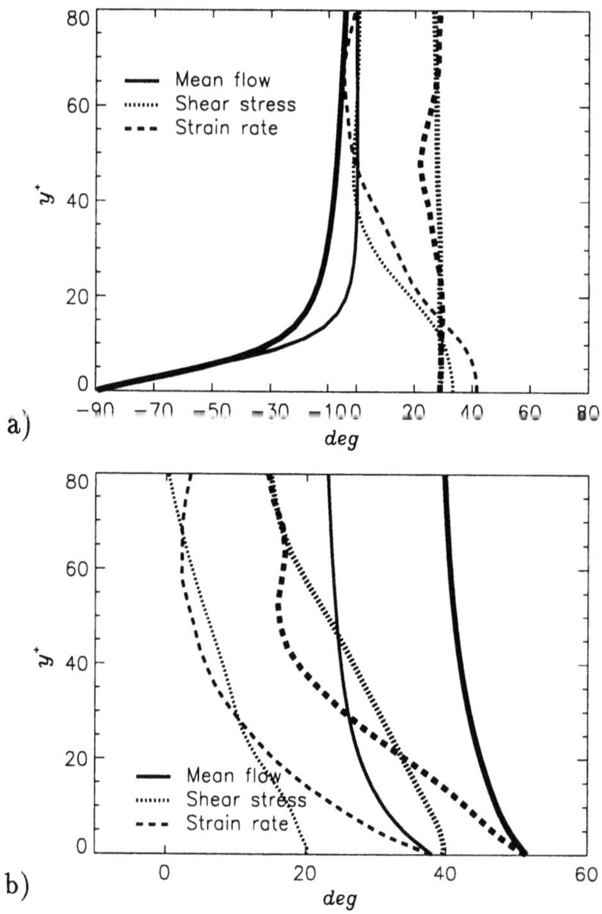


*Figure 3.* Flow angles of the mean velocity $\gamma_u$, shear stress $\gamma_\tau$ and strain rate $\gamma_s$ for (a) the wall shear driven case at t=0.3 and (in thicker lines) at t=2.0 and (b) the pressure driven case at t=0.7 and at t=1.4 (in thicker lines).

is more clearly shown by plotting the variation of the maximum of the turbulence kinetic energy with time (see later).

An indication of the lack of equilibrium can be obtained from the relative angles of the velocity $\gamma_u = \arctan(\overline{W}/\overline{U})$, shear stress, $\gamma_\tau = \arctan(\overline{v'w'}/\overline{u'v'})$, and strain rate, $\gamma_s = \arctan(\partial\overline{w}/\partial y/\partial\overline{u}/\partial y)$. In an equilibrium flow the strain rate direction is equal to the shear stress direction. Figure 3 shows the distributions of these angles at two different stages in the skewing process. Both cases show the shear stress vector lagging behind the strain-rate vector in the near wall region. The magnitude of the lag for the pressure driven case is of the order of the maximum observed by Bradshaw and Pontikos (1985).

All the terms in the kinetic energy transport equations and in the individual Reynolds stress equations have been calculated. The initial response of the flow is a reduction in the turbulence production $P$ and a reorganisation of the dissipation, reducing at the wall and increasing slightly at $y^+ \approx 10$. The diffusion terms also show reductions in magnitude in the region $5 < y^+ < 30$. At later times the kinetic energy profiles tend towards their original shapes.

## 3. Modelling

Most practical engineering flows are currently being calculated by turbulence models in which the shear stress is modelled as the product between the strain rate and an eddy viscosity, $\overline{(u_i'u_j')} = \nu_t \partial\overline{u_i}/\partial x_j$. The eddy viscosity is modelled as $\nu_t = C_\mu f_\mu k^2/\epsilon$ where $C_\mu$ is a constant and $f_\mu$ a damping function. This has one clear problem in that the flow angles $\gamma_\tau = \gamma_s$ for all flows. This means that this type of modelling will not show any lag between the shear stress angle and the strain rate angles. The model equations are typically closed by solving transport equations for $k$ and $\epsilon$

$$\frac{Dk}{Dt} = P - \epsilon + \frac{\partial}{\partial y}\left[(\nu + \frac{\nu_t}{\sigma_k})\frac{\partial k}{\partial y}\right] + \pi_k \tag{1}$$

$$\frac{D\epsilon}{Dt} = C_{\epsilon 1} P\frac{\epsilon}{k} - C_{\epsilon 2} f_\epsilon \frac{\epsilon^2}{k} + \frac{\partial}{\partial y}\left[(\nu + \frac{\nu_t}{\sigma_\epsilon})\frac{\partial \epsilon}{\partial y}\right] + E + \pi_\epsilon. \tag{2}$$

Here, $P = \nu_t\left(\left(\frac{\partial\overline{u}}{\partial y}\right)^2 + \left(\frac{\partial\overline{w}}{\partial y}\right)^2\right)$ is production. Solution of the dissipation equation requires a boundary value for dissipation which is not zero at the walls. There are three solutions to this problem: (i) set a wall value for $\epsilon$, (ii) solve a transport equation for $\tilde{\epsilon}$ (the 'isotropic' dissipation) where $\tilde{\epsilon} = 0$ at $y = 0$ and use the non-isotropic part ($D = \epsilon - \tilde{\epsilon}$) to close the $k$ transport equation, or (iii) combine transport equations of $k$ and $\epsilon$ such that the new variable requires a zero boundary condition, i.e. $z(k, \epsilon) = 0$ at $y = 0$. The models of LS, CH and KK are all examples of models which use the second method for the closure of $\epsilon$. Functions and constants within these models are shown in Table 1. Where the turbulent Reynolds number is $Re_\tau = \frac{k^2}{\nu\tilde{\epsilon}}$.

An example of the third method of solving the $\epsilon$ problem is the $k - g$ model where $g = \sqrt{k/\epsilon}$. The $k - g$ model time scale equation is

$$\frac{Dg}{Dt} = P\frac{5g}{18k} + \frac{0.075}{2C_\mu g} + \frac{\partial}{\partial y}[(\nu + \frac{\nu_t}{\sigma_\epsilon})\frac{\partial g}{\partial y}] - (\nu + \nu_t\sigma_\epsilon)\frac{3}{g}\left(\frac{\partial g}{\partial y}\right)^2 \tag{3}$$

Further constants for the $k - g$ model are $\sigma_k = \sigma_\epsilon = 2.0$, $C_\mu = 0.09$ and $f_\mu = 1$.

The two-equation models are solved together with the $\overline{u}$ and $\overline{w}$ momentum equations using second order central difference in the wall normal

TABLE 1. Model constants and functions

| | Launder Sharma (LS) | Chien (CH) | Kawashima Kawamura (KK) |
|---|---|---|---|
| $D$ | $2\nu\left(\frac{\partial\sqrt{k}}{\partial y}\right)^2$ | $2\nu\frac{k}{y^2}$ | $2\nu\left(\frac{\partial\sqrt{k}}{\partial x_j}\right)^2$ |
| $C_\mu$ | 0.09 | 0.09 | 0.09 |
| $C_{\epsilon 1}$ | 1.44 | 1.35 | 1.44 |
| $C_{\epsilon 2}$ | 1.92 | 1.8 | 1.92 |
| $\sigma_k$ | 1.0 | 1.0 | $1-0.5\exp\left[-\left(\frac{y^+}{20}\right)^2\right]$ |
| $\sigma_\epsilon$ | 1.3 | 1.3 | $1-0.5\exp\left[-\left(\frac{y^+}{20}\right)^2\right]$ |
| $E$ | $2\nu P$ | $-2\nu\frac{\tilde{\epsilon}}{y^2}\exp\left[-0.5y^+\right]$ | $0.6\nu P$ |
| $\pi_\epsilon$ | – | – | $-\frac{\tilde{\epsilon}}{k}D$ |
| $f_\mu$ | $\exp\left[\frac{-3.4}{\left(1+\frac{Re_T}{50}\right)^2}\right]$ | $1-\exp\left[-0.0115y^+\right]$ | $1-\exp\left[-\frac{y^+}{415}-\left(\frac{y^+}{36}\right)^3\right]$ |
| $f_\epsilon$ | $1-0.3\exp\left[-Re_T{}^2\right]$ | $1-0.22\exp\left[-\frac{Re_T{}^2}{6}\right]$ | 1 |
| $\pi_k$ | 0 | 0 | $-0.5\nu\frac{\partial}{\partial y}\left[\frac{k}{\epsilon}\left(\frac{D}{\epsilon}\right)^2\frac{\partial D}{\partial y}\right]$ |

direction and fourth order Runge-Kutta for time advancement. A stretched grid of 81 points provided sufficient resolution in the wall normal direction.

## 4. Comparison of models and DNS

Figure 4 shows the variation of the maximum of the turbulent kinetic energy with time for each of the models compared with the DNS for the two flow types. From this figure it can be seen that none of the models are able to capture the effects observed in the DNS. The KK and CH models actually predict the wrong sign of the change in $k$. The LS model showed the right general trends and could be described in many respects as the least inaccurate of the models considered for these flow cases.

Figure 5 shows the production and dissipation for the three cases of steady flow, wall shear driven disturbance at $t = 0.3$ and pressure driven disturbance at $t = 1.4$. Comparing the models with the DNS for the steady channel flow (Figure 5 (a),(b)), it can be seen that the models in general cannot capture the dissipation wall value or inflexion point. Many models are successful due to the balancing out of equal and opposite inaccuracies in the modelling of the dissipation and diffusion. All the models show a good agreement for the turbulence production of the steady channel flow. There are problems with modelling production for the skewed flows how-

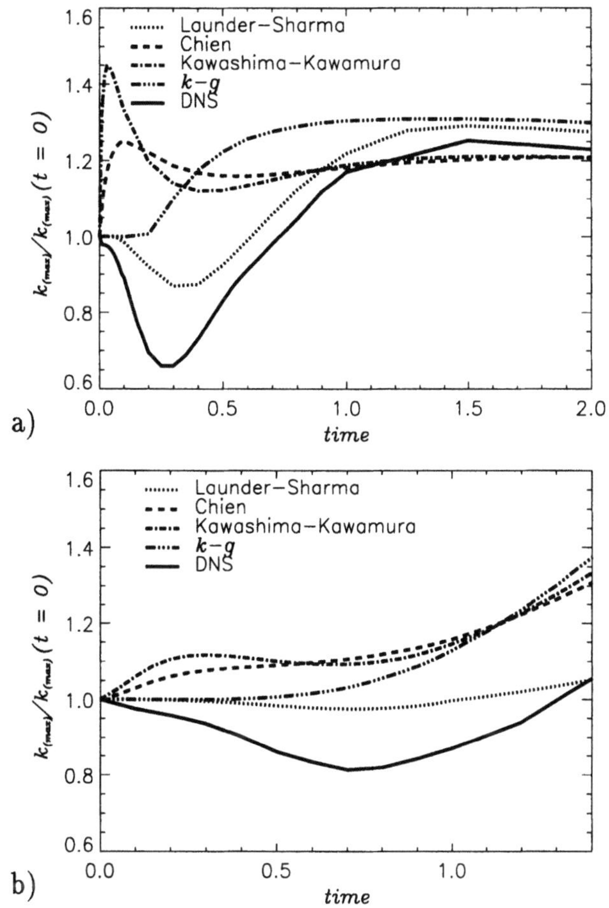


*Figure 4.* Variation in turbulence intensity with time for (a) wall shear driven and (b) pressure driven cases.

ever. None of the models are able to show the drop in production due to the disturbance.

The KK model showed good modelling of dissipation and diffusion for steady channel flow, since it was developed using DNS data for steady channel flow, but was unable to model skew disturbance. This demonstrates a common problem in two-equation modelling of matching a model too closely to one flow.

A common feature between the CH and KK models was their use of the wall normal distance $y^+$ as a modelling parameter. This means that any change in the flow is related to how far it is from the wall rather than flow quantities such as $k$ and $\epsilon$. Another problem with the use of distance from the wall occurs in wing body junction flows in which there is more than one wall and so there is some doubt as to what the wall normal distance is

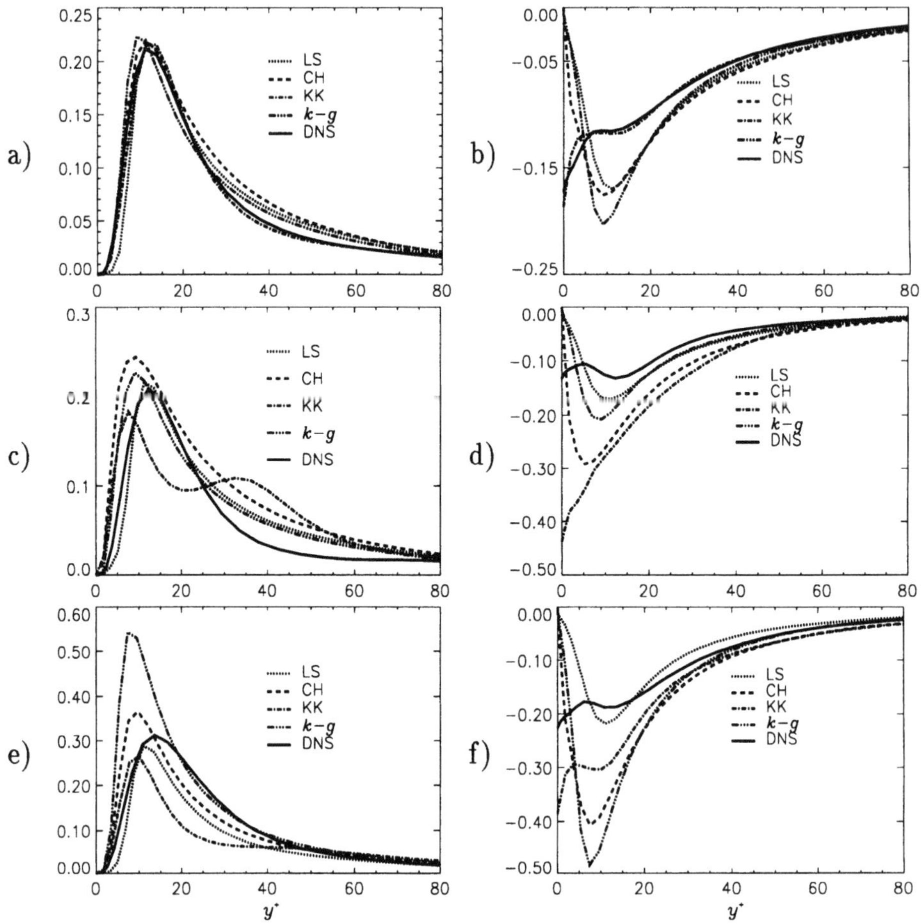


*Figure 5.* Selected terms in the kinetic energy budget for steady flow ((a) production, (b) dissipation), wall shear driven disturbance at t=0.3 ((c) production, (d) dissipation) and pressure driven disturbance at t=1.4 ((e) production, (f) dissipation).

and how it affects the flow. Although it may be a coincidence, it appears that models that have not used $y^+$ perform better than models that do.

The fundamental problem with all the models is most clearly shown in Figures 6 and 7 showing the angles between the velocity, shear stress and strain rate during the skewing process. From these figures it can be seen that there is a difference between the shear stress angle and strain rate angle. Figure 6 (a) shows that the models all collapse onto one curve for the shear stress/strain rate angle and this curve bears very little resemblance to the DNS shear stress or strain rate angles. The pressure driven disturbance flow is a less vigorously disturbed flow and at the earlier stage shown on Figure 6(b) the models agree quite closely with the DNS strain rate angle. At the later time shown on Figure 7 (b) the flow is more disturbed and the model

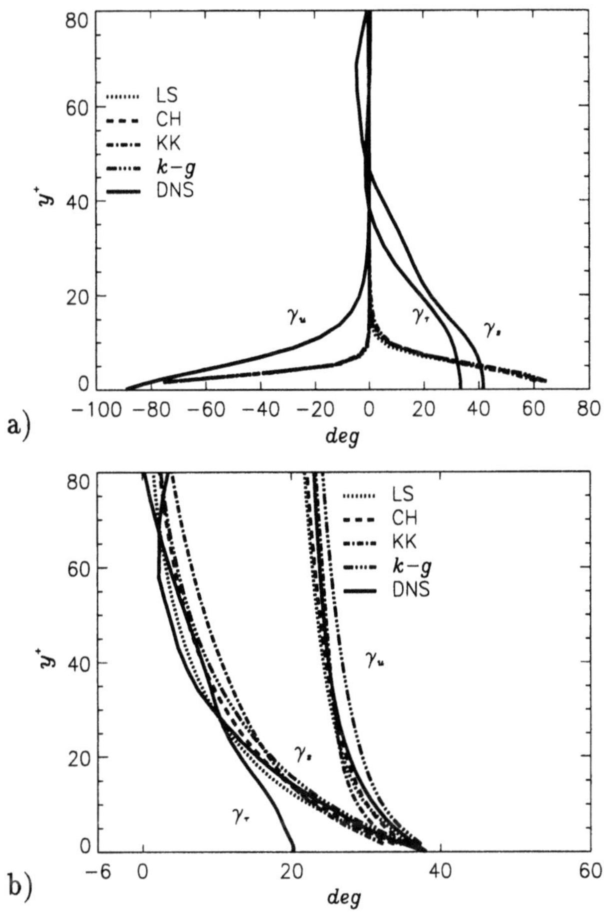


*Figure 6.* Flow angles of the velocity, shear stress and strain rate for (a) the wall shear driven ($t = 0.3$) and (b) the pressure driven ($t = 0.7$) cases.

results differ more widely. The LS model seems to underpredict the velocity and shear stress angles significantly. The CH, KK and $k - g$ models seem to perform slightly better. The models probably should be made to take account of the lag between the shear stress and strain rate if they are to model this type of flow more accurately.

## 5. Turbulence structure

The three dimensional structure of the turbulence is modified by the skew disturbance. However two-equation modelling does not include any information about turbulence structure. It is thus important to find any relationship between statistical quantities, used in two-equation modelling, and turbulence structure.

By observing the instantaneous flowfield from the DNS it is possible

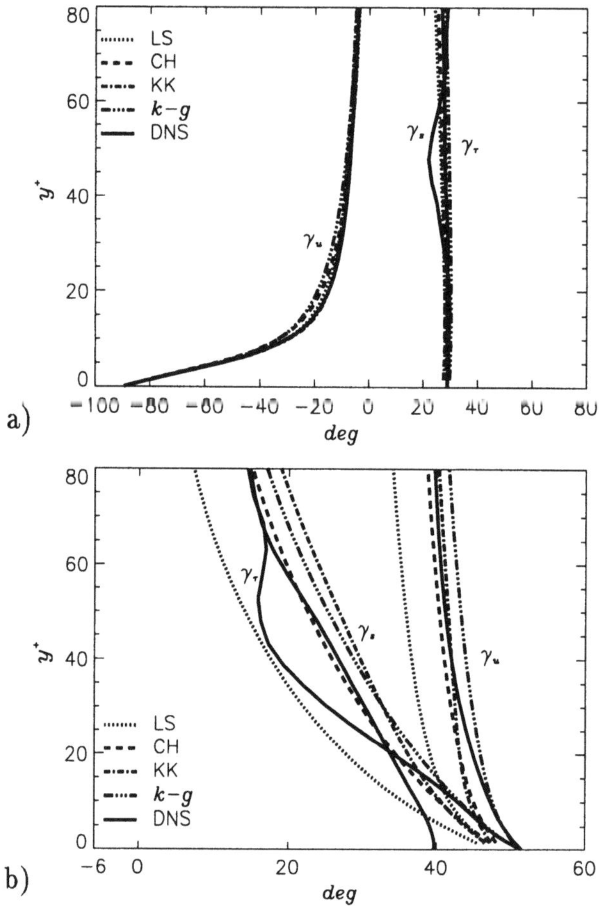


*Figure 7.* **Flow angles of the velocity, shear stress and strain rate for (a) the wall shear driven ($t = 2.0$) and (b) the pressure driven ($t = 1.4$) cases.**

to find out information about the turbulence structure. Figure 8 shows contours of streamwise velocity perturbation $u'$ at $y^+ = 9$ for the steady channel flow and during the wall shear driven disturbance. Figure 8 (a) shows the low speed streak structures (in black) as elongated coherent lumps aligned along the flow direction. In between the low speed streaks are faster moving high speed streaks. At $t = 0.3$ the turbulence intensity has reached a minimum (see Figure 4 (a)) and the resulting flow is shown in Figure 8 (b). It can be seen that the streak length has significantly reduced, they have become less organised and skewed. Looking at the later stage, $t = 2.0$, in Figure 8 (c) the streaks have reformed in the new direction and the flow structure is returning to the steady condition. From these observations it can be inferred that the coherence of the streak structures is related to the magnitude of the turbulence intensity. A key problem

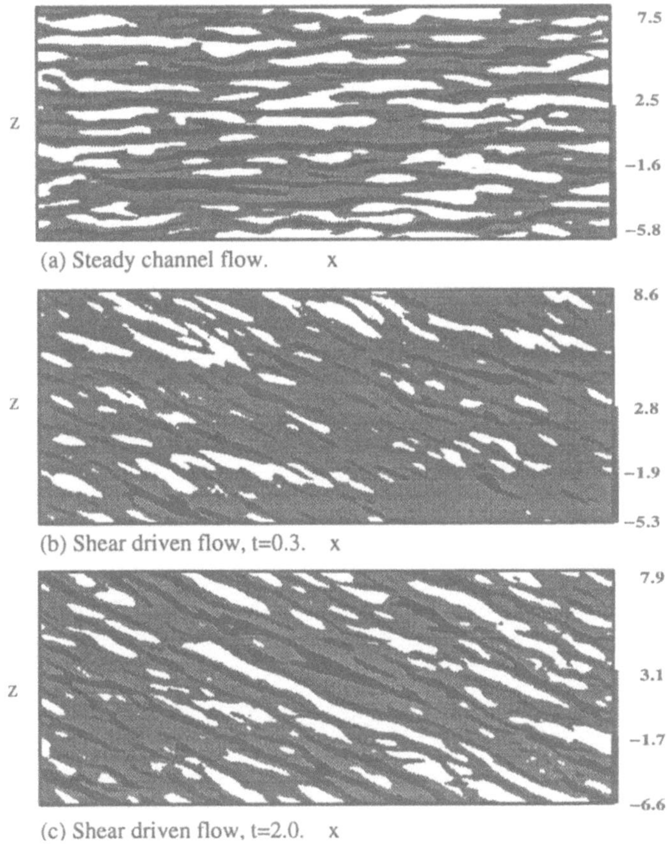


*Figure 8.* Streamwise velocity perturbations, $u'$, showing the response of the streak structures to the wall shear driven disturbance at $y_0^+ = 9$.

within two-equation turbulence modelling is that there is no involvement of the structures themselves so solutions must be found based on statistics. A more in depth analysis of the processes involved in this flow will be given elsewhere. However it is clear that these results are consistent with those observed by previous researchers.

## 6. Conclusions

A steady 2-D flow subjected to a spanwise disturbance shows an initial decrease in turbulent kinetic energy and other turbulence quantities, accompanied by a breakdown in the near wall low speed streak structures. There is also a lag between the angles of the shear stress and strain rate.

Of the four models investigated, the Kawamura & Kawashima (1995) and Chien (1982) models actually gave increases in turbulence intensity and friction velocity while the $k-g$ model (Kalitzin *et al.* (1996)) showed no initial change. The Launder & Sharma (1975) model underestimated the drop in turbulence intensity, but managed to show some of the correct trends.

Two-equation modelling is unable to show any lag between the strain rate and shear stress during a spanwise flow deformation as the eddy viscosity relation is non-directional. Correct modelling of these flows appears to require a directional eddy viscosity.

## 7. Acknowledgements

This research is funded by British Aerospace (operations) Ltd. (Sowerby Research Centre). Time on the Cray T3D is provided by EPSRC under grants GR/K 43902 and GR/K 43957.

## References

Bradshaw, P. and Pontikos, N.S.: Measurements in the turbulent boundary layer on an 'infinite swept wing', *J. Fluid Mech.* **159** (1985), 105-130.

Canuto, C., Hussaini, M.Y., Quarteroni, A. and Zang, T. : *Spectral Methods in Fluid Dynamics*,(1988), Springer.

Chien, K.-Y.: Predictions of channel and boundary layer flows with a low Reynolds-number turbulence model, *AIAA J.* **20** (1982), No.1, 33-38.

Coleman, G.N., Kim, J. and Le, A.-T.: A numerical study of 3-D boundary layers, *Tenth Symp. Turbulent Shear Flows* Penn. State, (1995), Aug 14-16.

Kalitzin, G., Gould, A.R.B. and Benton, J. : Application of two-equation tubrulence models to aircraft design, *34th Aerospace sciences meeting, AIAA-96-0327,* Reno,(1996), Jan 15-18.

Kawamura, H. and Kawashima, N.: A proposal of $k-\tilde{\epsilon}$ model with relevance to the near wall-turbulence, *Int. Symp. on Turbulence, Heat and Mass Transfer*, (1995), Lisbon.

Kleiser, L. and Schumann, U.: Treatment of incompressibility and boundary conditions in 3-D numerical spectral solutions of plane channel flows, In *Proc. 3rd GAMM Conf. on Numerical Methods in Fluid Mechanics*, (Ed. E.H. Hirschel), (1980), 165-173, Vieweg.

Launder, B.E. and Sharma, B.I.: Application of the energy-dissipation model of turbulence to the calculation of flow near a spinning disc, *Letters in Heat and Mass transfer*, **1** (1975), 131-138.

Moin, P., Shih, T.-H., Driver, D. and Mansour, N.N.: Direct numerical simulation of a three-dimensional turbulent boundary layer, *Phys. Fluids A* **2** (1990), 1846-1853.

Sandham, N.D. and Howard, R.J.A.: Statistics databases from DNS of fully developed turbulent channel flow, QMW-EP-1106, unpublished, 1995.

Sendstad, O. and Moin, P. : On the mechanics of 3-D turbulent boundary layers, *Eighth Symp. Turbulent Shear Flows* Munich, (1991), Sept 9-11.

# INFLUENCE OF A SPANWISE ROTATION UPON THE COHERENT-STRUCTURE DYNAMICS IN A TURBULENT CHANNEL FLOW

E. LAMBALLAIS, O. MÉTAIS AND M. LESIEUR
*LEGI-IMG*
*URA CNRS 1509*
*Institut National Polytechnique de Grenoble*
*and*
*Université Joseph Fourier, Grenoble*
*BP 53, 38041 Grenoble Cedex 9, France*

**Abstract.** We introduce a spectral-dynamic model, which is based on Kraichnan's spectral eddy-viscosity (Kraichnan, 1976) with an arbitrary spectral exponent $m$ for the kinetic-energy spectrum at the cutoff. This model is validated for LES of a turbulent channel flow at low and high Reynolds numbers. Afterwards we look at the influence of a rotation $\Omega$ of spanwise axis at a high rotation rate. We find that the following results, concerning the anticyclonic side of the channel and compared with the non-rotating case : a) the low- and high-speed streaks are reduced ; b) the longitudinal vorticity is increased ; c) the mean velocity profile is linear and equal to $2\Omega$; d) the organized vortices are more coherent, and preferentially inclined at 25° (instead of 45°) to the wall.

## 1. Introduction

We use large-eddy simulation (LES) to investigate the influence of spanwise rotation on a turbulent channel flow (see figure 1). Such a study finds numerous industrial applications in turbo-machinery and oceanography. In previous works, the rotating channel flow problem has been studied by means of direct numerical simulation (DNS) at low Reynolds number (Kristoffersen & Andersson, 1993; Lamballais *et al.*, 1996a,b). It has been



*J.-P. Chollet et al. (eds.), Direct and Large-Eddy Simulation II*, 225-236.

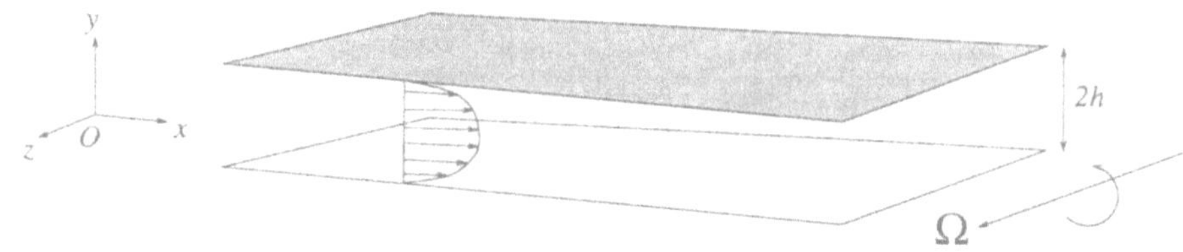


*Figure 1.* Schematic view of the rotating channel
$h$ : half-channel height ; $U_m$ : bulk velocity ; $Re = U_m 2h/\nu$ : Reynolds number
$Ro_g = 3\ U_m/\Omega 2h$ : global Rossby number
$x, y, z$ are respectively the streamwise, transverse and spanwise directions

shown that rotation can modify considerably the vortex topology of the channel flow. Previous authors (Kim, 1983; Tafti & Vanka, 1991; Piomelli & Liu, 1995) have already used LES successfully to predict the principal modifications brought by a weak rotation upon turbulent statistics. Note that it was Piomelli & Liu (1995), using a variant of the dynamic procedure applied to Smagorinsky's model, who obtained the best agreement with experimental data (Johnston *et al.*, 1972). Our aim in the present paper is to complete these studies by a study of vortex organization, and to reach higher rotation rates and Reynolds numbers.

## 2. The spectral-dynamic model

We use a dynamic version of the spectral eddy-viscosity model (Kraichnan, 1976; Chollet & Lesieur, 1981). Thus, the eddy-viscosity is normalized by $\sqrt{E(k_c,t)/k_c}$, where $E(k_c,t)$ is the kinetic-energy spectral density at the cutoff wave number $k_c$ :

$$\nu_t(k,k_c,t) = \nu_t^+(k/k_c)\left[\frac{E(k_c,t)}{k_c}\right]^{\frac{1}{2}} \quad . \tag{1}$$

Here, we have chosen to express $\nu_t^+(k/k_c)$ as Chollet (1985) :

$$\nu_t^+(k/k_c) = \nu_t^{+\infty}(1 + 34.5\ \mathrm{e}^{-3.03(k_c/k)}) \quad . \tag{2}$$

Until very recently, $\nu_t^{+\infty}$ (the value of the normalized eddy viscosity when $k \ll k_c$) was determined assuming that the spectrum at $k_c$ follows a Kolmogorov law

$$E(k) = C_k \varepsilon^{\frac{2}{3}} k^{-\frac{5}{3}}, \tag{3}$$

where $C_k$ is the Kolmogorov constant. Such an assumption yields $\nu_t^{+\infty} = 0.267$ if one takes $C_k = 1.4$.

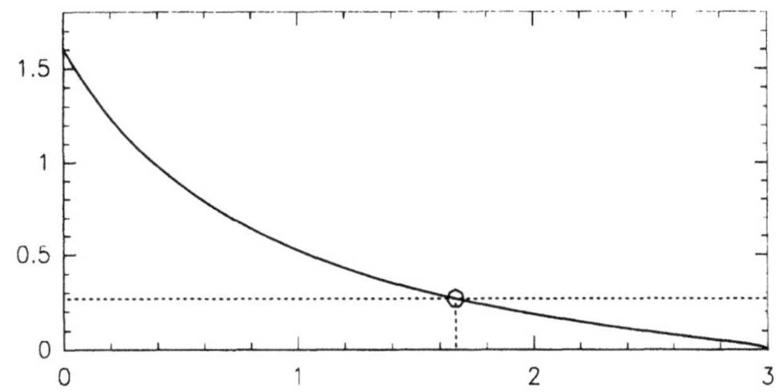


*Figure 2.* Variation of $\nu_t^{+\infty}$ with $m$, the slope of the kinetic-energy spectrum in the neighborhood of the cut-off wave number $k_c$. The symbol corresponds to the value obtained when considering a Kolmogorov law ($m = 5/3$).

We present in this paper an improvement of this model, in order to take into account spectra decreasing differently than $-5/3$ at the cutoff. Instead of a Kolmogorov law, we assume that the cutoff $k_c$ lies in the middle of an $E(k) \propto k^{-m}$ kinetic-energy spectrum. Métais & Lesieur (1992) proposed, on the basis of non-local spectral expansions within the frame of the E.D.Q.N.M theory, that $\nu_t^{+\infty}$ should be expressed as

$$\nu_t^{+\infty} = \frac{0.31}{C_k^{\frac{3}{2}}} \frac{(5-m)}{(m+1)} (3-m)^{\frac{1}{2}} \tag{4}$$

for $0 < m \leq 3$. We propose to use the expression (4) in our subgrid-scale model, and to take a zero eddy-viscosity for $m > 3$. The sensitivity of $\nu_t^{+\infty}$ to $m$ is illustrated in figure 2. It shows clearly that $\nu_t^{+\infty}$ decreases strongly when $m$ goes from 0 to 3. Note that when $m > 2.2$, $\nu_t^{+\infty}$ computed with (4) is reduced by more than 50 % with respect to its value for $m = 5/3$. This behaviour is particularly interesting for the two following cases :

- when the Reynolds number is not sufficient to permit the development of an inertial zone,
- for flows in the near-wall region, where spectra are steeper than a Kolmogorov one, independently of the Reynolds number.

In the following, we will consider flows which will present simultaneously these two characteristics.

## 3. Validation of the Model

We first present LES of the non-rotating turbulent channel flow at low Reynolds number ($Re = 6666$). We use mixed spectral (in the streamwise and spanwise directions) and compact (in the transverse direction)

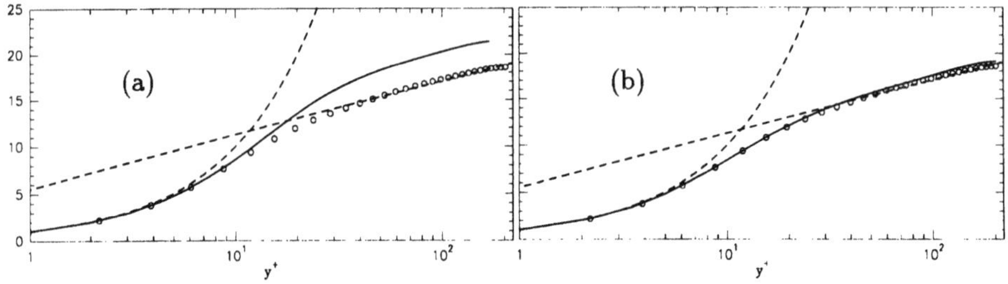


*Figure 3.* Mean velocity profiles in wall units.
—, present, $Re = 6666$ ; - - - , $u^+ = y^+$ and $u^+ = 2.5 \ln y^+ + 5.5$ ;
o, Piomelli (1993), $Re = 6500$
(a) : $m = 5/3$ ($h^+ = 181$) ; (b) : dynamic evaluation of $m(y,t)$ ($h^+ = 204$)
Size of the computational domain . $(L_x, L_y, L_z) = (2\pi h, 2h, \pi h)$
Grid : 64 × 65 × 32

schemes. In order to determine $E(k_c, t)$, we have chosen to use 2D spectra $E(k_{2D}, y, t)$, computed by averaging in $xz$ planes (parallel to the walls). The spectral formulation is used only in horizontal directions ($\nu_t(k_{2D}, k_c, t)$)

To emphasize the interest of considering the dynamically-adjusted eddy-viscosity, we compare now the turbulent statistics obtained from two different LES :

- a first one, where $m$ is maintained at the fixed value : $m = 5/3$
- a second one, where $m$ is computed using $E(k_{2D}, y, t)$ through a least-square fit between the highest wave numbers (for results presented in this paper, the spectral range $k_c/2 < k_{2D} < k_c$ have been used).

Mean velocity and turbulent intensity profiles are presented on figures 3, 4 for these two simulations. One observes a clear improvement of these statistics in the second case, where the results are in good agreement with a previous calculation using a dynamic model (Piomelli, 1993). For instance,

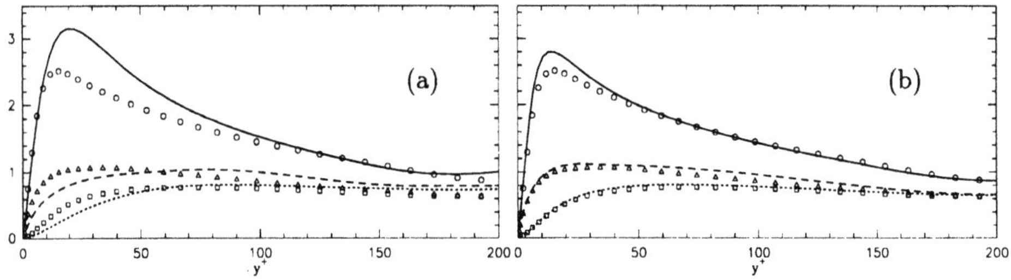


*Figure 4.* R.m.s. velocity fluctuations in wall units :
—, o, $\sqrt{\langle \overline{u}'^2 \rangle}/u_\tau$; ...,□, $\sqrt{\langle \overline{v}'^2 \rangle}/u_\tau$; - - -, △, $\sqrt{\langle \overline{w}'^2 \rangle}/u_\tau$; ($h^+ = 204$).
lines : present, $Re = 6666$ ; symbols : Piomelli (1993), $Re = 6500$
(a) : $m = 5/3$ ($h^+ = 181$) ; (b) : dynamic evaluation of $m(y,t)$ ($h^+ = 204$)
Same computation as in figure 3

(a)

(b)

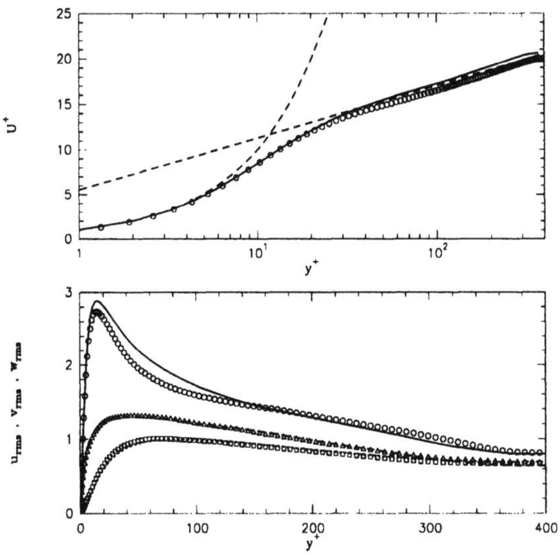

*Figure 5.* (a) : Mean velocity profile ; (b) : R.m.s. velocity fluctuations (wall units) : —, o, $\sqrt{\langle \overline{u}'^2 \rangle}/u_\tau$; ...,□, $\sqrt{\langle \overline{v}'^2 \rangle}/u_\tau$; - - -, △, $\sqrt{\langle \overline{w}'^2 \rangle}/u_\tau$.
lines : present ($h^+ = 389$, $Re = 14000$) ;
symbols : Antonia *et al.* (1992), $h^+ = 395$, unfiltered DNS results.

our new model permits a better prediction of the wall turbulent shear-stress which is close to experimental correlation (Dean, 1978). Thus, the deviation from the empirical law $h^+ = 0.0956\, Re^{7/8}$ is less than 4% with the new model instead of 15% when $m = 5/3$ is assumed. The intensity and the location of the maximum in the streamwise turbulence fluctuations are also in better agreement with the accepted values $u'^+_{max} \approx 2.7$ and $y^+_{max} \approx 12$. We emphasize the fact that none of these simulations uses damping function to reduce the near-wall eddy-viscosity. Thus, we conclude that the correction given by the expression (4) seems to be sufficient to obtain a good near-wall behaviour for statistics presented here, without use of any "hand-tuned" constant in the model or self-calibration procedure as in dynamic models in physical space.

## 4. Application to a Rotating Channel

We present now LES of a turbulent channel flow subjected to spanwise rotation at a higher Reynolds number ($Re = 14000$). The case without rotation is presented for comparison. The Rossby number, 6, is low compared with most experimental and numerical studies. For both simulations, we use the spectral-dynamic eddy-viscosity. In the non-rotating case, turbulent statistics are in very good agreement with DNS results of Antonia *et al.* 1992 (see figure 5).

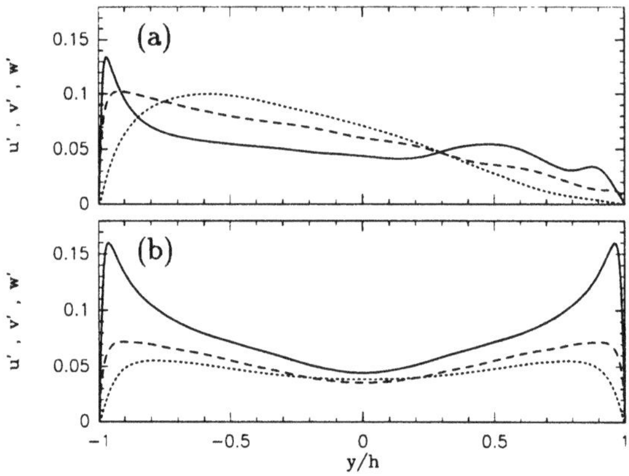


*Figure 6.* R.m.s of the velocity fluctuations ($Re = 14000$) : —, $\sqrt{\langle \overline{u}'^2 \rangle}/U_m$; ..., $\sqrt{\langle \overline{v}'^2 \rangle}/U_m$; - - -, $\sqrt{\langle \overline{w}'^2 \rangle}/U_m$; (a) $Ro_g = 6$ ; (b) $Ro_g = \infty$.

### 4.1. STATISTICAL RESULTS

Figure 6 shows profiles of the r.m.s velocity fluctuations for the two simulations considered here (non-rotating and rotating cases). A tendency similar to what was observed previously in DNS (Lamballais *et al.*, 1996a,b; Kristoffersen & Andersson, 1993; Métais *et al.*, 1995) is recovered : when rotation is applied, the cyclonic region of the flow is stabilized while turbulence remains very active in the anticyclonic region. The term cyclonic (anticyclonic) designates a situation where the mean vorticity vector $\vec{W} = (0, 0, -d\langle \overline{u} \rangle / dy)$ is parallel (antiparallel) to the rotation vector $\vec{\Omega} = (0, 0, \Omega)$. One sees that, in the anticyclonic region, rotation reduces the longitudinal velocity fluctuations and increases the transverse ones, which indicates a tendency for the low-and high-speed streaks of the non-rotating case to transform into transverse and spanwise oscillations. The study of this mechanism, already observed via DNS (Lamballais *et al.*, 1996b), will be the subject of a future work.

An important characteristic of the rotating turbulent channel flow is the existence of a linear zone in the mean velocity profile, where the (constant) mean velocity gradient is equal to $2\Omega$. This particular behaviour of the mean velocity profile is clearly observed in our LES (see figure 7). As we have shown previously with the aid of DNS at low Reynolds number (Lamballais *et al.*, 1996a,b), such a range is dynamically very important because it is associated with longitudinal vorticity production by absolute-vortex stretching mechanisms. The effect of this stretching can be observed in figure 8, where the $x$-component of the r.m.s vorticity fluctuations clearly

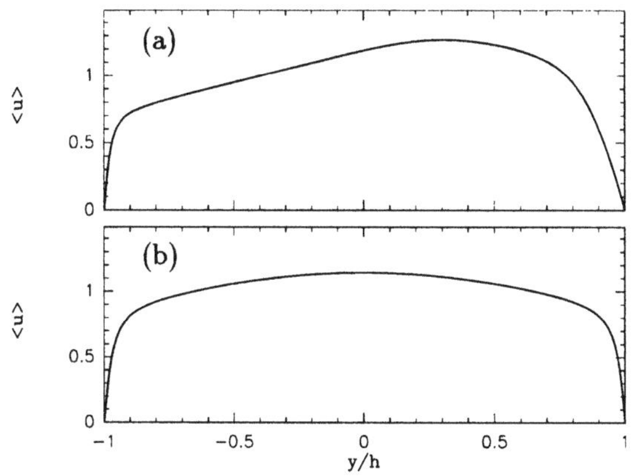


*Figure 7.* Mean velocity profiles $\langle u \rangle / U_m$ ($Re = 14000$) : (a) $Ro_g = 6$ ; (b) $Ro_g = \infty$.

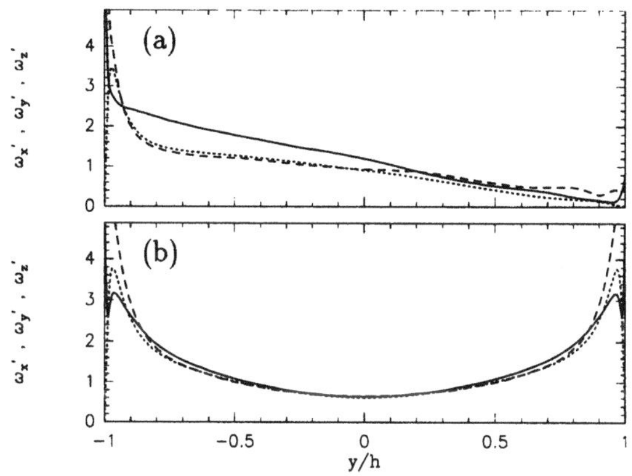


*Figure 8.* R.m.s of the vorticity fluctuations ($Re = 14000$) : : —, $\sqrt{\langle \overline{\omega}_x'^2 \rangle}/(U_m/h)$; ..., $\sqrt{\langle \overline{\omega}_y'^2 \rangle}/(U_m/h)$; - - -, $\sqrt{\langle \overline{\omega}_z'^2 \rangle}/(U_m/h)$; (a) $Ro_g = 6$ ; (b) $Ro_g = \infty$

dominates the two other components in the region where the mean velocity profile is linear. The ability for LES technique to account for such phenomena already observed via DNS is an important point.

### 4.2. COHERENT-STRUCTURE DYNAMICS

In figure 10, we compare isosurfaces of the vorticity modulus $\omega = 4.5\ U_m/h$ obtained for each simulation. Note first that we can observe numerous vortex structures (tube-like objects of high vorticity) in the two LES. The presence of such well organized vortical structures indicates that the co-

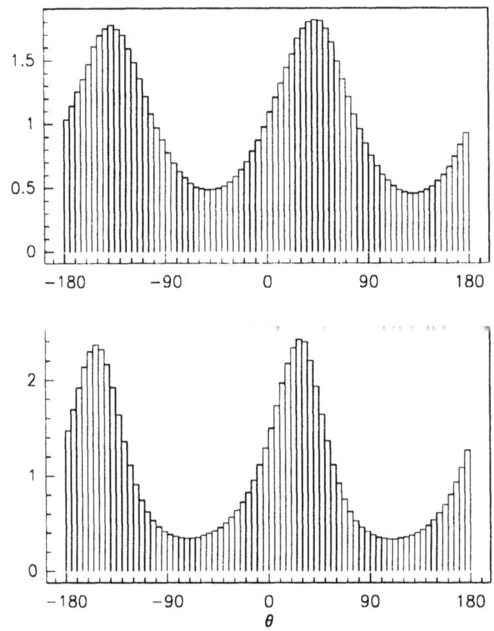


*Figure 9.* P.d.f of $\theta$ ($Re = 14000$) at $y = -0.6\ h$: $Ro_g = \infty$ (top) ; $Ro_g = 6$ (bottom).

herence of motion is preserved. We have verified that, if no subgrid-scale model is used (or a model with insufficient dissipation), the vorticity field become unrealistic (spurious oscillations, no vortex structure), while the statistics quality (mean velocity and turbulent intensities) is less dramatically affected.

In the rotating case, the modification of the vortex topology in the anticyclonic region is similar to than for lower Reynolds number (Lamballais *et al.*, 1996a,b) : compared with the case without rotation, the vortical structures (hairpin vortices) are more inclined with respect to the wall and more organized. This double tendency can be quantified by examining the p.d.f of the angle $\theta = tan^{-1}(\omega_y/\omega_x)$, representing the angle formed by the wall and the projection of the vorticity vector on the $(x, y)$-plane. Each contribution to the distribution is weighted by the magnitude of the vorticity projection, in order to give more weight to high vorticity regions in the statistics. It can be observed on figure 9 that in a $y$-location where hairpin vortices are present ($y = -0.6\ h$), the p.d.f peaks in the rotating case shift to lower values (more inclined vortex structure) and become narrower (more organized motions), in a similar manner to DNS results (Lamballais *et al.*, 1996a,b). Note that the most probable values obtained in DNS (at

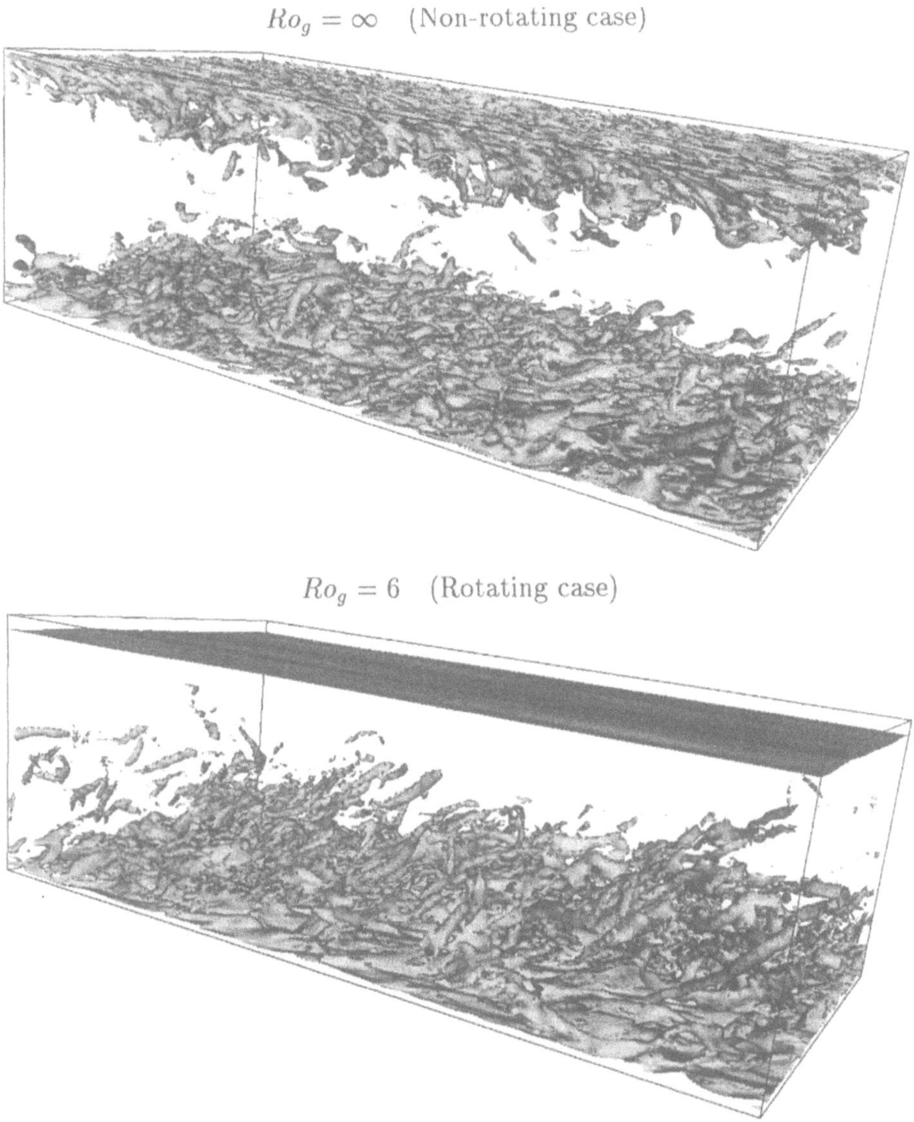


*Figure 10.* Isosurfaces of vorticity modulus $\omega = 4.5\ U_m/h$ ($Re = 14000$)
Size of the computational domain : $(L_x, L_y, L_z) = (2\pi h, 2h, \pi h)$
Grid : $128 \times 97 \times 64$

low Reynolds number) and in LES are identical (45° without rotation and nearly 25° for $Ro_g = 6$).

One remarks also, with respect to lower Reynolds number DNS or LES, a very intense small-scale variability showing that our subgrid-scale model is able to produce high Reynolds number effects.

## 5. Conclusion

Our general conclusion is that the spectral-dynamic model is able to capture the principal modifications of the flow due to rotation. We have shown that LES preserves the coherence of the turbulent motion and can give reliable information on the influence of the rotation on the dynamics of the vortical structures. These LES ($\approx$ 10 hours CPU time on a C90) are hundred times cheaper than DNS for the same Reynolds number. Furthermore, and because Reynolds-averaged Navier-Stokes calculations have some difficulties to reproduce accurately strong rotation effects in channel flows, LES is today the only technique available for predicting this flow at high Reynolds number.

## 6. Acknowledgments

We would like to thank J. Ferziger, P. Comte and J. Silvestrini for fruitful discussions. We are indebted to P. Bégou for the boost he gave to our numerical codes. Computing time was freely provided by IDRIS, the CNRS computing center.

## References

Antonia, R. A., Teitel, M., Kim, J., & Browne, L. W. B. 1992. Low-Reynolds-number effects in a fully developed turbulent channel flow. *J. Fluid Mech.*, **236**, 579–605.

Chollet, J. P. 1985. Two-point closure used for a sub-grid scale model in large eddy simulations. *Pages 62–72 of:* Bradbury, L. J. S. (ed), *Turbulent Shear Flows*. Springer.

Chollet, J. P., & Lesieur, M. 1981. Parameterization of small scales of the three-dimensional isotropic turbulence utilizing spectral closures. *J. Atmos. Sci.*, **38**, 2747–2757.

Dean, R. B. 1978. Reynolds Number Dependence of Skin Friction and Other Bulk Flow Variables in Two-Dimensional Rectanguler Duct Flow. *J. Fluids Engine.*, **100**, 215–223.

Johnston, J. P., Halleen, R. M., & Lezius, D. K. 1972. Effects of spanwise rotation on the structure of two-dimensional fully developed turbulent channel flow. *J. Fluid Mech.*, **56**, 533–557.

Kim, J. 1983. The effect of rotation on turbulence structure. *Pages 6.14–6.19 of: Proc. 4th Symp. on Turbulent Shear Flows.*

Kraichnan, R. H. 1976. Eddy viscosity in two and three dimensions. *J. Atmos. Sci.*, **33**, 1521–1536.

Kristoffersen, R., & Andersson, H. I. 1993. Direct simulations of low-Reynolds-number turbulent flow in rotating channel. *J. Fluid Mech.*, **256**, 163–197.

Lamballais, E., Lesieur, M., & Métais, O. 1996a. Effects of spanwise rotation on the vorticity stretching in transitional and turbulent channel flow. *Int. J. Heat and Fluid Flow*, **17**(3), 324–332.

Lamballais, E., Lesieur, M., & Métais, O. 1996b. Influence of a solid-body rotation upon coherent vortices in a channel. *C. R. Acad. Sci.*, 95–101. Série II b.

Métais, O., & Lesieur, M. 1992. Spectral large-eddy simulation of isotropic and stably stratified turbulence. *J. Fluid Mech.*, **239**, 157–194.

Métais, O., Flores, C., Yanase, S., Riley, J. J., & Lesieur, M. 1995. Rotating free shear flows Part 2 : Numerical simulations. *J. Fluid Mech.*, **293**, 41–80.

Piomelli, U. 1993. High Reynolds number calculations using the dynamic subgrid-scale stress model. *Phys. Fluids A*, **5**, 1484–1490.

Piomelli, U., & Liu, J. 1995. Large-eddy simulation of rotating channel flows using a localized dynamic model. *Phys. Fluids A*, **7**(4), 839–848.

Tafti, D. K., & Vanka, S. P. 1991. A numerical study of the effects of spanwise rotation on turbulent channel flow. *Phys. Fluids A*, **3**(4), 642–656.

# DIRECT NUMERICAL SIMULATION OF TURBULENT CHANNEL FLOW WITH PARTICLES

B.A. VAN HAARLEM AND F.T.M NIEUWSTADT
*Laboratorium voor Aëro- en Hydrodynamica,*
*Delft University of Technology,*
*Rotterdamseweg 145, 2628 AL Delft, The Netherlands*

**Abstract.** A direct numerical simulation (DNS) of a turbulent channel flow with a Reynolds number of 360 (based on channel height and friction velocity) has been performed. Particles with dimensionless response times of 10, 100 and 200 were introduced in the flow. For each particle type 30,000 particles were tracked for a number of eddy turnover times. The computed particle statistics are compared with the solution of Swailes and Reeks (1994) of the kinetic equation proposed by Reeks (1990) as a model for this flow.

## 1. Introduction

Particle-laden turbulent flows can be encountered in a wide variety of applications ranging from process engineering and combustion to geophysical flows. Interest from such diverse fields has led to many different approaches to the modelling of particle-laden flow (for a review, see Elgobashi 1994).

Here we are interested in the case of very dilute systems. The effect of the particles on the flow can then be assumed to be negligible and the particles as non-interacting. This means that the fluid- and the disperse- or particle-phase are de-coupled. The turbulent flow field is simulated by Direct Numerical Simulation (DNS). For the particles we use a Lagrangian approach in which individual particles are tracked in the flow field by integrating their equation of motion. This method has been used with success in the past (e.g. McLaughlin, 1989 and Elgobashi, 1992) and has proven to be a useful way of simulating particle laden turbulent flows with a minimal amount of modelling.



*J.-P. Chollet et al. (eds.), Direct and Large-Eddy Simulation II,* 237-246.

However the use of DNS restricts this approach to simple geometries and low Reynolds numbers. For higher Reynolds numbers and more complex flows a higher degree of modelling will be necessary. A promising model is the statistical description of the dispersed phase by means of a kinetic equation as proposed by Reeks (1990). Solving this equation yields the ensemble averaged joint-pdf (probability density function) for particle positions and velocities. With this information all quantities which depend on the particle concentrations and velocities can be determined.

The motivation for doing the work reported here is to test the merits of the kinetic equation model by making a comparison of this equation with results from particles tracked in a DNS. The case used here is the dispersion of particles in turbulent channel flow. A solution of the kinetic equation for this case has been obtained by Swailes and Reeks (1994). In the following sections we will discuss some aspects of this solution and also the details of our simulation before presenting our results and comparisons.

## 2. The kinetic equation solution

Starting point for the derivation of the kinetic equation (Reeks, 1990) is the equation of motion for a single particle. Apart from the deterministic forces like gravity, the forces due to the random turbulent motion of the fluid phase are modelled as a stochastic forcing term. The kinetic equation that is derived for an ensemble of particles takes the form of a modified Fokker Planck equation for the particle pdf.

Swailes and Reeks (1994) solve the kinetic equation in a form normalised by the particle response time $\tau$ and the particle mean free path $\ell$. The particle response time is defined as,

$$\tau = \frac{2}{9}\frac{a^2\rho}{\nu} \tag{1}$$

with $a$ the particle radius, $\rho$ the fluid-particle density ratio and $\nu$ the fluid kinematic viscosity. The response time can be interpreted as the time-scale which is needed for the particle to respond to sudden changes in the ambient flow. The mean free path is defined as,

$$\ell = \tau v_p\text{'} \tag{2}$$

with $v_p$' the particle rms velocity. The mean free path is the length-scale over which the particle is insensitive to inhomogeneities of the flow.

To solve the kinetic equation for the case of turbulent channel flow certain simplifying assumptions are made. Firstly the particle pdf is assumed to be stationary and a function of the distance from the wall $z$ and the particle velocity $v_p$ only. A second simplification is the fact that only particles with high inertia are considered, i.e. large values of $\tau$ with respect to the turbulence time scale. This implies that the fluid turbulence may be modelled as homogeneous for all $z$. Any spatial inhomogeneity of the particle pdf is thus a result of the imposed boundary conditions. A final simplification is the fact that the particles experience only a drag force, which is modelled as Stokes drag. We note that the latter two assumptions are inconsistent. For Stokes drag to be valid $Re_p = 2a|v_p - v_f|/\nu < 1$ should hold. However the term $|v_p - v_f|$, which is the magnitude of the particle-fluid velocity difference, increases with increasing particle inertia. Thus, the assumption of Stokes drag is in general invalid for heavy particles.

The boundary conditions used are a constant flux of particles with a prescribed velocity distribution at $z = D/2$ ($D$ is the full channel hight). The boundary conditions for particle velocities that are imposed at the wall range from adsorption to perfect reflection.

## 3. Description of the simulation

For the DNS of the flow field the Navier Stokes equation and the continuity equation in the incompressible form are discretised using a finite volume method on a staggered grid. In the streamwise and spanwise direction the grid used is uniform. Normal to the wall we use a slightly non-uniform grid. Towards the wall the grid is refined in such a way that we have three grid points in the viscous sublayer for each velocity component. Time integration of the discretised equations is done using the Adams Bashforth scheme. We use the pressure correction method for solving the discretised equations. At the wall we use no-slip boundary conditions; the boundary conditions in the streamwise and spanwise direction are periodic. The coordinate system used and the size of the domain is illustrated in Fig. 1. The Reynolds number of the simulation $Re = u_* D/\nu$ is 360 ($u_*$ is the wall shear stress velocity). The number of grid points used in is $225 \times 120 \times 156$ in the $x, y$ and $z$ direction respectively. The grid spacing is 8 and 9 wall units in the $x$ and $y$ direction, while in the $z$ direction it varies between 1.5 and 2.5 wall units. The Kolmogorov length scale based on the volume averaged dissipation is approximately 2 wall units so that we may conclude that all relevant flow scale are sufficiently resolved.

The simulation was started from random initial conditions and reached statistical equilibrium in approximately $9t^*$ ($t^* = tu_*/D$), at which point 50 data fields were collected at an interval of $0.1t^*$. The statistics calculated from these data were in good agreement with the DNS results of Kim *et al.* (1987).

Once the channel flow simulation has reached equilibrium, we start with

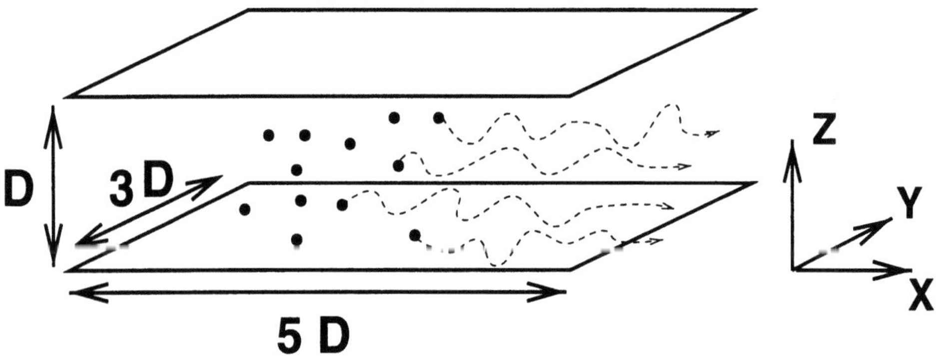


*Figure 1.* Geometry of the simulation. The mean flow is in the $x$-direction.

the particle tracking. Like Swailes and Reeks (1994) we assume that the particles experience only a Stokes drag force. The particle equation of motion for each particle thus reads,

$$\frac{d\underline{v}_p}{dt} = \frac{1}{\tau}(\underline{v}_f - \underline{v}_p)$$
$$\frac{d\underline{x}_p}{dt} = \underline{v}_p \tag{3}$$

with $\underline{x}_p$,$\underline{v}_p$ the particle position and velocity and $\underline{v}_p$ the fluid velocity at the particle position. The latter value is calculated by means of a quadratic interpolation over the 27 closest gridpoints. The time integration of these particle equations is performed using the Adams Bashforth scheme. As for the flow, periodic boundary conditions are used for the particles in the streamwise and spanwise direction. At the wall we use adsorbing boundary conditions. The omission of the terms describing added mass, lift and Basset history force which appear in the full equation of motion (see e.g Maxey and Riley, 1983) from equation (3) is justified by the assumption of high particle inertia. Note that there is no gravity in the equations.

Three different types of particles with dimensionless response times $\tau^+ = \tau u_*^2/\nu = 10, 100$, and 200 respectively, are used in the simulation. In choosing the particle response times of the particles to be tracked we wish to satisfy two constraints. Firstly, they should satisfy the high inertia limit, which implies particles with a response time greater than the integral time scale of the flow. Secondly, the particles should still be influenced in some way by the turbulence for the results to be of interest. This latter condition is met by demanding that the particle mean free path does not exceed the channel width $D$. This leads to the following restriction on the choice of the particle response time,

$$\mathcal{T}^+ < \tau^+ < \mathrm{Re}$$

with $\mathcal{T}^+$ the integral time scale of the turbulence in wall units. For the latter restraint ($\tau^+ < Re$) we have assumed $u_*/v_p' \approx 1$, which is a valid approximation for the particles considered here. The particles with $\tau^+ = 100, 200$ satisfy both these conditions, the particles with $\tau^+ = 10$ are included for comparison. For each particle type 30,000 particles are homogeneously distributed over the flow domain at the start of the simulation. All particles have zero initial velocity. The simulation is then run for $2.5t^*$. In wall units this is equal to 910 $t^+$ ($t^+ = tu_*^2/\nu$), thus approximately 4.5 times the response time of the heaviest particles. During this time the particles have become independent of their initial conditions. Thereafter the simulation is run for another 1.1 $t^*$ ($\approx 400t^+$) during which the particle positions and velocities are sampled at an interval of approximately 0.3 $t^+$

## 4. Results

To calculate the statistical quantities presented in this section the entire domain (shown in Fig. 1) is divided into 200 bins which are equally spaced in the $z$-direction and which span the entire $x, y$-plane. Averaging is performed over all particles in a certain bin and over all time samples.

Before discussing the results we wish to point out that we are faced with a difficulty when comparing results from our simulation with the solution of the kinetic equation. This difficulty is caused by the fact that the solution of the kinetic equation represents a steady state resulting from a constant flux of particles at $z = D/2$. However in our simulation the total number of particles will decrease due to deposition at the wall. Only if new particles are introduced during the simulation will it be possible to mimic the steady state solution of the kinetic equation. This however is not possible as for these new particles initial conditions, which are unknown, will have to be specified. To see to what extent our results may be compared to a steady state solution we have calculated the mean particle flux as a function of the distance from the wall ($z$) during the sampling interval. The flux calculated for the $\tau^+ = 100$ particles shows a linear dependence on $z$, increasing from the centre of the channel to the wall. For the particles with $\tau^+ = 200$ the behaviour of the flux is similar apart from a decrease of the flux close to

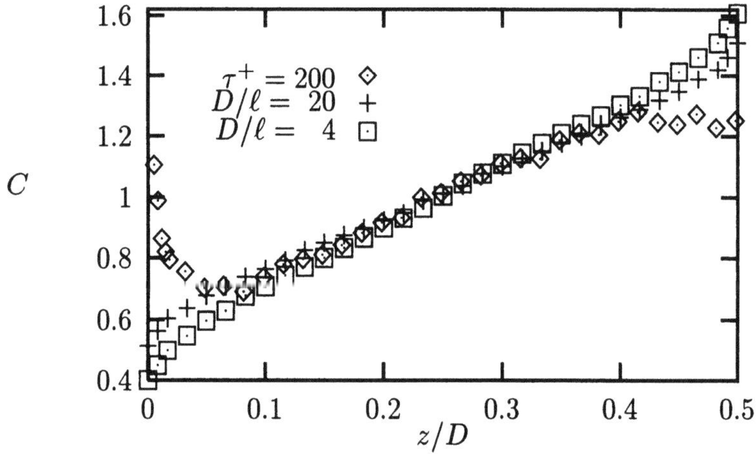


*Figure 2.* Particle concentration as a function of the distance from the wall. Results of the simulation are shown for $\tau^+ = 200$ which corresponds to $D/\ell \approx 6$. Results of the kinetic equation are shown for $D/\ell = 4, 20$

the wall. A linear dependence of the particle flux on $z$ indicates that the concentration gradient $(\partial C/\partial z)$ is independent of time. Hence we may say that (apart from a small region close to the wall for the heaviest particles) the shape of the concentration distribution is preserved during sampling, thus justifying that at least a qualitative comparison with the steady state is possible.

In Figure 2 we show the particle concentration as a function of the distance from the wall for the heaviest particles used in our simulation, $\tau^+ = 200$. Using the centre plane rms particle velocity we can determine the particle mean free path from equation (2). For the case of $\tau^+ = 200$ we find $D/\ell \approx 6$. Also shown are results of Swailes and Reeks (1994) for various values of $D/\ell$. We see that the kinetic equation result for $D/\ell = 4$ compares well with our results for $D/\ell \approx 6$ over a large portion of the domain. Near the wall $(z/D = 0)$ our simulation shows a slight increase of the particle concentration. This may be attributed due to a turbophoretic effect, i.e. migration of particles to regions of lower turbulence intensity (see e.g. Reeks 1983) It is interesting to see that these heavy particles are still influenced by inhomogeneity of the near-wall turbulence. On the other hand it is no surprise that the kinetic equation does not capture this effect as it assumes homogeneity of the turbulence up to the wall. However our results

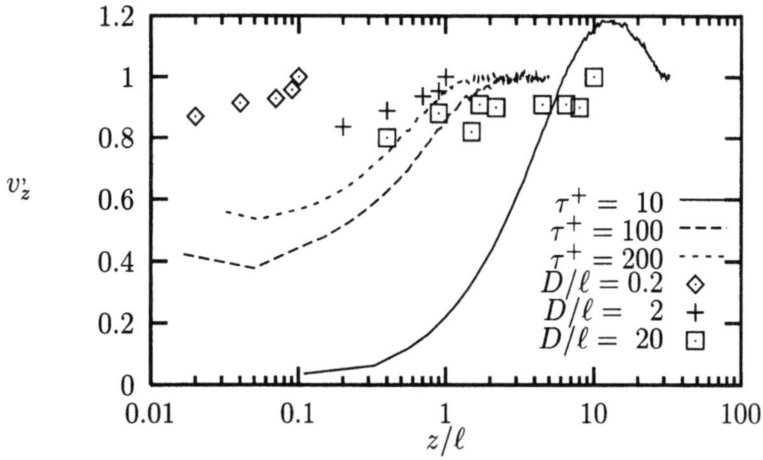


*Figure 3.* Root mean square values of the normal to the wall velocity fluctuations normalised by the centre plane value. Lines represent the results from our simulation while points represent the results from the kinetic equation. The particle response times of $\tau^+ = 10, 100, 200$ correspond to $D/\ell \approx 66, 10, 6$

show that this assumption may not be justified. For the light particles with $\tau^+ = 10$ (not shown in the figure) we find a very strong segregation leading to near-wall particle concentrations of roughly ten times the centre plane value. The particles with $\tau^+ = 100$ show a concentration profile that differs only very slightly from that shown for $\tau^+ = 200$.

In Figure 3 we show the root-mean-square values of the particle-velocity fluctuations for all three types of particles used in the simulation. For the two heavier particle types we see that the rms values are roughly constant over a large part of the domain starting at the centre. Towards the wall the velocity fluctuations are strongly damped. The light particles, $\tau^+ = 10$ which are also shown in the figure follow the shape of the fluid rms but with slightly lower values. The results of the kinetic equation are also shown in the figure for three different values of $D/\ell$. The difference in values of $D/\ell$ makes a comparison of the results difficult. However we can clearly see that the particle rms values are not damped enough in the near-wall region. Furthermore the roughly constant value of the rms in the neighbourhood of the centre plane found in the simulations is not found in the kinetic equation result, where a slight increase of the velocity fluctuations towards the channel centre is observed. We also note that the lighter particles show a local maximum in the rms values which is a result of the turbulence

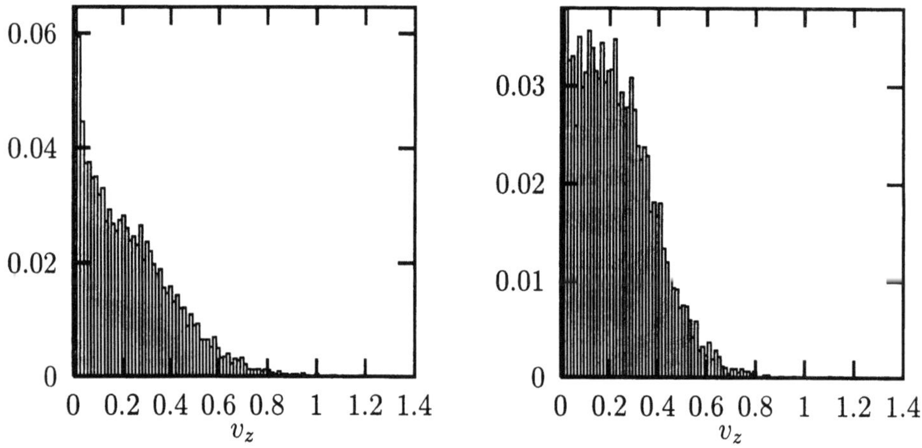


*Figure 4.* Probability distribution for the normal to the wall velocity $v_z$ (normalised by the wall shear stress velocity) of the particles depositing during the sampling time. The left figure shows the distribution for $\tau^+ = 100$, on the right the distribution for $\tau = 200^+$

inhomogeneity and therefore this effect cannot be captured in the kinetic equation solution.

The former results indicate that even for the high inertia particles that we use in our simulation there is still some influence due to the presence of the layer adjacent to the wall. As a result the particles do not travel through this layer totally undisturbed. For the particles with $\tau^+ = 100$ this effect is of course stronger than for the $\tau^+ = 200$ particles. This can be seen clearly in Fig. 4., where we show the probability distribution of the velocity normal to the wall for particles that have deposited at the wall. Results are given for the two different heavy particle types $\tau = 100, 200$. Both pdf's show a rather wide distribution with a strong contribution of particles with a small velocity. Thus a large part of the particles experience a strong damping resulting in small velocities near the wall; this effect is neglected in the kinetic equation solution. The difference between the two velocity distributions is the relatively strong peak near zero velocity for the particles with $\tau^+ = 100$ and thus an influence of the inhomogeneity of turbulence near the wall.

From our data we have also calculated the deposition rate of particles, $k$, which relates the deposition flux to the particle concentration in the following way

$$\frac{dN_d}{dt} = k\langle C\rangle \tag{4}$$

with $N_d$ the number of depositing particles per unit area and $\langle C\rangle$ the volume average concentration of particles. We find for the particles with $\tau^+ = 100$ and 200 the values of $k^+ = k/u_* = 0.15$ and 0.17 respectively. These values are somewhat higher than the approximately constant value of 0.09 that Swailes and Reeks (1994) find for the particles with these response times. So even though the particles in our simulation experience stronger damping in the near-wall region the deposition rate is higher than the value obtained from the kinetic equation. This effect should probably be attributed to the dynamics of the turbulence in the near-wall region.

## 5. Conclusions

We have performed a direct numerical simulation of turbulent channel flow in which the trajectories of high inertia particles have been calculated. The results have been compared with a solution to the kinetic equation (Swailes and Reeks 1994). The comparison is made somewhat difficult by the fact that only a steady state solution to the kinetic equation is available. Nevertheless, our results are approximately similar to a steady state. Results from our simulation indicate that high inertia particles with a mean free path appreciably smaller than the channel height are still influenced by inhomogeneity of the fluid turbulence at the wall. This contradicts one of the assumptions used to find a solution to the kinetic equation. Deposition rates have also been compared and were found to differ considerably. Contradictory to what may be expected the kinetic equation under-predicts the value found from the DNS.

## Acknowledgement

The first author gratefully acknowledges support by the Netherlands Geosciences Foundation (GOA) with financial aid from the Netherlands Organisation for Scientific Research (NWO) under grant number 750.195.14. The authors also want to thank the Dutch National Computer Facilities (NCF) for the availability of the Cray C98/4256 supercomputer.

## References

Elgobashi, S., On predicting particle-laden turbulent flows., *Appl. Sci. Res.* **52** (1994), 309–329.

Elgobashi, S., Truesdell, G.C. Direct simulation of particle dispersion in a decaying isotropic turbulence., *J. Fluid Mech.* **242**, 655–700.

Kim, J., Moin, P., Moser, R., Turbulence statistics in fully developed channel flow at low Reynolds numbers. *J. Fluid Mech.* **177** (1987), 133–166.

Maxey, M.R., Riley, J.J., Equation of motion for a small rigid sphere in a nonuniform flow. *Phys. Fluids* **26** (1983), 833–889.

McLaughlin, J.B., Aerosol deposition in numerically simulated channel flow. *Phys. Fluids A 1* **7** (1989), 1211–1224.

Reeks, M. W., The transport of discrete particles in inhomogeneous turbulence., *J. Aerosol Sci.* **14** (1983), 729–739.

Reeks, M. W., On a kinetic equation for the transport of particles in turbulent flows, *Phys. Fluids* **3** (1990), 446–456.

Swailes, D.C., Reeks, M.W., Particle deposition from a turbulent flow. 1. A steady state model for high inertia particles, *Phys. Fluids* **6** (1994), 3392–3403.

# DIRECT NUMERICAL SIMULATION OF TURBULENT OBSTACLE FLOW

R. KESSLER
*DLR, Institute for Fluid Mechanics,*
*D-37073 Göttingen, Germany*

AND

K.S. YANG
*Dept. of Mechanical Engineering, INHA University,*
*Inchon, 402-751, Korea*

**Abstract.** A direct numerical simulation (DNS) of the turbulent flow in a channel with a square obstacle mounted on a wall was carried out. The Reynolds number was 6820 based on the bulk velocity in the channel and the channel height. The main objective of the work was to evaluate statistical data from the time-dependent numerical solution. The influence of numerical and statistical errors on these data was examined. A comprehensive data base including 290 correlations was set up for testing and improving turbulence models for this complex, separated flow.

## 1. Introduction

Despite the progress in numerical algorithm and computer power, reliable numerical predictions of complex turbulent flows is still a challenge. Direct numerical simulations (DNS) of these turbulent flows require much computing time and cannot be applied as a standard engineering tool. Therefore, the Reynolds averaged equations or, for special cases, large eddy simulation have to be used together with appropriate turbulence models to do practical flow field predictions. However, even the most advanced turbulence models have shortcomings in predicting complex separated flows. Therefore, the further improvement of these models is one of the most important tasks in computational fluid dynamics. Statistical data derived from direct nu-



*J.-P. Chollet et al. (eds.), Direct and Large-Eddy Simulation II, 247-256.*

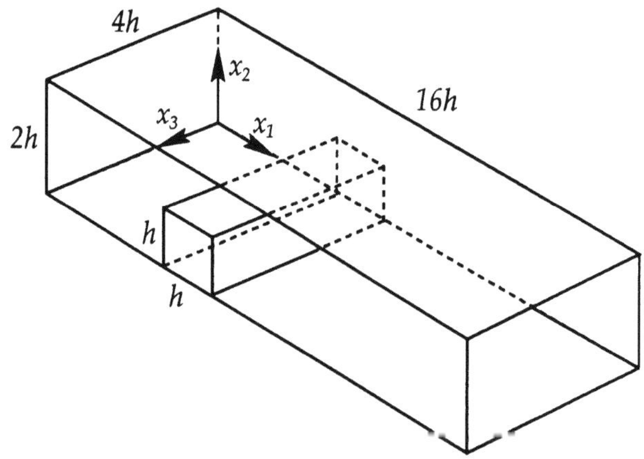


*Figure 1.* Geometry of computation domain

merical simulations of turbulent flows are good tools for comparing directly closure formulae with the terms being modeled. Such DNS data bases can provide guidelines for model developers and model testing. For simple developed channel flow, such data basis were created by Mansour *et al.* (1988) or Gilbert & Kleiser (1991). More recently, a data base for separated flows was created by Le *et al.* (1993).

The paper presents a direct numerical simulation of a turbulent channel flow over a two-dimensional square obstacle. The physical problem and the numerical method are illustrated in chapter 2. In chapter 3, the mean flow is shown and compared with experiments of Dimaczek *et al.* (1989). Special emphasis is laid on the error estimation for the statistical data in chapter 4. The budget of Reynolds stresses is presented in chapter 5 and conclusions can be found in chapter 6.

## 2. Physical problem and numerical method

The turbulent flow over rectangular obstacles has been studied by several authors (e. g. Tropea & Gackstatter (1985), Dimaczek *et al.* (1989), Werner & Wengle (1989)). In spite of the simple cartesian geometry, the flow topology is very complex with primary and secondary separation regions around the obstacle (see streamlines in Fig. 3a). The computational domain is shown in Fig. 1. Its length, height and width was set to $16h$, $2h$ and $4h$ respectively, where $h$ is the height of the square obstacle. No-slip boundary conditions were employed at the walls and periodic boundary conditions imposed in both the homogeneous ($x_3$) and the streamwise direction ($x_1$). Periodicity in the streamwise direction was chosen to avoid the expensive generation of input data and problems with appropriate outflow conditions. Moreover, the length of the computational domain could

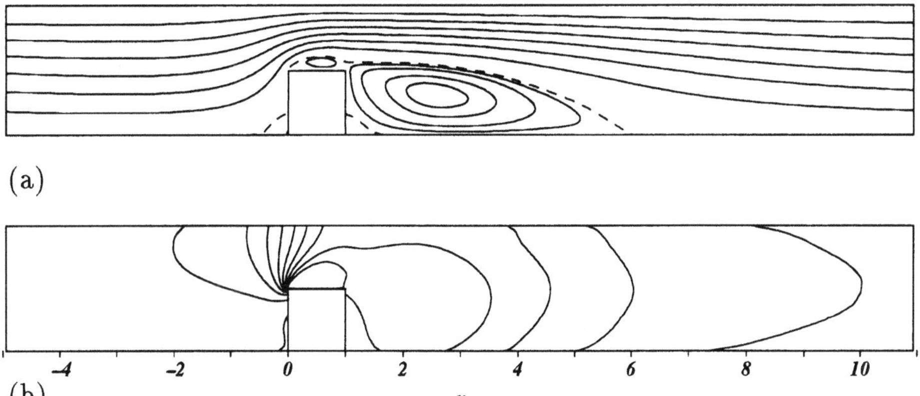


*Figure 2.* (*a*) Streamlines and (*b*) isobars of the mean flow.

be much shorter than using in/outflow conditions. However, when applying periodicity in the $x_1$-direction, we actually simulate the flow in an infinite long channel over a periodic array of obstacles rather than the flow over a single obstacle. The effect of this approach will be discussed in chapter 3. The Reynolds number based on the bulk velocity $U_B$ in the channel and the channel height $H = 2h$ was 6820.

The incompressible Navier-Stokes equations were solved by applying a finite volume method on a staggered grid. The method is of second order accuracy in space and time. The vectorized code was optimized and run on a NEC-SX3 at 1.5 GFLOPS. A detailed description of the numerical scheme is given in Yang & Ferziger (1993). Two runs on two non-equidistant grids ($204\times72\times36$ and $384\times168\times72$) were performed to check out the influence of the numerical resolution. The wall nearest grid points of the fine grid are located typically one wall unit apart the wall.

The numerical simulation and the data evaluation were done in succession. After an initial transient period the velocity and pressure fields were stored every 80th time step during an interval of 130 time units ($h/U_B$). This time interval corresponds to more than 9 flow-through times. On completion of the simulation all statistical data were calculated by averaging over the homogeneous direction ($x_3$) and the time interval mentioned above.

## 3. Mean flow results

The calculated streamlines of the time averaged flow are shown in Fig. 2a. The separation and reattachment points are connected by dashed lines. Three primary recirculation zones at the front, top and rear face of the obstacle and one secondary recirculation can be seen. Two of the recirculation zones are connected by a free stagnation point near the trailing edge of the obstacle. Isolines of the pressure are plotted in Fig. 2b. The pressure

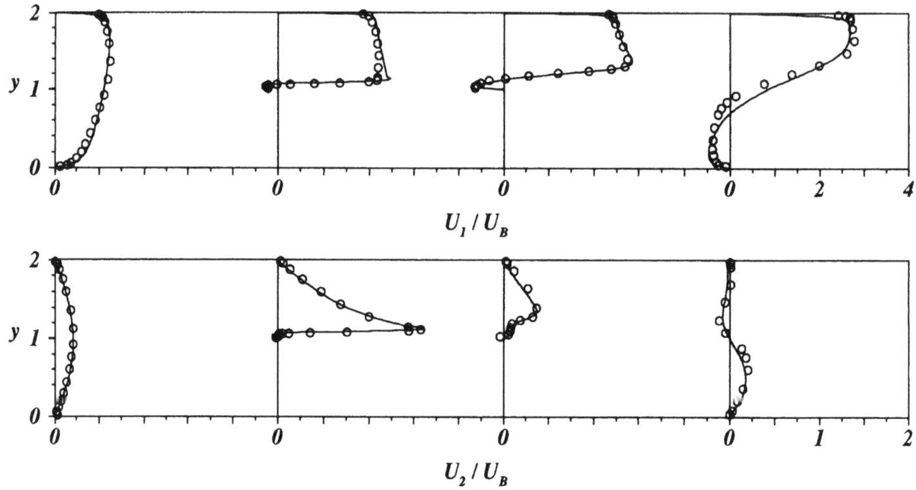


*Figure 3.* Profiles of normalized velocity components $U_1$ and $U_2$ at $x = -1.2h$, $x = 0.08h$, $x = 0.48h$ and $x = 2h$; —— DNS, o experiments (Dimaczek *et al.*, 1989)

drop caused by the obstacle can be observed up to $3h$ upstream of the constriction. Profiles of the mean velocities at four sections are shown in Fig. 3 together with experimental data of Dimaczek *et al.* (1989). The experiment was performed at a Reynolds number of 84000 with a fully developed channel flow upstream of the obstacle. An asymmetric $U_1$ profile, caused by the displacement effect of the obstacle can be seen at $x = -1.2h$. The development of the thin recirculation zone at the top wall of the obstacle is indicated by the negative $U_1$ values at $0.08h$ and $0.48h$. Although the experimental flow conditions are quite different to those applied in our simulation, the predicted and measured data are in astonishing good agreement at the sections upstream and above the obstacle. The mean profiles in this region seem to be not affected by the upstream flow or Reynolds number effects. An assessment of the reliability of the simulation data can hardly be derived from this comparison. At the fourth section, one $h$ downstream the obstacle, the predicted $U_1$ profile slightly differs from the measured values. The predicted shear layer expands more rapidly than the measured one, which may be attributed to the different Reynolds number of simulation and experiment.

## 4. Error estimation

Experimental data for obstacle flows at low Reynolds are rare. Moreover, no experimental data are available for pressure correlations, dissipation or other higher order terms. Therefore, an error estimation is essential for the assessment of the calculated data. The data derived by the statistical evaluation can be affected by the model used, statistical errors and numerical

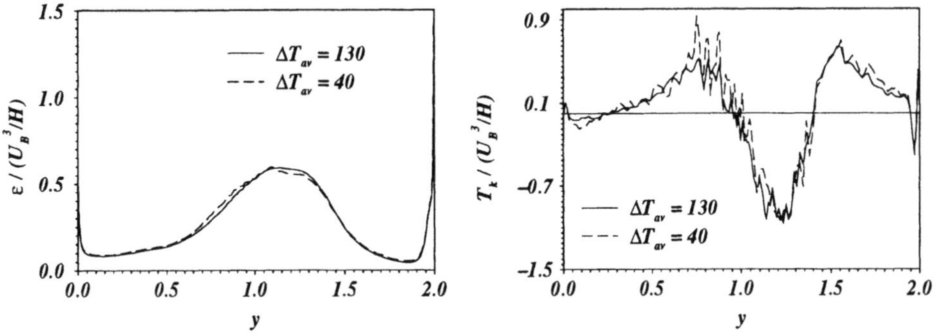


*Figure 4.* Profiles of the normalized dissipation rate $\varepsilon$ and the turbulent diffusion of $k$ for different averaging intervals $\Delta T_{av}$ at $x = 3h$.

errors of the simulation or the evaluation process.

## 4.1. STATISTICAL ERRORS

The statistical errors decrease with an increasing number of statistically independent data used for the averaging process. We applied data of about 800 time sheets for the evaluation. A second evaluation was performed with a reduced averaging interval (250 time sheets) to check out the influence of statistical errors.

The mean flow is less affected by statistical errors. Even in regions with a high turbulence level, the disagreement did not exceed 5% for the mean values. A different behavior can be observed for some higher order correlations. For example, the profiles of the dissipation rate $\varepsilon$ and the turbulent diffusion $T_k = (u_i u_i u_k)_{,k}/2$ of the turbulent kinetic energy $k$ are shown in Fig. 4 calculated with averaging intervals $\Delta T_{av}$ of 130 and 40 time units, respectively. The dissipation is smooth and little difference can be seen between the two curves. By contrast, the profiles of the triple correlations are superposed by fluctuations. High amplitudes of these fluctuations occur in regions with high turbulence levels. An increasing time averaging interval $\Delta T_{av}$ results in a decrease of the waviness.

A possible explanation for this behavior can be found in the different time scales of small and large scale elements. Mainly small scale motions contribute to the dissipation rate. The time scale of these elements is smaller than the time difference between two succeeding time sheets. Therefore, all the data are statistically independent and small statistical errors can be achieved for the dissipation rate. The time scale for large turbulent elements exceeds the time difference between two succeeding time sheets. Therefore, the number of statistically independent data is reduced and larger statistical errors occur if large scale motion contributes to a correlation.

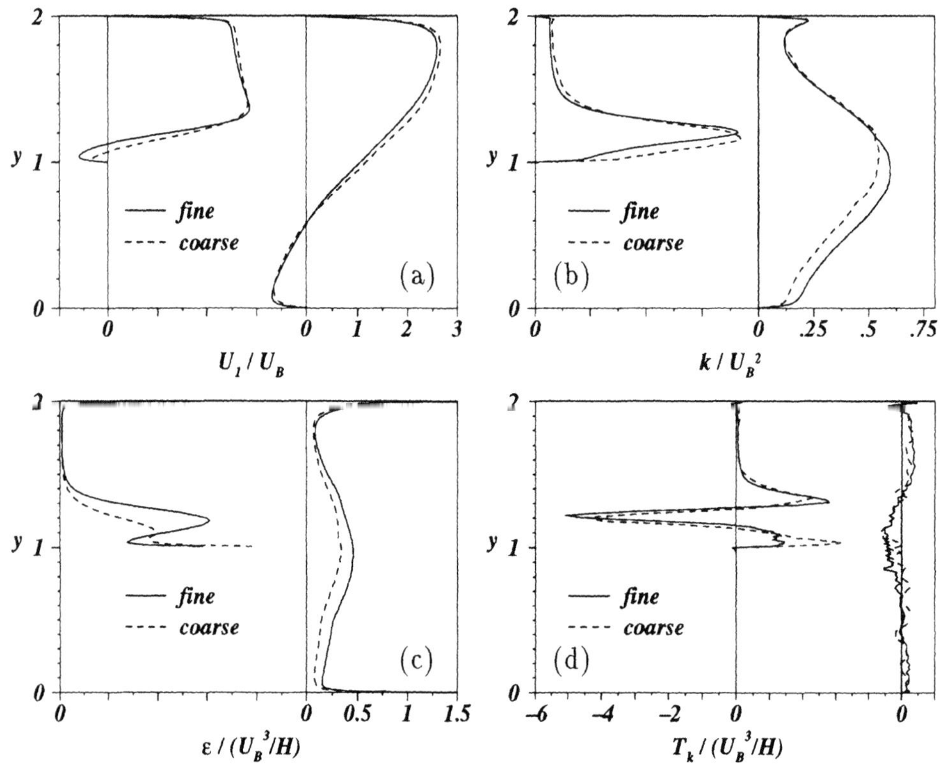


*Figure 5.* Comparison of coarse and fine grid solutions, (a) mean velocity $U_1$, (b) turbulent kinetic energy $k$, (c) dissipation rate $\varepsilon$ and (d) turbulent diffusion of $k$ at $x = 0.5h$ and $x = 3h$

Statistical errors deteriorate the smoothness of some correlations but generally they do not change the principal shape of the profiles. To ensure smooth profiles for all correlations, the time averaging interval of the present simulation should be extended by a factor of 5. Unfortunately, this was not possible because the cpu time resources were limited.

## 4.2. NUMERICAL ERRORS

An extra run was performed with approximately half the grid points of the fine grid DNS in each direction. A complete evaluation of the resulting data was carried out subsequently. The comparison of the results of both simulations can give a rough idea which parts of the results are mostly affected by an insufficient numerical resolution.

Four flow quantities were shown in Fig. 5 to illustrate the results of the comparison. Profiles were plotted at the midpoint of the obstacle ($x = 0.5h$) and downstream of the obstacle ($x = 3h$) near the center of the large recirculation zone.

The profiles of the mean velocities of coarse and fine grid simulation agree well in the whole computational domain with the exception of the thin separation region at the top wall of the obstacle (Fig. 5a). For the coarse grid simulation, the width of the shear layer grows more rapidly. In contrast to the experimental results and the fine grid solutions, the flow reattaches near the trailing edge of the obstacle. Profiles of turbulent fluctuations, represented by the turbulent kinetic energy $k$ can be seen in Fig. 5b. Differences of less than 10% can be observed, mostly concentrated near the separation zones. More pronounced deviations can be detected in Fig. 5c for the dissipation rate, which is underpredicted throughout the flow field by the coarse grid simulation. Differences up to 60% occur mainly in the regions of separation zones. Astonishingly, many of the higher order correlations were well predicted even by the coarse grid simulation. An example is given in Fig. 5d. Profiles of the turbulent diffusion $T_k$ resulting from the coarse and fine grid simulation are close together even in the "critical" region around the obstacle.

Two conclusions can be drawn from the comparison. Firstly, the numerical resolution is clearly insufficient near the top wall of the obstacle for the coarse grid simulation and probably also for the fine grid DNS. The remaining flow field seems to be well resolved at least with the fine grid simulation, and numerical errors seem to have negligible effects on the correlations away from the obstacle. Secondly, some of the turbulent quantities, above all the dissipation rate, are strongly affected by numerical errors, whereas other correlations are less sensitive and can be well predicted even by the coarse grid simulation. The differences between coarse and fine grid solutions were assumed to be an upper bound for the numerical errors of the fine grid DNS.

## 5. Statistical quantities

### 5.1. REYNOLDS STRESSES

Predicted Reynolds stresses at four different sections are shown in Fig. 6. At the first section, $2h$ upstream the obstacle, the flow is characterized by a decaying turbulence. The asymmetry is nearly balanced but all turbulent fluctuations are one order of magnitude higher than the values known from fully developed channel flow. At the next section ($x = -0.5$) the flow in the upper part of the channel is strongly accelerated due to the displacement effect of the obstacle. As a consequence, the streamwise fluctuations $u_1u_1$ decrease below the values of the wall normal component $u_2u_2$. Above the obstacle strong peaks of all components of the Reynolds stress tensor produced by the shear layer can be seen in Fig. 6c ($x = 0.5h$). Whereas the streamwise fluctuations dominate in the shear layer region, the spanwise and wall-normal fluctuations exceed them in the core region above.

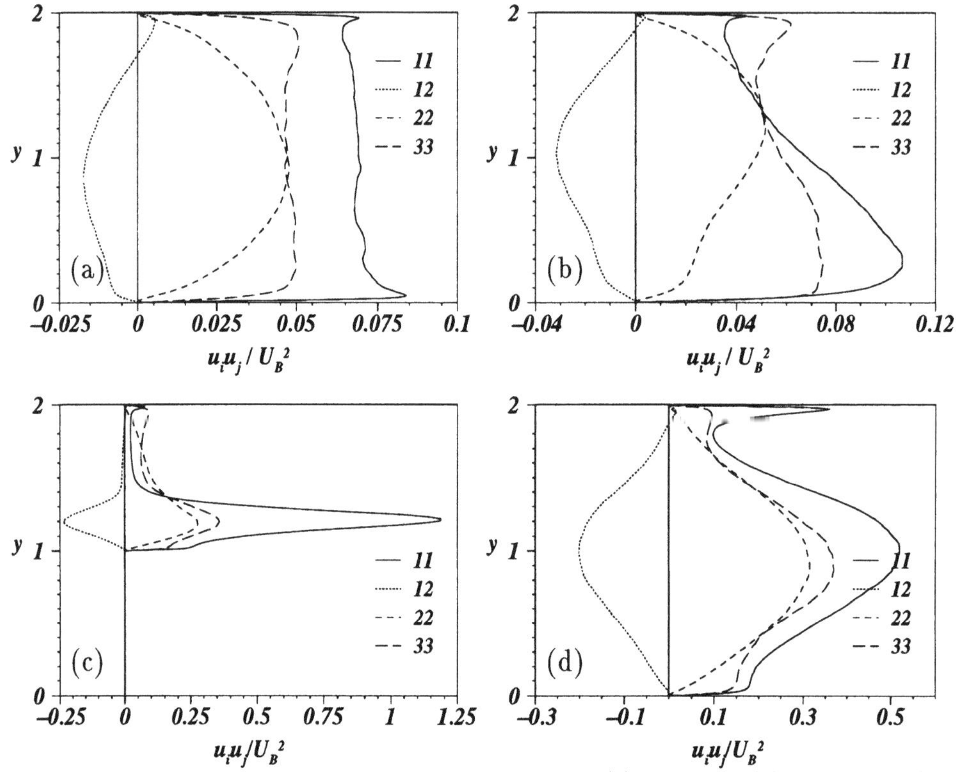


*Figure 6.* Reynolds stresses at various cross sections, (a) $x = -2h$, (b) $x = -0.5h$, (c) $x = 0.5h$ and (d) $x = 3h$.

The kink in the curves of all normal stresses in the vicinity of the obstacle wall was observed in related experiments too (Dimaczek *et al.*, 1989) and is therefore attributed to physical effects rather than to a numerical deficiency. The profiles displayed in the last figure are typical for separated flows. Similar shapes can be found in backward facing step flows. However, the turbulence level of the obstacle flow is clearly higher than that of many other separated flows.

### 5.2. BUDGET FOR REYNOLDS STRESSES

Our evaluation includes all terms appearing in the transport equations for the Reynolds stresses $u_iu_j$, the dissipation rate $\varepsilon$ and the vorticity correlation $\omega_i\omega_i$. Due to limited space, we only present budgets of the streamwise Reynolds stress $u_1u_1$. The transport equation for this correlation reads

$$\frac{D}{Dt} = P_{ij} + T_{ij} + D_{ij} + \Pi_{ij} + \phi_{ij} - \varepsilon_{ij}$$

with the production $P_{ij}$, the turbulent diffusion $T_{ij}$, the molecular diffusion $D_{ij}$, the pressure diffusion $\Pi_{ij}$, the pressure strain term $\phi_{ij}$ and the dissi-

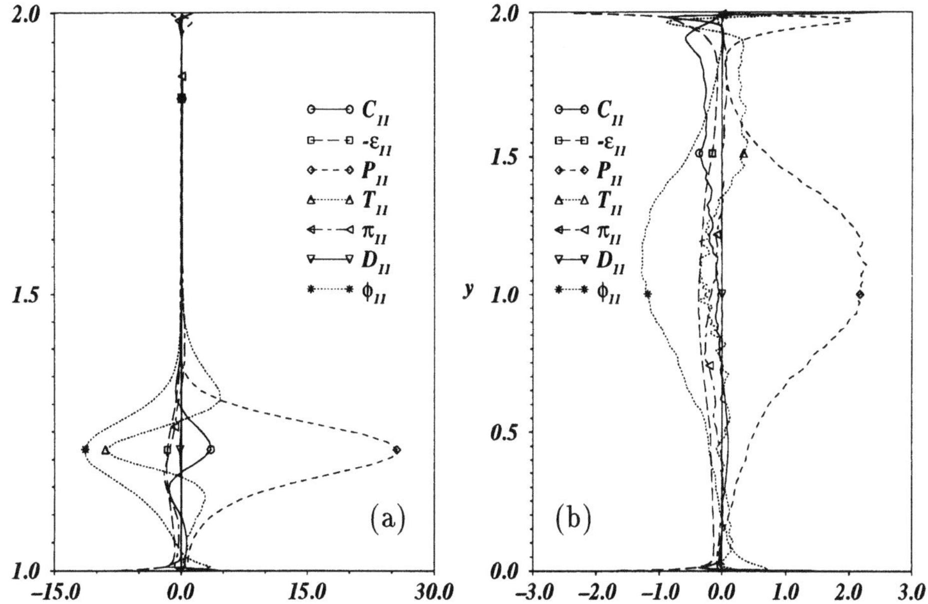


*Figure 7.* Budgets of the Reynolds stress $u_1u_1$, (a) $x = 0.5h$ and (b) $x = 3h$.

pation $\varepsilon_{ij}$. The complete formulae for all terms can be found for example in Mansour *et al.* (1988).

The budgets of the Reynolds normal stress $u_1u_1$ above the obstacle and at the center of the recirculation zone behind the obstacle are plotted in Fig. 7. At the first section, the budget is mainly determined by the thin shear layer. The production term is balanced by the pressure strain term and the turbulent diffusion. The latter is responsible for the transport of energy from the center of the shear layer to its borders, whereas the pressure strain term transfers energy to the other normal stresses. The pressure diffusion as well as the dissipation term act as a sink throughout the shear layer.

The residue of the budget has a positive peak at the center of the shear layer. We mainly attribute this to a strong underprediction of the dissipation rate in this region.

Fig. 7b shows the budget downstream of the obstacle at $x = 3h$. The production term is biggest at the center and the top wall of the channel. However, in the lower part of the channel the production is very small and even switches to negative values near the wall. The most important sink in the budget is again the pressure strain correlation. The turbulent diffusion transports energy from the center of the channel to the upper channel half. The dissipation rate has a local maximum near the center of the channel and grows up to higher values in the vicinity of both walls. The residue does not exceed 5% of the maximum value of the production term, which means numerical error are expected to be small at this section.

## 6. Summary and conclusions

The direct numerical simulation of a turbulent channel flow over a square obstacle was presented. Around the obstacle, the predicted mean velocity profiles are in very good agreement with experimental data, although Reynolds number and upstream flow conditions applied for the simulation are quite different from those realized in the experiment.

By comparing the coarse and fine grid solutions, we found the numerical resolution of the present DNS to be sufficient upstream and downstream the obstacle. Above the obstacle, not all the physically relevant scales are resolved and more grid points would be needed to reduce the numerical errors in this region.

The statistical error introduced by the finite time average interval is highest in the center of the shear layer, where the turbulence level is high. Dissipation is little affected by this error, whereas turbulent diffusion, pressure diffusion and the convective transport term are strongly influenced by statistical errors.

An estimation of the effect of numerical errors on statistical data was derived by comparing results of the coarse and fine grid solutions. The dissipation term was found to be most sensitive to numerical errors of the simulation. However, the mean flow and also some higher order correlations can be calculated satisfactorily even with a coarse grid simulation.

The data base derived from the present simulation will be a helpful tool for testing and improving advanced turbulence models. The obstacle flow itself remains a challenge for both direct numerical simulations and turbulence modelers.

## References

Dimaczek, G., Kessler, R., Martinuzzi, R. and Tropea, C., "The flow over Two-dimensional, surface-mounted obstacles at high Reynolds numbers", *Proc. of 7th Symposium on Turbulent Shear Flows*, Stanford, Aug. 21-23, 1989.

Gilbert, N. and Kleiser, L., "Turbulence model testing with the aid of direct numerical simulation results", *Proc. of 8th Symposium on Turbulent Shear Flows*, Paper 26-1, Munich, Sept. 9-11, 1991.

Le, H., Moin, P. and Kim, J., "Direct numerical simulation of turbulent flow over a backward-facing step", *Proc. of 9th Symposium on Turbulent Shear Flows*, Paper 13-2, Kyoto, Aug. 16-18, 1993.

Mansour, N.N., Kim, J. and Moin, P., "Reynolds-stress and dissipation-rate budgets in a turbulent channel flow", *J. Fluid Mech.*, vol. **194**, pp. 15-44, 1988.

Tropea, C. and Gackstatter, R., "The flow over two-dimensional surface-mounted obstacles at low Reynolds numbers", *J. Fluid Eng.*, vol. **107**, pp. 489-494, 1985.

Werner, H. and Wengle, H., "Large-eddy simulation of turbulent flow over a square rib in a channel", *Proc. of 7th Symp. on Turbulent Shear Flows*, Stanford, Aug. 21-23, 1989.

Yang, K.S. and Ferziger, J.H., "Large-eddy simulation of turbulent obstacle flow using a dynamic subgrid-scale model", *AIAA Journal*, vol. **31**, pp. 1406-1413, 1993.

# NON-UNIQUE SOLUTIONS IN TURBULENT CURVED PIPE FLOW

B. J. BOERSMA & F.T.M. NIEUWSTADT
*Delft University of Technology,*
*Laboratory for Aero and Hydrodynamics,*
*Rotterdamseweg 145, 2628 AL Delft, The Netherlands*

**Abstract.** The mean flow and turbulent statistics obtained from the direct numerical simulation of a fully developed turbulent flow through a curved pipe with circular cross-section are reported. The Reynolds number based on the bulk velocity and the pipe diameter is approximately 5500, the radius of curvature is equal to five pipe diameters. In the cross-section of the pipe a strong secondary motion is observed. It is shown that in the turbulent case multiple time mean solutions of governing equations exist.

## 1. Introduction

Direct Numerical Simulation (DNS) is nowadays a powerful tool to study turbulent flow at rather low Reynolds numbers. In the past several direct numerical simulations have been carried out. Most of these simulations were restricted to very simple Cartesian geometries like channels (e.g., Kim, Moin & Moser 1987) and ducts (e.g., Gavrilakis 1992). More recently simulations in cylindrical geometries have been carried out by for instance Eggels *et al.* (1994), and Orlandi & Fatica (1995). These simulations give detailed insight in the turbulent structures which are, for instance, present in pipe- or jet flow. Furthermore its shown that the agreement between DNS and experiment is good, e.g., Eggels *et al.* (1994). Therefore, the conclusion seems justified that DNS can be used to generate accurate databases of a certain type of flow. These databases contain more information than an experimental dataset.

In this paper we will use DNS to study turbulent flow in a pipe bend. This flow is of large practical importance, because almost any flow appliance or equipment contains some pipes with bends. Data on the parameters for such curved pipes are required for a number of applications. For instance,



*J.-P. Chollet et al. (eds.), Direct and Large-Eddy Simulation II, 257-266.*

the pressure loss due to a bend must be know in order to calculate the total loss of a pipeline. Curved pipes are also frequently used in heat exchangers where the secondary flow which occurs in the bend is used to transport heat from the pipe wall to the core of the pipe. It is obvious that a good knowledge of the structure and strength of the secondary motion is crucial for the design of such a cooler.

A (turbulent) flow field under influence of a centrifugal force is subjected to a remarkable change in flow patterns due to the generation of secondary motions in the direction of the centrifugal force. The secondary motion can be characterised by the Dean number of the flow which we define as:

$$K = Re\sqrt{\frac{D}{R}}, \tag{1}$$

where $Re$ is the bulk Reynolds number, $D$ the diameter of the pipe, and $R$ is the radius of curvature. The square of this number gives the ratio of the centrifugal force to the viscous force. For loosely coiled pipes, i.e. $R$ is large compared to $D$, the Dean number $K$ determines the flow completely. For pipes with small $R$ the flow must be characterised by both Dean and Reynolds numbers.

To give the organisation of the paper, in the next section we will introduce the coordinate system. In section three we will describe the numerical method, the results will be discussed in section four and some conclusions will be given in section five.

## 2. The coordinate system

The natural choice for a coordinate system is a toroidal system, which can be found in Morse & Feshbach (1953). In Figure 1 we have sketched this coordinate system. The orthogonal metric for this system is given by Germano (1989):

$$\mathbf{dx} \cdot \mathbf{dx} = dr^2 + (rd\theta)^2 + (1 + \kappa r \sin\theta)^2 ds^2, \tag{2}$$

where $dr$, $d\theta$, and $ds$ are the infinitesimals in the radial, circumferential and axial direction, and $\kappa$ stands for the inverse radius of curvature, i.e. $\kappa = D/R$. With the metric given by equation (2) one can obtain the scale factors $h_1, h_2, h_3$ (see Appendix 2 of Batchelor 1967).

$$h_1 = 1, \qquad h_2 = r, \qquad h_3 = 1 + \kappa r \sin\theta. \tag{3}$$

With the help of these scale factors one can express the continuity and Navier-Stokes equations in terms of this coordinate system (e.g., Boersma & Nieuwstadt 1996b). The equations are nondimensionalised with the pipe

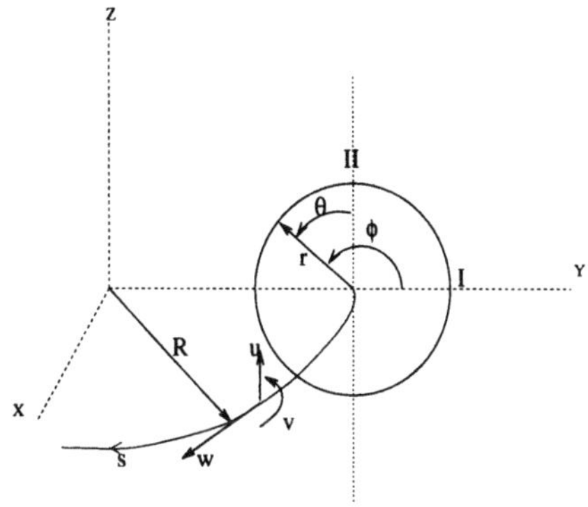


*Figure 1.* The coordinate system used, $u, v, w$ denote the velocity components in the radial, circumferential and axial direction, and $R$ is the radius of curvature. Plane I and II denote the planes in which we will present most of the results.

diameter $D$, the mean friction velocity $U_*$, and time $D/U_*$ as scale factors. The mean friction velocity $U_*$ is defined as:

$$U_* = \frac{1}{2\pi}\int_0^{2\pi} u_*(\theta)d\theta, \tag{4}$$

where $u_*$ is the local friction velocity at some position $\theta$ along the pipe circumference, which is by definition equal to the square root of the local wall shear stress divided by the mass density of the fluid.

## 3. Outline of the numerical method, boundary conditions and forcing

In this section we will describe the numerical method used to solve the Navier-Stokes equation in a toroidal system. The numerical method employed to solve these equations is basically the same as used by Boersma & Nieuwstadt (1996a). The continuity and Navier-Stokes equation are spatially discretized with the finite volume method on a staggered grid resulting in a second order accurate spatial discretisation. The integration in time has been done with the second order Adams-Bashforth and Crank-Nicolson for the convective and viscous terms. The pressure correction method is used to ensure the incompressibility of the flow. In the pressure correction step a discrete Poisson equation for the pressure correction has to be solved. The Poisson equation can be decoupled in the axial direction with a Fast Fourier Transform (e.g., Swarztrauber 1977). The remaining two dimensional Helmholtz equations are solved with a Conjugate Gradient method.

At the pipe wall ($r = D/2$) no slip boundary conditions for all three velocity components are imposed. The in- and outflow boundary conditions are taken periodic. These boundary conditions are only chosen for computational convenience. Periodic in and outflow conditions can only be justified if the distance between in- and outflow is considerably larger than the integral length scale of the turbulence. The latter can be checked a postiori by calculating the longitudinal correlation coefficient between variables at the entrance and halfway the computational domain.

The flow is forced in the axial direction by means of a constant pressure gradient $-\nabla P/\rho$. For a stationary flow this forcing should be balanced by the wall shear stress, i.e.

$$-\frac{1}{\rho}\nabla P = -\frac{1}{\rho}\frac{\Delta P}{\Delta s} = 4\rho U_*^2/D, \tag{5}$$

where $\Delta s$ is the distance along the centreline of the pipe.

## 4. Results for Turbulent flow

In this section we will investigate turbulent flow in curved pipe. The radius of curvature of the pipe is equal to five diameters and the Reynolds number based on the mean frictional velocity $U_*$ is equal to 460.

All the direct numerical simulations are started from the following velocity profile for straight pipe:

$$w = 2.5U_*\ln\left(\frac{yU_*}{\nu}\right) + 5.5 \quad \text{if} \quad yU_*/\nu \geq 11.6$$
$$w = yU_*^2/\nu \quad \text{if} \quad yU_*/\nu < 11.6.$$

On this velocity profile small random fluctuations are superimposed. After approximately four time scales $D/U_*$ the flow becomes fully developed. We note that for standard pipe flow it takes approximately $8D/U_*$ before the flow becomes fully developed. This difference can be attributed to the additional mixing by the secondary motion, which removes the transients due to the initial conditions rapidly.

In Table 1 we have summarised some important quantities with respect to the DNS of curved pipe flow (simulations A and B) and with respect to a DNS of standard pipe flow (simulation C). The computational grids in simulations A and B are non-uniform in the direction towards the wall, with the grid points slightly concentrated near the wall to resolve the near-wall gradients. For the standard pipe flow a uniform grid was used and therefore more grid points are required to get the same resolution in the near wall region.

The statistics for the standard pipe flow simulation (run C) were taken over 40 independent data fields with spatial averages calculated along the homogeneous tangential and axial direction. For the curved pipe flow simulations only spatial averages along the homogeneous axial direction can be used and therefore many more data fields are needed to obtain independent and stable statistics. Therefore for the curved pipe flow we used 80 data fields which probably is still too few to get statistically independent fields.

TABLE 1. Some important quantities for a DNS of a curved pipe (A and B) and a straight pipe (C): $N$ denotes the number of grid points, $Re_*$ the Reynolds number based on the mean friction velocity and pipe diameter, $R$ the radius of curvature, $K$ is the Dean numbers, and $L/D$ the length of the computational domain normalised with the pipe-diameter.

| | A | B | C |
|---|---|---|---|
| $N(r \times \theta \times s)$ | $90 \times 160 \times 350$ | $90 \times 160 \times 350$ | $96 \times 128 \times 256$ |
| $Re_*$ | 460 | 460 | 360 |
| $R$ | 5 | 5 | 0 |
| $Re_b$ | 5470 | 5520 | 5300 |
| $K$ | 2445 | 2470 | 0 |
| $L/D$ | 7.5 | 7.5 | 5 |

Simulations A and B are performed at exactly the same $Re_*$ and $R$ however with different initial conditions. Simulation B has been started from a velocity profile for straight pipe flow, on which random fluctuations are superimposed. The variances obtained from simulation B were found not to be in agreement with earlier results obtained by Boersma & Niewstadt (1996a, 1996b). Therefore more energy, in the form of random noise, was added to the initial conditions and this resulted in a second stationary solution, i.e. simulation A. The latter simulation shows the same behaviour as reported in Boersma & Nieuwstadt (1996a). This result shows that turbulence in a curved pipe may lead to more than one ensemble averaged flow field. Such a non-unique solution was already known for laminar flow in a curved pipe (e.g. Ito 1987) but to our knowledge nobody has reported non-unique solutions for turbulent flow fields. In the sequel we will present some results for both stationary solutions and we will compare these results with results for standard pipe flow.

### 4.1. MEAN FLOW QUANTITIES

In Fig. 2 we show the mean velocity profiles obtained from the simulation A and B (curved pipe flow) and simulation C (standard pipe flow). The results are presented in the planes I and II of Fig. 1. The difference in axial velocity profiles between a curved pipe flow and standard pipe flow is striking. The explanation for this difference is that the centrifugal force acting on the flow forces the rapid flowing parts of the flow to the outside. Consequently the velocity is low at the inside and high at the outside of the bend. Furthermore there is a significant difference between the axial velocity profiles for simulations A and B near the outside of the bend. The velocity profiles in plane II are symmetric and have their maxima close to the pipe wall. This velocity maximum near the pipe wall in plane II is

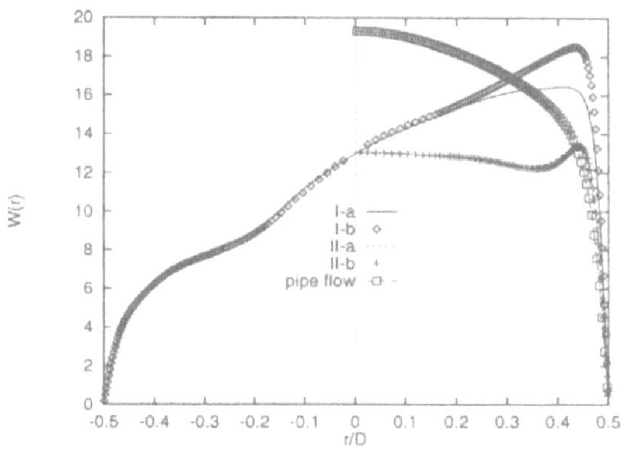


*Figure 2.* The axial velocity profiles obtained from simulation A and B (curved pipe flow, and D (standard pipe flow). The outside of the curvature is to the right.

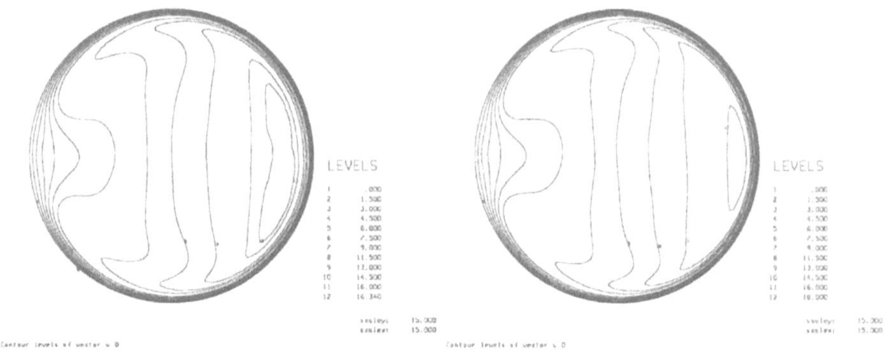


*Figure 3.* A contour plot of the axial velocity profile normalised with $U_*$. The left Figure is for case A and the right Figure for case B (in both case the outside of the curvature is to the left).

caused by the secondary motion which advects the velocity maximum at the outside of the bend somewhat along the wall.

In Fig. 3 contour plots of the axial velocity profiles obtained from simulation A and B. This figure shows again clearly that the centrifugal force, forces the rapid flowing parts of the fluid to the outside of the pipe.

### 4.2. SECONDARY MOTION

In the previous sections we have already noted that there is a secondary circulation in the cross plane of the pipe. In this section we will study this motion in some detail. The secondary motion can be characterised in terms of a stream function, which for the cylindrical geometry used here can be

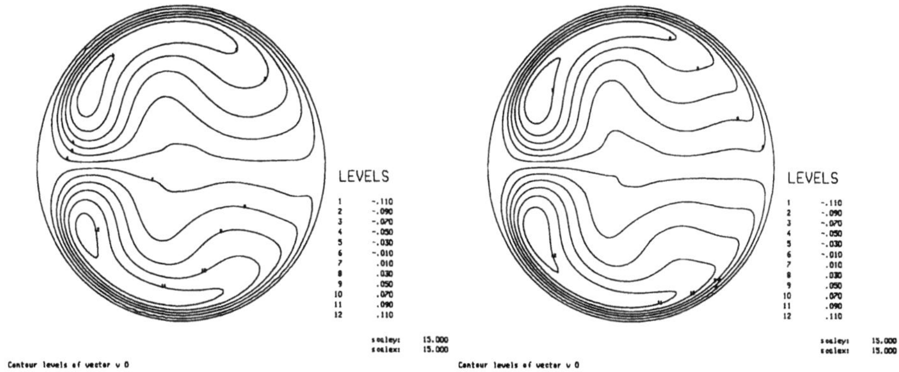


*Figure 4.* The secondary motion in the cross plane of the pipe defined by equation (6). The left figure shows the results obtained from simulation A and the right figure the results obtained from simulation B. In both case the outside of the curvature is to the right).

found from:

$$\psi(r,\theta) = -\int_r^{D/2} v(r,\theta)dr \tag{6}$$

The stream function $\psi$ is zero at the wall. In Fig. 4, this stream function has been plotted, for simulations A and B. The distance between the streamlines is a measure for the magnitude of the secondary velocity. The figure shows that the secondary velocity is strong near the pipe wall close to plane II. Furthermore, it follows that the secondary velocities are much larger at the inside of the bend than at the outside.

The strong vortices which are apparent in Fig. 4 near the inside of the bend are regular Dean vortices which are also present in laminar curved pipe flow (e.g., Collins & Dennis 1975). The contour plots of the secondary motion obtained from simulation A and B give a small difference in the core region of the bend. This difference can qualitatively explained as follows: The pressure gradient in the radial direction, in the symmetry plane (I) is approximately equal to:

$$\frac{\partial P}{\partial r} \approx \frac{\rho w^2}{R+r}. \tag{7}$$

From Fig. 2 it follows that the velocity for case B near the outside of the bend is approximately 10% larger than for case A. Consequently the pressure gradient $\partial P/\partial r$ will be larger for case B which also implies that the radial velocity for case B is smaller than for case A, which explains more or less the difference in secondary flow patterns.

## 4.3. TURBULENT STATISTICS

In this section we will investigate how the curvature, i.e., the centrifugal force influences the turbulent statistics. Our point of departure is the

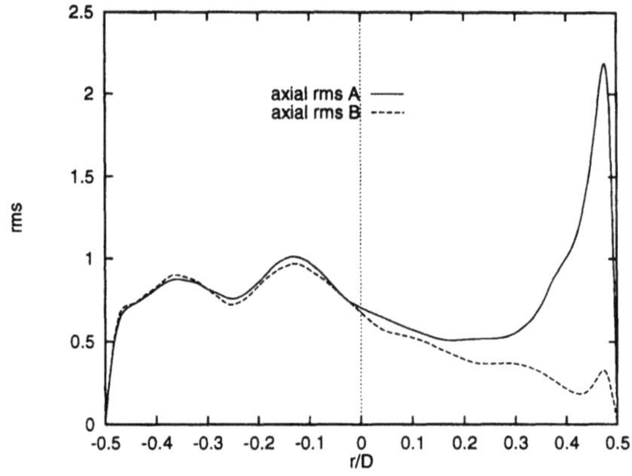


*Figure 5.* The axial root mean square profiles obtained from simulation A and B, plotted in plane I. The outside of the curvature is to the right.

Reynolds decomposition

$$u = \overline{u} + u', \quad v = \overline{v} + v', \quad w = \overline{w} + w', \quad \text{and} \quad p = \overline{p} + p', \tag{8}$$

where a bar denotes an average in the homogeneous axial direction and in time, and a prime a deviation of this average. In Fig. 5 we show the axial root mean square profiles in plane I. This figure again clearly illustrates the difference in the two solutions A and B. Simulation A is fully turbulent near the outside, while simulation B is almost laminar near the outside. When we apply the Reynolds decomposition given above to the Navier Stokes equations in toroidal coordinates we obtain the following equation for the mean flow quantities in the axial direction:

$$\begin{aligned} &\frac{\partial}{\partial r}[rh_3(\overline{u}\,\overline{w} + \overline{u'w'})] + \frac{\partial}{\partial \theta}[h_3(\overline{v}\,\overline{w} + \overline{v'w'})] + r\kappa\overline{w}(\overline{u}\sin\theta + \overline{v}\cos\theta) = \\ &4rU_*^2/D + \frac{\partial r h_3 \overline{\tau}_{rs}}{\partial r} + \frac{\partial h_3 \overline{\tau}_{\theta s}}{\partial \theta} + r\kappa(\overline{\tau}_{rs}\sin\theta + \overline{\tau}_{\theta s}\cos\theta). \end{aligned} \tag{9}$$

In which $h_3 = 1 + \kappa r \sin\theta$ and $\overline{u}, \overline{v}, \overline{w}$ the mean velocity components, and $\overline{\tau}_{ij}$ the mean stress tensor components. In the symmetry plane some terms vanish, because:

$$v = \frac{\partial \overline{u}}{\partial \theta} = \frac{\partial \overline{w}}{\partial \theta} = 0, \tag{10}$$

and equation (9) reduces in this plane to:

$$\begin{aligned} rh_3\overline{u}\frac{\partial \overline{w}}{\partial r} + \frac{\partial}{\partial r}[rh_3\overline{u'w'}] + \frac{\partial}{\partial \theta}[h_3\overline{v'w'}] + r\kappa(\overline{u}\,\overline{w} + \overline{u'w'}) = \\ 4rU_*^2/D + \frac{\partial}{\partial r}[rh_3\overline{\tau}_{rs}] + r\kappa\overline{\tau}_{rs}, \end{aligned} \tag{11}$$

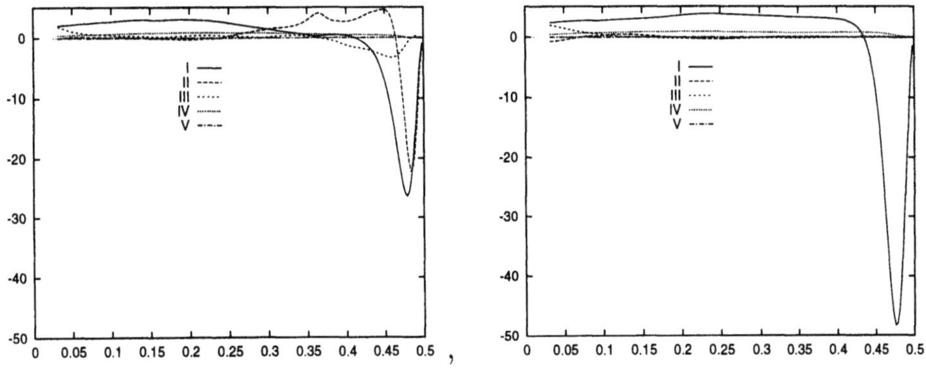


*Figure 6.* The terms at the right hand side of equation (12). Left simulation A and right simulation B. I : $rh_3\overline{u}\frac{\partial \overline{w}}{\partial r}$, II: $r^{-1}\frac{\partial}{\partial r}[rh_3\overline{u'w'}]$, III:$r^{-1}\frac{\partial}{\partial \theta}[h_3\overline{v'w'}]$, IV :$\kappa\overline{u}\,\overline{w}$, V: $\kappa\overline{u'w'}$

where we have used the continuity equation to rewrite some of the partial derivatives in equation (9). In the core region of the pipe we can neglect the viscous terms in equation (11) and we get the following equation for the mean variables

$$h_3\overline{u}\frac{\partial \overline{w}}{\partial r} + \frac{1}{r}\frac{\partial}{\partial r}[rh_3\overline{u'w'}] + \frac{1}{r}\frac{\partial}{\partial \theta}[h_3\overline{v'w'}] + \kappa\overline{u}\,\overline{w} + \kappa\overline{u'w'} = 4U_*^2/D. \quad (12)$$

In Figure 6a and b we have plotted the five terms on the right hand side of the equation (12), normalised with $U_*^2/D$ The sum of the five terms in Figure 6a and 6b is almost constant in the core region and equal to $4U_*^2/D$. In both cases $\overline{u}\partial\overline{w}/\partial r$ is by far the largest term. This term mainly balances the imposed pressure gradient in the axial direction. This is in contrast to standard pipe or channel flow where the pressure gradient is balanced by the gradient of the Reynolds stress, i.e. $r^{-1}\partial(r\overline{u'w'})/\partial r$. In principle this observation is only valid in the symmetry plane. However, similar behaviour will be observed in the other planes.

The balance between $\overline{u}\partial\overline{w}/\partial r$ and the imposed pressure gradient partly explains the large difference between parallel flow and mildly curved flow (e.g., Bradshaw 1987), i.e., a small curvature ratio will give also a small radial velocity, but $\overline{u}\partial\overline{w}/\partial r$ can be large ($\overline{u}\partial\overline{w}/\partial r \gg \tau_{ij}$). Figure 6b furthermore shows that the derivatives of the Reynolds stresses $\overline{u'w'}$ and $\overline{v'w'}$ are small near the outside of the bend, this in contrast to Figure 6a. Thus in case A the momentum $\partial P/\partial s$ is balanced by the Reynolds stresses and $\overline{u}\partial\overline{w}/\partial r$ while in case B the momentum is balanced by $\overline{u}\partial\overline{w}/\partial r$ only.

## 5. Conclusions

In this paper we have presented some results obtained from a DNS of curved pipe flow. The Reynolds and Dean numbers are 5500 and 2450 respectively. It is shown that at this low Reynolds number multiple solutions of the

Navier-Stokes equations can exist. One of these solutions is a fully turbulent one, i.e. the flow is turbulent in the whole cross-section of the pipe. In the other simulation the flow becomes almost laminar near the outside of the bend and turbulent near the inside of the bend. It is expected that this second solution can only be found for a small range of Reynolds and Dean numbers. The existence of a solution which is partly laminar and partly turbulent is maybe related to the gradual transition from laminar to turbulent flow which is observed in experiments (e.g., Ramshankar & Sreenivasan 1988).

## Acknowledgement

The first author gratefully acknowledge the Dutch Foundation of Technology (STW) for their financial support under grant DTN22.2646. The authors also want to thank the Dutch National Computer Facilities (NCF) for the availability of the Cray C98/4256 supercomputer.

## References

Batchelor, G.K., 1967, *An Introduction to Fluid Dynamics*, Cambridge University Press, Cambridge

Boersma, B.J. & Nieuwstadt, F.T.M.,1996a, Large Eddy simulation of turbulent flow in a curved pipe, *J. Fluids Eng.*

Boersma, B.J. & Nieuwstadt, F.T.M., 1996b, Direct Numerical Simulation of the flow in a bend, In the Proceedings of: 3rd International Symposium on Engineering Turbulence Modelling and Measurements, Elsevier.

Bradshaw, P., 1987, Turbulent secondary flows, *Ann. Rev. Fluid Mech.*, **19**, 53-74.

Eggels, J.G.M., Unger, F., Weiss, M.H., Westerweel, J., Adrian, R.J., Friedrich R., & Nieuwstadt, F.T.M., 1994, Fully developed turbulent pipe flow: a comparison between direct numerical simulation and experiment. *J. Fluid Mech.* **268**, 175-209

Collins, W.M., & Dennis, S.C.R., 1975, The steady motion of a viscous fluid in a curved tube, *Q. J. Mech. Appl. Math.*, **28**, 133-156.

Germano, M., 1989, The Dean equations extended to a helical pipe flow. *J. Fluid Mech.*, **203**, 289-305.

Gavrilakis, S., 1992, Numerical simulation of low Reynolds-number turbulent flow through a straight square duct, *J. Fluid Mech.*, **244**, 101-129.

Ito, H., 1987, Flow in curved pipes, *JSME International Journal*, **30**, 543-552.

Kim, J., Moin, P., and Moser, R., 1987 Turbulence statistics in fully developped channel flow at low Reynolds number, *J. Fluid Mech.*, **177**, 133-166.

Morse, P.M., & Feshbach, H., 1953, *Methods of Theoretical Physics*, McGraw-Hill, New York

Orlandi, P. & Fatica, M., 1995, Direct simulation of a turbulent pipe rotating along the axis, *Submitted to : J. Fluid Mech.*.

Ramshankar, R. & Sreenivasan, K.R., 1988, A paradox concerning the extended Stokes series solution for the pressure drop in coiled pipes *Phys. Fluids*, **31**, 1339-1347.

Swarztrauber,P., 1977, The methods of cyclic reduction, Fourier analysis and the FACR algorithm for the discrete solution of Poisson's equation on a rectangle, *SIAM Review*, **19**,490-501.

Taylor, G.I., 1929, The criterion for turbulence in curved pipes, *Proc. R. Soc. London Ser. A.*, **124**, 243-249. (also in: The collected paper of G.I. Taylor).

# Large Eddy Simulation of Two Dimensional and Rectangular Impinging Jets

T. Cziesla, N. K. Mitra
Ruhr-University Bochum
Institut für Thermo- und Fluiddynamik
44780 Bochum, Germany

## Abstract

Large eddy simulations of two dimensional and rectangular jets impinging on a flat surface have been performed using a localized dynamic subgrid stress model. The 3D jet flow is characterized by backflow at the exit. Comparison of LES with direct numerical simulation shows good agreement and confirms the accuracy of the model and the boundary conditions used.

## Introduction

Impinging jet flows are of great interest both in fundamental as well as in applied fluid mechanics. In technical applications such flows are often used for heating, cooling and drying of materials. There are two main forms of impinging jet flows, namely round or circular nozzles and slot or rectangular nozzles (Viskanta, 1993). In the case of round nozzles the fluid is ducted by a tube before impinging near the tube exit. So there is a stagnation point on the plate. In slot nozzles the fluid flows through a channel. In this case a stagnation line may appear at the impingement plate. For enhancement of global transfer rates in technical applications often banks of round or slot nozzles are used. A single nozzle can be treated as a periodic element of such a configuration.

This geometry causes a very complex flow. Martin (1977) studied experimentally the influence of variation of parameters like nozzle-to-plate-distance, nozzle height, nozzle width and width of a periodic nozzle element on the heat transfer at the impingement plate. He found that after impingement two oppositely directed wall jets develop which seperate at the periodic sides because of the impact of the wall jets of two neighboring elements. Martin (1977) could show that corner vortices appear at each lower corner. Other authors (Viskanta, 1993; Martin, 1977; Gardon, 1966) pointed out, that depending on the Reynolds number and nozzle height the flow below the nozzle is laminar first (stagnation zone) and then becomes turbulent during the wall jet development. Our numerical investigations for laminar flow show the appearance of large helical vortex structures in each quarter of the domain (Laschefski, 1996). We found, that for $Re_{2B}<400$ the flow is laminar. Here the Reynolds number is defined with the double slot width, which can be understood as the hydraulic diameter of the feed tube (Martin, 1977; Laschefski, 1996). Furthermore backflow across the exit plane appears [see Fig. 1], for which a special boundary condition treatment is necessary (Laschefski, 1996; Hoffmann, 1994).

The purpose of this work is to investigate the flow for higher Reynolds numbers ($Re_{2B}=600$), where transition and turbulence can occur. So not only turbulent but also



*J.-P. Chollet et al. (eds.), Direct and Large-Eddy Simulation II*, 267-278.

laminar flow must be considered. Because of the strong streamline curvatures it is not evident if Reynolds-averaged models can give sufficently correct results. Besides, these models are geometry dependent and need special tuning of their model coefficients for each flow geometry (damping functions). Hence Large Eddy Simulation (LES) or Direct Navier Stokes Simulation (DNS) seems to be the most appropriate tool for the numerical simulation of impinging jets. The accuracy of DNS is limited by the grid size. It has been observed that the larger scales carry the main part of the turbulent kinetic energy and that the smaller scales are universal and can be assumed as isotropic. So the main advantage of LES is that it is more general in principle and less geometry dependent. So far only simple geometries were handled, because LES based mostly on the Smagorinsky Lilly subgrid stress (SGS) model (Smagorinsky, 1963) suffered from the following limitiations:

- optimal model constant must be changed in different flows
- the model does not have the correct limiting behavior near to the wall
- the constant does not vanish in laminar flow
- backscatter, i.e. energy transport from small to larger scales, cannot be considered.

Germano (1991) and later Lilly (1992) proposed the "dynamic" subgrid model which improves capability of the model, removes the above mentioned limitations and makes LES more universal. The basic idea of this model is the dynamic calculation of the model constant during the simulation. The main problem of this procedure is the stabilization of the calculated model parameter values, i. e. the dynamic model has been so far applied for geometries with at least one homegeneous direction, in which the model coefficient is averaged for stabilization. Since there is no homogeneous direction for the impinging jet flow we investigated here (see. Fig. 1), we use a local averaging procedure suggested by Zang (1993). Because the spatially-varying coefficient is inconsistent with the model derivation (Germano, 1991; Lilly, 1992), we implemented a modification presented by Piomelli (1994), which avoids this mathematical inconsistency.

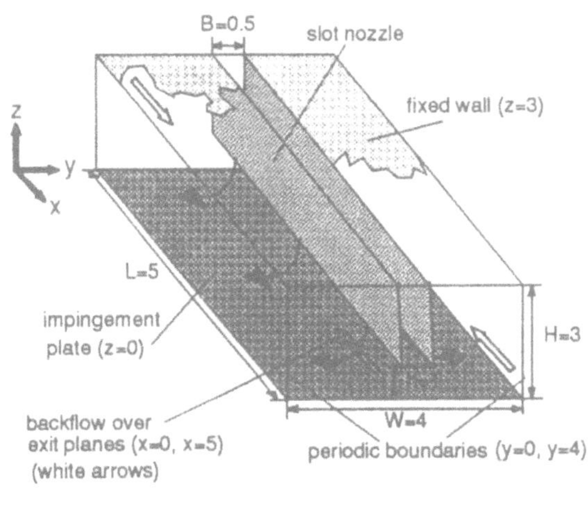


Fig. 1: Schematic of an periodic element of a slot nozzle bank

The purpose of our work is LES of an impinging jet with a dynamic SGS for both two dimensional and three dimensional impinging jets. Figure 1 shows the computational domain for a 3D jet. For the 2D case we use periodic condition on the boundaries of y=0 and y=4 and average the flowfield in y-direction. For comparison some direct numerical simulations of turbulent flow have also been done.

## Numerical Method

The LES is based on the application of a filter operation (denoted by an overbar) to the 3D time-dependent Navier-Stokes equations. The turbulent scales can be distinguished at scales larger and smaller than the filter width. For finite difference methods the filter width is defined by the grid size. The large scales (denoted by an overline) are solved directly, the small or subgrid scales have to be modelled. The basic equations are the continuity and momentum equations, whose conservative, nondimensional and incompressible form for the large scales can be written as follows:

$$\frac{\partial \overline{u}_i}{\partial x_i} = 0 \tag{1}$$

$$\frac{\partial \overline{u}_i}{\partial t} + \frac{\partial \overline{u}_i \overline{u}_j}{\partial x_j} = - \frac{\partial \overline{p}}{\partial x_i} + \frac{1}{Re_{2B}} \frac{\partial^2 \overline{u}_i}{\partial x_j^2} - \frac{\partial \tau_{ij}}{\partial x_j} \tag{2}$$

The Reynolds number $Re_{2B}$ is based on the nozzle inlet velocity $u_{in}$ and the hydraulic diameter of the geometry, which can be assumed as the double slot width 2B.
All variables are arranged on a co-located grid. The discretization scheme is second order in space (central differences). For the convective terms the Adams-Bashforth scheme is used for second order time discretization. The diffusive terms are discretized by the Crank-Nicolson scheme.
The algorithm is based on the fractional-step method of Kim and Moin (1985). First an intermediate velocity field $u^*$ is calculated

$$\frac{u_i^* - u_i^n}{\Delta t} = - \frac{3}{2} \, conv^{\,n} + \frac{1}{2} \, conv^{\,n-1} + \frac{1}{2\,Re} \nabla^2 (diff^{\,n} + diff^{\,n+1}) \tag{3}$$

where

$$u_i^* = u_i^{n+1} + \Delta t \, \nabla p^{n+1} \tag{4}$$

Applying continuity to Eq. 4 one obtains the following Poisson equation for the pressure field

$$\nabla^2 p^{n+1} = \frac{1}{\Delta t} \nabla u_i^* \tag{5}$$

Once Eq. 5 is solved, the intermediate velocity field $u^*$ is corrrected by Eq.4 in order to obtain the final velocity field.

$$u_i^{n+1} = u_i^* - \Delta t \, \nabla p^{n+1} \tag{6}$$

**Subgrid Scale Closure**

In (2) the last term $\tau_{ij}$, given by

$$\tau_{ij} = \overline{u_i u_j} - \overline{u}_i \, \overline{u}_j \tag{7}$$

represents the unresolved subgrid scales. Since the small scales tend to be more isotropic than the large scales, they can be modelled by the eddy viscosity model of Smagorinsky (1963)

$$\tau_{ij} - \frac{1}{3} \tau_{kk} \, \delta_{ij} = - 2 \, \nu_t \, \overline{S}_{ij} = - 2 \, C \, \overline{\Delta}^2 \, |\overline{S}| \, \overline{S}_{ij} = - 2 \, C \, \beta_{ij} \tag{8}$$

where $\delta_{ij}$ is the Kronecker delta, $|\overline{S}| = (2 \, \overline{S}_{ij} \, \overline{S}_{ij})^{\frac{1}{2}}$ and

$$\overline{S}_{ij} = \frac{1}{2} \left( \frac{\partial \overline{u}_i}{\partial x_j} + \frac{\partial \overline{u}_j}{\partial x_i} \right) \tag{9}$$

is the large scale strain rate tensor. In finite difference procedures the length scales, defined by the filter, are given by the grid size

$$\overline{\Delta} = ( \overline{\Delta}_1 \, \overline{\Delta}_2 \, \overline{\Delta}_3 )^{\frac{1}{3}} \tag{10}$$

The value of $C$ in (8) depends on the flow type under consideration. For different geometries the value must be determined again. This is what made LES limited to simple geometries like channel flows so far and unapplicable to complex flows of technological importance. Germano (1991) and later Lilly (1992) suggested a method to calculate the value of $C$ for each time step and grid point dynamically from flow field data.

The basic idea of the dynamic subgrid model is the definition of a coarser test filter and to sample the scales in the range between this test filter and the model filter. The background is the assumption that the scales which are only a little bit larger than the subgrid scales (= test filter scales) behave in a similar manner in comparison to the subgrid scales, so that analogous expressions for their definition can be used. The test filter stresses (denoted by a hat or [...]^)

$$T_{ij} = [ \, \overline{u_i u_j} \, ]^{\wedge} - \hat{\overline{u}}_i \, \hat{\overline{u}}_j \tag{11}$$

are expressed by using the Smagorinsky closure again

$$T_{ij} - \frac{1}{3} T_{kk} \delta_{ij} = \overline{\tau}_{ij} = -2\, C\, \hat{\Delta}^2\, |\hat{\overline{S}}|\, \hat{\overline{S}}_{ij} = -2\, C\, \alpha_{ij} \tag{12}$$

$$\text{where} \quad \hat{\overline{S}}_{ij} = \frac{1}{2}\left(\frac{\partial \hat{\overline{u}}_i}{\partial x_j} + \frac{\partial \hat{\overline{u}}_j}{\partial x_i}\right), \quad |\hat{\overline{S}}| = \left(2\, \hat{\overline{S}}_{ij}\, \hat{\overline{S}}_{ij}\right) \tag{13}$$

$$\text{and} \quad \hat{\Delta} = (\hat{\Delta}_1\, \hat{\Delta}_2\, \hat{\Delta}_3)^{\frac{1}{3}}; \quad \hat{\Delta}_1 = 2\Delta_1,\ \hat{\Delta}_2 = \Delta_2,\ \hat{\Delta}_3 = 2\Delta_3 \tag{14}$$

A relation for the resolved scales of motion between the test and the grid scale can be obtained by substracting (8) from (12).

$$L_{ij} = T_{ij} + \hat{\overline{\tau}}_{ij} = [\,\overline{u}_i\, \overline{u}_j\,]^{\wedge} - \hat{\overline{u}}_i\, \hat{\overline{u}}_j = -2\, C\, \alpha_{ij} + 2\, [\, C\, \beta_{ij}\,]^{\wedge} \tag{15}$$

There are five independent equations (see Eqs. 8, 12 and 15) which cannot be explicitly solved for the model constant $C$ because it appears in a filter operation (last term of Eq. 15). Some authors (Germano, 1991; Lilly,1992; Zang, 1993; Yang, 1993) assumed

$$[\, C\, \beta_{ij}\,]^{\wedge} = C\, \hat{\beta}_{ij} \tag{16}$$

so that (15) can be written as

$$E_{ij} = L_{ij} + 2\, C\, \alpha_{ij} - 2\, C\, \hat{\beta}_{ij} \tag{17}$$

where $E_{ij}$ is the residual of (10). By application of the least squares approach to minimize the residual, the following equation for $C$ can be obtained

$$C(y,t) = -\frac{1}{2} \frac{\langle L_{ij}\, (\alpha_{ij} - \hat{\beta}_{ij}) \rangle}{\langle (\alpha_{mn} - \hat{\beta}_{mn})\, (\alpha_{mn} - \hat{\beta}_{mn}) \rangle} \tag{18}$$

Here <> means plane-averaging for stabilization of the solution for $C$ during the calculation. It is mentioned (Hoffmann, 1994; Germano, 1991; Lilly,1992; Zang, 1993; Yang, 1993) that this averaging represents a great problem of the dynamic model procedure. Since the distribution of $C$ can change rapidly in space and time it is difficult to avoid numerical errors. Besides negative values of $C$ can appear and lead to in-

stabilities. Negative values of $C$ can be interpreted as backscatter (Lilly, 1992), i.e. energy transfer upscale, because in that case the turbulent viscosity $\nu_t$ in (8) is negative. For totally 3D flows like the rectangular impinging jet there is no homogeneous space direction. This is why we use instead of plane averaging a local average over the test filter cell, which was also used by Zang (1993) for recirculating flows.
Additionally we used a modification of (16) suggested by Piomelli (1994). From a mathematical point of view Eqn. (16) has an inconsistency, because it results from neglecting that $C$ is a function of space (Piomelli, 1994; Ghosal, 1993). But this is finally the goal of the dynamic model procedure, to get a distribution of $C$ depending on space and time. Piomelli (1994) pointed out that Ghosal et al. (1993) developed a consistent procedure without any least square approach. For calculating $C$ after this method an additional iteration is necessary which leads to a higher computational effort. Piomelli suggested a simpler approach which is not only mathematically consistent but also needs no iteration procedure.
Eqn. (15) is written as

$$C\,\alpha_{ij} = L_{ij} - 2\,[\,C^*\,\beta_{ij}\,]^\wedge \tag{19}$$

On the right hand side an estimate of the coefficient $C$, expressed by $C^*$ and assumed to be known, is used. In that case minimization of the sum of the squares results

$$C\,(x,y,z,t) = -\frac{1}{2}\,\frac{(\,L_{ij} - [\,C^*\,\beta_{ij}\,]^\wedge\,)\,\alpha_{ij}}{\alpha_{mn}\,\alpha_{mn}} \tag{20}$$

This is the equation for calculating the model coefficient $C$. Piomelli (1994) pointed out that there is no significant difference between zeroth- and first-order- approximation for estimating $C^*$. We used zeroth-order-approximation

$$C^* = C^{n-1} \tag{21}$$

During the calculation numerical errors by high peak values of $C$ appeared, although we used the local averaging procedure of Zang (1993). A limit of $C$ after local averaging was necessary.

$$0. \leq C \leq 0.04 \tag{22}$$

[Note: the minimum $C$=0 is necessary to avoid negative turbulent viscosity; the maximum $C$=0.04 means for the Smagorinsky coefficient $C_S$=0.2, so that $C$ ranges up to twice $C_s$=0.1, which was often used (Braun, 1994; Cziesla, 1996)].

## Boundary conditions

At the jet exit constant velocity (top-hat) profile and at the fixed walls no-slip condition are assumed. At the lower (impingement) plate the no-slip condition for the diffusive terms is implemented in two different ways. On the one hand the near wall area is calculated directly by grid refinement, so that a linear velocity profile for the wall gradients could be assumed (as it is done for laminar calculations). Therefore the distance of the centre of the near wall grid cells to the impingement plate is smaller then $y^+=yRe_\tau=1$, where $Re_\tau$ is the Reynolds number based on the friction velocity $u_\tau$. A second calculation was done using the logarithmic law of the wall (Braun, 1994; Cziesla, 1996) for calculation of the time averaged (denoted by <>) wall shear stress distribution. Applying the Schumann assumption (Schumann, 1975)

$$c = \frac{\tau_w}{\overline{u}_1|_w} = \frac{<\tau_w>}{<\overline{u}_1|_w>} \tag{23}$$

the instantaneous wall shear stresses can be obtained. At y=0 and y=4 periodic boundary condition are used. At the exit planes (x=0, x=5) there is the problem of inflow from the ambient into the computational domain due to entrainment effect by jet flows (Laschefski,1996; Hoffmann, 1994). In contrast to the periodic sides, where the handling of backflow makes no problem because of the periodic conditions which guarantee mass conservation, the amount of the backflow at the exit planes is unkown. So the exit boundary conditions have to be derived by a numerical algorithm to get mass conservation (Hoffmann, 1994). Integration of Eqn. (5) gives

$$\iiint_G (\nabla^2 \overline{p}^{\,n+1})\, dV = \frac{1}{\Delta t} \iiint_G (\nabla\, \overline{u}_i^*)\, dV \tag{24}$$

Application of the theorem of Gauss leads to

$$\iint_{\partial G} (\frac{\partial \overline{p}^{\,n+1}}{\partial n})\, dS = \frac{1}{\Delta t} \iiint_G (\nabla \overline{u}_i^*)\, dV \tag{25}$$

It is easy to show that the common pressure boundary condition, zero first derivative, cannot be used for jet flows. Since the intermediate velocity field $u^*$ is only the solution of the momentum equations and do not fulfil the global continuity in contrast to flows without backflow over the exit plane (e.g. channel flows), the term on the left hand side in (25) need not to be equal zero either. Our investigations for laminar impinging jet flows gave good results with p=const at the exit plane. But because of the turbulent nature we implemented a condition presented by Childs and Nixon (1986) and Grinstein et al. (1987), so that pressure fluctuations across the exit planes are possible.

$$\overline{p}_{exit\ plane} = 0.7\ \overline{p}_{interior} + 0.3\ \overline{p}_{ambient} \tag{26}$$

The pressure at the exit plane is an interpolated value between the pressure of the last interior cell of the domain and the ambient pressure, which is known ($p_{ambient}=0$ in case of nondimensional pressure).
Vanishing first derivative conditions for the velocities and $u^*$ are assumed.

**Results**

The computations have been performed on an IBM RISC workstation. For the statistical results 50 samples over 65 nondimensional time-units are used. For the LES with grid refining near the impingement plate instead of wall functions 82x82x54 (=363096) grids are used. The LES using the logarithmic law of the wall are done on 82x66x50 (=270600) grids. Besides we made a DNS (analogous numerical method as mentioned above) on 102x82x68 (568752) grids.

## 2D Results

Figure 2 shows the instantaneous contour of constant lines of the model coefficent $C$ for the 2D jet for $Re_B$=10000. On the boundaries $C$ becomes small as expected. This non-constant distribution of $C$ guarantees the damping of turbulent fluctuations in the sublayer. At the slot outlet (z=20) the flow is laminar, hence vanishing $C$. Similar results of $C$ have been found for the 3D jet. Figure 3 shows instantaneous velocity vectors for the 2D flow with $Re_B$=10000. Here z=20 is the slot outlet and z=0 is the location of the impingement plate. The jet generates vortices as the flow changes from laminar at the slot exit to turbulent on the impingement plate. The vortices are asymmetric except on the plate. The entrainment of the flow from ambient can also be seen. Figure 4 compares the normalized jet axial velocity $w/w_m$ distribution against x/B at different distances (z) from the impingement plate. The experimental result is given by Namer et al. (1988). We notice here the self-similar development of the axial velocity between z=12 and z=6. For larger x (x=0 at the jet axis) the incoming flow from ambient causes the disagreement with experiment.

## 3D Results

Figure 5 shows the flow struture for $Re_{2B}$=600 based on the nozzle exit velocity and double slot width B. Streamlines are plotted in half of the domain for the case of LES with grid refining at the lower wall. In the other half, vector velocity plots are shown for yz-planes at x=0 and x=5 (exit planes). The fluid leaving the slot nozzle ($1.75 \le y \le 2.25$, z=1) impinges directly on the lower plate, so that the stagnation line (y=2, z=0) appears. Then two wall jets develop in opposite directions. At the periodic sides of the domain, fluid separates and some part reattaches again on the lower plate exactly at the periodic boundaries (y=0 and 4, z=0). So corner vortices develop. Upwash fountains at the periodic boundaries (y=0, y=4) are obtained. Resulting from the geometrical symmetry of the periodic nozzle element, helical main flow structures appear, i.e the movement of the fluid towards the exit planes (x=0, x=5) can be described as a longitudinal vortex. In each quarter of the domain such structures appear. In Fig. 5 only two oppositely directed structures are shown (left half of the domain). Since jet flows entrain ambient fluid there is inflow over the two exit planes. Our results demonstrate, that the dynamic subgrid model used together with the exit boundary conditions is able to treat a complex flow with strong streamline curvature, because no damping of the helical structures appeared.

Figure 6 shows the time averaged pressure distribution on the impingement surface. In Fig. 7 the surface pressure distribution averaged over the x-direction for different calculations (LES with grid refinement, LES with wall function, DNS) is shown. Below the nozzle (1.75≤y≤2.25) the peak values caused by the stagnation appear. In the directions towards the periodic sides the pressure decreases because of the wall jet flow acceleration. The minimum values at about y=1.2 and 2.8 characterize the beginning of the seperation zones. Finally there are local maxima at the reattachement zones at y=0 and 4. The pressure decreases from the middle towards the exit planes (x=0, x=5) below the slot results from fluid acceleration directed parallel to the x-direction. This effect has been found by Martin (1977) also. Figure 7 shows that the surface pressure distributions obtained from the LES, with or without the wall function, agree very well with DNS. Only at the separation zones the values from the simulation using the wall function differ from the other two calculations. These differences are expressed by the global averaged surface pressure also. Figure 8 shows the dimensionless wall shear stress distribution, where the values are also averaged in x-direction. The LES with grid refining shows good agreement with the DNS data. In contrast to that the LES with wall function shows significant differences. Not only the value at the stagnation line (y=2) is lower then the ones of the other cases, but also the distribution between y=0.8 and 3.2, which is the wall jet region, gives lower wall shear stresses.

## Conclusion

LES using a localized dynamic subgrid model of a rectangular impinging jet for $Re_{2B}$=600 were done. Results confirm that the used localized subgrid model is able to simulate the complex helical flow structure. No damping of the helical flow structures appeared. No empirical data are necessary for this model. The coefficient is calculated by flow field data only. The problem of ill-conditioning of the model coefficient is described and suggestion for solution presented. Since the calculations show satisfactory flow field results the implemented condition describes the physical flow behavior (backflow) very well. LES with grid refinement near the impingement surface agrees very well with DNS data. LES using a wall function shows larger differences, so that the LES with grid refining should be preferred.

**Acknowledgement:** We thank DFG for financial support for this work.

## References

Braun, H., Fiebig, M., Mitra, N.K. (1994) Large-Eddy Simulation of Seperated Flow in a Ribbed Duct. *AGARD, Chania, Greece,* 18-21, April 1994, Proceedings, pp. 9/1-9

Childs, R. E., Nixon, D. (1986) Unsteady Three Dimensional Simulations of a VTOL Upwash Fountain. *AIAA-Paper, 86-0212*

Cziesla, T., Braun, H., Biswas, G., Mitra, N.K. (1996) Large Eddy Simulation in a Channel with Exit Boundary Conditions. *ICASE (NASA Langley) Report* 198304, No. 96-18

Gardon, R., Akfirat, J.C, (1966) Heat Transfer Characteristics of Impinging Two-Dimensional Air Jets. *Journal of Heat Transfer*, February 1966, pp. 101-108

Germano, M., Piomelli, U., Moin, P., Cabot, W. H. (1991) A Dynamic subgrid-scale

eddy viscosity model. *Phys. of Fluids*, A3 (7), pp. 1760-1765
Ghosal, S., Lund, T.S., Moin, P. (1993) A local dynamic model for LES. *CTR Annual Research Briefs*, Center for Turbulence Research, Stanford University, 3
Grinstein, F.F., Oran, E.S., Boris, J.P. (1987) Direct Numerical Simulation of Axisymmetric Jets. *AIAA-Journal*, Vol. 25, No. 1
Hoffman, G., Benocci, C. (1994) Numerical Simulation of Spatially-Developing Planar Jets. *AGARD, Chania, Greece*, 18-21, April 1994, Proceeding, pp. 26/1-6
Kim, J., Moin, P. (1985) Application of a fractional step method to incompressible Navier-stokes equations. *J. of Comp. Phys.*, 59, pp. 308-323
Laschefski, H. et al. (1996) Numerical Investigation of Heat Transfer by Rows of Rectangular Impinging Jets. *Numerical Heat Transfer*, Vol. 30, pp. 87-101
Lilly, D. K. (1992) A proposed modification of the Germano subgrid-scale closure method. *Phys. of Fluids* A4 (3), pp. 633-635
Martin, H. (1977) Heat and Mass Transfer between Impinging Gas Jets and Solid Surfaces. *Advances in Heat Transfer*, 13, pp. 1-60
Namer, I., Ötügen, M.V. (1988) Velocity measurements in a plane turbulent air jet at moderate Reynolds numbers. *Exp. in Fluids,* Vol. 6, pp. 387-399
Piomelli, U., Junhui, L. (1994) Large-Eddy Simulation of rotating channel flows using a localized dynamic model. *AGARD, Chania, Greece*, 18-21 April 1994, Proceedings, pp. 3/1-9
Schumann, U. (1975) Subgrid-Scale Model for Finite Difference Simulations of Turbulent Flows in a Plane Channel and Annuli. *J. of Comp. Phys.*, 18, pp. 376-404
Smagorinsky, J.S. (1963) General Circulation Experiments with Primitive equations. I. The Basic Experiment. *Monthly Weather Review*, 91
Viskanta, R. (1993) Heat Transfer to Impinging Isothermal Gas and Flame Jets. *Exp. Thermal and Fluid Science*, 6, pp. 111-134
Yang, K.S., Ferziger, J.H. (1993) Large-Eddy Simulation of Turbulent Obstacle Flow Using a Dynamic subgrid-Scale Model. *AIAA Journal*, Vol. 31, No. 8, pp. 1406-1413
Zang, Y. et al. (1993) A dynamic mixed subgrid-scale model and its application to turbulent recirculating flows. *Phys. Fluids*, A5, 12, pp. 3186-3196

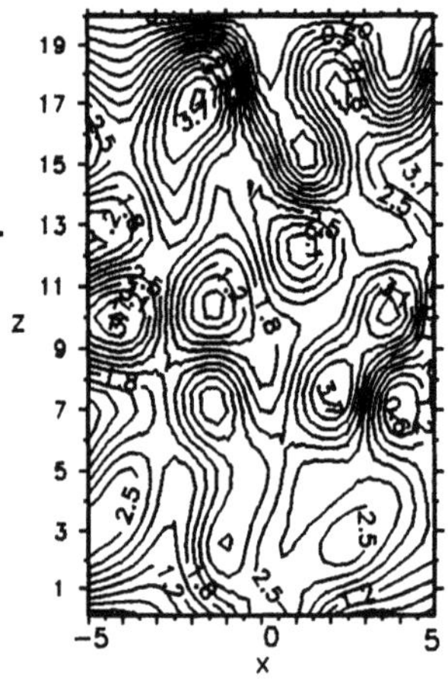


Fig. 2:
Contourplot of model coefficent $C$, instantaneous

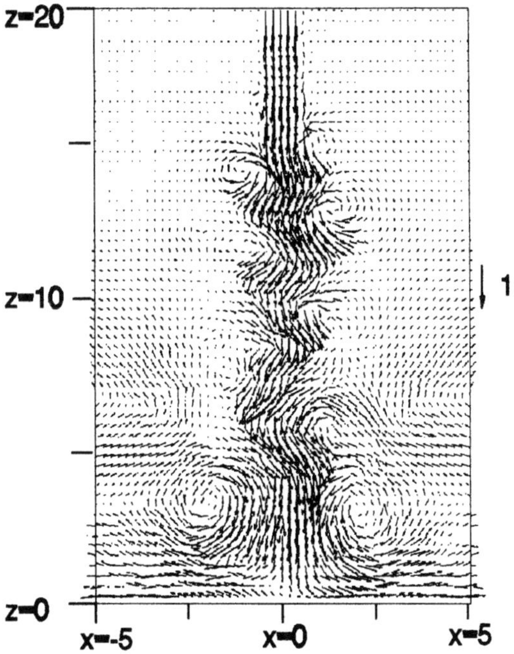


Fig.3: Velocity vector plot, instantaneous

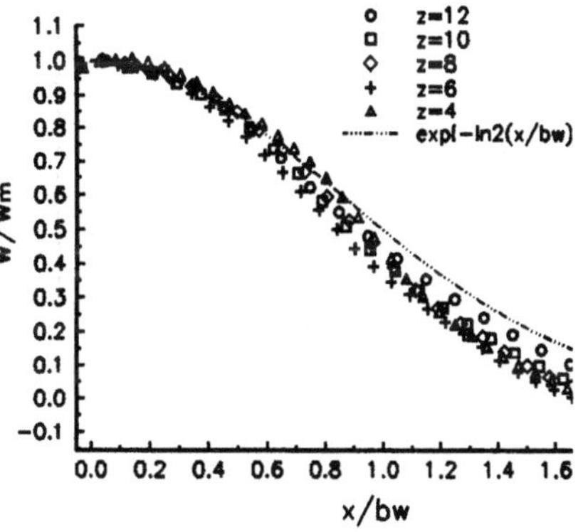


Fig. 4: Normalized jet axial velocity distribution

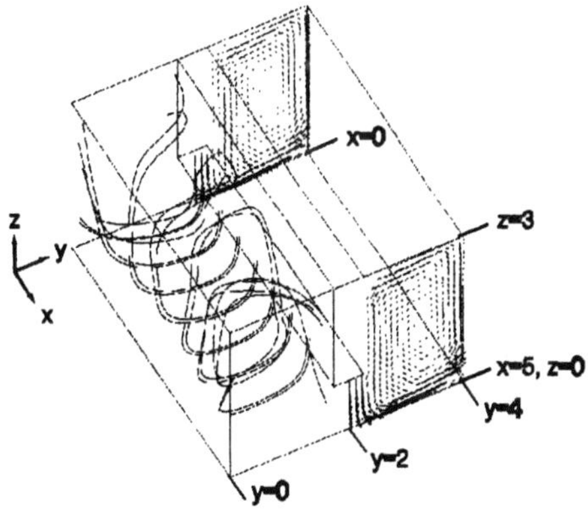


Fig. 5: Streamlines and vectorplots of different yz-planes, time-averaged

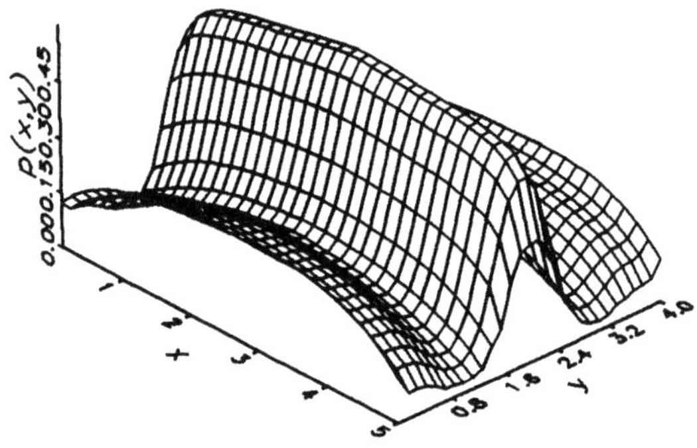


Fig. 6: Nondimensional surface pressure distribution, time-averaged

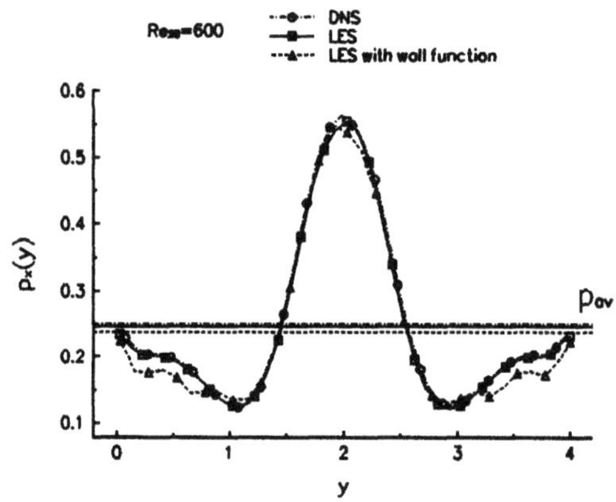


Fig. 7 : Comparison of nondimensional surface pressure distribution, averaged in x-direction

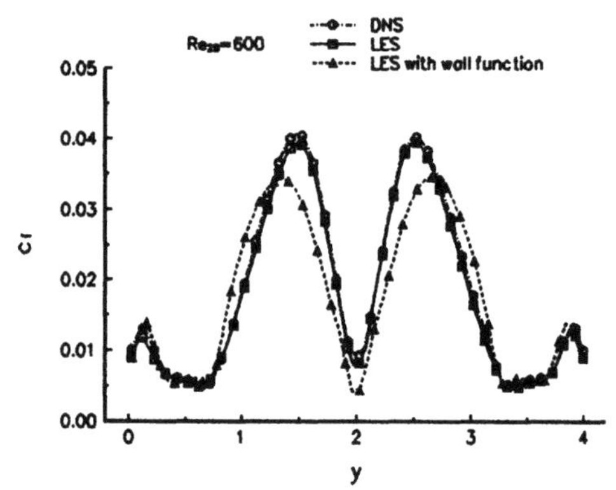


Fig. 8: Comparison of nondimensional shear stress distribution, averaged in x-direction

# LES PREDICTION OF GAS DIFFUSION NEAR BUILDING USING DYNAMIC MIXED SGS MODEL BASED ON A COMPOSITE GRID SYSTEM

A. MOCHIDA*, Y. TOMINAGA* AND S. MURAKAMI**
* *Niigata Institute of Technology, Niigata, Japan*
** *I.I.S., University of Tokyo, Tokyo, Japan*

**Abstract** The turbulent diffusion of gaseous contaminant near a building model is predicted by Large Eddy Simulations (LES) using a Dynamic Mixed SGS model (DM model ) with the aid of the composite grid technique. The results using a composite grid show better agreement with the experiment than does the case using a single grid. Some discrepancies from the experimental results are observed in the results of the standard Smagorinsky model (S model) with a composite grid. These discrepancies are reduced remarkably by using the DM model.

## 1. Introduction

This paper reports on the results of LES of a turbulent diffusion of gaseous contaminant near a building model using the dynamic subgrid-scale(SGS) model. In recent years, the present authors have carried out the computations of a turbulent flow past a bluff body, such as a cube, two-dimensional square cylinder, etc. using various types of SGS models ; i.e., the conventional standard Smagorinsky model (S model) , Dynamic Smagorinsky model (DS model)(Germano et al.(1991), Lilly(1992)) and Dynamic Mixed model (DM model)(Zang et al.(1993)), etc. In our experience, results of DS and DM models

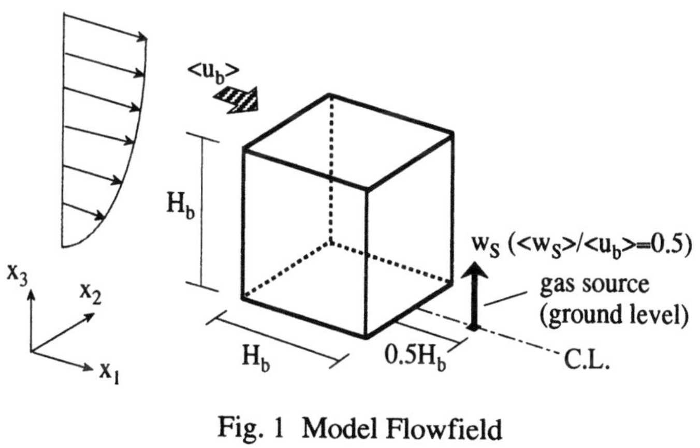


Fig. 1 Model Flowfield



*J.-P. Chollet et al. (eds.), Direct and Large-Eddy Simulation II*, 279-290.

showed better agreement with the experimental data in comparison with the static type of S model. Furthermore, the DM model provided more accurate results than did DS model in some respects (Murakami et al.(1996)).

In this study, the DM model is applied to the analysis of a turbulent diffusion field near a building model. Here, a localization technique proposed by Piomelli et al. is employed in order to avoid instability in proceeding with the calculation (Piomelli and Liu(1995)). Furthermore, a composite grid technique supported by a fortified solution algorithm is adopted (Delsem, and Steger (1986), Mochida et al. (1993)). Computations based on the S model are also carried out for comparison. Accuracy of these computations is examined precisely by comparing the numerical results with experimental data measured by the present authors (Tominaga et al. (1992), Murakami et al. (1995)).

## 2.Outline of Computations

### 2.1 Flow and diffusion fields analyzed

Fig. 1 illustrates the flow situation analyzed here. A cubical model is located in the turbulent boundary layer expressed as $<u_1 (x_3)> \propto x_3^{1/4}$. Here, $u_1$ is streamwise velocity and $x_3$ is vertical direction. The angle brackets denote time-averaged values. The square-shaped gas source is set at ground level in the recirculation region behind the cube. The side length of the gas source is $0.025H_b$ where $H_b$ is the height of the building model. In this study, tracer gas is treated as a passive scalar ($\rho_s/\rho_a=1.0$, $\rho_s$ : gas density, $\rho_a$ : ambient air density). Exit gas speed $<w_S>$ is $0.5<u_b>$, where $u_b$ is $<u_1>$ value at the inflow of the computational domain at height $H_b$.

### 2.2 Computed cases

Table 1 lists the three cases compared : two different grid systems, and two different SGS models. A composite grid technique (Delsem and Steger (1986), Mochida et al. (1993)) is adopted in cases 2 and 3. Two types of SGS models (S

Table 1 Computed cases

| case | SGS model | grid system | number of grid points | computational domain |
|---|---|---|---|---|
| 1 | S model | single | 116,025<br>$65(x_1)\times51(x_2)\times35(x_3)$ | $15.7(x_1)\times9.7(x_2)$<br>$\times5.2(x_3)$ |
| 2 | S model | composite | 112,458<br>Grid A : $25(x_1)\times25(x_2)\times12(x_3)$<br>Grid B : $63(x_1)\times49(x_2)\times34(x_3)$ | Grid A : $0.5(x_1)\times0.5(x_2)$<br>$\times0.25(x_3)$<br>Grid B : $15.7(x_1)\times9.7(x_2)$<br>$\times5.2(x_3)$ |
| 3 | DM model | composite | | |

S model : Smagorinsky model, DM model : Dynamic Mixed model

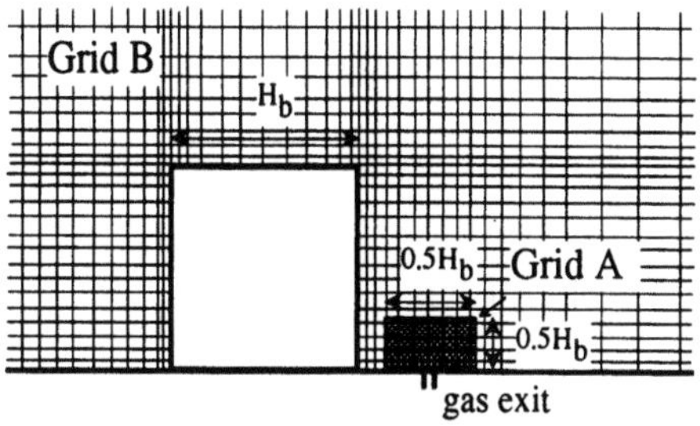


Fig. 2 Grid layout for cases 2 and 3
[Composite grid,
Grid A is embeded in grid B]

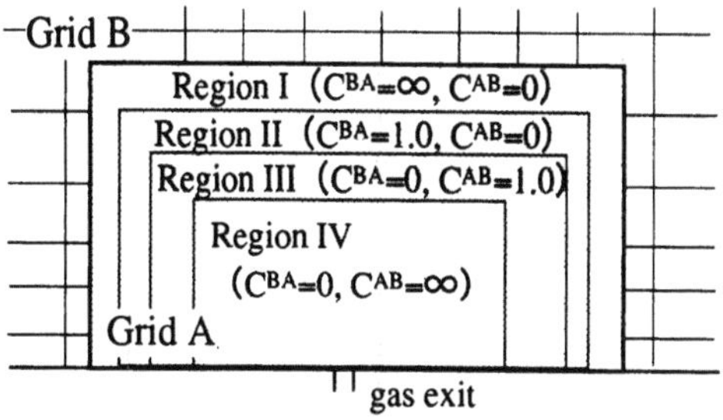


Fig. 3 Distributions of switching parameters

model and DM model) are compared using the composite grid technique for cases 2 and 3. Computation using a usual rectangular structural grid (case 1) is also carried out and compared with the results of computations on the composite grid.

**2.3 Grid system used**

Fig. 2 illustrates the grid layout employed for cases 2 and 3. Grid B is used to cover the whole computational domain ($15.7H_b$ ($x_1$) $\times$ $9.7H_b$ ($x_2$) $\times$ $5.2H_b$ ($x_3$)) and grid A is applied in the region near the gas exit ($0.5H_b$ ($x_1$) $\times$ $0.5H_b$ ($x_2$) $\times$ $0.25H_b$ ($x_3$)) . In order to avoid serious numerical instability caused by inappropriate connection at the connecting regions, the composite grid technique is supported by the fortified Navier-Stokes approach (Delsem, and Steger (1986) , Mochida et al. (1993)) . Details of the fortified Navier-Stokes approach are given in the appendix.

The total number of grid points is 112,458 ( 63 ($x_1$) $\times$ 49 ($x_2$) $\times$ 34 ($x_3$) for grid B and 25 ($x_1$) $\times$ 25 ($x_2$) $\times$ 12 ($x_3$) for grid A) for cases 2 and 3. The smallest mesh size on grid A is $0.005H_b$ at the area just near the gas exit in these cases. In case 1, the usual single grid system is used. In this case, the total number of grid points is 116,025 (=65 ($x_1$) $\times$ 51 ($x_2$) $\times$ 35 ($x_3$)) , almost the same as for cases 2 and 3. The smallest mesh size in case 1 is $0.025H_b$ at the area near the gas exit.

**2.4 SGS models compared**

Two types of SGS models, the static type of conventional Smagorinsky model (S model, cases 1 and 2 in Table 1) and the dynamic mixed SGS model (DM model, case 3) are compared.

In the S model, the anisotropic part of the SGS stress $\tau_{ij}$ is modeled as follows :

$$\tau_{ij}^{a} = -2C\,(f_{\mu}\overline{\Delta})^{2}\,|\overline{S}|\overline{S}_{ij}\,. \qquad (1)$$

The superscript "a" denotes the anisotropic part of the tensor. C is the model coefficient, $\overline{\phantom{x}}$ denotes the grid-filtered value, $\overline{\Delta}$ is the width of the grid-filter, and $\overline{S}_{ij}$ is the resolved-scale strain rate tensor,

$$\bar{S}_{ij}=\frac{1}{2}\left(\frac{\partial\bar{u}_i}{\partial x_j}+\frac{\partial\bar{u}_j}{\partial x_i}\right),\quad |\bar{S}|=(2\bar{S}_{ij}\bar{S}_{ij})^{1/2}. \tag{2}$$

In the S model (cases 1 and 2), C is treated as a constant. The value of 0.0144 is selected in cases 1 and 2. This value corresponds to 0.12 of the so-called Smagorinsky constant $C_S$. Previous studies by the present authors reported that the appropriate value of $C_S$ was 0.12 in a flowfield around a cube (Murakami et al. (1992)). $\bar{\Delta}$ is multiplied by the Van Driest type wall damping function $f_\mu$, 1-exp $(-x_n^+/25)$, in order to account for the near wall effect in the S model, while $f_\mu$ is not necessary in the dynamic SGS models.

In dynamic SGS models, the model coefficient C is determined dynamically. Following Germano et al. (1991), a test filter (denoted as $\hat{\ }$ in this paper) is introduced to derive an expression for C. The test filter width is taken to be twice the grid filter width. Germano et al. defined the resolved turbulent stress as follows :

$$\mathcal{L}_{ij}=\widehat{\bar{u}_i\bar{u}_j}-\hat{\bar{u}}_i\hat{\bar{u}}_j\,. \tag{3}$$

$\mathcal{L}_{ij}$ can be related to the SGS stress $\tau_{ij}$ and the subtest-scale stress $T_{ij}=\widehat{\overline{u_iu_j}}-\hat{\bar{u}}_i\hat{\bar{u}}_j$ by the Germano identity

$$\mathcal{L}_{ij}=T_{ij}-\hat{\tau}_{ij}. \tag{4}$$

The DM model (Zang et al. (1993)) employs a linear combination of the dynamic Smagorinsky model and the scale similarity model (Bardina et al. (1981). In the DM model, the anisotropic part of the SGS stresses $\tau_{ij}$ and subtest-scale stresses $T_{ij}$ are expressed as :

$$\tau_{ij}^a=-2C\beta_{ij}+b_{ij}, \tag{5}$$

$$T_{ij}^a=-2C\alpha_{ij}+B_{ij}, \tag{6}$$

$$\text{where}\quad \alpha_{ij}=\hat{\bar{\Delta}}|\hat{\bar{S}}|\hat{\bar{S}}_{ij},\ \beta_{ij}=\bar{\Delta}|\bar{S}|\bar{S}_{ij}, \tag{7}$$

$$b_{ij}=\overline{\bar{u}_i\bar{u}_j}-\bar{\bar{u}}_i\bar{\bar{u}}_j,\ B_{ij}=\widehat{\overline{\bar{u}_i\bar{u}_j}}-\hat{\bar{u}}_i\hat{\bar{u}}_j\,. \tag{8}$$

The first and the second terms in the right-hand of eqs. (5) and (6) derive from the Smagorinsky model and the scale similarity model respectively. Substituting eqs. (5) and (6) into eq. (4) yields

$$\mathcal{L}_{ij}^a=-2C\alpha_{ij}+\widehat{2C\beta_{ij}}+\gamma_{ij}, \tag{9}$$

$$\text{where}\ \gamma_{ij}=B_{ij}-\hat{b}_{ij}=\widehat{\bar{\bar{u}}_i\bar{\bar{u}}_j}-\hat{\bar{\bar{u}}}_i\hat{\bar{\bar{u}}}_j, \tag{10}$$

$\mathcal{L}_{ij}^a$ indicates the anisotropic part of $\mathcal{L}_{ij}$.

Eq. (9) cannot be solved explicitly for C, since C appears inside a filtering operation in the right-hand side in eq. (9). Previous authors (Germano et al. (1991), Zang et al. (1993), Lilly (1992)) assumed $\widehat{C\beta_{ij}}=C\hat{\beta}_{ij}$ to avoid this problem. However, this assumption is obviously inconsistent with the concept of dynamic LES, in which C is treated as a function of space and time. In this study, a localization technique proposed by Piomelli and Liu (1995) is employed.

Piomelli and Liu (1995) developed an approximate localization technique based on recasting the expression in eq. (9) in the form

$$-2C\alpha_{ij} = \mathcal{L}_{ij}^{a} - \gamma_{ij} - 2\widehat{C^{*}\beta_{ij}}. \tag{11}$$

Here, C* is the model coefficient inside the test filter, which is treated as the known value shown below. Eq. (11) is a tensor equation with six independent components. Hence, we need a procedure to specify the scalar value of C from the tensor equation. We use a least-square method suggested by Lilly (1992), in which the sum of the squares of the residual of eq. (11) is minimized. Using this method, the expression for a scalar value of C is deduced from eq. (11) (Piomelli and Liu (1995)):

$$C = -\frac{1}{2}\frac{(\mathcal{L}_{ij}^{a} - \gamma_{ij} - 2\widehat{C^{*}\beta_{ij}})\,\alpha_{ij}}{\alpha_{kl}^{2}}. \tag{12}$$

Since C* must be treated as a known value in order to solve eq. (12), we assume $C^{*}=C^{n-1}$ for C* at time step *n* following Piomelli and Liu (1995).

The dynamic SGS model can be extended easily to the modeling of a SGS scalar flux $(h_3 = \overline{u_3 c_g} - \bar{u}_3\bar{c}_g)$.

In the S model, $h_j$ is modeled as follows:

$$h_j = -\frac{\nu_{SGS}}{Sc_{SGS}}\frac{\partial \bar{c}_g}{\partial x_j} = -\frac{C\bar{\Delta}^2}{Sc_{SGS}}|\bar{S}|\frac{\partial \bar{c}_g}{\partial x_j}, \tag{13}$$

where $Sc_{SGS}$ is the SGS Schmidt number defined as follows:

$$Sc_{SGS} = \frac{\nu_{SGS}}{K_{SGS}}. \tag{14}$$

$K_{SGS}$ is the SGS eddy diffusivity. In this study, $Sc_{SGS}$ is treated as a constant 0.5 (Antonopoulos-Domis (1981)) in the S model.

In the DM model, $h_j$ is expressed by

$$h_j = -\frac{\nu_{SGS}}{Sc_{SGS}}\frac{\partial \bar{c}_g}{\partial x_j} + b_{jc}, \tag{15}$$

$$b_{jc} = \overline{\bar{u}_j\bar{c}_g} - \bar{\bar{u}}_j\bar{\bar{c}}_g. \tag{16}$$

$b_{jc}$ corresponds to $b_{ij}$ in eq. (5), which derives from the scale similarity model. We employed the localization technique here, in the same manner as for estimating the momentum flux:

$$\frac{1}{Sc_{SGS}} = \frac{\left(\mathcal{L}_{jc} - \gamma_{jc} - 2\widehat{\dfrac{\beta_{jc}}{Sc^{*}_{SGS}}}\right)\alpha_{jc}}{\alpha_{lc}^{2}}, \tag{17}$$

where $b_{jc} = \overline{\bar{u}_j\bar{c}_g} - \bar{\bar{u}}_j\bar{\bar{c}}_g$, $\quad \gamma_{jc} = \widehat{\overline{\bar{u}_j\bar{c}_g}} - \widehat{\bar{\bar{u}}}_j\widehat{\bar{\bar{c}}}_g$ (18)

$$\alpha_{jc} = C\hat{\Delta}^2 |\hat{\bar{S}}| \frac{\partial \hat{\bar{c}}_g}{\partial x_j}, \quad \beta_{jc} = C\bar{\Delta}^2 |\bar{S}| \frac{\partial \bar{c}_g}{\partial x_j}. \tag{19}$$

$Sc_{SGS}$ is determined dynamically by eq. (17) . In this study, negative values of C and $Sc_{SGS}$ appeared even though the localization technique was used. Therefore we used the clipping technique, i.e., the coefficients C and $Sc_{SGS}$ were set equal to zero wherever they were estimated to be negative by eqs. (12) and (17) .

### 2.5 Numerical methods and boundary conditions

For the computations conducted here, a second-order centered difference scheme was adopted for the spatial derivatives. For time advancement, the second order Adams-Bashforth scheme was used for the convection terms and the Crank-Nicolson scheme for the diffusion terms. The interval for time advancement is $1.0 \times 10^{-3}$ for all cases in a non-dimensional time scale based on $\langle u_b \rangle$ and $H_b$.

At the inflow boundary, the time history of velocity fluctuations $u_i'$ (t) in the boundary layer of a fully developed channel flow predicted by LES were utilized. These velocity fluctuations were adjusted in advance by multiplying by scaling parameters such that the turbulence intensities were the same as those of the wind tunnel experiment. The adjusted fluctuations were superimposed on the mean profile expressed as $\langle u_1 (x_3) \rangle \propto x_3^{1/4}$ which corresponds well with the measured profile in the experiment. The value of turbulent energy k at height $H_b$ at inflow boundary is $0.03\langle u_b \rangle$. For the boundary condition at the solid walls, Werner and Wengle's approach (Werner and Wengle (1991)) was adopted, in which a linear or 1/7 power law distribution of the instantaneous velocity is assumed :

$$\frac{\bar{u}}{u_*} = x_n^+ \quad ( x_n^+ \leq 11.81 ), \tag{20}$$

$$\frac{\bar{u}}{u_*} = 8.3\, x_n^{+1/7} \quad ( x_n^+ > 11.81 ). \tag{21}$$

where u* is the friction velocity and $x_n^+$ is the distance from the wall ($x_n^+ = x_n u^*/\nu$) .

## 3. Results and Discussion

### 3.1 Influence of mesh spacing near gas exit : Comparison between results on single grid system (case 1) and composite grid system (case 2) with S model

Firstly, results from the S model using a single grid (case 1) and a composite grid (case 2) are compared. In Fig 4, the distributions of the time-averaged gas concentration $\langle c_g \rangle$ in the center section are shown. The values are normalized by the reference gas concentration $\langle c_{g0} \rangle$ ($=q/\langle u_b \rangle/H_b^2$, q is gas emission rate) . In the result of the experiment, tracer gas discharged from the ground is advected upstream by the reverse flow in the recirculation region behind the cube, and thus the value of $\langle c_g \rangle$ is relatively large in the area between the leeward face of the cube and the gas exit (Fig. 4 (1)) .

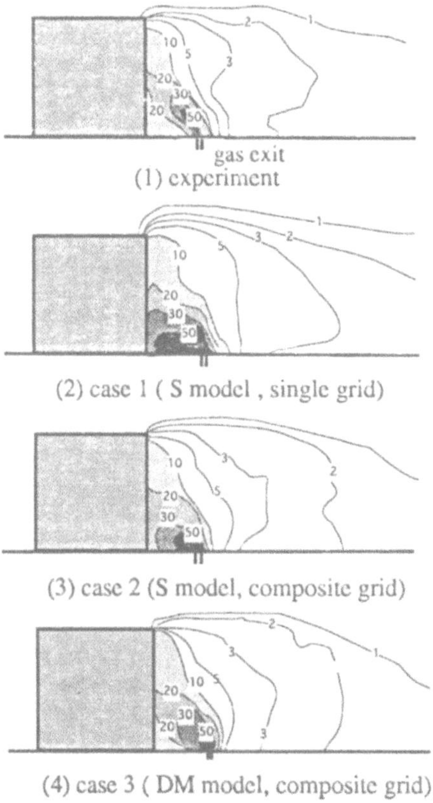


Fig. 4 $\langle c_g \rangle / \langle c_{g0} \rangle$ (at the center section)

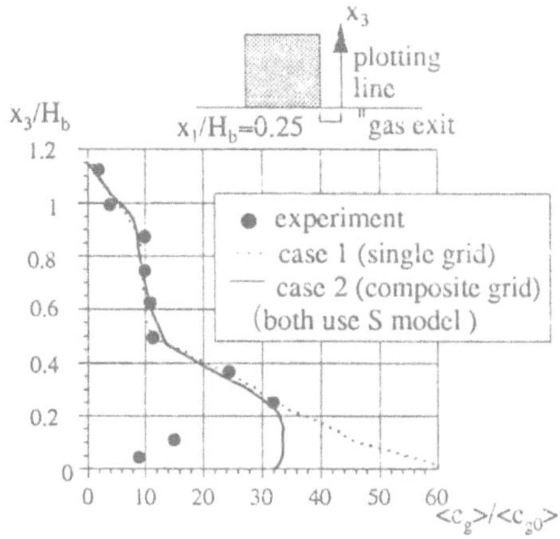


Fig. 5 Vertical profiles of $\langle c_g \rangle / \langle c_{g0} \rangle$
(comparison between cases 1 and 2)

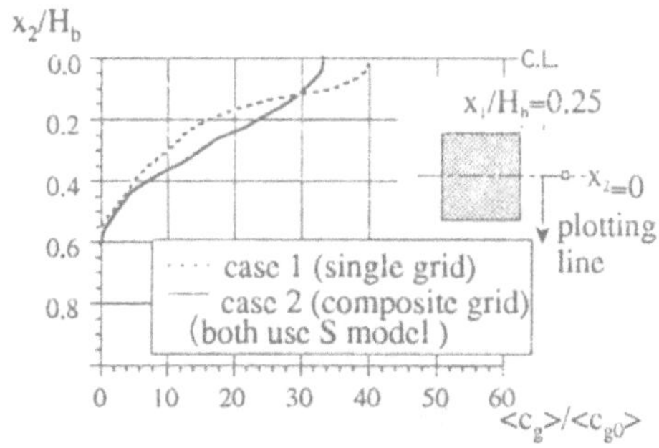


Fig. 6 Lateral profiles of $\langle c_g \rangle / \langle c_{g0} \rangle$
($x_3/H_b$=0.15)

As shown in Fig. 4, $\langle c_g \rangle$ is overestimated in the region near the leeward face of the cube in the results of cases 1 and 2. However, the result from case 2 (composite grid) shows better agreement with the experiment than does case 1 (single grid) (cf. Figs. 4 (1) ~ (3)) . The better agreement for the distribution of $\langle c_g \rangle$ in case 2 is due to the fine mesh spacing in the vicinity of the gas exit in this case.

Fig. 5 shows the vertical profiles of $\langle c_g \rangle$ in cases 1 and 2. In the region near the leeward face of the cube, $\langle c_g \rangle$ given from the experiment decreases rapidly as the measuring point approaches ground level. This tendency is not reproduced at all in case 1 (single grid) . In comparison with the result of case 1, the result of case 2 agrees well with the experimental data. Fig. 6 compares lateral profiles of $\langle c_g \rangle$ behind the cube at $x_3/H_b = 0.15$. The value of $\langle c_g \rangle$ in case 2 becomes smaller than that in case 1 near the centerline. This is because the diffusion effect in the lateral direction is well reproduced in case 2 by using the finer mesh spacing near the gas exit.

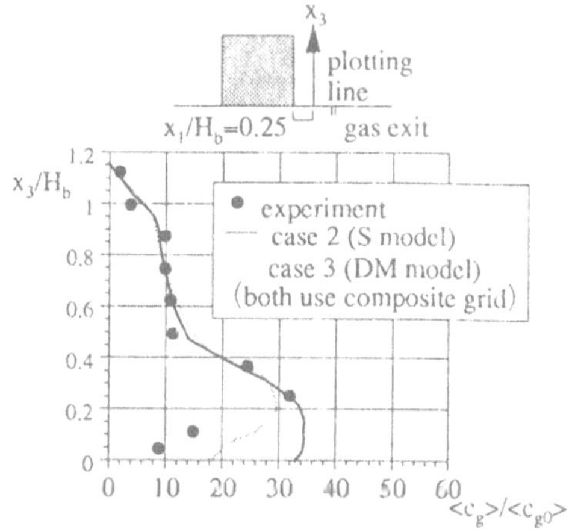


Fig. 7 Vertical profiles of $<c_g>/<c_{g0}>$ (comparison between cases 2 and 3)

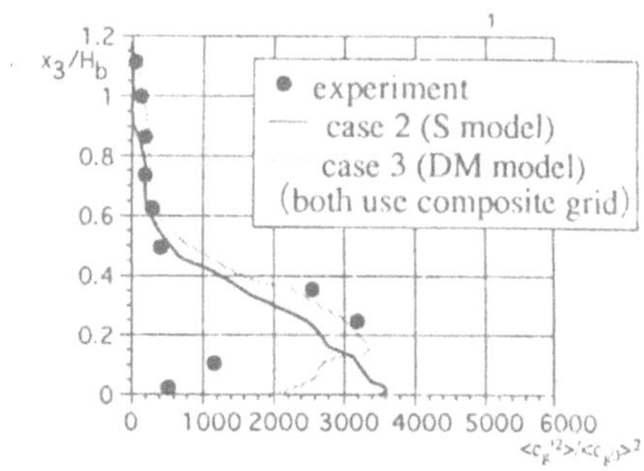


Fig. 9 Vertical profiles of $<c_g'^2>/<c_{g0}>^2$ (comparison between cases 2 and 3)

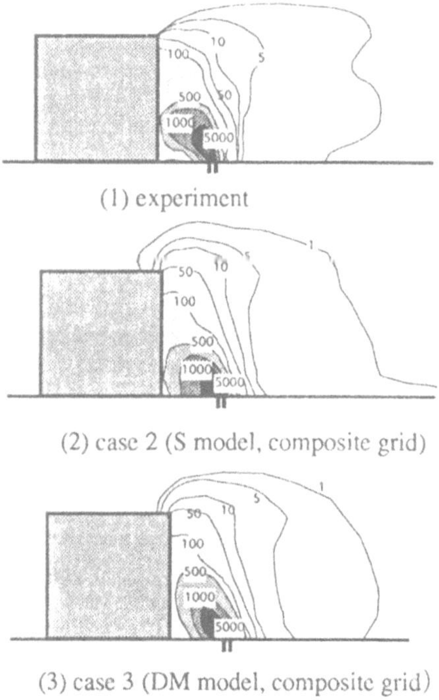


Fig. 8 $<c_g'^2>/<c_{g0}>^2$ (at the center section)

## 3.2 Influence of SGS models : Comparison between results with S model (case 2) and DM model (case 3) on composite grid system

### 3.2.1 Distributions of time-averaged gas concentration $<c_g>$ (Figs. 4(3) and (4), Fig. 7)

The result of case 3 (DM model) shows rather good agreement with the experiment (Figs. 4 (1) and (4)) . The result of the experiment shows a maximum peak of $<c_g>$ at $x_3/H_b \doteqdot 0.25$ just behind the cube, as is shown in Fig. 7. The result given by case 3 (DM model) well reproduces the decrease of $<c_g>$ near the ground. On the other hand, the decrease of $<c_g>$ near the ground is not reproduced well in case 2 (S model) .

### 3.2.2 Distributions of concentration fluctuation $<c_g'^2>$ (Figs. 8 and 9)

Fig. 8 illustrates the distributions of $<c_g'^2>$. Here, the prime denotes the deviations

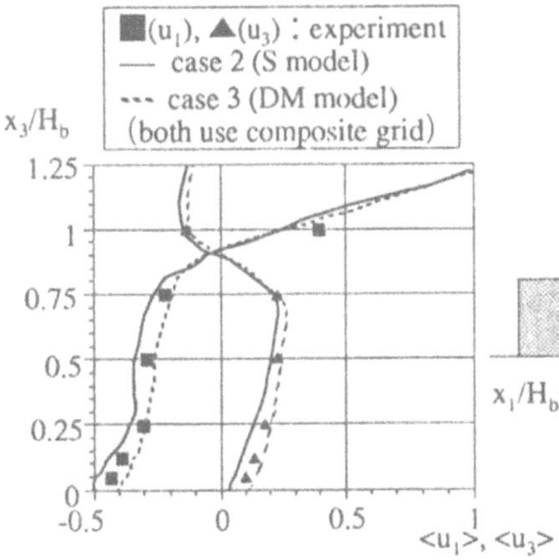


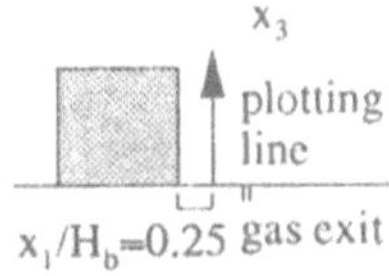


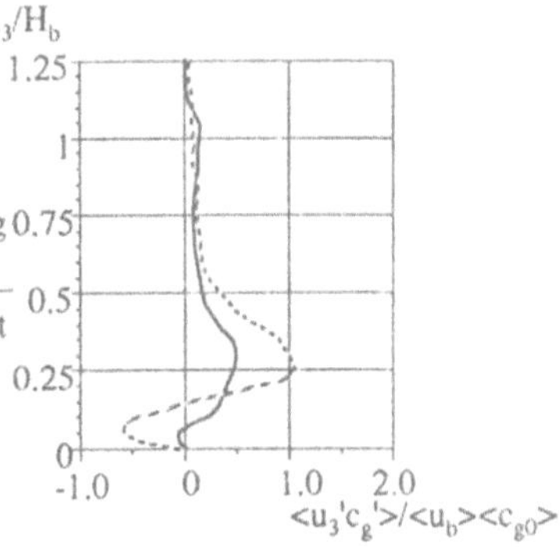


Fig. 10 Vertical profiles of $<u_1>$ and $<u_3>$

Fig. 11 Vertical profiles of $<u_3'c_g'>/<u_b><c_{g0}>$ ( total turbulent scalar flux )

from the time-averaged value. It can be seen that the patterns of the distributions of $<c_g'^2>$ are similar to those of $<c_g>$ for each case. Fig. 9 shows the vertical profiles of $<c_g'^2>$ behind the cube. Case 3 (DM model) shows much better agreement with the experiment than that of case 2 (S model) .

**3.2.3. Distributions of time-averaged velocities $<u_1>$ and $<u_3>$ (Fig. 10)**

Fig. 10 shows a comparison of the vertical distributions of time-averaged velocities $<u_1>$ and $<u_3>$ behind the cube in the center section. In this region, values from both cases agree rather well with those from the experiment. However, the S model overestimates the absolute value of $<u_1>$ near the ground (in this region, $<u_1>$ is negative) and underestimates the value of $<u_3>$. With this decrease of $<u_3>$ in this region in case 2 (S model) , advection of highly contaminated air from the area near the ground to the upper area also decreases. Thus, the distribution of $<c_g>$ in case 2 (S model) corresponds less favorable with the experiment than does the result from case 3 (DM model) , as is seen in Figs. 4 and 7.

**3.2.4 Distributions of $<u_3'c_g'>$ (Fig. 11)**

Fig. 11 compares the vertical profiles of $<u_3'c_g'>$, the total turbulent scalar flux (Grid Scale + SGS components) , behind the cube. The absolute value of $<u_3'c_g'>$ in case 3 (DM model) is much larger than that in case 2 (S model) . The difference between the results of cases 2 and 3 is significant at $x_3/H_b \fallingdotseq 0.25$, where a large positive value of $<u_3'c_g'>$ is observed in the result of case 3. According to this large turbulent flux $<u_3'c_g'>$, concentration near the ground decreases in case 3, and consequently the distribution of $<c_g>$ given by case 3 shows much better agreement with the experiment in comparison with that given by case 2, as is shown in Fig. 7.

## 4. Conclusions

1) The results using a composite grid show better agreement with the experiment than do the results using a single grid. This indicates that the fineness of mesh spacing near the gas exit is very important for reproducing the early process of gas diffusion.

2) However, some discrepancies from the experimental results are observed in the distributions of $<c_g>$ and $<c_g'^2>$ near the ground in the results of case 2, which employed the S model with a composite grid. These discrepancies are reduced remarkably by using the DM model (case 3).

3) The DM model reliably reproduces the distributions of the mean concentration $<c_g>$ and the concentration fluctuation $<c_g'^2>$.

## Appendix Fortified Navier-Stokes approach

In this approach, Navier-Stokes equations and the transport equation for the gas concentration are modified to include a forcing term. As an example, the modified Navier-Stokes equation for the streamwise velocity component for the fortification from grid B to grid A is shown in Eq. (21).

$$\frac{\partial U^A_{I,J,K}}{\partial t} + Cx^A_{I,J,K} = -Px^A_{I,J,K} + Dx^A_{I,J,K} + \underline{C^{BA}_{I,J,K}(U^B_{(I,J,K)} - U^A_{I,J,K})} \quad (21)$$

Here, I, J and K denote spatial positions at $x_1$, $x_2$ and $x_3$ directions respectively. $U^A_{I,J,K}$ indicates streamwise velocity U, defined at the nodal point (I,J,K) on grid A and $U^B_{(I,J,K)}$ is the interpolated value onto the nodal point (I,J,K) using velocities defined at the nodal points on grid B. Cx, Px and Dx in eq. (21) are convection, pressure gradient and diffusion terms, respectively.

The underlined term on the right-hand side in eq. (21) is the added forcing term. $C^{BA}$ in the forcing term is a switching parameter. The superscript $(\ )^{BA}$ indicates the fortification from grid B to grid A.

The fortified Navier-Stokes equations for fortification from grid A to grid B can easily be derived in the same manner as that for eq. (21). In this case, the switching parameter is expressed as $C^{AB}$. $C^{BA}$ and $C^{AB}$ are varied from 0 to infinity in the whole computational domain. When $C^{BA}$ is 0, eq. (21) becomes the usual Navier-Stokes equation for the streamwise velocity component, since the forcing term vanishes. When $C^{BA}$ is infinite, eq. (21) becomes equivalent to $U^A_{I,J,K}=U^B_{(I,J,K)}$, since all other terms in eq. (21) become negligibly small except for the forcing term. This means that velocity value $U^A_{I,J,K}$ at the nodal point on grid A is completely replaced by the interpolated value $U^B_{(I,J,K)}$ given from grid B. Fortification from grid B to grid A is performed in this manner, when $C^{BA}=\infty$. When $C^{BA}$ takes a finite value, $U^A_{I,J,K}$, the velocity defined at the nodal point on grid A, blends with $U^B_{(I,J,K)}$, the interpolated value of velocity from grid B. Thus,

$U^{A}_{I,J,K}$ becomes the weighted average between velocity values on grids A and B. Through this blending, smooth fortification is attained in the connecting region.

Regions I ~IV in Fig. 3 illustrate connecting regions set in the region of grid A. Here, regions II and III are the blending regions. Distribution of the switching parameters is also indicated in Fig. 3.

## References


Antonopoulos-Domis, M., Large-eddy simulation of a passive scalar in isotropic turbulence, J. Fluid Mech., vol.104, pp.55-79 (1981)

Bardina, J., J. H. Ferziger and W. C. Reynolds, Improved Subgrid-scale Models for Large-Eddy Simulation, AIAA paper-80, 1357 (1981)

Delsem, W.R.V. and J.L. Steger, The fortified Navier-Stokes approach, Workshop on Computational Fluid Dynamics, University of California, Davis, June 17-18 (1986)

Germano, M., U. Piomelli, P. Moin and W.H.Cabot : A dynamic subgrid-scale eddy viscosity model, Phys. Fluids, A3, pp.1760-1765 (1991)

Lilly, D.K., A proposed modification of the Germano subgrid-scale closure method, Phys. Fluids A4, pp.633-635 (1992)

Mochida, A. , S. Murakami and Y. Ishida, Numerical Study on Flowfield around Structures with Oblique Wind Angle based on Composite Grid System, The 7th U.S. National Conference on Wind Engineering (1993)

Murakami, S., A. Mochida, Y. Hayashi and S. Sakamoto, J. Wind Eng. Ind. Aerodyn., 41-44, pp.2841-2852 (1992)

Murakami, S., A. Mochida, Y. Tominaga, Numerical simulation of turbulent diffusion in cities, Wind Climate in Cities, Kluwer Academic Publishers,pp.681-701 (1995)

Murakami, S., A. Mochida, Y. Tominaga, S. Iizuka, Large Eddy Simulation of Turbulent Flow past 2D Square Cylinder using Dynamic SGS Models, Proceeding of the Second ERCOFTAC Workshop on Direct and Large-Eddy Simulation (1996)

Piomelli, U. and J. Liu , Large-eddy simulation of rotating channel flows using a localized dynamic model, Phys. Fluids 7 (4) pp.839-848 (1995)

Tominaga, Y., S. Murakami, A. Mochida, A. Shibuya and Y. Noguchi, Wind tunnel test on turbulent diffusion and concentration fluctuation of buoyant gas near building, 12th Japan National Symp. on Wind Eng., pp.119-124 (in Japanese) (1992)

Werner, H and H. Wengle, Large-eddy simulation of turbulent flow over and around a cube in a plate channel, 8th Symp. on Turbulent Shear Flows, 19-4 (1991)

Zang, Y., R. L. Street and J.R. Koseff : A dynamic mixed subgrid-scale model and its application to turbulent recirculating flows, Phys. Fluids A5, pp.3186-3196 (12) , (1993)

**Nomenclature**

$x_i$ : three components of the spatial coordinate
(i = 1, 2, 3 : streamwise, lateral, vertical)

$u_i$ : three components of the velocity vector

$f$ : instantaneous value of a quantity

$\bar{f}$ : grid-filtered value of f

$\langle f \rangle$ : time-averaged value of f

$\hat{\bar{f}}$ : test-filtered value of f

$c_g$ : gas concentration

$\langle c_{g0} \rangle$ : the reference gas concentration $(=q/\langle u_b \rangle/H_b^2)$

$q$ : gas emission rate

$\rho_s$ : gas density

$\rho_a$ : ambient air density

$x_n^+$ : distance from the wall $(x_n^+ = x_n u^*/\nu)$

$H_b$ : height of model

$u_b$ : $\langle u_1 \rangle$ value at inflow of computational domain at height $H_b$

When values are made dimensionless, representative length scale $H_b$, velocity scale $\langle u_b \rangle$ and air density $\rho_a$ are used.

# ON THE EXTENSION OF THE DYNAMIC MODELLING PROCEDURE TO TURBULENT REACTING FLOWS

M. GERMANO
*Politecnico di Torino*
*Dip. di Ing. Aeronautica e Spaziale, Torino, Italy*

A. MAFFIO AND S. SELLO
*CISE SpA*
*Divisione Sistemi e Modelli, Milano, Italy*

AND

G. MARIOTTI
*ENEL SpA*
*Centro Ricerca Termica, Pisa, Italy*

**Abstract.** The dynamic modelling procedures are based on the use of the information extracted at a filtered level. These procedures have been applied to the subgrid scale stresses in incompressible flow, but their extension to reacting flows poses a lot of problems. In this paper a preliminary study of the dynamic modelling procedure applied to the turbulent transport and the turbulent production of a scalar has been performed.

## 1. Introduction

Modelling procedures particularly fitted to the Large Eddy Simulations of turbulent reacting flows are currently the object of a lot of studies. In particular we notice that the transport equations for the mass fractions $C_\alpha$ of the reacting chemical species

$$\frac{\partial(\varrho C_\alpha)}{\partial t} + \frac{\partial(\varrho C_\alpha u_j)}{\partial x_j} = \frac{\partial}{\partial x_j}\left(\varrho \mathcal{D} \frac{\partial C_\alpha}{\partial x_j}\right) + P_\alpha \tag{1}$$



*J.-P. Chollet et al. (eds.), Direct and Large-Eddy Simulation II*, 291-300.

produce when averaged terms very difficult to model, the turbulent transport terms

$$\langle \varrho C_\alpha u_j \rangle_f$$

and the averaged chemical production terms

$$\langle P_\alpha \rangle_f$$

where the following operatorial notation $\langle \cdots \rangle_f$ that describes a LES as a particular representation of the turbulent field produced by a filtering operator $\mathcal{F}$ has been adopted, Germano (1992). In the simple case of two reactants and one product and if we can assume that the reacting scalars have no influence on the turbulence, (no heat release and density effects), the basic quantities to model are the following generalized central moments

$$\tau_f(C_\alpha, C_\beta) = \langle C_\alpha C_\beta \rangle_f - \langle C_\alpha \rangle_f \langle C_\beta \rangle_f \tag{2}$$

$$\tau_f(C_\alpha, u_j) = \langle C_\alpha u_j \rangle_f - \langle C_\alpha \rangle_f \langle u_j \rangle_f \tag{3}$$

The arguments that stand at the basis of the turbulent modelling of these terms are numerous. From the physical point of view we notice that the interactions between the flow and the chemistry can be interpreted in terms of typical length and time scales, see Chollet *et al.* (1994), and the important point is to represent these unresolved scales at the computational grid level. In the framework of a research program started by the ENEL-CRT (Italian Electricity Board-Centre for Thermal Research) in collaboration with CISE devoted to testing and developing subgrid scale models particularly fitted to industrial codes the main aim of this work is to examine simple modelling procedures. In particular the turbulent production term (2) has been analyzed on the basis of the similarity modelling procedure applied by Cook *et al.* (1994) to the decay of an isotropic reacting turbulent flow while a simple extension of the dynamic modelling procedure previously proposed by Moin *et al.* (1991) has been examined for the turbulent transport term (3). We remark that the work is in progress and that the possible applications in the field of combustion are much beyond the field of this paper not only because of the *no heat release - constant density* frame of assumptions but also because of the flame features. In fact turbulent flames may exhibit inner and outer cut-off length scales that might reduce similarity effects to a very narrow range of grid meshes particularly in the case of *hard* combustion. The main results of this preliminary study are briefly exposed in the following sections.

## 2. The extension of the dynamic modelling procedure to the production terms.

A simple similarity model

$$\tau_f(C_\alpha, C_\beta) = k_{fg}\tau_g(\langle C_\alpha\rangle_f, \langle C_\beta\rangle_f) \tag{4}$$

has been explored in order to test the possibility of representing the unresolved production term

$$\tau_f(C_\alpha, C_\beta) = \langle C_\alpha C_\beta\rangle_f - \langle C_\alpha\rangle_f\langle C_\beta\rangle_f \tag{5}$$

by means of the resolved similarity term

$$\tau_g(\langle C_\alpha\rangle_f, \langle C_\beta\rangle_f) = \langle\langle C_\alpha\rangle_f\langle C_\beta\rangle_f\rangle_g - \langle C_\alpha\rangle_{fg}\langle C_\beta\rangle_{fg} \tag{6}$$

extracted by a similarity filter $\mathcal{G}$. This model has been originally proposed by Cook *et al.* (1994) and a suggested value of the similarity constant $k_{fg}$ for a Damköhler number of $\infty$ and a grid ratio $l_{fg} = 2l_f$ is 1, where $l_{fg}$ and $l_f$ are the grid lengths at the different resolution levels. The problem is obviously to test the value of the similarity constant $k_{fg}$ and the study has been conducted both numerically and theoretically. From the numerical point of view the similarity constant has been determined by means of *a priori* tests through statistical analysis. The code and the databases that have been utilized for such experimental study of the similarity constant refer to the DNS of a 3-D temporal mixing layer generated by two streams carrying two separated reactants. The simulations are performed using an extended version, Brusa *et al.* (1994), of a hybrid semianalytic pseudospectral code, Banerjee *et al.* (1993), previously used for the determination of the fluidynamic characteristic of the flow, Andreola *et al.* (1995). As regards the DNS the initial Reynolds number based on the vorticity thickness was $Re = 92$, the initial conditions are random for the velocity field and three different Damköhler numbers, $Da = 0.02, 0.4, 1$, have been explored by using a $64 * 64 * 129$ mesh.

The Fig. 1 shows the total production rate for different Damköhler numbers. The roll up time is at $T = 0.022$ and the pairing time is nearly at the end of the computation, $T = 0.034$. We notice an initial decrease due to the high initial product generation rate due to overlap in initial species concentration, Park *et al.* (1994). After the roll up time the flow starts to develop tridimensionally, with a rapid increase of the turbulent kinetic energy, see Fig. 2, and a net increase on the total product generation.

*A priori* different LES have been simulated by filtering the DNS databases with different meshes, and a similarity test filter $\mathcal{G}$ has been successively applied in the fixed grid ratio 2 : 1. The results show a significative dependence of the similarity constant on the specific cases, see table 1.

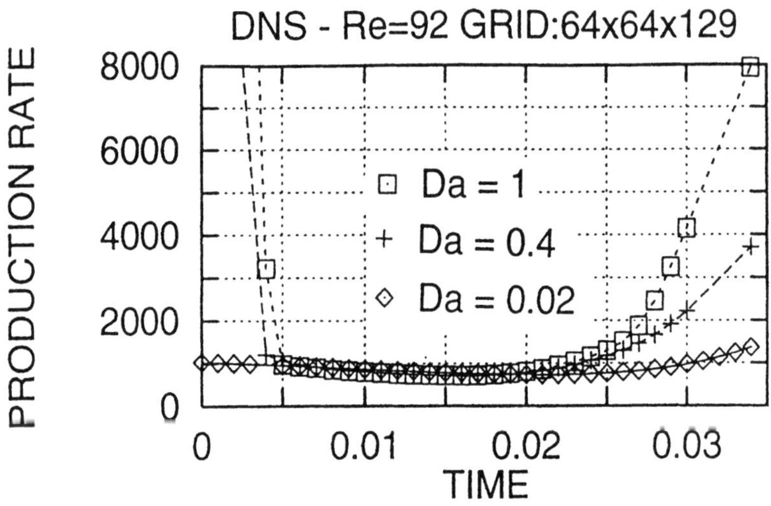


*Figure 1.* Production rate

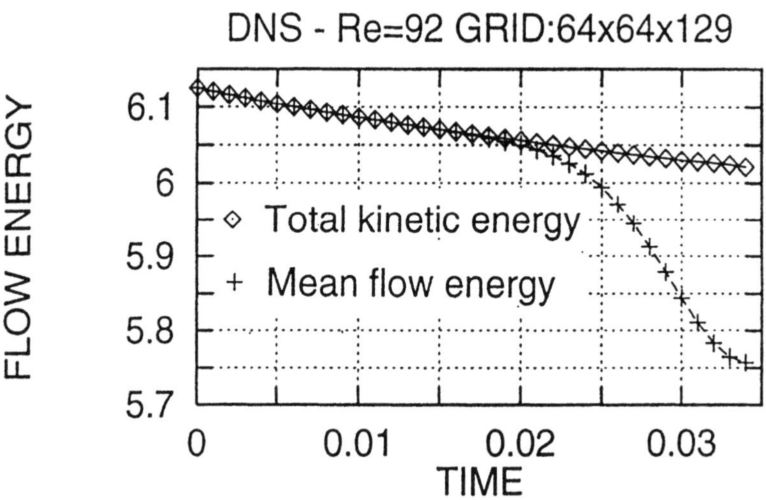


*Figure 2.* Total kinetic energy and mean flow energy.

In particular we notice that both the values of $k_{fg}$ and the related statistical correlations coefficients $r$ calculated in different planes of simmetry at different times increase when we increase the Damköhler number whereas decrease when we decrease the mesh size. In order to shed some light on the functional dependence and on the order of magnitude of the similarity constant $k_{fg}$ we have performed the following preliminary theoretical analysis. Let us assume as a first approximation that the filtered quantity

TABLE 1.

| | $\Delta_{LES} = 2\Delta_{DNS}$ | $\Delta_{LES} = 4\Delta_{DNS}$ | $\Delta_{LES} = 8\Delta_{DNS}$ |
|---|---|---|---|
| $Da = 0.02$ | $k_{fg} = \cdots$ $r = \cdots$ | $k_{fg} = 0.065$ $r = \cdots$ | $k_{fg} = \cdots$ $r = \cdots$ |
| $Da = 0.4$ | $k_{fg} = 0.004$ $r = 0.062$ | $k_{fg} = 0.15$ $r = 0.44$ | $k_{fg} = 0.28$ $r = 0.52$ |
| $Da = 1$ | $k_{fg} = 0.014$ $r = 0.14$ | $k_{fg} = 0.15$ $r = 0.20$ | $k_{fg} = 0.30$ $r = 0.42$ |

$\langle C_\alpha \rangle_f$ can be represented by the box average in one dimension

$$\langle C_\alpha \rangle_f = \frac{1}{2l_f} \int_{x-l_f}^{x+l_f} C_\alpha(x')dx' \tag{7}$$

extended to a finite interval of length $l_f$ and let us introduce the *ensemble* average $\mathcal{E}$ that in the case of spatially homogeneous turbulence is equivalent to the infinite length average

$$\langle C_\alpha \rangle_e = \lim_{l \to \infty} \frac{1}{2l} \int_{x-l}^{x+l} C_\alpha(x')dx' \tag{8}$$

We notice that in this particular case we have

$$\langle \langle \cdots \rangle_f \rangle_e = \langle \cdots \rangle_{fe} = \langle \cdots \rangle_e \tag{9}$$

so that the identity that relates the generalized central moments, Germano (1992), at the different levels $\mathcal{F}$ and $\mathcal{FE}$

$$\tau_{fe}(C_\alpha, C_\beta) = \langle \tau_f(C_\alpha, C_\beta) \rangle_e + \tau_e(\langle C_\alpha \rangle_f, \langle C_\beta \rangle_f) \tag{10}$$

can be written as

$$\tau_e(C_\alpha, C_\beta) = \langle \tau_f(C_\alpha, C_\beta) \rangle_e + \tau_e(\langle C_\alpha \rangle_f, \langle C_\beta \rangle_f) \tag{11}$$

We can apply the same considerations at the $\mathcal{FG}$-level. Let us assume that in a first approximation $\langle C_\alpha \rangle_{fg}$ can be represented as

$$\langle C_\alpha \rangle_{fg} = \frac{1}{2l_{fg}} \int_{x-l_{fg}}^{x+l_{fg}} C_\alpha(x')dx' \tag{12}$$

We notice that also in this case we have

$$\langle \langle \cdots \rangle_{fg} \rangle_e = \langle \cdots \rangle_{fge} = \langle \cdots \rangle_e \tag{13}$$

so that the identity that relates the generalized central moments at the different levels $\mathcal{FG}$ and $\mathcal{FGE}$

$$\tau_{fge}(C_\alpha, C_\beta) = \langle \tau_{fg}(C_\alpha, C_\beta) \rangle_e + \tau_e(\langle C_\alpha \rangle_{fg}, \langle C_\beta \rangle_{fg}) \tag{14}$$

also in this case can be written as

$$\tau_e(C_\alpha, C_\beta) = \langle \tau_{fg}(C_\alpha, C_\beta) \rangle_e + \tau_e(\langle C_\alpha \rangle_{fg}, \langle C_\beta \rangle_{fg}) \tag{15}$$

If we now introduce the cross covariance coefficient $\varrho_{\alpha\beta}(\xi)$ given by

$$\varrho_{\alpha\beta}(\xi) = \frac{\tau_e(C_\alpha(x), C_\beta(x+\xi))}{\tau_e(C_\alpha, C_\beta)} \tag{16}$$

where

$$\tau_e(C_\alpha(x), C_\beta(x+\xi)) = \langle C_\alpha(x) C_\beta(x+\xi) \rangle_e - \langle C_\alpha(x) \rangle_e \langle C_\beta(x+\xi) \rangle_e \tag{17}$$

we can write at the $\mathcal{F}$-level, Papoulis (1965), Tennekes and Lumley (1972),

$$\frac{\tau_e(\langle C_\alpha \rangle_f, \langle C_\beta \rangle_f)}{\tau_e(C_\alpha, C_\beta)} = I_{\alpha\beta}(l_f) \tag{18}$$

where

$$I_{\alpha\beta}(l_f) = \frac{1}{l_f} \int_0^{2l_f} \left(1 - \frac{\xi}{2l_f}\right) \varrho_{\alpha\beta}(\xi) d\xi \tag{19}$$

On the basis of the same considerations we can also write at the $\mathcal{FG}$-level that

$$\frac{\tau_e(\langle C_\alpha \rangle_{fg}, \langle C_\beta \rangle_{fg})}{\tau_e(C_\alpha, C_\beta)} = I_{\alpha\beta}(l_{fg}) \tag{20}$$

where now

$$I_{\alpha\beta}(l_{fg}) = \frac{1}{l_{fg}} \int_0^{2l_{fg}} \left(1 - \frac{\xi}{2l_{fg}}\right) \varrho_{\alpha\beta}(\xi) d\xi \tag{21}$$

Let us now apply the basic identity in the form

$$\tau_{ge}(\langle C_\alpha \rangle_f, \langle C_\beta \rangle_f) = \langle \tau_g(\langle C_\alpha \rangle_f, \langle C_\beta \rangle_f) \rangle_e + \tau_e(\langle C_\alpha \rangle_{fg}, \langle C_\beta \rangle_{fg}) \tag{22}$$

Owing to the fact that also in this case we can write

$$\langle \langle \cdots \rangle_g \rangle_e = \langle \cdots \rangle_{ge} = \langle \cdots \rangle_e \tag{23}$$

we obtain

$$\tau_e(\langle C_\alpha \rangle_f, \langle C_\beta \rangle_f) = \langle \tau_g(\langle C_\alpha \rangle_f, \langle C_\beta \rangle_f) \rangle_e + \tau_e(\langle C_\alpha \rangle_{fg}, \langle C_\beta \rangle_{fg}) \tag{24}$$

and on the basis of the relations (11), (18),(20) and (24) we can finally write

$$k_{fg} = \frac{1 - I_{\alpha\beta}(l_f)}{I_{\alpha\beta}(l_f) - I_{\alpha\beta}(l_{fg})} \tag{25}$$

where we have assumed that

$$\langle \tau_f(C_\alpha, C_\beta)\rangle_e = k_{fg}\langle \tau_g(\langle C_\alpha\rangle_f, \langle C_\beta\rangle_f)\rangle_e \tag{26}$$

Let us consider now the particular case in which the cross covariance $\varrho_{\alpha\beta}(\xi)$ can be approximated as

$$\varrho_{\alpha\beta}(\xi) = \exp\left(-\frac{\xi}{l_e}\right) \tag{27}$$

where $l_e$ is an integral scale. We obtain

$$I_{\alpha\beta}(l_f) = \frac{1}{r_{fe}} - \frac{1 - \exp(-2r_{fe})}{2r_{fe}^2} \tag{28}$$

where $r_{fe}$ is the ratio between the finite length scale $l_f$ and the integral scale $l_e$

$$r_{fe} = \frac{l_f}{l_e} \tag{29}$$

In the limit $r_{fe} \to 0$ we obtain

$$I_{\alpha\beta}(l_f) = \frac{1}{l_f}\int_0^{2l_f}\left(1 - \frac{\xi}{2l_f}\right)\varrho_{\alpha\beta}(\xi)d\xi = 1 - \frac{2l_f}{3l_e} \tag{30}$$

and if we can also assume that

$$I_{\alpha,\beta}(l_{fg}) = \frac{1}{l_{fg}}\int_0^{2l_{fg}}\left(1 - \frac{\xi}{2l_{fg}}\right)\varrho_{\alpha\beta}(\xi)d\xi = 1 - \frac{2l_{fg}}{3l_e} \tag{31}$$

by making use of the relation (25) we can finally write

$$k_{fg} = \frac{1}{\frac{l_{fg}}{l_f} - 1} \tag{32}$$

and we notice that in the case $l_{fg} = 2l_f$ the predicted value of $k_{fg}$ is 1.

## 3. The extension of the dynamic modelling procedure to the turbulent transport of the chemical mass fractions.

One largely used model for the turbulent transport terms

$$\tau_f(C_\alpha, u_j) = \langle C_\alpha u_j\rangle_f - \langle C_\alpha\rangle_f\langle u_j\rangle_f \tag{33}$$

is based on the diffusive assumption

$$\tau_f(C_\alpha, u_j) = -\mathcal{D}_f \frac{\partial \langle C_\alpha \rangle_f}{\partial x_j} \tag{34}$$

where now $\mathcal{D}_f$ is a turbulent diffusion coefficient at the $\mathcal{F}$-level. This model has been largely criticized from the physical point of view, and here we notice another difficulty peculiar to the extension of the dynamical approach to this modelling procedure. As a matter of fact if we directly apply to every mass fraction the dynamic procedure suggested by Moin *et al.* (1991) for the turbulent transport of a massive scalar we obtain different dynamic turbulent diffusion coefficient $\mathcal{D}_{f\alpha}$, one for each chemical component. This obviously will produce a lot of paradoxes similar to those examined by Pope (1983). In particular we notice that a different set of scalars defined by the linear transformation

$$C^*_\beta = L_{\beta\alpha} C_\alpha \tag{35}$$

produce scalar fluxes that commute following the same rule

$$\tau_f(C^*_\beta, u_j) = L_{\beta\alpha} \tau_f(C_\alpha, u_j) \tag{36}$$

and the only possible parameterization in terms of an eddy diffusivity that satisfies this rule is given by a constant $\mathcal{D}_f$. Different procedures can be conceived in order to remove this difficulty and to extend the dynamic modelling procedure to the case of a multicomponent mixture. The simplest one is the parameterization in terms of an eddy viscosity $\nu_f$ and an imposed turbulent Schmidt number $\mathcal{S}c_f$. In this case we simply have

$$\mathcal{D}_f = \frac{\nu_f}{\mathcal{S}c_f} \tag{37}$$

where $\nu_f$ is obtained by applying some dynamic procedure to the turbulent velocity field. Obviously it would be better to derive a turbulent Schmidt number more appropriate to the real turbulent field. As for the turbulent Prandtl number we know that the turbulent Schmidt number varies along the flow and near the walls, so that it is interesting to explore modelling procedures that give an appropriate Schmidt number. One possibility to derive a dynamic Schmidt number consistent with the linear transformation rule could be to extract a single dynamic eddy diffusivity by some least square method. If we write the basic identity

$$\tau_{fg}(C_\alpha, u_j) = \langle \tau_f(C_\alpha, u_j) \rangle_g + \tau_g(\langle C_\alpha \rangle_f, \langle u_j \rangle_f) \tag{38}$$

in terms of a diffusivity coefficient $C_D$

$$\tau_f(C_\alpha, u_j) \quad = \quad -C_D l_f^2 S_f \frac{\partial \langle C_\alpha \rangle_f}{\partial x_j}$$

$$\tau_{fg}(C_\alpha, u_j) = -C_D l_{fg}^2 S_{fg} \frac{\partial \langle C_\alpha \rangle_{fg}}{\partial x_j} \tag{39}$$

where $l_f$ and $l_{fg}$ are the filter scales,typically the grid lengths, and

$$\begin{aligned} S_f &= (2\langle s_{lm}\rangle_f \langle s_{lm}\rangle_f)^{1/2} \qquad \langle s_{ij}\rangle_f = \frac{1}{2}\left(\frac{\partial \langle u_i\rangle_f}{\partial x_j} + \frac{\partial \langle u_j\rangle_f}{\partial x_i}\right) \\ S_{fg} &= (2\langle s_{lm}\rangle_{fg} \langle s_{lm}\rangle_{fg})^{1/2} \qquad \langle s_{ij}\rangle_{fg} = \frac{1}{2}\left(\frac{\partial \langle u_i\rangle_{fg}}{\partial x_j} + \frac{\partial \langle u_j\rangle_{fg}}{\partial x_i}\right) \end{aligned}$$

we obviously obtain different values of $C_D$ for different scalars. If we now indicate with $M_{\alpha j}$ the difference

$$M_{\alpha j} = \tau_{fg}(C_\alpha, u_j) - \langle \tau_f(C_\alpha, u_j)\rangle_g - \tau_g(\langle C_\alpha\rangle_f, \langle u_j\rangle_f) \tag{40}$$

we can extract by a least square procedure applied to the contraction

$$M = M_{\alpha j} M_{\alpha j} \tag{41}$$

a single diffusivity coefficient $C_D$.

Another possibility is to apply the statistical formulation of the dynamic model recently proposed by Germano (1996). We notice that the production of the total scalar variance due to the eddy diffusivity is given by, Shih *et al.* (1990),

$$P_f(C_\alpha, C_\alpha) = -2\tau_f(C_\alpha, u_j)\frac{\partial \langle C_\alpha\rangle_f}{\partial x_j} \tag{42}$$

where also in this case the contraction must be applied on both the indices $j$ and $\alpha$. If we now introduce the test filter $\mathcal{G}$ and the ensemble statistical average $\mathcal{E}$ we can impose that

$$P_{fe}(C_\alpha, C_\alpha) = P_{fge}(C_\alpha, C_\alpha) \tag{43}$$

The meaning of this constraint is evident. It forces the different LES obtained at the $\mathcal{F}$-level and at the $\mathcal{FG}$-level to produce the same statistical representation, and the idea is now to derive an eddy coefficient $C_D$ consistent with this constraint. We can write the following chain of identities

$$\begin{aligned} \tau_{fe}(C_\alpha, u_j) &= \tau_e(\langle C_\alpha\rangle_f, \langle u_j\rangle_f) + \langle \tau_f(C_\alpha, u_j)\rangle_e \\ \tau_{fge}(C_\alpha, u_j) &= \tau_e(\langle C_\alpha\rangle_{fg}, \langle u_j\rangle_{fg}) + \langle \tau_{fg}(C_\alpha, u_j)\rangle_e \end{aligned} \tag{44}$$

that introduced in the statistical constraint (43) produce finally the equation

$$\tau_{fe}(C_\alpha, u_j)\frac{\partial \langle C_\alpha\rangle_{fe}}{\partial x_j} = \tau_{fge}(C_\alpha, u_j)\frac{\partial \langle C_\alpha\rangle_{fge}}{\partial x_j} \tag{45}$$

Future *a priori* tests will give some indications on the better choice.

## 4. Conclusions

The extension of the dynamic modelling procedures to turbulent reacting flows has been examined both numerically and theoretically. As regards the turbulent production term of a scalar the similarity model has been studied with particular attention to the determination of the similarity constant. As regards the turbulent transport of a scalar three modelling procedures (37), (41), (45) for the the turbulent diffusivity in a multicomponent mixture have been proposed. More theoretical and numerical work seems necessary in order to assess the feasibility of these modelling procedures in industrial codes.

## References


Andreola M., Maffio A., Sello S., Germano M., & Mariotti G., 1995: LES of a temporal mixing layer. *10th Symposium on Turbulent Shear Flows*, The Pennsylvania State University, August 14-16, 1995.

Banerjee S., Brusa L., Roy C.M., & Za S., 1993: A hybrid semianalytic spectral method for direct numerical simulation of turbulent flows. *Proc. 5th Int. Symp. on Refined Flow Modelling and Turbulence Measurements*, Paris, September 7-10, 1993.

Brusa L., Maffio A., & Sello S., 1994: Nuova versione del programma MILCH per la simulazione di strati di mescolamento 3-D turbolenti reagenti chimicamente. *R.T. CISE-SMA* **94-14**.

Chollet J.P., Si Ameur M., & Vallcorba M.R., 1994: Direct and Large Eddy Simulations of chemically reacting flows. *Direct and Large Eddy Simulation I*, Kluwer, Dordrecht, 375-385.

Cook A.W.,& Riley J.J., 1994: A subgrid model for equilibrium chemistry in turbulent flows. *Phys.Fluids* **6**, 2868-2870.

Germano M., 1992: Turbulence : the filtering approach. *J.Fluid Mech. bf 238*, 325-336.

Germano M., 1996: A statistical formulation of the dynamic model. *Phys. Fluids* **8**, 565-570.

Moin P., Squires K., Cabot W., & Lee S., 1991: A dynamic subgrid scale model for compressible turbulence and scalar transport, *Phys.Fluids* **A3**, 2746-2757.

Papoulis A., 1965: Probability, Random Variables and Stochastic Processes. McGraw-Hill

Park K., Metcalfe R.W., & Hussain F., 1994: Role of coherent structures in an isothermally reacting mixing layer *Phys.Fluids* **6**, 885-902.

Pope S.B., 1983: Consistent modelling of scalar in turbulent flows. *Phys. Fluids* **26**, 404-408.

Shih T.H., Lumley J.L., & Chen J.Y., 1990: Second order modeling of a passive scalar in a turbulent shear flow. *AIAA Journal* **28**, 610-617.

Tennekes H., & Lumley J.L. 1972: A first course in turbulence. MIT Press, Cambridge MA.

# SUBGRID-SCALE SEGREGATION OF CHEMICALLY REACTIVE SPECIES IN A NEUTRAL BOUNDARY LAYER

J.P. MEEDER AND F.T.M. NIEUWSTADT
*J.M. Burgers Centre, Laboratory of Aero- and Hydrodynamics*
*Rotterdamseweg 145*
*2628 AL Delft*
*The Netherlands*

**Abstract.** We study the binary reaction between an ambient species and a species emitted from a point source in the form of a plume. The concentration of both species is simulated in a neutral boundary layer with help of large-eddy simulation. The contribution of subgrid-scale segregation of the reactants is taken into account by using second-order closure technique for the subgrid-scale covariance. For the slow reaction case the model results agree with wind tunnel experiments. The calculations reveal that subgrid-scale segregation is important for relatively fast reactions, for example the reaction between NO and $O_3$. Regions of high subgrid-scale segregation are located in the vicinity of the source and near the plume edge.

In general, the subgrid contribution to the segregation reduces the mean reaction rate.

## 1. Introduction

The influence of atmospheric turbulence on dispersion and chemistry of air pollution is one of the main topics in the present research field of air pollution. Areas of application are e.g. photochemical smog. A relatively modern technique to study this type of problem is Large-Eddy Simulation (LES), see e.g. Schumann, 1989; Meeder and Nieuwstadt, 1994 and Sykes *et al.*, 1992. A challenge in LES of turbulent reacting flows is the parameterisation of the effects of chemistry on smaller scales which cannot be resolved by the numerical resolution. For relatively fast reactions, the chemically reactive species are not well mixed even at subgrid-scale (SGS). Although it has been proven in early studies (Toor, 1969; Donaldson and Hilst, 1972)



*J.-P. Chollet et al. (eds.), Direct and Large-Eddy Simulation II,* 301-310.

that the influence of inhomogeneous mixing on the reaction rate can be significant, the LES studies mentioned above do not account for this SGS "unmixedness" or segregation.

In principle, the segregation effect depends on the ratio of the time scales of turbulence and chemistry. For example, let us consider the reaction A + B $\rightarrow$ 2P between initially non-premixed species A and B with a reaction rate coefficient $k$. The reaction can only take place in a zone near the interface between the reacting species, where the two species can mix on a molecular level. For the limit of $k$ to infinity, the volume fraction of this zone decreases to zero and the species are completely segregated. The reaction rate is in that case completely determined by diffusion processes and no longer by the chemical reaction rate. For finite but still large reaction rates, the reaction zone is so small that it cannot be resolved explicitly. In that case, the SGS contribution to the segregation cannot be neglected. This case is characterised by a high value of the Damköhler number, which is defined as

$$Da_{SGS} = \frac{\tau_t}{\tau_c}. \tag{1}$$

with $\tau_t$ representing the subgrid turbulence time scale and $\tau_c$ denoting the time scale of the chemical reaction.

For present LES, $\tau_t$ is estimated by

$$\tau_t = \frac{\Delta}{\sqrt{e}} \sim O(100)\mathrm{s}, \tag{2}$$

with $\Delta$ representing the characteristic mesh size and $e$ representing the subgrid energy. The time scale of the chemistry follows from

$$\tau_c = (kC)^{-1} \tag{3}$$

where $C$ denotes a concentration scale, for example the mean concentration of the two reactants. Considering the reaction between NO and $O_3$, $k = 0.4\,\mathrm{ppm}^{-1}\mathrm{s}^{-1}$ and $C \sim O(1)$ ppm. This yields $Da_{SGS} \sim O(100)$, which means that the SGS contribution to the segregation cannot be neglected. Beets *et al.* show how the SGS segregation can be approximated for $k \rightarrow \infty$.

In this study a subgrid model for the SGS-segregation is proposed and applied to a chemically reactive plume in the neutral boundary layer. We consider a second-order reaction between species A emitted from a point source and ambient species B for different Damköhler number. For a relatively slow reaction, the results of the simulation will be compared to wind-tunnel data provided by Builtjes (1983). The high Damköhler case can be associated with the reaction between nitrogen-oxide (NO) and ozone ($O_3$) in the atmosphere.

## 2. Numerical model

### 2.1. GOVERNING EQUATIONS

We consider a neutral atmospheric flow including chemically reactive species which undergo a second order reaction $A+B \rightarrow 2P$. The basic set of equations used for the LES of this case read

$$\frac{\partial \overline{u}_i}{\partial x_i} = 0, \tag{4}$$

$$\frac{\partial \overline{u}_i}{\partial t} + \frac{\partial(\overline{u}_j \overline{u}_i)}{\partial x_j} = -\frac{1}{\rho_0}\frac{\partial \overline{p}}{\partial x_i} + f\varepsilon_{ij3}\overline{u}_j + \frac{\partial}{\partial x_j}\left(\nu\frac{\partial \overline{u}_i}{\partial x_j} - \overline{u_i'' u_j''} + \frac{2}{3}e\delta_{ij}\right), \tag{5}$$

$$\frac{\partial \overline{c}_m}{\partial t} + \frac{\partial \overline{u}_j \overline{c}_m}{\partial x_j} = \frac{\partial}{\partial x_j}\left(\gamma\frac{\partial \overline{c}_m}{\partial x_j} - \overline{u_j'' c_m''}\right) - k\overline{c_A c_B}, \tag{6}$$

where $u_i$ is the velocity vector, $p$ the pressure, $\rho_0$ the constant reference density, $f$ is the vertical Coriolis component, $\nu$ is the kinematic viscosity and $\gamma$ denotes the scalar diffusivity. The $c_m$ represents concentration, with the index $m$ denoting either species A or B. The over-bar represents the average over one grid volume, denoted as resolved, and the double prime represents the deviation from this average, denoted as subgrid.

### 2.2. SGS STRESS AND TRANSPORT

The SGS stress and transport terms in the equations are parameterised by introducing an eddy-viscosity and diffusivity, which are given by

$$K_m = C_1 \Delta e^{1/2}, \tag{7}$$

$$K_h = C_2 \Delta e^{1/2}, \tag{8}$$

where $C_1$ and $C_2$ are taken here to be equal to 0.12 and 0.36 respectively. The SGS kinetic energy is calculated from an additional prognostic equation. The SGS stress for momentum and SGS transport terms of species are now approximated by

$$\overline{u_i'' u_j''} - \frac{2}{3}e\delta_{ij} = -K_m\left(\frac{\partial \overline{u}_i}{\partial x_j} + \frac{\partial \overline{u}_j}{\partial x_i}\right), \tag{9}$$

$$\overline{u_j'' c_m''} = -K_h \frac{\partial \overline{c}_m}{\partial x_j}. \tag{10}$$

### 2.3. SGS SEGREGATION

The last term at the right-hand side of equation (6), which is the chemistry term, can be decomposed into a resolved part and a subgrid contribution, according to

$$\overline{c_A c_B} = \bar{c}_A \bar{c}_B + \overline{c''_A c''_B} = \bar{c}_A \bar{c}_B (1 + I_{SGS}), \tag{11}$$

where the intensity of SGS segregation is defined as

$$I_{SGS} = \frac{\overline{c''_A c''_B}}{\bar{c}_A \bar{c}_B}. \tag{12}$$

In general, $I_{SGS} < 0$ and this implies that subgrid effects effectively reduce the mean reaction rate for non-premixed species. $I_{SGS} = 0$ refers to the state that species A and B are completely mixed and the species are completely segregated for $I_{SGS} = -1$.

In analogy with second-order closure models (see e.g. Sykes *et al.*, 1994), a closure hypothesis for the SGS segregation can be based on the evolution equation for the second-moment species correlation. This reads

$$\begin{aligned}
\underbrace{\frac{\partial \overline{c''_A c''_B}}{\partial t}}_{(a)} + \underbrace{\bar{u}_j \frac{\partial \overline{c''_A c''_B}}{\partial x_j}}_{(b)} &= \underbrace{-\overline{u''_j c''_A} \frac{\partial \bar{c}_B}{\partial x_j} - \overline{u''_j c''_B} \frac{\partial \bar{c}_A}{\partial x_j}}_{(c)} \\
&\quad \underbrace{- \frac{\partial \overline{u''_j c''_A c''_B}}{\partial x_j} + \gamma \frac{\partial^2 \overline{c''_A c''_B}}{\partial x_j^2}}_{(d)} \underbrace{- 2\gamma \overline{\frac{\partial c''_A}{\partial x_j} \frac{\partial c''_B}{\partial x_j}}}_{(e)} \\
&\underbrace{- k(\bar{c}_A \overline{c''_A c''_B} + \bar{c}_B \overline{c''^2_A} + \overline{c''^2_A c''_B}) - k(\bar{c}_B \overline{c''_A c''_B} + \bar{c}_A \overline{c''^2_B} + \overline{c''_A c''^2_B})}_{(f)}.
\end{aligned} \tag{13}$$

where the following terms can be distinguished: the temporal change of covariance (a), advection by resolved velocities (b), production by resolved concentration gradients (c), turbulent and molecular diffusion (d), destruction by molecular diffusion (e) and the chemical terms (f).

Sykes *et al.* (1994) have studied different parameterisations for the additional triple correlation terms $\overline{c''^2_A c''_B}$ and $\overline{c''_A c''^2_B}$ in (f). They showed that for second-order turbulence closure schemes applied to surface diffusion of reactive species in the convective boundary layer, the results are not very sensitive to the exact formulation of these triple correlations. In the present study, we will utilise one of the models proposed by Sykes *et al.* (1994), namely

$$\begin{aligned}
\overline{c''^2_A c''_B} &= -(\overline{c''^2_A} + I_{SGS}\, \bar{c}^2_A)\bar{c}_B, \\
\overline{c''_A c''^2_B} &= -(\overline{c''^2_B} + I_{SGS}\, \bar{c}^2_B)\bar{c}_A.
\end{aligned} \tag{14}$$

In fact, this model implies that the reaction term (f) vanishes for all values of $I_{SGS}$.

Using this closure model and the SGS transport model described in section 2.2, we obtain from (13)

$$\frac{\partial \overline{c_A''c_B''}}{\partial t} + \overline{u}_j \frac{\partial \overline{c_A''c_B''}}{\partial x_j} = 2K_h \frac{\partial \overline{c}_A}{\partial x_j}\frac{\partial \overline{c}_B}{\partial x_j} + \frac{\partial}{\partial x_j}\left((\gamma + K_h)\frac{\partial \overline{c_A''c_B''}}{\partial x_j}\right) - C_3\overline{c_A''c_B''}\frac{\sqrt{e}}{\Delta}, \quad (15)$$

where $C_3$ is chosen 1.49.

Notice that the molecular viscosity and diffusivity can be assumed negligible compared with the turbulent stress and transport.

## 2.4. DISCRETISATION AND BOUNDARY CONDITIONS

The LES equations are discretised on an equidistant rectangular grid. The advective terms of momentum, SGS energy and SGS covariance are discretised using the method of Piacsek and Williams (1970). The Leap-Frog scheme, with an Asselin filter to prevent decoupling of odd and even time-levels, is used for the time integration. Advection of species is predicted by the Multi-dimensional Flux Corrected Transport method (MFCT), proposed by Zalesak (1979). This scheme uses a first order approximation in time at steep concentration gradients and is second order accurate in time at smooth parts of the concentration distribution. As a result, the method leads to a non-oscillating, positive definite solution for the concentration field. Therefore, it is well suited for dispersion problems with sharp concentration gradients, where second-order methods lead to oscillations. For the chemical sink term we use an explicit time integration scheme with variable time step, which can be considerably smaller than the time step used for the advection terms. The Poisson equation for the pressure is solved by applying Fast Fourier Transformation (FFT) technique.

The computations are performed on a domain with dimensions of 4000 m, 2000 m and 1500 m in the $x$, $y$ and $z$ direction, respectively, where according to meteorological convention $z$ is taken as vertical perpendicular to the surface. The domain is divided into $50 \times 50 \times 50$ grid volumes. The geostrophic wind components are chosen $U_g = 10.0$ ms$^{-1}$ and $V_g = 0.0$ ms$^{-1}$ and the surface roughness is $z_0 = 0.10$ m. Table 1 summarises the characteristic data for the source configuration and the chemistry. These are the source area $A_s$, the source height $h_s$, the initial source concentration $c_A^0$, the initial ambient concentration $c_B^0$ and the reaction rate coefficient. $I_{SGS} = 0$ means that no SGS segregation is taken into account, for the cases where $I_{SGS} < 0$ the SGS model described in section 2.3 is applied.

For velocity, periodic boundary conditions are imposed in the horizontal directions. At the lower boundary the vertical velocity is set to zero, Monin-Obukhov similarity is used as a wall function and zero gradient conditions are prescribed for the concentrations. At the upper boundary zero gradient and stress free conditions are imposed. Finally, inflow/outflow conditions for the species are used.

TABLE 1. Parameter settings related to source configuration and chemistry

| Case | $A_s$ ($m^2$) | $h_s$ (m) | $c_A^0$ (ppm) | $c_B^0$ (ppm) | $k$ $ppm^{-1}s^{-1}$ | $I_{SGS}$ (-) |
|---|---|---|---|---|---|---|
| N1 | 1200 | 105 | 9.0 | 0.35 | 0.011 | 0 |
| N2 | 1200 | 105 | 9.0 | 0.35 | 0.011 | <0 |
| N3 | 1200 | 105 | 9.0 | 0.35 | 0.400 | 0 |
| N4 | 1200 | 105 | 9.0 | 0.35 | 0.400 | <0 |

## 3. Results

The calculations carried out in this study corresponds to simulations of a neutral boundary layer above farmland. Point source simulations are performed for a relatively slow reaction (cases N1 and N2) as well as a fast reaction (cases N3 and N4). Both reactions read A + B $\rightarrow$ 2P. Statistics are based on time averages of about 10,000 s, starting from a situation with fully developed turbulence and a plume fully developed in the boundary layer.

The parameters of the simulations corresponding to the slow reaction, i.e. N1 and N2, are chosen to be comparable to the wind-tunnel experiments performed by Builtjes (1983). The ratio of the time scales of processes in the wind tunnel and in the atmosphere is approximately 1:35, and for the length scale ratio it is about 1:700. From now on, we shall express the wind-tunnel results in atmospheric dimensions with the help of these conversion factors. We notice that the source area is significantly larger for the simulations than used in the experiments, because the area in the computations is limited by the dimensions of the grid volumes. For this reason we can compare the LES results only with the wind tunnel data at distances larger than about 1000 m from the source.

Fig. 1 (left panel) shows the mean concentration, denoted by the square brackets, of the source species added to half of the product species ($\overline{c}_A + \overline{c}_P$) at the plume centreline as a function of the distance from the source. It

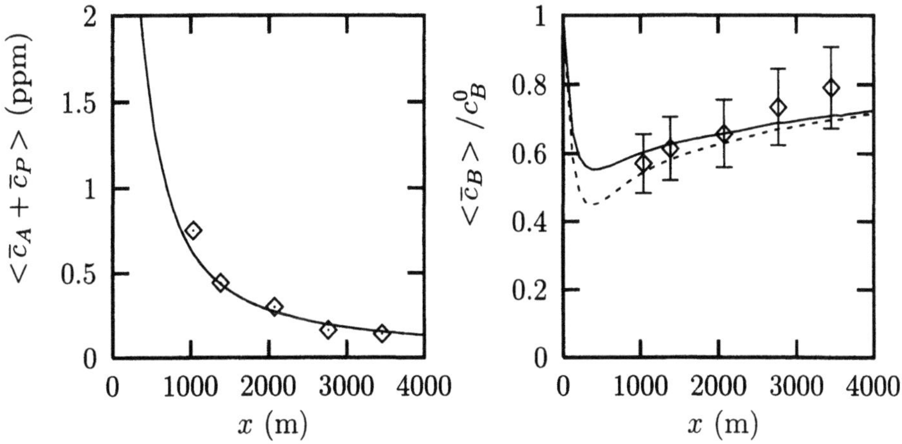


*Figure 1.* Average concentration of species A + P (left panel) and species B (right panel) at plume centreline as a function of downstream distance. The solid line shows the results of case N2, the dashed line the results of case N1 and the diamonds represent the wind tunnel results of Builtjes (1983).

can be shown by using the conservation equations for the species (6), that $\bar{c}_A + \bar{c}_P$ can be associated with a conserved scalar. Dispersion of this passive scalar is predicted quite accurately by the LES results. The correspondence of the model results with the wind tunnel data is not as perfect for the centreline concentration of reactant B, according to Fig. 1 (right panel). However, the difference between the experimental results and the calculations are less then the reproduction error of the measurements, which is about $\pm$ 15% (P.J.H. Builtjes, personal communication). In Fig. 1 we show the LES data for the cases with and without a subgrid model for $I_{SGS}$. We find that due to SGS segregation the reaction rate is reduced and $\bar{c}_B$ increases. However, this effect is only significant close to the source, where $Da_{SGS} > O(1)$.

For the higher Damköhler case it is expected that the SGS segregation effect becomes more important. To illustrate this we give in Fig. 2 the instantaneous plume concentration and SGS intensity of segregation in the vertical plane through the plume centreline. According to equation (13), production of SGS covariance and segregation is connected to resolved concentration gradients. This means that most of the production of SGS covariance or segregation takes place close to the source or at the plume edge, where the gradients are large. This is confirmed by the contour plots. Fig. 3 shows the total intensity of segregation of the time averaged concentration on the plume centreline:

$$I_s = \frac{<\bar{c}'_A \bar{c}'_B> + <\overline{c''_A c''_B}>}{<\bar{c}_A><\bar{c}_B>}. \tag{16}$$

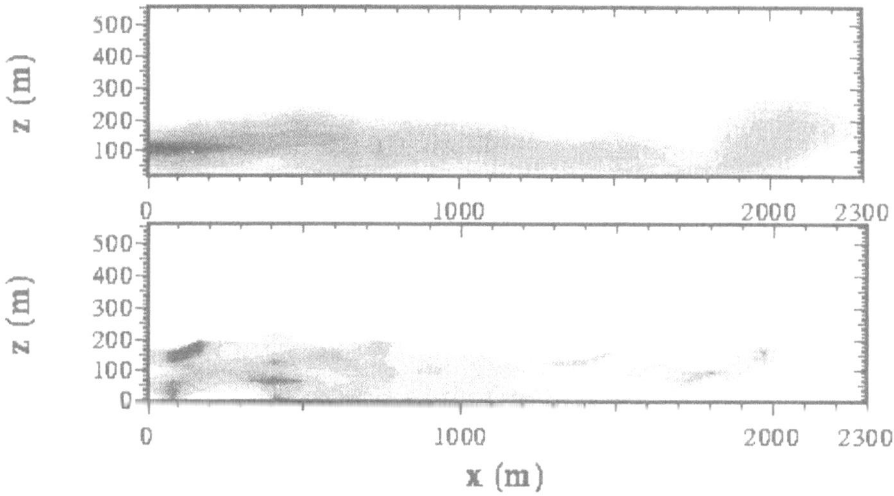


*Figure 2.* Instantaneous plume concentration (top panel) and SGS intensity of segregation (bottom panel) obtained from simulation N4. The concentration ranges from 0.0 ppm (white) to 5.0 ppm (black) and $I_{SGS}$ ranges from 0.0 (white) to -1.0 (black).

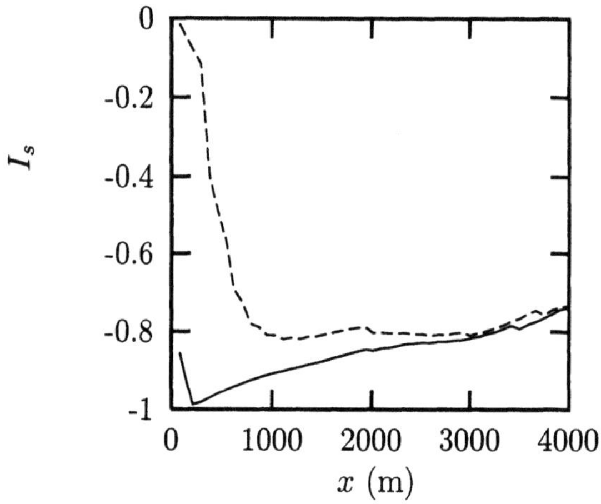


*Figure 3.* Intensity of segregation at plume centreline as a function of downstream distance. The solid line shows the results of case N4 and the dashed line the results of case N3.

Here the square brackets denote the time average, and the prime represents the deviation from this average. $I_s = 0$ refers to the situation without reaction ($Da_{SGS} = 0$), $I_s = -1$ indicates that the reacting species are completely segregated ($Da_{SGS} \to \infty$). It is observed that $I_s$ is strongly affected by the influence of SGS segregation. Due to SGS segregation the total segregation is increased and the reaction is slowed down, which is reflected in Fig. 4.

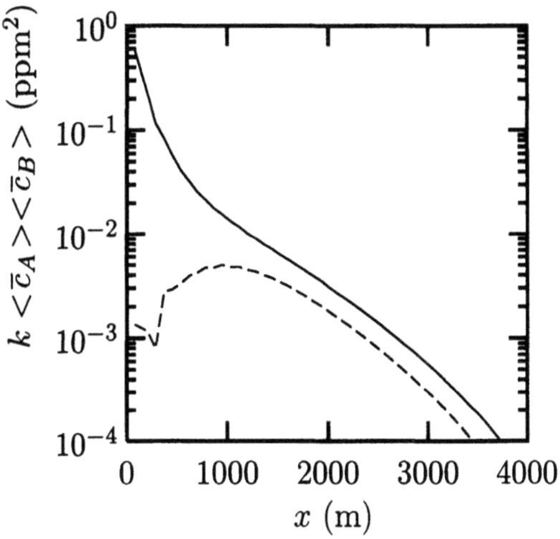


*Figure 4.* Mean reaction rate at plume centreline as a function of downstream distance. The solid line shows the results of case N4 and the dashed line the results of case N3.

## 4. Conclusions

We have carried out LES of plumes in a neutral atmospheric boundary layer, subjected to a reaction classified by an intermediate Damköhler number and a high Damköhler number. The LES model was modified for SGS segregation of the reacting species. It was found that the effect of SGS segregation on the conversion rate of the reactants is significant for fast reactions, for example $NO + O_3 \rightarrow NO_2 + O_2$. Due to SGS segregation, the reaction is slowed down and the segregation of the plume with respect to time averaging is increased.

## 5. Acknowledgements

This work was financially supported by the Nederlandse Organisatie voor Wetenschappelijk Onderzoek (Netherlands Organization for Scientific Research, NWO) and sponsored by the Stichting Nationale Computerfaciliteiten (National Computing Facilities Foundation, NCF) for the use of supercomputer facilities.

## References

Andren, A., Brown, A.R., Graf, J., Mason, P.J., Moeng, C.H., Nieuwstadt, F.T.M., and Schumann, U. (1994) Large-eddy simulation of a neutrally stratified boundary layer: A comparison of four computer codes, *Q.J.R. Meteorol. Soc.*, **120**, pp. 1457-1484.

Beets, C., Duynkerke, P.G., Meeder, J.P. and Nieuwstadt, F.T.M., Large-eddy simulation of chemically reactive species in a convective atmospheric boundary layer, *submitted to J. Appl. Meteor.*

Builtjes, P.J.H. (1983) A comparison between chemically reacting plume models and wind tunnel experiments, *Air Pollution Modelling and its Application II* (edited by de Wispelaere, C.), pp. 59-84. Plenum Press, N.Y.

Donaldson, C. du P. and Hilst, G.R. (1972) Effect of inhomogeneous mixing on atmospheric photochemical reactions, *Environ. Sci. Technol.*, **6**, pp. 812-816.

Meeder, J.P., Bouwmans, I. and Nieuwstadt, F.T.M. (1994) The large-eddy simulation of dispersion of passive and chemically reactive pollutants in a convective atmospheric boundary layer, *Direct and Large-Eddy Simulation I* (edited by Voke, P.R., Kleiser, L. and Chollet, J.P.), pp. 179-188.

Nieuwstadt, F.T.M., Mason, P.J., Moeng, C.H. and Schumann, U. (1991) Large-eddy simulation of the convective boundary layer: A comparison of four computer codes, *Proc. 8th Symposium on Turbulent Shear Flows.*

Piacsek, A.P. and Williams, G.P. (1970) Conservation properties of convection difference schemes, *J. of Comp. Phys.*, **6**, pp. 392-405.

Schumann, U. (1989) Large-eddy simulation of turbulent diffusion with chemical reactions in the convective boundary layer, *Atmos. Environ.*, **23** (8), pp. 1713-1727.

Sykes, R.I., Henn, D.S. and Parker, S.F. (1992) Large-eddy simulation of a turbulent reacting plume, *Atmos. Environ.*, **26A** (14), pp. 2565-2574.

Sykes, R.I., Parker, S.F., Henn, D.S. and Lewellen, W.S. (1994) Turbulent mixing with chemical reaction in the planetary boundary layer, *J. Appl. Meteor.*, **33**, pp. 825-834.

Toor, H.L. (1969) Turbulent mixing of two species with and without chemical reactions, *IEC Fundamentals*, **8**, pp. 655-659.

Zalesak, S.T. (1979) A fully multidimensional flux-corrected transport algorithm for fluids, *J. of Comp. Phys.*, **31**, pp. 335-362.

# DYNAMIC SUBGRID PDF MODELING FOR NONPREMIXED TURBULENT COMBUSTION

J. RÉVEILLON AND L. VERVISCH
*LMFN-INSA de Rouen*
*URA - CNRS 230 - CORIA,*
*76821 Mont St. Aignan Cédex,*
*FRANCE*

**Abstract.**

Large Damköhler number nonpremixed combustion is considered and subgrid fluctuations of mixture fraction are accounted for through a subgrid probability density function (SGPDF). First, the response of the LES dynamic model to heat release induced effects is studied by using a presumed form for the SGPDF. Then, a dynamic closure for diffusive processes is proposed. It provides a subgrid mixing time scale and a solution for the SGPDF transport equation. To estimate the validity of these closures, LES and DNS have been performed in 3D homogeneous and freely decaying turbulence.

## 1. Introduction

One of the basic question arising when one wishes to use LES for combustion is linked to the possible dependence of the large scales of the flow on the small scale detail. In this spirit, two issues of large scale turbulent combustion modeling are addressed in the following. The effects of heat release on the large scales motion along with the description of subgrid diffusive processes (mixing) are discussed in the context of the subgrid probability density function approach.

The dynamic formalism proposed by Germano et al., (1991) has been retained. By applying a double filtering operation to the turbulent flow, the flux of energy transferred between two characteristic length scales of the turbulent motion is estimated. This allows for a dynamic determination of the parameters of the subgrid model which then depend on the local prop-



*J.-P. Chollet et al. (eds.), Direct and Large-Eddy Simulation II,* 311-320.

erties of the flow. The dynamic LES model has been previously developed and used in the case of nonreactive flows, in both constant and variable density situations (Moin et al., (1991), Ghosal and Rogers, (1994), Friedrich, (1995)). This dynamic formalism is very attractive for flames, indeed model parameters can a priori be sensitive to the modifications of the flow resulting from heat release. In addition to the basic dynamic approach, use is made in this work of a procedure developed by Meneveau et al., (1994) to minimize, over flow pathlines, the error in the determination of the parameters of the model. This approach has a large potential for complex flows because it extends the field of application of classical averaging dynamic LES techniques to flows that do not possess any homogeneous direction.

Nonpremixed turbulent flames are essentially controlled by the rate of mixing between fuel and oxidizer, and it is usually believed that chemical reactions are completed at the small scales (Bray and Peters, (1994)). In the following, for the case of large Damköhler number turbulent combustion, it is illustrated how the properties of the large scales of the turbulent flame can be captured by introducing the subgrid probability density function (SGPDF) of the mixture fraction.

First a scale similarity assumption is formulated to get the statistical properties of the subgrid mixture fraction field. This permits to extract information from the resolved part of the spectrum, in order to estimate the fluctuations of the mixture fraction at the subgrid level and, then, to presume the SGPDF of the mixture fraction. In the second part of this work, a dynamic approach is proposed for subgrid mixing modeling, this closure for small scale diffusion allows for solving the transport equation for SGPDF.

Both fully compressible direct numerical simulations and large eddy simulations have been independently performed. DNS fields are used as reference databases to evaluate the dynamic large eddy turbulent combustion modeling possibilities. Then, nonpremixed turbulent flames are computed by using dynamic LES. It is found that the response of the dynamic model to different levels of heat release reproduces DNS observations. Moreover, the proposed dynamic subgrid mixing model correctly represents SGPDF evolution along with subgrid mixing time scales.

## 2. Dynamic SGPDF modeling

The objective of large eddy simulation is to exactly simulate the large scales of the turbulent motion, while the smaller scales are modeled. All spatial fluctuations smaller than a length $\overline{\Delta}$ are damped by using a filtering operation defined from a convolution between a turbulent signal $Y(\underline{x},t)$ and a normalized filter $G(\underline{x})$, whose width is $\overline{\Delta}$. $\overline{Y}$ is defined as the $\overline{\Delta}$ level filtered value of $Y$. Favre averaged filtering is also introduced as $\widetilde{Y} = \overline{\rho Y}/\overline{\rho}$.

To account for the chemical activity, it is assumed that the heat release is controlled by the rate of mixing between fuel and oxidizer. This hypothesis restricts the scope of the study to flows in which chemical reactions are infinitely fast. However, this choice allows for testing the possibility of coupling dynamic LES and nonpremixed turbulent flame modeling under well defined demanding assumptions. The mixture fraction $Z$ is introduced ($Z$ is equal to unity in the pure fuel stream and to zero in the oxidizer stream). Then, the instantaneous fuel mass fraction field may be expressed as a function of $Z$, in the form $Y_F(\underline{x},t) = Y_F^{eq}(Z(\underline{x},t))$, where $Y_F^{eq}$ is fully determined for large Damköhler flames (Bray and Peters, (1994)). The filtering operation leads to

$$\begin{aligned} \widetilde{Y}_F(\underline{x},t) &= \frac{1}{\overline{\rho}} \int_{-\infty}^{+\infty} \rho(Z(\underline{x}',t)) Y_F^{eq}(Z(\underline{x}',t)) G(\underline{x}' - \underline{x}) d\underline{x}' \\ &= \int_0^1 Y_F^{eq}(Z^*) \widetilde{P}_{sg}(Z^*;\underline{x},t)\, dZ^* \end{aligned}$$

$\widetilde{Y}_F(\underline{x},t)$ appears as the large eddy fuel mass fraction determined from an infinitely fast chemistry assumption. $\widetilde{P}_{sg}(Z^*;\underline{x},t)$, the probability density function of the mixture fraction $Z$, is introduced to capture the statistics at the subgrid level. $\widetilde{P}_{sg}(Z^*;\underline{x},t)$ may be presumed or computed from its transport equation. For both approaches, it is shown in the remaining part of the paper that information can be collected from the large eddy field to help in the dynamic determination of $\widetilde{P}_{sg}(Z^*;\underline{x},t)$.

## 2.1. SUBGRID DYNAMIC MODELING

More details about the basis of dynamic subgrid modeling may be found in Germano et al., (1991). The introduction of variable density follows the strategy of Moin et al., (1991) and Yoshizawa, (1986), while the parameters of the closure are computed along flow pathlines according to the ideas of Meneveau et al., (1994). The modeled subgrid stress tensor may be written as

$$\begin{aligned} \tau_{ij} &= \overline{\rho}\left(\widetilde{u_i u_j} - \widetilde{u}_i \widetilde{u}_j\right) \\ &= -2C_s \overline{\Delta}^2 \overline{\rho} \mid \widetilde{S} \mid \left(\widetilde{S}_{ij} - \frac{1}{3}\delta_{ij}\widetilde{S}_{kk}\right) + \frac{1}{3}\delta_{ij}\left(2C_I \overline{\Delta}^2 \overline{\rho} \mid \widetilde{S} \mid^2\right) \end{aligned} \quad (1)$$

where $\widetilde{S}_{ij} = \left(\partial\widetilde{u}_i/\partial x_j + \partial\widetilde{u}_j/\partial x_i\right)/2$ and $\mid \widetilde{S} \mid = \sqrt{2\widetilde{S}_{ij}\widetilde{S}_{ij}}$. In Eq. (1) the parameters $C_s$ and $C_I$ need to be determined. In nonpremixed turbulent flames, where chemical reactions inducing heat release are found, it is difficult to expect that the entire flow could be characterized by unique values for $C_s$ and $C_I$.

3-D DNS and LES of inert and nonpremixed reactive flows have been performed. The compressible Navier Stokes equations are solved in physical space by using a DNS numerical code previously developed by Guichard et al., (1995) following the methodology of Poinsot et al., (1995).

The initial conditions for the velocity field and the mixture fraction field are generated in spectral space from a prescribed energy spectrum (3-D homogeneous flows) (Réveillon and Vervisch, (1996a)). From this initial configuration the turbulence is allowed to decay. In the DNS, the initial turbulent Reynolds number based on the integral length scale is about 70, while the Reynolds number based on the initial Taylor microscale is about 50. The chosen initial PDF of mixture fraction leads to a segregated $Z$ field featuring pockets of fuel and oxidizer (Réveillon and Vervisch, (1996b)). The LES simulations are performed on a $33^3$ grid while the DNS correspond to a $129^3$ mesh. The initial conditions for the LES is the result of the filtering operation (size $\overline{\Delta}$) applied to the initial DNS field.

## 2.2. COUPLING DYNAMIC SUBGRID MODEL AND FAST CHEMISTRY

In this section, it is assumed that the shape of $\widetilde{P}_{sg}(Z^*; \underline{x}, t)$ can be presumed from the knowledge of $\widetilde{Z}$ and $\widetilde{Z'^2}$, the first and second subgrid moments of mixture fraction. The filtered value of the mixture fraction $\widetilde{Z}$ is computed from a transport equation closed by using the dynamic model. The subgrid variance $\widetilde{Z'^2}$ needs to be evaluated; following Cook and Riley, (1994) a scale similarity assumption is formulated to estimate $\widetilde{Z'^2}$ from the known field at the test grid level. This approach allows for focussing on heat release induced effects without solving an additional transport equation requiring more closures. A beta function (Borghi, (1988)) is used as the presumed shape of the pdf, $\widetilde{P}_{sg}(Z^*; x, t)$, with given $\widetilde{Z}$ and $\widetilde{Z'^2}$.

In the present study, the impact on the parameters $C_s$ and $C_I$ of geometrical properties of burners cannot be addressed. However, the effect of heat release on turbulence activity can be investigated. In other words, at the level of the large eddy field, the modifications of the turbulent flow induced by heat release should lead to different values of the parameters $C_s$ and $C_I$.

The subgrid energy flux may be estimated as $\tau_{ij}\widetilde{S}_{ij}$, where $\tau_{ij}$ is given by Eq. (1). This flux may be extracted from filtered DNS in its exact

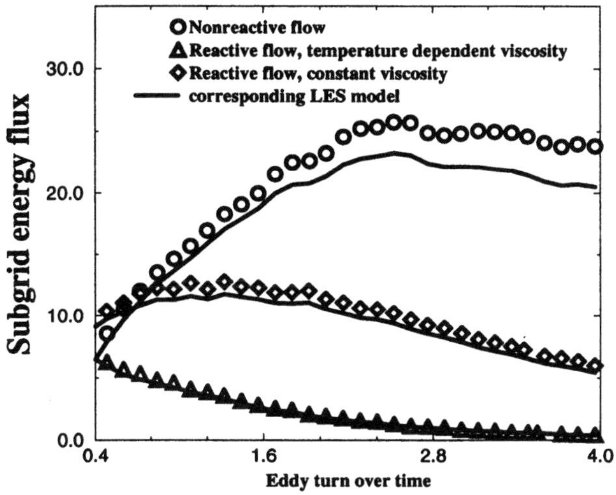


*Figure 1.* Flux of energy toward the small scales, DNS versus LES.

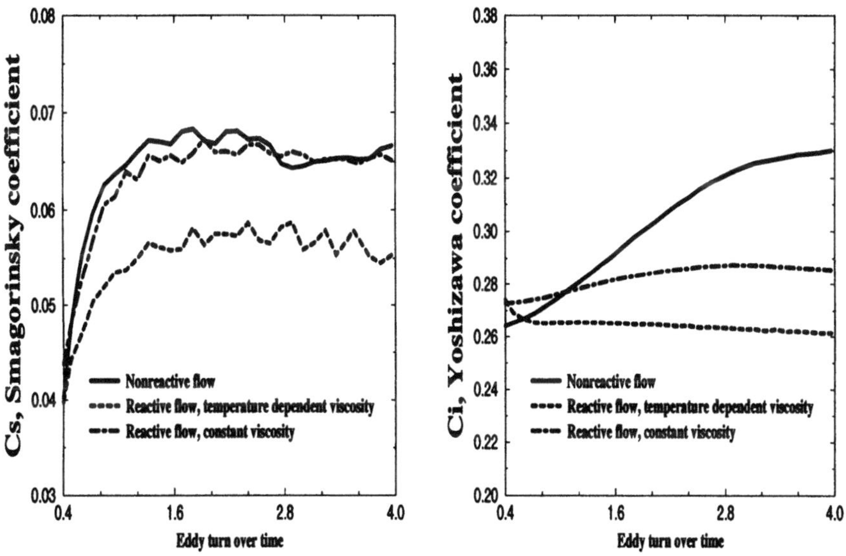


*Figure 2.* Time evolution of $C_s$ (left) and $C_I$ (right).

form $-(\widetilde{u_i u_j} - \widetilde{u}_i\ \widetilde{u}_j)\,\widetilde{S}_{ij}$ and compared against the closure. For these fluxes the agreement between LES and DNS is found to be acceptable (fig. 1). Simulations with constant laminar viscosity are performed to isolate the effects related to the decrease in density, and it is observed (fig. 2, left) that the Smagorinsky coefficient $C_s$, involved in the anisotropic part of the subgrid stress tensor, takes values close to those of the inert case when it is evaluated for a nonpremixed turbulent flame with constant laminar viscosity. However, the parameter $C_I$, linked to the compressible part of the subgrid stress tensor (Eq. (1)), is expected to be sensitive to heat release, even for constant laminar viscosity. This is confirmed by fig. 2 (right), where time evolutions of $C_I$ are presented for the different cases. It is observed that in opposition to $C_s$, $C_I$ is directly modified by variations in density promoted by combustion.

To conclude, comparisons with DNS indicate that variations in density resulting from heat release is properly accounted for when estimating the subgrid flux of energy with the lagrangian dynamic model.

### 2.3. SUBGRID MIXING MODELING, A DYNAMIC APPROACH

The objective of this section is to investigate possibilities of constructing a dynamic model for the unclosed diffusive (mixing) term found in the transport equation for the SGPDF. To focus on the mixing part of the problem, a field $Z(\underline{x},t)$ homogeneous and isotropic in mean (Reynolds average) and evolving in a constant density turbulent flow is considered. The transport equation for the SGPDF may be written (Gao and O'Brien, (1993))

$$\begin{aligned}\frac{\partial \overline{P}_{sg}(Z^*;\underline{x},t)}{\partial t} = & - \frac{\partial}{\partial Z^*}\left[\left(\overline{-\mathrm{u}.\nabla Z(\underline{x},t)\Big|Z(\underline{x},t)=Z^*}\right)\overline{P}_{sg}(Z^*;\underline{x},t)\right] \\ & - \frac{\partial}{\partial Z^*}\left[\left(\overline{D\nabla^2 Z(\underline{x},t)\Big|Z(\underline{x},t)=Z^*}\right)\overline{P}_{sg}(Z^*;\underline{x},t)\right] \quad (2)\end{aligned}$$

where the terms of the form $\left(\overline{A\Big|Z(\underline{x},t)=Z^*}\right)$ denote unclosed conditional filtered mean values. In particular, the diffusive term may be written

$$\begin{aligned}\left(\overline{D\nabla^2 Z(\underline{x},t)\Big|Z(\underline{x},t)=Z^*}\right)\overline{P}_{sg}(Z^*;\underline{x},t) &= \overline{D\nabla^2 Z(\underline{x},t)\delta(Z(\underline{x},t)-Z^*)} \\ &= \overline{\mathcal{H}}(Z^*;\underline{x},t)\end{aligned}$$

The basic idea of the proposed dynamic mixing model is to estimate the unknown quantity $\overline{\mathcal{H}}(Z^*;\underline{x},t)$ from the test mixing term

$$\widehat{\mathcal{H}}(\overline{Z}^*;\underline{x},t) = \widehat{D\nabla^2\overline{Z}(\underline{x},t)\delta(\overline{Z}(\underline{x},t)-\overline{Z}^*)}$$

which is known from the resolved large eddy field. $\widehat{Y}$ corresponds to a test filtered quantity, with $\widehat{\Delta} > \overline{\Delta}$.

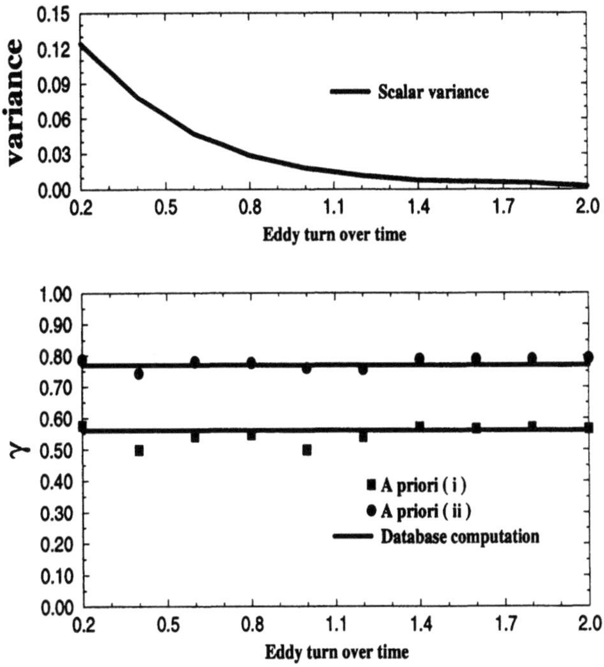


*Figure 3.* Top: Time evolution of $\left\langle (\langle Z\rangle - Z(\underline{x},t))^2\right\rangle$ computed from DNS. Bottom: Comparison of the time evolution of $\gamma(\overline{\Delta},\widehat{\Delta})$ computed from DNS (symbols) and from the $\gamma$ database (solid). Two cases are considered, (i) $\overline{\Delta}=2\delta_x$, $\widehat{\Delta}=4\delta_x$ and (ii) $\overline{\Delta}=4\delta_x$, $\widehat{\Delta}=8\delta_x$, where $\delta x$ is the uniform grid spacing.

The DNS field $Z(\underline{x},t)$ is filtered, and the resulting large eddy quantities are averaged over the computational domain. The large eddy mixing problem is then characterized by the time evolution of the mean of the SGPDF $\left\langle \overline{P}(Z^*,t)\right\rangle = \left\langle \overline{P}_{sg}(Z^*;\underline{x},t)\right\rangle$, which is expressed by $\partial\left\langle \overline{P}(Z^*,t)\right\rangle/\partial t = -\partial\left\langle \overline{\mathcal{H}}(Z^*,t)\right\rangle/\partial Z^*$.

Mean values of the subgrid mixing term $\left\langle \overline{\mathcal{H}}(Z^*,t)\right\rangle$ and of the test mixing term $\left\langle \widehat{\mathcal{H}}(\overline{Z}^*,t)\right\rangle$ have been extracted from $65^3$ DNS. As expected (Leonard and Hill, (1991)), the functions $\left\langle \overline{\mathcal{H}}(Z^*,t)\right\rangle$ and $\left\langle \widehat{\mathcal{H}}(\overline{Z}^*,t)\right\rangle$ feature maximum and minimum values which are, in the inert case, symmetrically located with respect to the means $Z^* = \langle Z(\underline{x},t)\rangle$ and $\overline{Z}^* = \left\langle \overline{Z}(\underline{x},t)\right\rangle$.

DNS analysis of the two functions $\left\langle \overline{\mathcal{H}}(Z^*,t)\right\rangle$ and $\left\langle \widehat{\mathcal{H}}(\overline{Z}^*,t)\right\rangle$ leads to the following expected results: 1) Different amplitudes for the extrema of the functions are found, 2) When the test filtering operation is applied, the location of these extrema in $Z^*$ and $\overline{Z}^*$ sample spaces are different for the two functions. In particular, these extrema are moved towards the mean when the test filtering operation is applied. These observations are explained by the fact that the filtering process globally decreases gradients of the mixture fraction, this effect being inversely proportional to filter sizes.

The basic idea of the proposed closure is to reconstruct the unknown subgrid function $\left\langle\overline{\mathcal{H}}(Z^*,t)\right\rangle$ from the computable function $\left\langle\widehat{\mathcal{H}}(\overline{Z}^*,t)\right\rangle$ by using a reference field $Z^r(\underline{x})$.

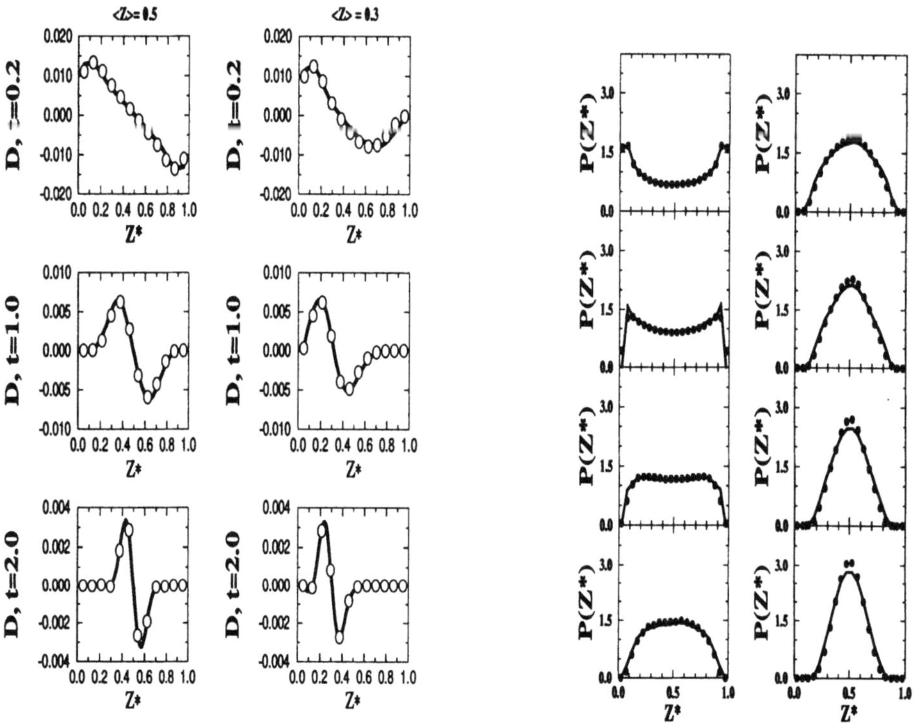


*Figure 4.* Comparison between DNS and subgrid diffusion model (symbol : exact from DNS, solid : modelisation). Left: mean subgrid function $\left\langle\overline{\mathcal{H}}(Z^*,t)\right\rangle$. Right: time evolution of the subgrid PDF.

One may write

$$\left\langle\overline{\mathcal{H}}(Z^*,t)\right\rangle = \frac{1}{\gamma(\overline{\Delta},\widehat{\Delta})}\left\langle\widehat{\mathcal{H}}(\mathcal{F}(\overline{Z}^*;\overline{\Delta},\widehat{\Delta}),t)\right\rangle \quad , \tag{3}$$

The function $\gamma(\overline{\Delta},\widehat{\Delta})$ relates the amplitude of the unknown function $\left\langle\overline{\mathcal{H}}(Z^*,t)\right\rangle$ to the computable quantity $\left\langle\widehat{\mathcal{H}}(\overline{Z}^*,t)\right\rangle$. $\mathcal{F}(\overline{Z}^*;\overline{\Delta},\widehat{\Delta})$ links both sample space variable $Z^*$ and $\overline{Z}^*$.

A database of $\gamma(\overline{\Delta},\widehat{\Delta})$ is constructed from the given random scalar field $Z^r(\underline{x})$ featuring a wide range of values $|\nabla Z^r(\underline{x})|$. Figure 3 (bottom) shows the time evolution of $\gamma$ computed from DNS for two different filter sizes, along with the corresponding values extracted from the $\gamma$ database. Despite the dramatic evolution of the statistical properties of the mixture fraction field $Z$ (fig. 3 top), $\gamma$ can be considered as only a function of the size of the filters.

The mapping function $\mathcal{F}$ such as $Z^* = \mathcal{F}(\overline{Z}^*; \overline{\Delta}, \widehat{\Delta})$ is also tabulated from the reference $Z^r(\underline{x})$ field.

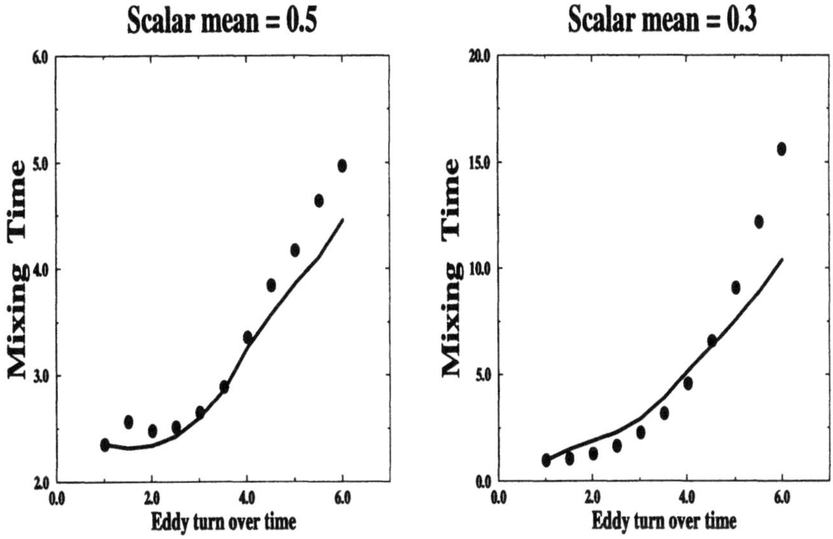


*Figure 5.* Time evolution of the subgrig mixing time scale (solid: subgrid diffusion model, symbol: DNS).

Figure 4 (left) displays the time evolution of $\left\langle \overline{\mathcal{H}}(Z^*, t) \right\rangle$ extracted from DNS and presumed by the model. Similarly fig. 4 (right) shows the relaxation of the SGPDF during mixing obtained from DNS and from the solution of its transport equation, in which the proposed model has been used to close the mixing term. Results are very encouraging.

The subgrid mixing time can be extracted from $\left\langle \overline{\mathcal{H}}(Z^*, t) \right\rangle$. Following the basic LMSE (or IEM) mixing model (Dopazo, (1994)), the slope of $\left\langle \overline{\mathcal{H}}(Z^*, t) \right\rangle$ at the position $Z^* = \overline{Z}$ is the inverse of the characteristic subgrid mixing time. Comparisons of subgrid mixing time evolution from DNS and provided by the proposed dynamic diffusion model are displayed in fig. 5.

Because the turbulence is decaying, after 4 eddy turn over times most of the subgrid energy has been dissipated and a deviation between the modeled time scale and the computed one is observed. A good agreement is found, however, before the subgrid energy is comsumed. This result suggests that this dynamic mixing closure can be used not only to solve the SGPDF transport equation, but it is also of great interest to estimate precisely the subgrid mixing time for mixture fraction.

It is concluded that dynamic mixing closures can be constructed for solving SGPDF transport equations.

## 3. Acknowledgments

The authors have benefited from discussion with Prof. Cesar Dopazo and Prof. Roland Borghi. Cray C-98 cpu time has been provided by IDRIS-CNRS (Institut du Développement et des Ressources en Informatique Scientifique).

## References

Borghi, R. (1988). Turbulent combustion modelling. *Prog. Energy Combust. Sci.*, 14.

Bray, K. and Peters, N. (1994). Laminar flamelets in turbulent flames. In Libby, P. and Williams, F., editors, *Turbulent Reacting Flows*, pages 63–113. Academic Press London.

Cook, A. W. and Riley, J. J. (1994). A subgrid model for equilibrium chemistry in turbulent flows. *Phys. Fluids*, 8(6):2868–2870.

Dopazo, C. (1994). Recent developments in pdf methods. In Libby, P. and Williams, F., editors, *Turbulent Reacting Flows*, pages 375–474. Academic Press London.

Friedrich, R. (1995). Dns and les of incompressible turbulent flow. In 4$^{eme}$ *Ecole de Printemps de mécanique des fluides numérique, Carcan-Maubuisson, France*. COST-MFN, GDR MFN, CNRS.

Gao, F. and O'Brien, E. E. (1993). A large-eddy simulation scheme for turbulent reacting flows. *Phys. Fluids*, 5(6):1282–1284.

Germano, M., Piomelli, U., Moin, P., and Cabot, W. H. (1991). A dynamic subgrid-scale eddy viscosity model. *Phys. Fluids*, 7(3):1760–1765.

Ghosal, S. and Rogers, M. (1994). Large-eddy simulation of a plane wake. *Center for Turbulence Research, Annual Research Briefs, Stanford*, pages 127–141.

Guichard, L., Vervisch, L., and Domingo, P. (1995). Two-dimensional weak choc-vortex interaction in a mixing zone. *AIAA journal*, 33(10):1797–1802.

Leonard, A. D. and Hill, J. C. (1991). Scalar dissipation and mixing in turbulent reacting flows. *Phys. of Fluids*, 3(5):1286–1299.

Meneveau, C., Lund, T. S., and Cabot, W. (1994). A lagrangian dynamic subgrid-scale model of turbulence. *Center for Turbulence Research, Proceeding of the 1994 Summer Program*, pages 271–299.

Moin, P., Squires, K., Cabot, W., and Lee, S. (1991). A dynamic subgrid-scale model for compressible turbulence and scalar transport. *Phys. Fluids*, 3(11):2746–2757.

Poinsot, T., Candel, S., and Trouvé, A. (1995). Direct numerical simulation of premixed turbulent combustion. *Prog. Energy Combust. Sci.* submitted.

Réveillon, J. and Vervisch, L. (1996a). Dynamics of isoconcentration surfaces in weak chock, turbulent mixing interaction. In *AIAA-960517, 34th Aerospace Sciences Meeting and Exhibit*, Reno, NV.

Réveillon, J. and Vervisch, L. (1996b). Response of the dynamic les model to heat release induced effects. *Phys. Fluids (in print)*.

Yoshizawa, A. (1986). Statistical theory for compressible turbulent shear flows, with the application to subgrid modeling. *Phys. Fluids*, 7(29):2152–2164.

# DIRECT NUMERICAL SIMULATION ANALYSIS OF THE G-EQUATION IN PREMIXED COMBUSTION

J. PIANA, D. VEYNANTE, S. CANDEL
*Laboratoire EM2C*
*Ecole Centrale Paris and CNRS*
*Grande voie des vignes*
*92295 CHATENAY-MALABRY Cedex*
*FRANCE*

AND

T. POINSOT
*IMFT - CERFACS*
*42, Av G. Coriolis*
*31057 TOULOUSE Cedex*
*FRANCE*

**Abstract.** This study is aimed at the development of a method applicable to the description of turbulent premixed flames in large eddy simulation framework. The basic idea is to use a transport equation to describe the propagation of the flame contour and thus avoid the calculation of the inner flame structure on the LES coarse grid. The so called $G$-equation is used for that purpose. One isolevel of the $G$ variable $G = G_0$ defines the flame location. While this idea has received considerable attention, it has been mainly used in constant density ("thermodiffusive") simulations. Extension to variable density flows of the kind encountered in combustion introduces serious difficulties. A novel method: the Forward Estimation of Temperature (FET) is devised in this investigation to deal with these problems. Flame/vortex and flame/turbulence interactions are calculated using the new model and compared with direct numerical simulations relying on simple chemistry representation of the reaction rate. Results obtained with the $G$-FET method are satisfying and this technique is now ready to be tested in a real LES reactive calculation.



*J.-P. Chollet et al. (eds.), Direct and Large-Eddy Simulation II,* 321-330.

## 1. Introduction

Flamelet models are now commonly used to describe turbulent premixed combustion and provide a convenient framework for subgrid scale model developments for large eddy simulations of reacting flows. Under the assumption of a short chemical reaction time and small flame thickness relative to the turbulence time and length scales respectively, combustion may be assumed to take place in thin reactive sheets separating unburnt and fully burnt gases. Then, the dynamics of the flame front may be described as the propagation of an infinitely thin surface using the $G$-equation (Kerstein *et al.*, 1988),

$$\frac{\partial G}{\partial t} + \boldsymbol{u} \cdot \nabla G = w \left| \nabla G \right| \tag{1}$$

where $G$ is a scalar field whose level surfaces represent interfaces. The flame front corresponds to one of the surface iso-levels $G = G_0$ of the scalar $G$ and moves with the propagation speed $w$. Points where $G < G_0$ define the region of burnt gases whereas $G > G_0$ corresponds to the unburnt mixture.

The combustion problem is replaced by the determination of the propagation speed $w$. This approach may also be viewed as an attractive tool for Large Eddy Simulation (LES) in premixed turbulent combustion. The filtered equation remains formally the same as Eq. (1) but the displacement speed $w$ of the flame front is replaced by the turbulent flame speed $S_T$ which needs to be modelled. While the principle is straightforward, application leads to various theoretical and practical difficulties.

One first notices that in most studies conducted with the $G$-equation (Kerstein *et al.*, 1988; Ashurst *et al.*, 1994), a constant density flow is assumed, which is clearly unsatisfactory in combustion situations. In the present work, Eq. (1) is used, where $w$ is the displacement speed of the local iso-level $G$ and depends on the local density $\rho$ to take into account the thermal expansion due to heat release (see section 2).

The aim of the present investigation is to devise a method which may be combined with the $G$-equation to accurately describe a premixed flame in a turbulent flow field. The novel method is explored by carrying direct simulations in well controled situations. Two sets of direct numerical simulations (DNS) of flame/vortex and flame/turbulence interactions are used to this purpose (see section 3). The code NTMIX (Baum *et al.*, 1993) is used as a plateform for this comparison. In the first set of DNS, the flame is described by a simple one step chemical reaction. The second one is based on the $G$-equation. The coupling between the propagation speed of the flame front $w$ and the heat release rate is investigated. A flame/vortex interaction is used to validate the ability of the $G$-equation to describe the evolution and the heat release rate of the flame front and to take into account flame front curvature effects. Turbulent flame propagation is then considered with a flame/turbulence interaction calculation.

## 2. G-Equation used for 1D flame description

### 2.1. VARIABLE DENSITY G-EQUATION

The $G$-equation is used in this section to model the propagation of a 1D flame at the speed $S_l$, stabilized by an inflow of premixed gases at the same speed $u_u = S_l$. The first problem here is the definition of the local propagation speed $w$. Due to density effects, and especially thermal expansion, the local velocity of the flow increases inside the flame from $u_u$ in the fresh gases to $u_b$ in the burnt gases and $\rho_u u_u = \rho_b u_b$. If one assumes conservation of mass inside the flame, the local velocity is $u = u_u \rho_u / \rho$. As described above, the flame is represented by an iso-level of the variable $G$ satisfying equation (1), then assuming the steady state,

$$\frac{\partial G}{\partial t} + u \cdot \nabla G = w \left|\nabla G\right| \text{ and } \frac{\partial G}{\partial t} = 0. \tag{2}$$

yielding $w = u$. The self propagation speed $w$ is not constant and must be equal to the local velocity of the flow. Then, each iso-level of the variable $G$ (and in particular the flame iso-level) propagates at the local speed of the flow and remains steady. It is then natural to scale $w$ with $\rho_u/\rho$:

$$w = S_l \frac{\rho_u}{\rho} \tag{3}$$

### 2.2. NUMERICAL METHOD

It is useful to validate the previous argument by a numerical simulation. The compressible Navier-Stokes set of equations are solved and the variable density G-equation replaces the standard equation of species conservation. The flame propagation, described by $w$, is combined with a simple model for heat release valid for the 1D propagation. In this case, the consumption speed $S_c$ which corresponds to the fuel consumption rate (the reaction rate per unit of flame area), is equal to the propagation speed $S_l$. The amout of heat release due to the flame propagation is then $\delta Q = \rho_u Y_{F_u} S_l \, c_p (T_b - T_u)$. The $G$ variable is initialized with a hyperbolic tangent with $G = 1$ in the fresh gases and $G = 0$ in the burnt gases. The flame is defined as the iso-level $G = 0.5$. Energy release at the flame is represented by: $\dot{\omega} = \rho_u Y_{F_u} S_l \, c_p (T_b - T_u) \left|\nabla G\right|$. This model replaces the Arrhenius source and the corresponding set of balance equations may be written:

$$\frac{\partial \rho}{\partial t} + \nabla \cdot (\rho \boldsymbol{u}) = 0 \tag{4}$$

$$\frac{\partial \rho \boldsymbol{u}}{\partial t} + \nabla (\rho \boldsymbol{u}\boldsymbol{u}) + \nabla p = \nabla \cdot (\boldsymbol{\tau}) \tag{5}$$

$$\frac{\partial G}{\partial t} + \boldsymbol{u} \cdot \nabla G = w \, |\nabla G| \tag{6}$$

$$\frac{\partial}{\partial t} (\rho e_t) + \nabla \cdot [(\rho e_t + p) \, \boldsymbol{u}] = \nabla \cdot (\boldsymbol{u} \cdot \boldsymbol{\tau}) + \dot{\omega} \tag{7}$$

The left graph in figure (1) shows a converged solution of the steady state 1D problem of a premixed flame propagating towards a constant velocity field ($u_u = S_l$).
In this figure, the thickness of the $G$ profile is similar to that of the density

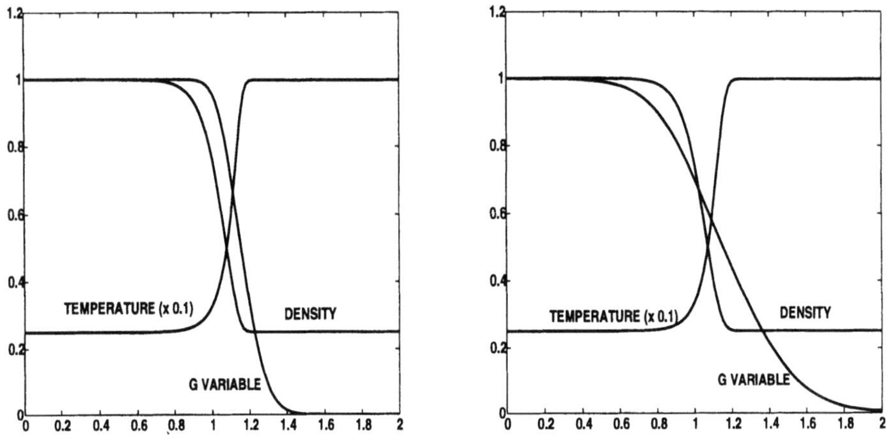


*Figure 1.* Converged solution of the propagation of a 1D flame with the G-equation with shape function (right) and without (left). left: $\dot{\omega} \propto w \; c_p \, (T_b - T_u) \, |\nabla G|$, right: $\dot{\omega} \propto w \; c_p \, (T_b - T_u) \, F \, (G) \, |\nabla G|$. The profile of $G$ is much larger than the density profile.

and temperature profiles. In an LES framework, a much wider $G$ profile must be used and its iso-level $G = G_0$ will accurately localize the flame front. This is done by modifying the shape of the energy release by a normalized shape function $F \, (G)$. The new energy source term is then :

$$\dot{\omega} = \rho_u Y_{F_u} w \; c_p \, (T_b - T_u) \, F \, (G) \, |\nabla G| \text{ where } \int_0^1 F \, (G) \, dG = 1. \tag{8}$$

with

$$F \, (G) = A \, \exp \left( \frac{(G - G_0)^2}{\delta^2 \, |\nabla G|^2} \right). \tag{9}$$

where $A$ is a normalizing constant and $\delta$ an adjustable width parameter. The energy release distribution can be made much narrower than the $G$

variable profile. The graph on the right of figure (1) shows the case where a shape function $F(G)$ is used to localize the energy release. By choosing the proper parameter $\delta$ it is possible to broaden the domain of $G$ arbitrarily to meet resolution requirements for a LES calculation.

## 3. Turbulent flame modeling

In this section, flame front wrinkling effects are investigated. This work is based on 2D direct numerical simulations. We are aware that three-dimensional simulations are clearly desirable as they provide a more complete physical description of turbulent flames. But, because of their computational cost, they cannot be used for parametric studies. Previous studies (see for example (Veynante D. and T., 1997)) have also shown that similiar results are achieved with 2D and 3D simulations for turbulence / premixed flames interactions. Accordingly, 2D DNS are well suited to investigate some difficulties induced by the $G$-equation.

### 3.1. CURVATURE EFFECTS

In configurations where the curvature is large, the flame propagation speed differs significantly from the consumption speed $S_c$ (Poinsot *et al.*, 1991). In this section, the curvature dependence on the flame consumption speed is illustrated and a relation between $S_c$ and $S_l$ is given. Consider the standard conservation equation for the reduced fuel mass fraction $Y = Y_F/Y_{F_u}$:

$$\frac{\partial Y}{\partial t} + u \cdot \nabla Y - D\nabla^2 Y = \dot{\Omega} \tag{10}$$

This expression differs from the $G$-equation by the diffusion term $D\nabla^2 Y$. Taking $G = Y$, a propagation at the speed $S_l$ is equivalent to

$$S_l\,|\nabla Y| = \dot{\Omega} + D\nabla^2 Y. \tag{11}$$

yielding

$$S_l = S_c + D\;\nabla\cdot\boldsymbol{n} + D\;\boldsymbol{n}\cdot\frac{\nabla\,|\nabla G|}{|\nabla G|} \tag{12}$$

where $\boldsymbol{n}$ is the vector normal to the flame front oriented from fresh to burnt gases and $\nabla\cdot\boldsymbol{n}$ is its curvature. This relation shows that the propagation speed $S_l$ combines reactive effects throught the fuel consumption speed $S_c$ and diffusive effects. Let us consider the converging propagation of a cylindrical flame. In this simulation, the last term (stretch in the tangent plane) is neglected and only the curvature term $D\nabla\cdot\boldsymbol{n}$ is considered. The left part of figure 2 shows the fuel mass fraction profiles at several times for a calculation where the consumption speed is kept constant and equal to

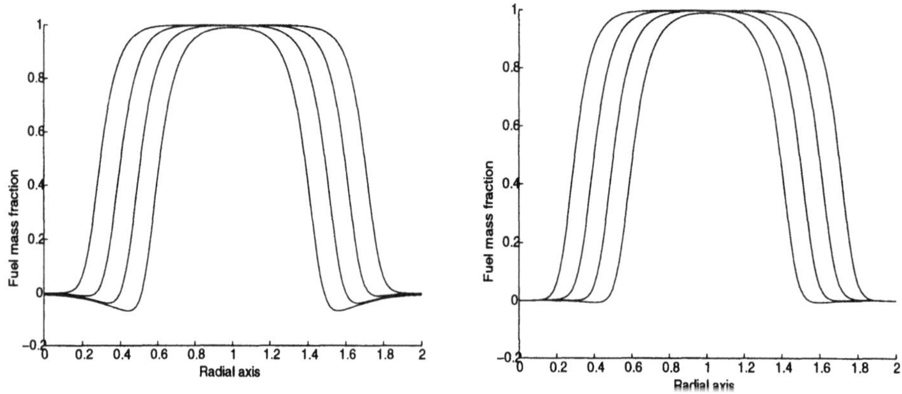


*Figure 2.* Propagation of a cylindrical flame with no curvature correction (left) and with curvature correction (right) where the consumption speed is better described.

the plane laminar flame speed: $S_c = S_l$. The right part of the same figure correspond to a consumption speed modified by curvature: $S_c = S_l - D\nabla \cdot \boldsymbol{n}$. Consumption is overestimated when curvature is not taken into account while it is well represented in the case of simple curvature with no tangential stretch. In more complicated cases such as flame-vortex interaction, the tangential stretch $D\boldsymbol{n} \cdot \nabla |\nabla G| / |\nabla G|$ has to be taken into account and the relation between propagation and consumption is less simple. To elude this problem, some techniques have been developed (Menon and Kerstein, 1992), (Im, 1995). These authors propose to write directly the temperature or the energy as a function of the variable $G$. Peters and Wirth (Wirth and Peters, 1992) use a front tracking method to accurately localize the iso-level $G = G_0$ and then separate the unburnt domain from the burnt one. In the following section, a novel method is devised for more complex flow situations.

### 3.2. FORWARD ESTIMATION OF TEMPERATURE (FET)

The FET method is based on a temperature prediction relying on the field estimates of $G$. The $G$-equation is first advanced in time, before the Navier-Stokes equations. The new $G$ field (at $t + dt$) defines the new location of the flame through its iso-level $G = G_0$. From this new flame location, one may deduce a forward estimate of the temperature field $T^*$ by imposing a suitable distribution. One may use for example a hyperbolic tangent function:

$$T^* = T_1 + \frac{1}{2}\left[\tanh\left(\frac{(G_0 - G)\pi}{2\delta}\right) + 1\right](T_b - T_u), \qquad (13)$$

where $\delta$ is an adustable width parameter allowing a variation of the stiffness of the presumed temperature $T^*$. Then the energy release term $\dot{\omega}$ is calculated from $T^*$ and the real temperature $T(t) : \dot{\omega} = \rho_u Y_{F_u} \, c_p (T^* - T)$. This energy source term is then injected in the energy equation and the Navier-Stokes equations are solved. In this model, the $G$-equation provides the location of the flame, and the constructed reaction rate is compatible with the propagation speed of the flame. By keeping an energy equation in the system, it is still possible to include in our model compressible effect or energy losses at boundaries of the domain. In the next section two test cases are presented and compared with the simple chemisty direct simulations, showing that this method gives reasonable results, and that it may be used in LES calculations where spatial resolution is low.

## 4. DNS comparisons

### 4.1. FLAME-VORTEX INTERACTION

The first test case is the interaction of a pair of counter-rotating vortices and a planar flame. The initial conditions are displayed in Figure (3). A

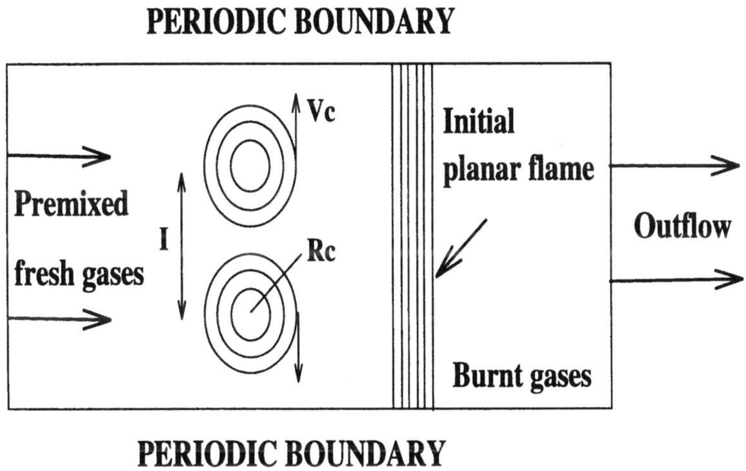


*Figure 3.* Initial condition for the flame-vortex interaction case

DNS calculation with simple chemistry is first carried out. In a second calculation, the G-equation and the FET model are combined to deal with the same flow configuration. The first FET+$G$-equation calculation is performed with a constant propagation speed $S_l$ while the second one uses a propagation speed $S_d$ corrected by the curvature (Markstein, 1964). The flame speed is a function of the laminar flame speed $S_l$, the curvature and the fresh to burnt temperature ratio. This relation uses the Markstein number Ma and for a unity Lewis number:

$$S_d = S_l (1 - \text{Ma} \; d_f \nabla \cdot \boldsymbol{n}) \tag{14}$$

where $d_f$ is the width of the flame estimated from the laminar flame speed $S_l$. Figure (4) shows the location of the flame calculated with the G-equation on top of the temperature profile of the DNS simple chemistry case. In the case of no curvature correction (left), the flame propagation is slightly im error in regions where the flame is curved. When the curvature is taken into account (right), the results are quite satisfactory: the location of the modeled flame exactly meets the simple chemistry case. The method

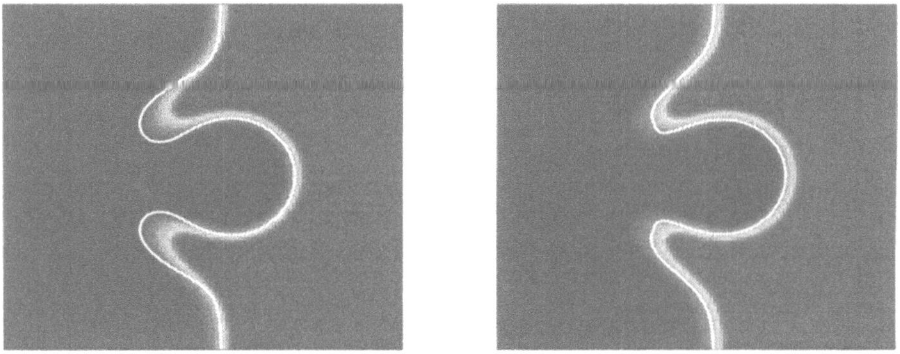

*Figure 4.* Flame/vortex interaction. Temperature field of the DNS-simple chemistry calculation with the white iso-level $G = G_0$ from the $G$-FET method. The propagation speed is correct (right) when the curvature correction of the flame speed is used.

gives good results for this case where the resolution is equal to the DNS resolution. Now the question is to know how does the model behaves when the resolution decreases. The calculation presented in figure (4) corresponds to a $121 \times 121$ grid. The grid has been reduced to a $31 \times 31$ grid for which the DNS could not be carried out because of the stiffness of the Arrhenius term while the G-equation method was still giving reasonable results (Fig. 5). This indicates that the $G$-FET method is suitable for low spatial resolution grids and therefore applicable to large eddy simulations.

## 4.2. FLAME-TURBULENCE INTERACTION

The second test case is the evolution of a flat flame embedded in a homogeneous isotropic turbulence. As before, simple chemistry simulations and FET+$G$-equation calculations are performed and comparisons are made. The turbulence characteristics (reduced variables) are summarized in table (1): $K_e$ and $K_d$ are the large and microscale wave numbers, $Re_t$ is the turbulent Reynolds number, $\varepsilon$ and $u$' are respectively the dissipation

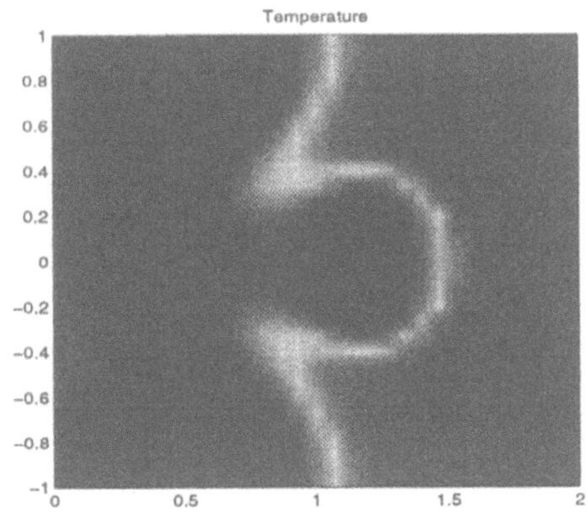


*Figure 5.* Flame/vortex interaction. Temperature field from the *G*-FET method. Calculation with a 31x31 grid, the G-equation method still holds and gives reasonable results. The DNS does not hold.

and the turbulence level. We use a von-Kármán Pao turbulent kinetic energy spectrum. For the homogeneous turbulence case, the model is rather

TABLE 1. Turbulence characteristics

| $K_e$ | $K_d$ | $Re_t$ | $\varepsilon$ | $u'$ |
|---|---|---|---|---|
| 1 | 50 | 100 | $3.210^{-4}$ | $6.510^{-2}$ |

representative of the DNS calculation. However, some errors in the local flame speed are made because of the difficulties to capture the curvature in some highly wrinkled regions. In order to limit spurious consequences of this non-physical behaviour, some extra diffusivity has been added in the G-equation. The *G*-FET method works well in this case.

## 5. Conclusions

This work constitutes a first step in the process of using the G-equation for combustion modeling in Large Eddy Simulations. A new compressible formulation has been tested in a 1D flame propagation configuration. A novel method designated as FET (Forward Estimation of Temperature) has been developed to decouple the energy release from the flame propagation and still keep them compatible. Comparisons with exact solutions obtained with DNS indicate that the *G*-FET method may be used to accurately

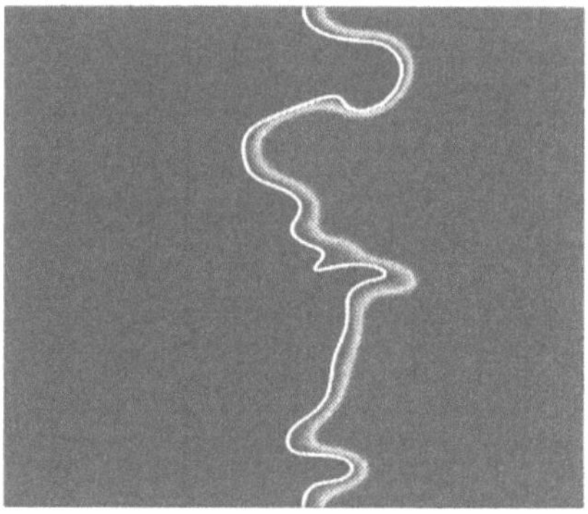

*Figure 6.* Flame/turbulence interaction. The position of the flame is not perfectly superimposed. Calculating the curvature for very low radii is a source of errors.

describe flame propagation. It is also shown that the method is suitable for the coarse grid calculations typically found in large eddy simulations.

## References


W.T. Ashurst, G.R. Ruetsch, and T.S. Lund. Passive turbulent flamelet propagation. In C. f. T. Research, editor, *Proceedings of the Summer Program*, pages 151–156, Stanford, 1994. Center for turbulence Research, Stanford.

M. Baum, D. Haworth, and T. Poinsot. Direct Simulation of turbulent flames with complex chemistry. *J. Fluid Mech.*, submitted, 1993.

H.T. Im. Study of turbulent premixed flame propagation using a laminar flamelet model. In C. f. T. Research, editor, *Annual Research Briefs*, pages 347–360, Stanford, 1995. Center for Turbulence Research, Stanford.

A.R. Kerstein, W. Ashurst, and F.A. Williams. Field equation for interface propagation in an unsteady homogeneous flow field. *Phys. Rev. A*, 37(7):2728–2731, 1988.

G.H. Markstein. Nonsteady flame propagation. AGARD MONOGRAPH 75, AGARD, 1964.

S. Menon and A.R. Kerstein. Stochastic simulation of the structure and propagation rate of turbulent premixed flames. In *24th Symposium (International) on Combustion*, pages 443–450, 1992.

T. Poinsot, T. Echekki, and M.G. Mungal. A study of the laminar flame tip and implications for premixed turbulent combustion. *Combust. Sci. Technol.*, 81(1-3):45, 1991.

Bray K.N.C. Veynante D., Trouvé A. and Mantel T. Gradient and counter-gradient scalar transport in turbulent premixed flames . *J. Fluid Mech.*, 332:263–293, 1997.

D. Wirth and N. Peters. Turbulent premixed combustion: a flamelet formulation and spectral analysis in theory and i.c. engine experiments. In *24th Symposium (International) on Combustion*, pages 493–501. The Combustion Institute, Pittsburgh, 1992.

# AUTOIGNITION OF NONPREMIXED TURBULENT MIXTURES: PARTIALLY PREMIXED COMBUSTION

P. DOMINGO AND L. VERVISCH
*LMFN-INSA de Rouen*
*URA - CNRS 230 - CORIA,*
*76821 Mont St. Aignan Cédex,*
*FRANCE*

**Abstract.**
Autoignition of nonpremixed turbulent mixtures undergoing a compression is studied by using DNS. The temporal evolution of the mean amount of heat released during ignition is used to distinguish between different ignition regimes. In particular, nonuniform distribution of mixture fraction dissipation rate along the stoichiometric line contributes to the development of partially premixed ignition fronts, in which the triple flame structure is observed.

## 1. Introduction, numerical procedure

The objective of this work is to seek out the basic properties of the reactive layers developing during autoignition of nonuniform turbulent mixtures undergoing a compression. In particular, the contribution of the partially premixed regime is investigated in detail.

Ignition of a mixing zone separating fuel and oxidizer is obtained by simulating an adiabatic compression. To mimic this compression, mass is added through a source term $\dot{\omega}_\rho$ acting in the continuity, momentum, energy and species transport equations. Viscosity is a function of temperature and the Lewis number is set to unity. Equations are solved by using a fully compressible DNS code developed by Guichard et al., (1995) following the idea of Poinsot et al., (1995). A sixth order accurate PADE Lele, (1992) spatial discretization is jointly used with a third order accurate Runge-Kutta method for time-stepping. Results are presented in stoichiometric premixed flame units. This choice being made because of the presence of



*J.-P. Chollet et al. (eds.), Direct and Large-Eddy Simulation II,* 331-341.

partially premixed fronts. The Zel'dovitch number $\beta$ is set to 8 and the heat release parameter $\alpha = (T_{burnt} - T_{fresh})/T_{burnt}$ is equal to 0.8. An Arrhenius law is used to express the burning rate of the one-step global chemistry.

The compression source term has been cast in two forms:

1) Two-dimensional simulations ($257^2$, laminar and turbulent) for which $\dot{\omega}_\rho(t) = Kt$, when $t \leq t_{comp}$, and, $\dot{\omega}_\rho(t) = 0$, when $t > t_{comp}$. This source term allows for simulating a compression. A very short compression time ($t_{comp}$) compared to the ignition time is considered.

2) Three-dimensional simulations ($129^3$, turbulent) in which compression occurs for $t < t_0$ and is followed by a decompression $\dot{\omega}_\rho(t) = K\rho^2(t_0 - t)$.

The second form of the compression term allows for mimicking diesel engine situations, where $t = t_0$ is the time at the top-dead center. It has been verified that results are not affected by the detail of the compression source term.

## 2. DNS of autoignition

Partially premixed combustion is observed when fuel and oxidizer are mixed prior to reaction. These flames may be of importance for instance in gas turbines in which stabilization is achieved through a fuel rich recirculation zone, or in diesel engines where high pressure injection systems are used to improve the premixing before ignition. The standard model for partially premixed combustion is the triple flame.

Triple flames exist in two generic forms. 1) A three layer system, where one stoichiometric diffusion flame and two premixed flames (lean and rich) coexist(fig. 1). 2) A premixed kernel along with a trailing diffusion flame. The stoichiometric premixed kernel evolves into a rich partially premixed flame in the direction of the fuel stream, while a lean partially premixed flame develops on the air side (fig. 2). In this case, the two premixed flames are curved because their self induced propagation velocities decrease when moving away from the stoichiometric point (Kioni et al., (1993)), ( Dold, (1989)), ( Liñán, (1994)), ( Ruetsch et al., (1995)).

Following Dold, (1989), the parameter $\mathcal{B} = \beta \ |\nabla Z|_{st}$ is introduced. $\beta$ is the Zel'dovitch number and $|\nabla Z|_{st}$ the magnitude of the mixture fraction gradient evaluated under stoichiometric conditions. ($Z$ is equal to unity in pure fuel and to zero in oxidizer stream). The $S$ curve ( Liñán and Williams, (1993)), describing the response of diffusion flames to variations in mixture fraction gradient, indicates that for a nonuniform distribution of $|\nabla Z|$ (or $\mathcal{B}$) autoignition will first occur at locations where $\mathcal{B}$ is small. Indeed, for too large values of $\mathcal{B}$ the reaction is not able to keep up with high level of heat losses. Therefore, for each particular instant in time when

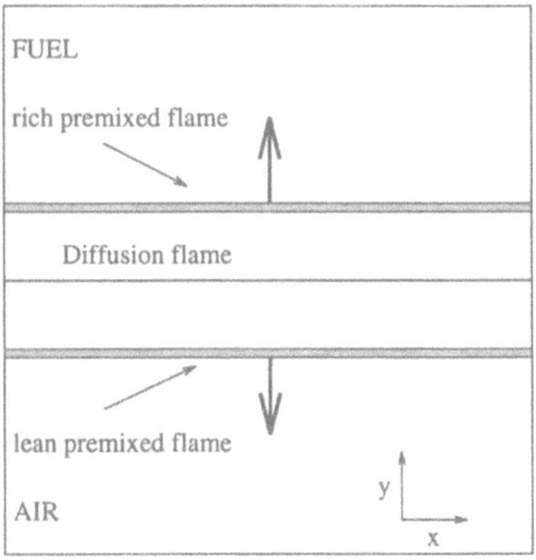


*Figure 1.* Sketch of three layers partially premixed structure.

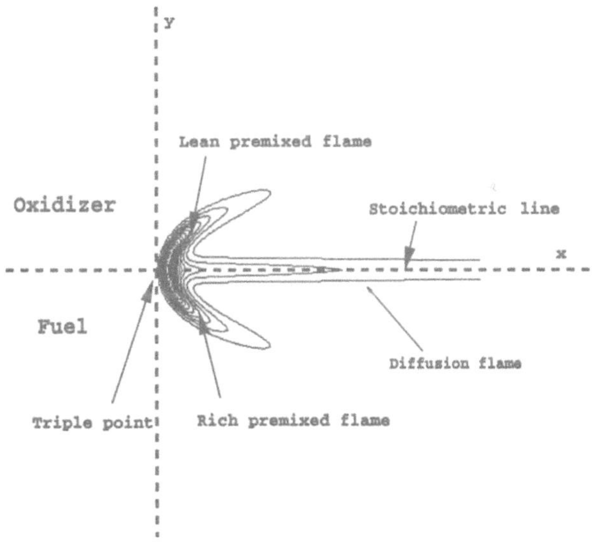


*Figure 2.* Sketch of a propagating triple flame.

ignition develops, one may introduce the value $\mathcal{B}_q$ from the $S$ curve, such that ignition is only observed at locations where $\mathcal{B} < \mathcal{B}_q$.

In the case of uniform $\mathcal{B}$ distribution, previous studies have shown that ignition develops according to the three layer system (fig. 1) ( Liñan and Crespo, (1976)), ( Thévenin and Candel, (1995)). However, when a nonuniform $\mathcal{B}$ distribution exists, conditions are found to produce propagating triple flamelets (fig. 2). Indeed, this propagating triple flame moves along the stoichiometric line from low values of $\mathcal{B}$ toward larger values of $\mathcal{B}$ ( Ruetsch et al., (1995)).

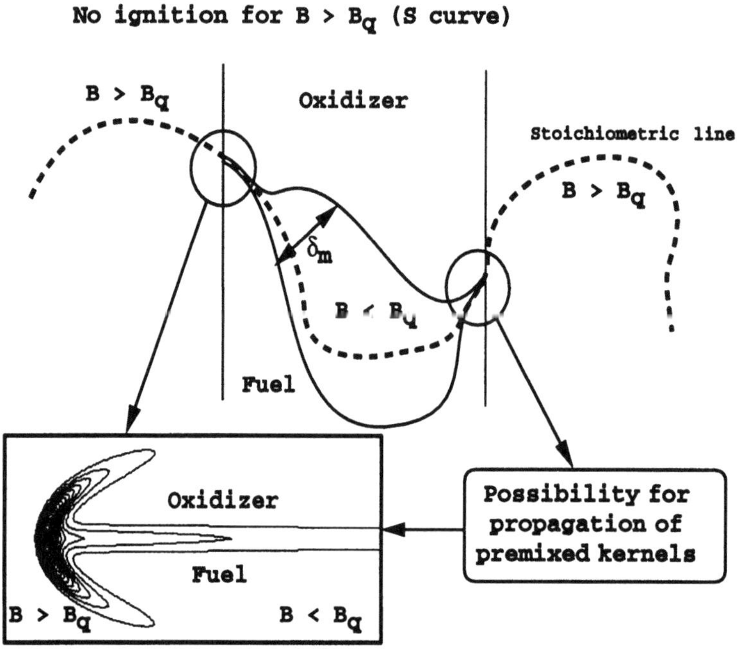


*Figure 3.* Sketch of the conditions allowing for the development of partially premixed ignition. $\mathcal{B} = \beta\ |\nabla Z|_{st}$, at this instant of the compression, autoignition occurs for $\mathcal{B} < \mathcal{B}_q$, $\delta_m$ is the local thickness of the mixing zone.

The rate of decay of the mean mixture fraction dissipation rate $\overline{\chi} = \overline{2D|\nabla Z'|^2}$ ($D$: diffusion coefficient, $Z'$: fluctuation of $Z$) evaluated under stoichiometric conditions (Bray and Peters, (1994)) is an important parameter of this study. When all the local characteristic mixing times ($\simeq 1/\chi$) are much smaller than the ignition time of the stoichiometric premixed mixtures, ignition is uniformly distributed. However, in the presence of turbulent mixing leading to possibilities of nonuniform distribution of ignition times, partially premixed combustion should strongly contribute to the propagation of ignition. Indeed, when traveling from low values of $\mathcal{B}$ toward the value $\mathcal{B}_q$, the size of the mixing layer decreases and the degree of premixing of the flow locally grows (fig. 3).

DNS results suggest that the temporal evolution of the mean amount of heat released can be used to distinguish between the different topologies for reactive layers described above. In particular, the turbulence is found to promote the propagation of ignition by allowing for autoignition through a local decrease of $\mathcal{B}$, or through a process of "parthenogenesis triple flamelets" as described in Veynante et al., (1994) and Domingo and Vervisch, (1996).

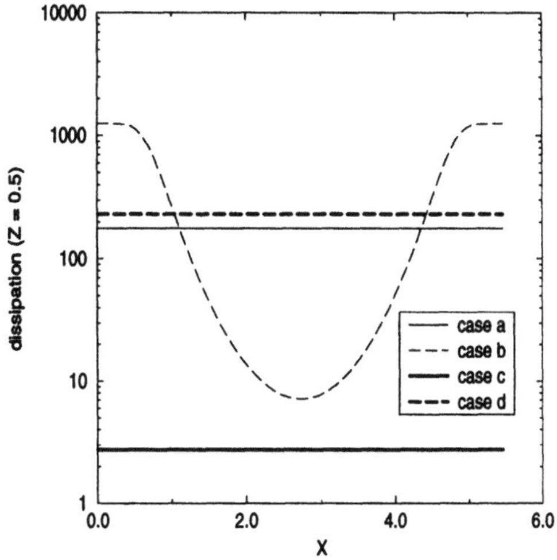


*Figure 4.* For the various initial conditions, distribution of $\chi = 2D\ |\nabla Z|^2$ along the stoichiometric line.

## 2.1. AUTOIGNITION WITHIN A LAMINAR FLOW

The aim of this section is to estimate the validity of the proposed topology (fig. 1 and fig. 2) for the flame structure at ignition. Simulations are performed with uniform and nonuniform $\chi$ (or $\mathcal{B}$) distribution (fig. 4).

The uniform $\mathcal{B}$ distribution leads to the three layer system (fig. 5). While in the case (b) (fig. 6) ignition first occurs within the zone of low level of $\mathcal{B}$ (or $\chi$), then two premixed flames, lean and rich, develop and migrate toward the edges of the initial low mixture fraction dissipation rate zone (fig. 6, t = 0.0955). The two generated premixed kernels contribute to the propagation of the ignition from the zone where the mixture fraction gradient is weak toward the zones where it is larger. Two propagating triple flames are then obtained (fig. 6, t = 0.118).

When triple flamelets are observed, two peaks are found in the time evolution of the mean amount of heat released (fig. 7). The combustion of the zone with a low level of $\mathcal{B}$ leads to the first peak. Then the burning rate is enhanced by the formation of the premixed kernels, leading to the second peak.

## 2.2. AUTOIGNITION IN A TURBULENT ENVIRONMENT

Simulations are carried out considering various initial thicknesses for the mixing zone separating fuel and oxidizer (cases (c) and (d) of fig. 4). A random homogeneous turbulent velocity field is prescribed. The turbulence

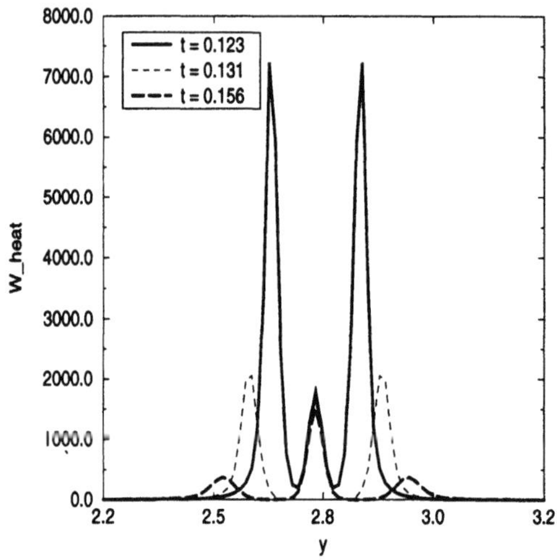


*Figure 5.* Case (a). Evolution of the heat release in the direction perpendicular to the stoichiometric line. The diffusion flame and the two premixed flames (lean and rich) are observed.

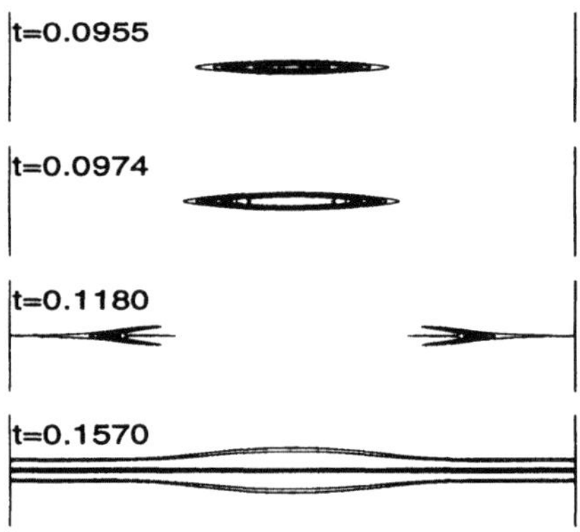


*Figure 6.* Case (b), laminar flow. Isocontour of heat release for four successive times, top to bottom.

will then decay freely. The initial rms turbulent velocity to laminar flame speed ratio is on the order of ten. The initial turbulent Reynolds number based on the integral length scale is about 70, while the Reynolds number

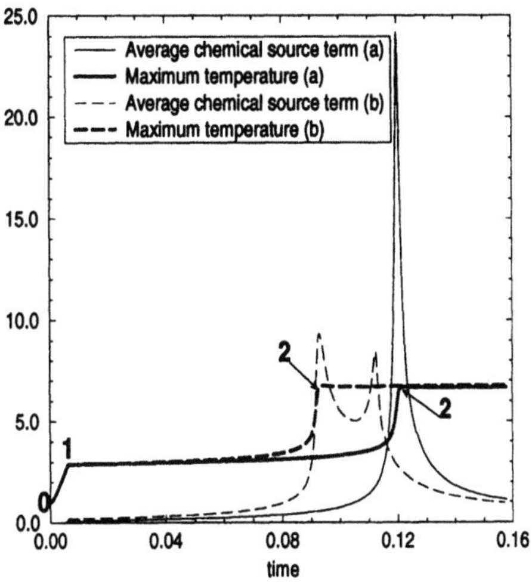


*Figure 7.* Cases (a) and (b), laminar flow. Time evolution of the maximum value of the temperature and of the average value of the heat release (chemical source term) over the computational domain.

based on the initial Taylor microscale is about 50. The initial condition gives a ratio between the integral length scale of the turbulent flow and the characteristic premixed laminar flame thickness of about 4.

Turbulent mixing generates nonuniform distributions of the mixture fraction dissipation rate (or $\mathcal{B}$); and for an initial condition with a low level of $\mathcal{B}$, case (c), the mechanism controlling the development of the flame is similar to the one described in fig. 6. Thus, one retrieves the double-peak in the temporal evolution of heat released (fig. 8). The corresponding propagating triple flames are displayed on fig. 9.

When a large value of $\mathcal{B}$, case (d), is imposed initially, both large ignition delay time along with possibilities for local extinction of diffusion flames are likely to occur. The propagating kernels are then more numerous (fig. 10) and only one peak is observed in the temporal evolution of heat released (fig. 8) The observed peak fully corresponds to the propagation of the partially premixed fronts. 3D simulations lead to similar results (fig. 11 and fig. 12).

## Conclusion

DNS suggests that the possible contribution of partially premixed combustion to autoignition of nonpremixed turbulent mixtures can be investigated from a simple physical picture. It is based on the appearance of nonuniform

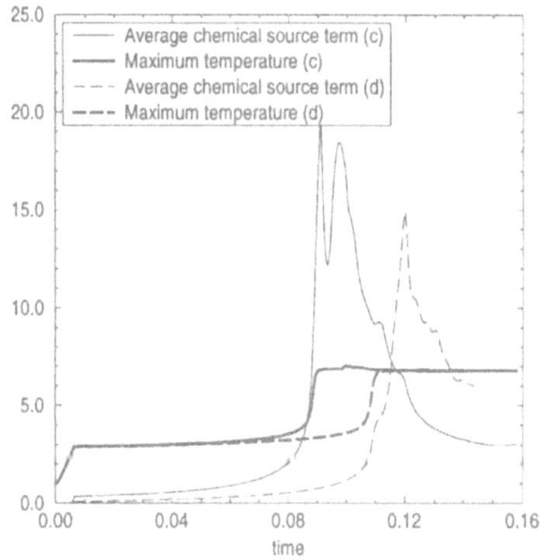


*Figure 8.* Case (c) and (d) ($\chi_c < \chi_d$), turbulent environment. Time evolution of the maximum value of the temperature and of the average value of the heat release (chemical source term) over the computational domain.

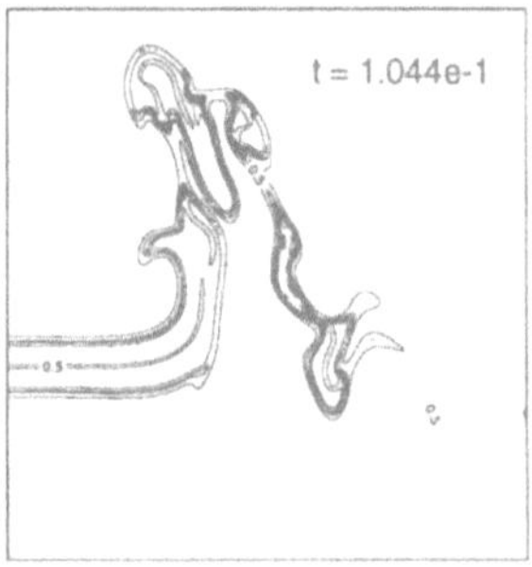


*Figure 9.* Case (c), turbulent environment. Isocontour of the heat release at t = 0.104.

ignition delay time along the stoichiometric line. This nonuniform ignition is directly linked to a nonuniform distribution of mixture fraction gradient.

During ignition, the time evolution of the mean heat release may include one or two peaks. The first peak corresponds to ignition in a diffusive-reactive system, for which ignition is first initiated at the stoichiometric premixed condition along the stoichiometric line. The second peak is representative of the propagation of ignition through partially premixed fronts. These two limiting regimes may lead to only one peak when one of the

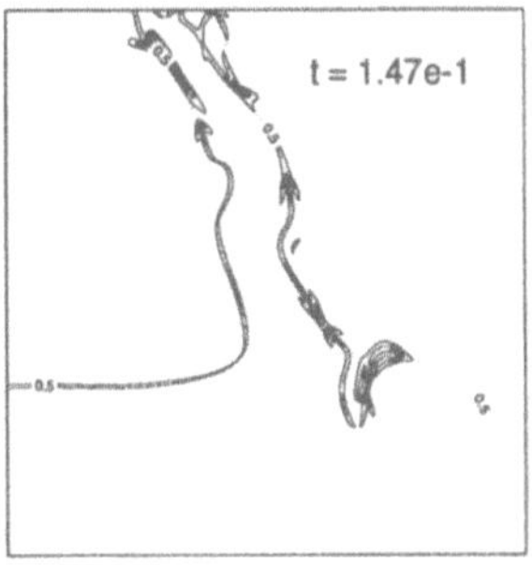


*Figure 10.* Case (d), turbulent environment. Isocontour of the heat release at t = 0.147

*Figure 11.* Case (c), three dimensional turbulent flow undergoing compression and relaxation. Reaction rate isosurface $\dot{\omega} = max(\dot{\omega})/10$, $t = 0.09$.

above processes is dominant.

For turbulent combustion modeling development, it is important to note that the propagation of ignition is mainly achieved through partially premixed fronts developing along the stoichiometric line and interacting with turbulence.

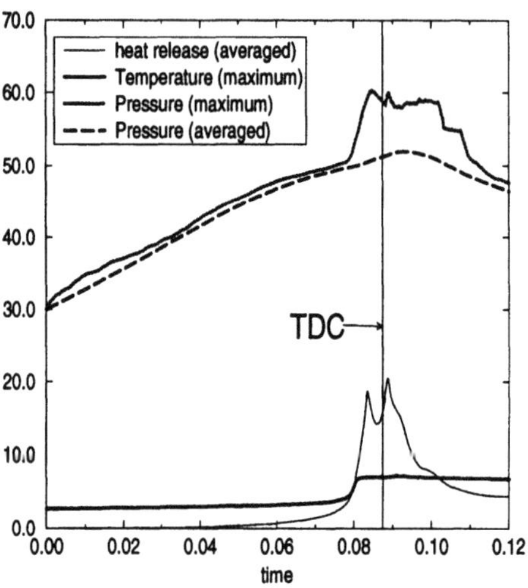


*Figure 12.* Case (c), three dimensional turbulent flow undergoing compression and relaxation. Time evolution of the mean heat release, maximum and average for temperature and pressure. TDC denotes and equivalent top dead center.

## 3. Acknowledgments

Cray C-98 cpu time has been provided by IDRIS-CNRS (Institut du Développ ment et des Ressources en Informatique Scientifique).

## References

Dold, J. W. (1989). Flame propagation in a nonuniform mixture: analysis of a slowly varying triple flame. *Comb. and Flame*, 76:71–88.

Guichard, L., Vervisch, L., and Domingo, P. (1995). Two-dimensional weak shock-vortex interaction in a mixing zone. *AIAA Journal*, 33(10).

Lele, S. K. (1992). Compact finite difference schemes with spectral like resolution. *J. Comput. Phys.*, (103):16–42.

Liñan, A. and Crespo, A. (1976). An asymptotic analysis of unsteady diffusion flames for large activation energies. *Comb. Sci. Tech.*, 95(14).

Liñán, A. (1994). Ignition and flame spread in laminar mixing layer. In Buckmaster, J., Jackson, T. L., and Kumar, A., editors, *Combustion in High Speed Flows*, page 461. Kluwer Academic Publishers.

Liñán, A. and Williams, F. A. (1993). *Fundamental aspects of combustion.* Oxford university press.

Poinsot, T., Candel, S., and Trouvé, A. (1995). Direct numerical simulation of premixed turbulent combustion. *Prog. Energy Combust. Sci.* submitted.

Ruetsch, G. R., Vervisch, L., and Liñán, A. (1995). Effects of heat release on triple flames. *Phys. of Fluids*, 7(6):1447–1454.

Thévenin, D. and Candel, S. (1995). Ignition dynamics of a diffusion flame rolled up in a vortex. *Phys. Fluids*, 2(7):434–445.

Veynante, D., Vervisch, L., Poinsot, T., A. Liñán, and Ruetsch, G. (1994). Triple flame structure and diffusion flame stabilization. In Moin, P. and Reynolds, W., editors, *Studying turbulence using numerical databases - V*, pages 55–73. Center for Turbulence Research.

Bray, K. and Peters, N. (1994). Laminar flamelets in turbulent flames. In Libby, P. and Williams, F., editors, *Turbulent Reacting Flows*, pages 63–113. Academic Press London.

Domingo, P. and Vervisch, L. (1996). Triple flames and partially premixed combustion in autoignition of nonpremixed mixtures. In *The Combustion Institute*, 26th Symp. on Comb.

Kioni, P. N., B.Rogg, Bray, K. N. C., and Liñán, A. (1993). Flame spread in laminar mixing layers: the triple flame. *Comb. and Flame*, 95:276.

# TWO-DIMENSIONAL DIRECT NUMERICAL SIMULATIONS OF TURBULENT DIFFUSION FLAMES USING DETAILED CHEMISTRY

D. THEVENIN, E. VAN KALMTHOUT AND S. CANDEL
*Laboratoire EM2C, Ecole Centrale Paris*
*Grande Voie des Vignes, 92295 Châtenay-Malabry, FRANCE*

**Abstract.** In order to make progress in the numerical investigation of practical turbulent propulsion systems, it is necessary to develop accurate Turbulent Combustion Models (TCM). A lot of work has been devoted to this task, particularly in the case of premixed combustion, but much remains to be done. Non-premixed systems are used in many industrial applications, e.g. in Diesel engines or in domestic burners, and it seems appropriate to study in detail turbulent non-premixed reacting flows. Apart from very complex experimental investigations, the only way to carry out such studies seems to be at present time Direct Numerical Simulation (DNS). Of course, an accurate simulation of reacting flows implies in many cases the use of detailed models for reactive and transport processes. We describe in this paper first results concerning such Direct Numerical Simulations of turbulent non-premixed flames using detailed chemistry.

## 1. Introduction

Numerical simulation of reactive flows has progressed impressively in recent years. This has brought considerable improvement in analysis, design and optimization tools applicable to practical systems. In some cases, the numerical predetermination of the flows has replaced experimental testing. There are nevertheless many cases where this is not possible. This state of affairs is mainly due to the complexity of the interaction between turbulent effects and reactive-diffusive processes. Available Turbulent Combustion Models fail in many configurations to give an acceptable answer. In order



*J.-P. Chollet et al. (eds.), Direct and Large-Eddy Simulation II,* 343-354.

to increase the impact of numerical predictions on industrial developments, the accuracy and reliability of these models must be greatly increased, while keeping numerical costs at a reasonable level.

One possible way is to try using Large-Eddy Simulations (LES) in place of more simple TCM. While LES is now widely employed to simulate complex turbulent flows, progress is still needed before LES can be used as an everyday tool to investigate reacting flows. The representation of the reactive layers on the coarse grid, as well as the coupling with subgrid-scale models, are still open questions at this point.

The other possible way is to improve existing TCM by introducing additional effects or new modelisation ideas. This requires a very good knowledge of the fundamental processes associated with turbulent combustion. Theoretical work is limited to basic configurations due to the complexity of this problem. Experimentation in combustion requires expensive testing facilities and complex diagnostics. It is often limited by the restricted access to the flame, the hostile high-temperature and pressure conditions prevailing in many cases and the abilities of current measurement methods. A possible complement or alternative is given by Direct Numerical Simulations, by which turbulent combustion can be investigated without introducing too many questionable models. It requires solving the full Navier-Stokes equations with transport and reaction terms, which can be cast in 3D using summation convention on indices $j$ and $k$ in the form (Williams 1985) :

$$
\begin{aligned}
\frac{\partial \rho}{\partial t} + \frac{\partial (\rho u_j)}{\partial x_j} &= 0 \\
\frac{\partial (\rho u_i)}{\partial t} + \frac{\partial (\rho u_i u_j)}{\partial x_j} + \frac{\partial p}{\partial x_i} &= \frac{\partial \tau_{ij}}{\partial x_j} \qquad (i=1,\ldots,3) \\
\frac{\partial e_t}{\partial t} + \frac{\partial [(e_t + p) u_j]}{\partial x_j} &= \frac{\partial (u_j \tau_{kj})}{\partial x_k} - \frac{\partial q_j}{\partial x_j} \qquad (1) \\
\frac{\partial (\rho Y_l)}{\partial t} + \frac{\partial (\rho Y_l u_j)}{\partial x_j} &= -\frac{\partial (\rho Y_l V_{Dl,j})}{\partial x_j} + W_l \dot{\omega}_l \qquad (l=1,\ldots,N_s)
\end{aligned}
$$

In this system, corresponding to $N_s$ chemical species, the variable $\rho$ represents the density, $p$ the pressure, $u_i$ the velocity in direction $i$, $q_j$ the heat flux along direction $j$, $Y_l$ the mass fraction of species $l$, $h_l$ its specific enthalpy, $V_{Dl,j}$ its diffusion velocity along direction $j$, $W_l$ its molar mass, $\dot{\omega}_l$ its molar production rate, and $e_t$ the total energy defined by :

$$
e_t = \rho \left( \frac{u_1{}^2 + u_2{}^2 + u_3{}^2}{2} + \sum_{l=1}^{N_s} (h_l Y_l) \right) - p \qquad (2)
$$

The accuracy required by such DNS and the corresponding cost lead to severe constraints on the code structure, which will be described in Section

2. Physical models used to represent reactive and diffusive processes are explained in Section 3, while results obtained with this code pertaining to the simulation of turbulent non-premixed flames are given in Section 4, and conclusions are listed in Section 5.

## 2. Numerical Techniques

The development procedure of the code *parcomb* has been described in detail in Thévenin *et al.* (1996), and will be only briefly summarized here. An estimation of the numerical effort necessary for two-dimensional as well as prospective future three-dimensional studies is also given in Thévenin *et al.* (1996).

### 2.1. BASIC CODE DESCRIPTION

The code we use for Direct Numerical Simulations of reacting flows is a finite-difference solver written in FORTRAN 77 and using domain-decomposition with message-passing for parallelization. It is at present time limited to two-dimensional configurations. It is well-known that many turbulent phenomena cannot be reproduced precisely with two-dimensional computations (Lesieur 1987), but it was not possible with currently available computer resources to implement detailed chemistry models in a three-dimensional code. Considering that many authors proved that reduced chemistry often leads to large errors in the obtained solution (e.g. Varma *et al.* (1986), Rhie *et al.* (1989), Sloane and Ronney (1991), Warnatz and Maas (1993)), we decided to implement detailed models for reaction processes and to restrain ourselves at this point to two-dimensional flows. It must nevertheless be kept in mind that the evolution of two-dimensional turbulence is different from that observed in real three-dimensional cases. This could modify at least quantitatively some of the obtained results, particularly for interactions with a considerable duration. The code could easily be extended to three dimensions if more powerful computers became available.

The space derivatives are computed using sixth-order central derivatives. In order to integrate the system of equations in time, a fourth-order fully explicit Runge-Kutta formulation is used. Implicit methods do not seem very beneficial in this case (Choi and Moin 1994), and are extremely difficult to parallelize with a good efficiency, which lead us to use this explicit technique. Moreover, it is planned to implement in this DNS code the Implicit Low-Dimensional Manifold method (ILDM, see Maas and Pope (1992)). This method removes the stiffest time-scales associated with reactive processes, and would therefore allow a considerable increase of the timestep.

In the present version, this timestep is controlled by three independent mechanisms. A Courant-Friedrichs-Levy (CFL) condition and a Fourier condition (for the diffusion terms) are tested to ensure the stability of the integration. Using timestep-doubling, the size of the timestep is also limited by a condition on the accuracy of the result. This generally leads to a timestep controlled by the chemical processes during the ignition and the extinction of the flame.

### 2.2. PARALLELIZATION

The parallelization of the code is based on a two-dimensional domain-decomposition, projected on a corresponding two-dimensional processor topology. Each node controls a subdomain of the global computational domain. The solution method consists then in the following repeated procedure : first, each processor carries out an integration step on its own subdomain completely independently from the others, then all processors communicate their new boundary values with their nearest neighbours. The initial grid-point distribution among the processors is computed by the code to minimize the length of inter-processor boundaries, leading to a minimum cost for communications, as described in Thévenin *et al.* (1994b). On all implemented architectures, we use either native asynchronous communications or PVM3.

As it was, the obtained parallel efficiency was not satisfactory, due to large differences in the effective work-load of the involved processors. In agreement with other authors, we found that a dynamic load-balancing procedure, which controls the distribution of the work among the processors during the run, is essential to obtain acceptable efficiency. We therefore implemented such a dynamic procedure (Thévenin *et al.* 1994a), by controlling the time needed by each processor to carry out an integration step. If the relative discrepancies between these measured times exceed a prescribed value, a grid-point redistribution procedure is activated. The nodes which presently need more computing time than the average send a part of their grid points to their nearest neighbour, on the side where the average load-per-processor is the smallest. This process is carried out independently in the $x$ and $y$-direction. In spite of its simplicity, this procedure was found to be robust and very efficient. In the best cases, it can lead to a reduction of over 80 % of the required wall clock time. The obtained efficiencies on all tested computers are then excellent (see Thévenin *et al.* (1996)). For example, a parallel efficiency of over 80% is achieved using 256 processors on a Cray T3D. This means that larger domains can be studied using parallel computers, which means larger values of the integral length scale of turbulence and of the turbulent Reynolds number. Alternatively, computing accurately three dimensional turbulent reacting flows should become possible in the near future.

## 2.3. BOUNDARY AND INITIAL CONDITIONS

The importance of very precise boundary conditions and their impact on the obtained solution is now widely recognized (Poinsot and Lele (1992), Carpenter *et al.* (1994), Sani and Gresho (1994)). We improved the Navier-Stokes Characteristic Boundary Conditions (Poinsot and Lele 1992) in order to take into account detailed models for chemistry and thermodynamics. Using this extended formulation, a precise and correct behaviour is obtained at the boundaries even for such difficult test-cases as propagating acoustic waves or flames leaving the numerical domain through the boundaries, as described in Baum *et al.* (1995).

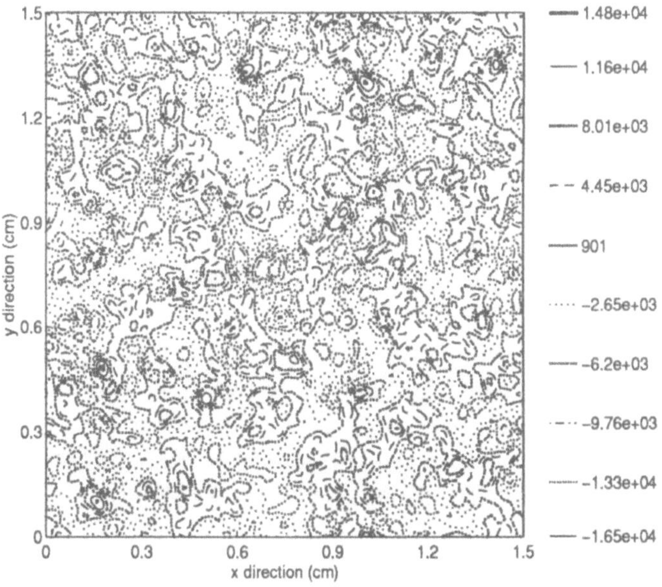


*Figure 1.* Example of initial vorticity field (1/s)

The initiation of the pseudo-turbulent velocity field is performed in the associated Fourier space. The parameters chosen by the user are the RMS velocity $u'$, the wave numbers $k_e$ associated with the maximum of turbulent kinetic energy and with the maximum of the dissipation, $k_d$. The distribution of turbulent energy versus wave number is given by a von Karman relation with Pao correction for near-dissipation scales in the form given in Hinze (1975) :

$$E(k) = A\frac{\left(\overline{u'^2}\right)^{5/2}}{\epsilon}\frac{(k/k_e)^4}{[1+(k/k_e)^2]^{17/6}}\exp\left(-\frac{3}{2}\alpha\,(k/k_d)^{4/3}\right) \tag{3}$$

In this equation, $A$ and $\alpha$ are model parameters and $\epsilon$ is the dissipation. The phases of the turbulent velocities in the Fourier space are considered to be uncorrelated and are therefore given by a random-number generator. The obtained turbulent velocity field is incompressible. An inverse FFT leads back to the values of the turbulent velocities in the physical space. An example is given in Fig.1, corresponding to $u' = 2.3$ m/s, an integral length-scale $\Lambda = 4.06$ mm, and turbulent Reynolds number $u'\Lambda/\nu$ of 397.

## 3. Physical models

It is known that many effects associated with turbulent combustion, like for example ignition, extinction or pollutant formation, can in general only be reproduced by using detailed models for the reactive and diffusive processes. For the results presented in this paper, we therefore employ a detailed reaction scheme to represent the combustion of hydrogen in air, using 9 species and 37 chemical reactions described by a set of Arrhenius parameters (Maas and Warnatz 1988). The multicomponent diffusion velocities are computed the standard way (Hirschfelder *et al.* 1954). Thermodynamical parameters are determined using fifth-order polynomial fits of experimental measurements.

For the practical computation of the chemical production terms, of the thermodynamical properties as well as of the molecular and thermal diffusion velocities, we use an existing library described for example in Maas and Warnatz (1989). Note that the Soret (or thermodiffusion) effect is also taken into account to compute the diffusion velocities $V_{Dl,j}$, as it is known to be important for hydrogen flames.

## 4. Results

For the results presented in this section, we use the initial turbulent velocity field presented in Fig.1 and described above. A laminar non-premixed flame of hydrogen diluted in nitrogen and air is first computed in a one-dimensional configuration with the same code. Resulting profiles are then used to initialize the two-dimensional computation by superimposing the turbulent velocity field, and the evolution of the resulting flame is observed. Periodicity is assumed along the $y$ direction, while both $x$ boundaries correspond to non-reflecting outflow boundary conditions with pressure re-

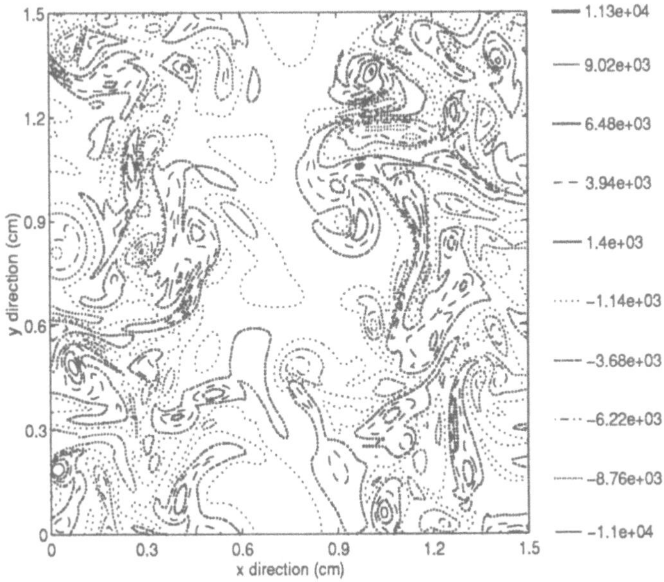


*Figure 2.* Field of vorticity (1/s)

laxation. A grid spacing of 50 $\mu$m is used in both directions for the case presented here.

The computation is carried out using 9 processors of an IBM SP2 parallel supercomputer. Total computing time on one node is slightly less than 40 hours. Results are presented for a time corresponding to one half of the large-eddy turn-over time $\Lambda/u'$. For this time, we show the vorticity, the mass fraction of the OH radical, the temperature and the heat released by the reactions respectively in Fig.2 to 5.

These pictures reveal some interesting features. The diffusion flame is highly contorted by the turbulent velocity field. The impact on the heat release as well as on the concentrations of some radicals is tremendous. In particular, the heat released by the chemical reactions is increased locally by a factor of almost 10 when compared to the laminar situation. Effects of curvature and strain appear to be essential to predict the evolution of the chemistry in such turbulent flows.

In order to give an idea of the coupling between heat release and detailed chemistry effects, the instantaneous correlation between heat release and $H_2O_2$ mass fraction is shown in Fig.6. In spite of the considerable dispersion of the points, this is the best correlation that could be obtained with any radical present in the flow. This shows that $H_2O_2$ could be used as a tracer for the reaction intensity.

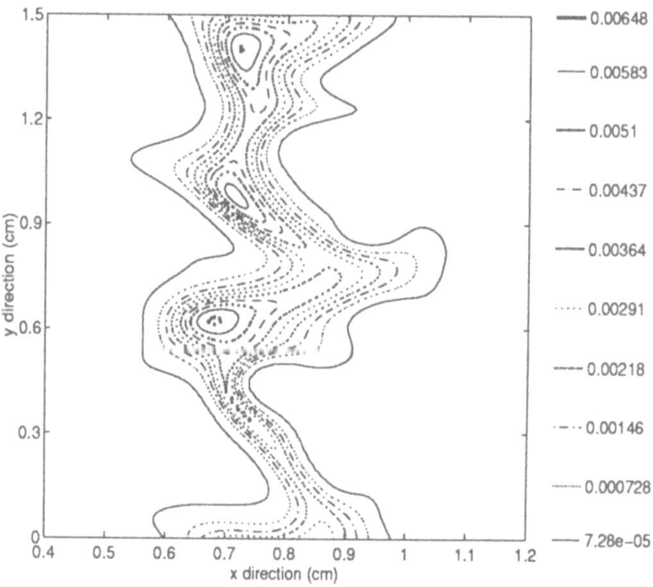


*Figure 3.* Field of OH mass fraction

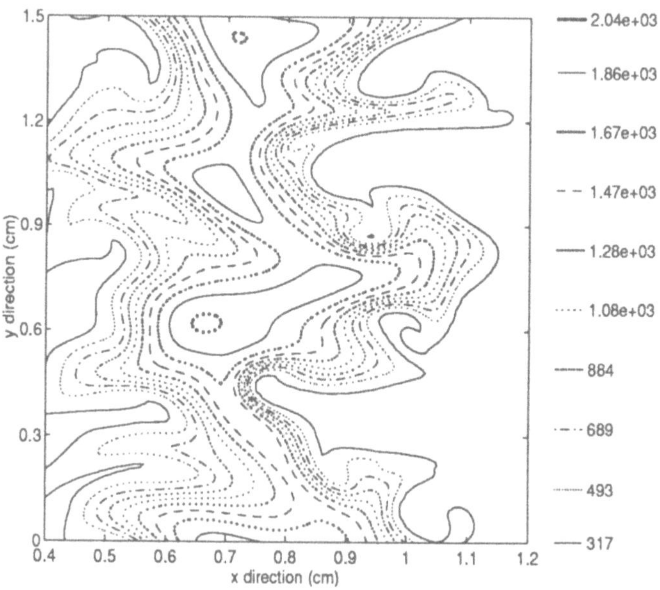


***Figure 4.*** **Field of temperature (K)**

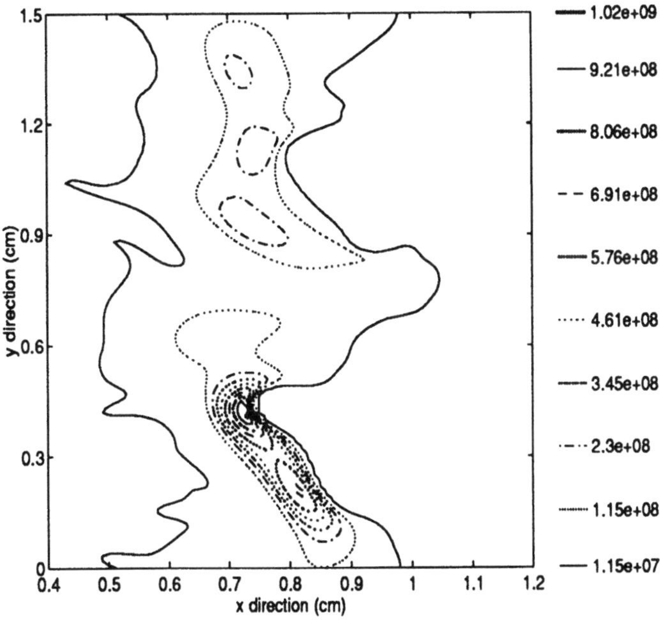


*Figure 5.* Field of heat release (J/m$^3$.s)

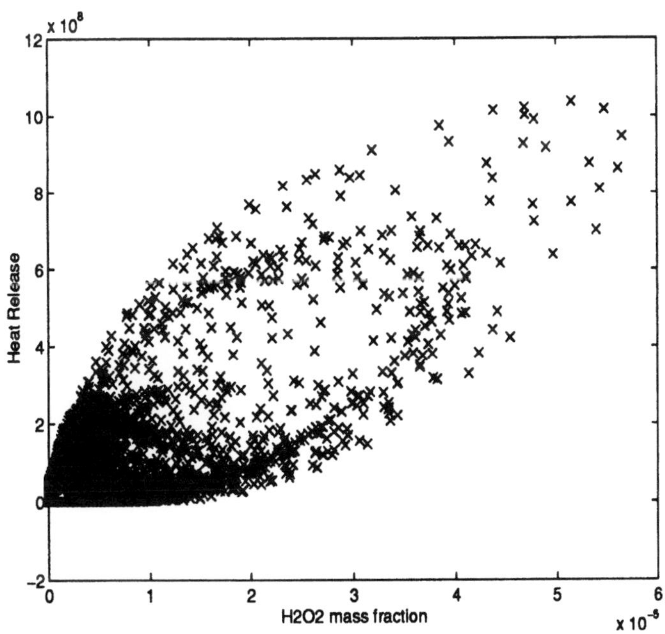


*Figure 6.* Correlation between heat release (J/m$^3$.s) and $H_2O_2$ mass fraction

## 5. Conclusions

A parallel code has been developed to investigate turbulent non-premixed flames using Direct Numerical Simulations. Detailed models for chemical and transport processes are implemented in order to represent accurately the interaction between turbulence and chemical reactions. Thanks to an efficient parallelization, results can be obtained in a reasonable computing time. Examples of such results have been presented. The coupling of the code with a detailed post-processing package is underway and will give access to numerous informations. This should lead to a considerable improvement in the understanding of turbulent combustion, which will help in developing more accurate Turbulent Combustion Models. In particular, results obtained by using DNS and TCM can be directly compared in the mean in order to identify ways of improvement and major deficiencies present in the turbulent combustion models.

## Acknowledgements

The authors would like to acknowledge the continuous support of IDRIS and CNUSC for accessing and using efficiently parallel supercomputers.

## References

Baum, M., Poinsot, T. and Thévenin, D. 1995 Accurate boundary conditions for multicomponent reactive flows, *J. Comput. Phys.*, **116**, pp 247-261.

Carpenter, M. H., Gottlieb, D. and Abarbanel, S. 1994 Time-stable boundary conditions for finite-difference schemes solving hyperbolic systems : methodology and application to high-order compact schemes, *J. Comput. Phys.*, **111**, pp 220-236.

Choi, H. and Moin, P. 1994 Effects of the computational time-step on numerical solutions of turbulent flow, *J. Comput. Phys.*, **113**, pp 1-4.

Hinze, J. O. 1975 *Turbulence*, McGraw Hill Book Company, 2nd Edition.

Hirschfelder, J. O., Curtiss, C. F. and Bird, R. B. 1954 *Molecular theory of gases and liquids*, John Wiley and Sons, Inc., New York.

Lesieur, M. 1987 *Turbulence in fluids*, Martinus Nijhoff.

Maas, U. and Pope, S. B. 1992 Simplifying chemical kinetics : intrinsic low-dimensional manifolds in composition space, *Combust. Flame*, **88**, pp 239-264.

Maas, U. and Warnatz, J. 1988 Ignition processes in carbon-monoxide-hydrogen-oxygen mixtures. In *Proc. Twenty-second Symposium (International) on Combustion*, The Combustion Institute, pp 1695-1704.

Maas, U. and Warnatz, J. 1989 Simulation of chemically reacting flows in two-dimensional geometries, *Impact Comput. Science Eng.*, **1**, pp 394-420.

Poinsot, T. J. and Lele, S. K. 1992 Boundary conditions for direct simulations of compressible viscous flows, *J. Comput. Phys.*, **101**, pp 104-129.

Rhie, C. M., Stowers, S. T. and Ebrahimi, H. B. 1989 Numerical analysis of reacting flows using finite rate chemistry models. In *Proc. AIAA 27th Aerospace Sciences Meeting*, AIAA 89-0459.

Sani, R. L. and Gresho, P. M. 1994 Résumé and remarks on the open boundary condition minisymposium, *Int. J. Numer. Methods in Fluids*, **18**, pp 983-1008.

Sloane, T. M. and Ronney, P. D. 1991 Comparison of ignition phenomena modelled with detailed and simple chemistry. In *Proc. Fourth International Conference on Numerical Combustion*, SIAM, Saint Petersburg, Florida.

Thévenin, D., Behrendt, F., Maas, U., Przywara, B. and Warnatz, J. 1996 Development of a portable parallel direct simulation code to investigate reactive flows, *Comput. Fluids*, **25**, 5, pp 485-496.

Thévenin, D., Behrendt, F., Maas, U. and Warnatz, J. 1994a Dynamic load balancing for parallel simulation of reacting flows. In *Proc. Parallel CFD'94*, (Ecer, A., Periaux, P. and Satofuka, N., Eds.), Elsevier.

Thévenin, D., Behrendt, F., Maas, U. and Warnatz, J. 1994b Parallel simulation of reacting flows using detailed chemistry. In *High-performance computing and networking*, (Gentzsch, W. and Harms, U., Eds.), Lecture Notes in Computer Science **796**, Springer-Verlag, pp 125-131.

Varma, A. K., Chatwani, A. U. and Bracco, F. V. 1986 Studies of premixed laminar hydrogen-air flames using elementary and global kinetics models, *Combust. Flame*, **64**, pp 233-236.

Warnatz, J. and Maas, U. 1993 *Technische Verbrennung*, Springer-Verlag.

Williams, F. E. 1985 *Combustion Theory*, 2nd Edition, Addison-Wesley.

# FLOW PAST A SQUARE CYLINDER: TEST CASE LES2

PETER R. VOKE
*Department of Mechanical Engineering*
*The University of Surrey, Guildford GU2 5XH, UK*

**Abstract.** A series of large-eddy simulations of flow past a square cylinder at $Re = 21400$ have been conducted by seven groups of researchers. A comparative report is presented of the time-average results with a view to establishing any clear advantage of particular subgrid scale modelling approaches, wall treatments, and numerical methods.

## 1. Background and Specification

The flow past a square cylinder at a Reynolds number of 21400 based on upstream velocity and cylinder side dimension was studied experimentally by Lyn and Rodi (1994) and Lyn *et al.* (1995). The flow is interesting as a test case for LES since it involves semi-coherent shedding of vortices from the cylinder, which is mounted transverse to the flow (Figure 1), breaking down into turbulence. It was selected by W. Rodi and J. Ferziger as a test case for a workshop held at Rottach-Egern, Germany in June 1995, reported by Rodi *et al.* (1996). The same flow was adopted at the First ERCOFTAC Workshop on Direct and Large-Eddy Simulation in March 1994 as test case LES2 to be studied for the Second Workshop reported here.

Since the flow involves coherent shedding of vortices from the cylinder, both the experimental data of Lyn *et al.* (1995) and the computations submitted to the Rodi-Ferziger workshop involved detailed phase averages at various phases through the shedding cycle. The phase averages constitute a large mass of data, and do not appear to add greatly to an understanding of the advantages and disadvantages of the various LES techniques (subgrid and wall models, resolution effects, etc.) over and above that provided by time-averaged data. In the present exercise, it was specified that only time-averaged data should be submitted for comparison.



*J.-P. Chollet et al. (eds.), Direct and Large-Eddy Simulation II,* 355-373.

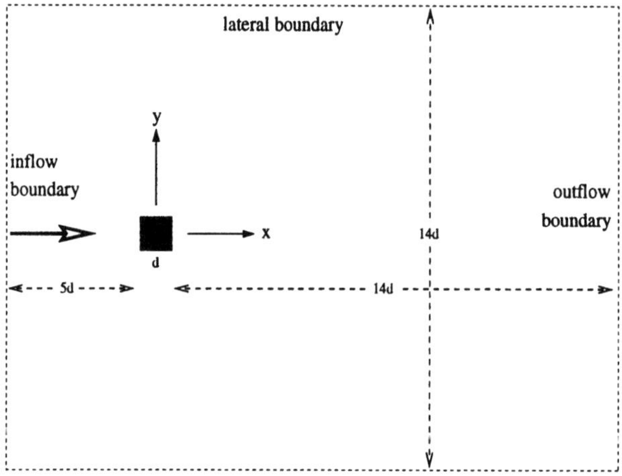


*Figure 1.* Geometry of the simulations.

## 2. Contributions to the Test Case

Seven groups have taken part in the test case exercise, though several of these have conducted more extended investigations by varying several aspects or parameters in their simulations, and have therefore submitted more than one dataset. The groups and datasets are as follows: Universität Karlsruhe (datasets UK1 to UK3); University of Groningen (GRO); University of Illinois (UOI); Institute of Technology, Niigata and University of Tokyo (NT1 to NT7); University of Tokyo Institute of Industrial Science (IS1 to IS3); Science University of Tokyo (ST2 to ST5); and the Tokyo Institute of Technology (TIT). This exercise is a continuation of that of Rodi and Ferziger, since all groups except GRO were participants in the Rottach-Egern workshop. The details of the simulations resulting in each of the datasets is laid out in Table 1. Here and elsewhere all distances are normalised by the cylinder dimension $d$ and velocities by the upstream velocity $U_0$.

The parameters given are: $N_x, N_y, N_z$, meshes in streamwise, normal and span directions; $L_x$, the domain length downstream from the cylinder centre to the ouflow plane; $\Delta t$, time step in terms of $d/U_0$; $T/T_s$, the statistical sampling time in the simulation divided by the shedding period $T_s$ (i.e. the number of cycles over which samples were taken, excluding any startup or conditioning run time); $\Delta_w$, the mesh dimension adjacent and normal to the cylinder surface, in terms of $d$; b.c. is the type of boundary condition applied at the wall of the cylinder.

Certain computational conditions were specified or suggested for the test case, notably the domain dimensions of $20d \times 14d \times 4d$. (In fact the groups used a variety of streamwise dimensions for their computational do-

TABLE 1. Simulations

| set | $N_x$ | $N_y$ | $N_z$ | $L_x$ | $10^3\Delta t$ | $T/T_s$ | $\Delta_w$ | b.c. | model |
|---|---|---|---|---|---|---|---|---|---|
| UK1 | 109 | 105 | 20 | 14 | 1 | 13.5 | 0.02 | power | $C_s = 0.1$ |
| UK2 | 146 | 146 | 20 | 14 | 1 | 8.3 | 0.01 | power | $C_s = 0.1$ |
| UK3 | 146 | 146 | 20 | 14 | 1 | 12.5 | 0.01 | no-slip | $C_s = 0.1$ |
| GRO | 280 | 210 | 64 | 20 | 1 | 3 | 0.005 | no-slip | none |
| NT1-5 | 104 | 68 | 10 | 14.5 | 1 | 8-2 | 0.022 | power | varied* |
| NT6 | 107 | 103 | 14 | 14.5 | 1 | 2 | 0.02 | power | $C_s = 0.1$ |
| NT7 | 140 | 103 | 32 | 14.5 | 1 | 8 | 0.02 | power | LDMM |
| UOI | 192 | 160 | 48 | 14 | 2 | 11 | 0.01 | no-slip | dynamic |
| IS1 | 82 | 63 | 16 | 21 | 2 | 13 | 0.04 | no-slip | $C_s = 0.13$ |
| IS2 | 82 | 63 | 32 | 21 | 2 | 13 | 0.04 | no-slip | $C_s = 0.13$ |
| IS3 | 112 | 104 | 32 | 20 | 2 | 6.7 | 0.02 | no-slip | dyn. mix. |
| TIT | 121 | 113 | 127* | 10 | 1 | 5 | 0.005* | no-slip | dynamic |
| ST2 | 109 | 78 | 20 | 7.7 | 0.5 | 16 | 0.024 | no-slip | $C_s = 0.1$ |
| ST3 | 125 | 78 | 20 | 11 | 0.5 | 15 | 0.024 | no-slip | numerical |
| ST4 | 107 | 103 | 20 | 15 | 0.5 | 12.5 | 0.017 | no-slip | $C_s = 0.1$ |
| ST5 | 107 | 103 | 20 | 15 | 0.5 | 14.5 | 0.017 | no-slip | $C_s = 0.1$ |

∗ Varies: see text.

mains downstream of the cylinder, but this variation does not appear to have affected the results in any obvious way.) This and other suggestions were made on the basis of experience from the Rottach-Egern workshop results; in fact one simulation from that meeting, the submission UKAHY1 from the University of Karlsruhe, was laid down as a baseline configuration for test case LES2, with the contributing groups encouraged to investigate variations and improvements of that computation. Nevertheless the specific numerical methods, subgrid-scale and wall models, mesh concentrations and stretching, boundary conditions and computational efficiency vary considerably in this exercise; we shall therefore now briefly summarise the key aspects of each simulation.

The Karlsruhe group provided the baseline computation UK1 which is identical to UKAHY1 of the Rottach-Egern workshop. UK2 is also included, which is essentially the same as UKAHY2 of the previous workshop, and has higher streamwise and lateral meshing than UK1 to allow a halving of the dimensions of the wall-adjacent meshes. Both UK1 and UK2 use the linear/one-seventh power law artificial boundary condition of Werner and Wengle (1993). The third computation UK3 is new, but differs from UK2 only in applying the no-slip boundary condition on the cylinder surface and in the sampling time which was 50% greater. The code used is non-staggered (cell-centred) with second-order central differencing and a standard explicit second-order time advancement and pressure correction scheme. The Smagorinsky model was used in all runs, with Van Driest damping functions close to the solid walls. UK3 required 10 hours per shedding cycle on a SNI (Fujitsu) S600/20.

The Groningen group (GRO) used a fourth-order finite-volume method which is detailed by Verstappen amd Veldman (1997) in this volume. Their grid is stretched using a sinh function, the ratio of the largest to the smallest grid being approximately 200 in $x$ and 100 in $y$. Standard inflow conditions, $u = 1, v = 0, w = 0$, were imposed at $x = -7.0$. Derivative outflow boundary conditions were imposed, $d^2v/dx^2 = 0, d^2w/dx^2 = 0$ and $dp/dx$ approximately zero, its value being determined at each time-step such that the mass inflow equalled the mass outflow. The outflow boundary was located beyond a buffer zone extending from $x = 15$ to 20, in which the Reynolds number was decreased. There was no subgrid model. Samples were taken every time step over three shedding cycles following a startup period of flow conditioning. The computation required 40 hours per shedding cycle on a Cray J916.

The group from the University of Illinois (UOI) performed six simulations as part of their study, of which the fourth is considered the most accurate and is included in this exercise. Further details of these simulations are given by Wang and Vanka (1996), and for the case included here (Wang and Vanka, 1997) later in this volume. The resolution in the spanwise direction is improved by reducing the spanwise domain dimension to $\pi d$ and using 48 cells. The inflow boundary conditions were standard, the outflow was a convective condition using a convection velocity of 0.7, but the lateral boundaries were moved further away so that the predicted $C_D$ had to be corrected for the different blockage; it is not known what other effects this different blockage may have had. The numerical scheme for the run utilised fifth-order upwind differencing for the convection terms and the fourth-order central scheme for other terms. A dynamic procedure based on the Smagorinsky model was used, stabilised by span averaging and bounding the total viscosity below by zero viscosity. The work was performed on a Thinking Machines CM-5.

The group co-operating between Niigata and Tokyo performed a high resolution simulation NT7 which will be included fully in the subsequent analysis, together with a series of six other test simulations with lower spanwise meshing (10 meshes) and box size ($2d$ instead of $4d$) and also with a very short sampling time (two shedding cycles out of a total run time of 4 cycles in each case, apart from NT1 where $T = 8T_s$). NT6 was slightly different; here the mesh of the baseline solution UKAHY1 from the Rottach-Egern workshop was closely reproduced, though the statistics were again gathered over two cycles. Since the implications of some of these supporting studies are of interest, the main features of the runs are summarised in Table 2. It is assumed that statistics gathered over two cycles can give no more than an indication of trends.

TABLE 2. Test Simulations NT1 to NT6

| set | Feature(s) |
|---|---|
| NT1 | Smagorinsky model with $C_s = 0.13$ and Van Driest damping function |
| NT2 | Dynamic Smagorinsky model; $C$ bounded by zero |
| NT3 | Dynamic mixed model; $C$ bounded below by zero |
| NT4 | Lagrangian dynamic mixed model |
| NT5 | As NT4 but on a staggered Cartesian grid |
| NT6 | Mesh and domain as UKAHY1, ($107 \times 103 \times 14$, domain width $4d$) Smagorinsky model with $C_s$=0.1 and damping function |

Most of the tests used collocated (centered) meshing. NT5 and NT7 used staggered meshing. The inflow and lateral boundary conditions were standard, the outflow convective, and the cylinder walls were subject to the boundary condition of Werner and Wengle (1993) in which a linear subayer is matched to a 1/7 power-law region above. In fact the wall-normal resolution of these simulations was in all cases not notably inferior to that of the other groups, and it is found that the linear law was being used in most regions of the upper and lower surface of the cylinder, exept near the corners. The numerical methods were second-order central differencing in all cases, and time advancement was by Adams-Bashforth for the convection terms and Crank-Nicolson for others. The tests that used the Smagorinsky model employed wall damping functions, $f = 1 - \exp(-y^+/25)$ for NT1 and $f = (1 - \exp(-y^+/25)^3)^{0.5}$ for NT6. The computations were carried out on a Hitachi S-3800 machine and required up to 2 hours per cycle for NT1 to NT6 and 9 hours per cycle for NT7.

The group from the IIS (Kogaki *et al.* (1997), following in this volume) performed three simulations with varying resolutions, numerical methods and subgrid-scale models. Their first two simulations, IS1 and IS2, differed only in the spanwise resolution (though IS1 was sampled every ten time steps and IS2 every step for statistics). Both simulations used a Smagorinsky model with $C_s = 0.13$ and standard Van Driest damping functions close to the walls. The differencing schemes were second order (QUICK for the convection terms with Adams-Bashforth time advancement, central for other derivatives with Crank-Nicolson advancement). The inflow was standard and the outflow boundary condition convective. The simulations were conditioned for 13 cycles before being sampled over a similar period. The cpu requirements were about 1.5 and 3 hours per shedding cycle on a Hitachi S-3800/480. The run IS3 was at higher resolution and required over 6 hours per shedding cycle on the same machine; as a result only half as many cycles were spent on both conditioning and sampling. The fifth-order upwind scheme was used for the convection terms and fourth-order central differences for diffusion terms. The same time-integration schemes were retained as in IS1 and IS2, but the run differed in utilising the dynamic mixed model of Vreman *et al.* (1994) and Zang *et al.* (1993), with the least-squares approach, stabilised by span averaging and bounding the viscosity below by the limit of zero viscosity.

The Science University group (Kawashima and Kawamura (1997), following in this volume) performed a set of four simulations, applying to this problem their convective schemes which act partially as subgrid models. The simulations differ slightly in the domain, meshing and sampling regimes. In several cases the convection schemes were used in combination with a Smagorinsky model. ST1 used central differencing and on the mesh employed was numerically unstable. In ST2 the UTOPIA convective scheme, which is a combination of fourth-order central and third-order upwind interpolation, was combined with the Smagorinsky model. ST3 omitted the Smagorinsky model entirely and relied on UTOPIA as a numerical model. ST4, with different lateral meshing, restricted the use of UTOPIA to the region $-3.5 < y/d < 3.5$; and ST5 utilised a different combination of fourth-order central and third-order interpolation in which the coefficient of the third order term, $\alpha$, depends on the ratio of the Smagorinsky and molecular viscosities, thus: $\alpha = \exp(-\nu_t/100\nu)$; this we call the variable-alpha scheme. Otherwise the simulations use standard methods and boundary conditions, though the outflow was governed by zero streamwise gradient rather than the recommended convective condition, and the streamwise domain lengths varied; ST2 was performed in a rather short domain.

The group of the Tokyo Institute of Technology performed a single LES of some interest using embedded meshes. The outermost mesh was $121 \times 114 \times 27$, the middle mesh $113 \times 97 \times 57$ and the inner mesh which surrounded the cylinder was $91\times91\times127$. Diagrams of the meshing are given subsequently in this volume (Nozawa and Tamura, 1997). By this means the resolution close to the upstream face of the cylinder was made very fine (0.005) though somewhat coarser near-wall resolution was used on the cylinder sides (0.02) and rear faces (0.033). A third-order upwind scheme for the convection terms was combined with second-order differencing for the other terms and a subgrid eddy viscosity generated by a dynamic procedure applied to the Smagorinsky model with the least squares approach. The computation required 20 hours per cycle on one processor of a Fujitsu

VX.

The groups were invited to estimate the magnitude of the numerical dispersion and dissipation of their codes, but little data was submitted on this aspect of the LES apart from the specific studies of interaction between subgrid models and convection treatments by the IIS and SUT groups. Some of the submissions constituted a demonstration of the mesh dependence of the results; it does not appear that anyone was able to demonstrate mesh independence convincingly.

## 3. Results

The principal integral parameters predicted by the simulations in the exercise are shown in Table 3. The columns give the dataset followed by the predicted recirculation length, Strouhal number of the shedding, the drag and lift coefficients of the cylinder (the latter should be zero by symmetry) and the r.m.s. variation of the lift and drag.

TABLE 3. Results: Integral Parameters

| set | $l_r$ | $St$ | $C_D$ | $C_L$ | r.m.s. $C_D$ | r.m.s. $C_L$ |
|---|---|---|---|---|---|---|
| **UK1** | 1.32 | 0.13 | 2.20 | -0.02 | 0.14 | 1.01 |
| UK2 | 1.46 | 0.13 | 2.30 | -0.04 | 0.14 | 1.15 |
| **UK3** | 1.44 | 0.13 | 2.23 | -0.05 | 0.13 | 1.02 |
| **GRO** | 1.61 | 0.133 | 2.09 | 0.005 | 0.18 | 1.45 |
| **NT7** | 1.39 | 0.131 | 2.05 | -0.05 | 0.12 | 1.39 |
| **UOI** | 1.20 | 0.13 | 2.03[b] | 0.04 | 0.18 | 1.29 |
| IS1 | 1.12 | 0.13 | 2.041 | -0.29[e] | 0.26 | 1.31 |
| IS2 | 1.20 | 0.13 | 2.067 | -0.0066 | 0.15 | 1.235 |
| **IS3** | 1.36 | 0.133 | 2.79 | -0.125 | 0.36 | 1.68 |
| **TIT** | 1.23 | 0.131 | 2.62 | 0.0093 | 0.23 | 1.39 |
| ST2 | 1.06 | 0.16 | 2.72 | 0.01 | 0.28 | 1.26 |
| **ST3** | 1.24 | 0.15 | 2.66 | -0.005 | 0.27 | 1.33 |
| ST4 | 1.12 | 0.139 | 2.74 | 0.012 | 0.29 | 1.36 |
| ST5 | 1.02 | 0.161 | 2.78 | 0.009 | 0.28 | 1.38 |
| LYN | 1.38 | 0.132 | 2.1 | | | |
| DUR | 1.33 | 0.138 | | | | |
| LEE | | | 2.05 | | 0.16–0.23 | |
| VIC | | | 2.05 | | | 0.68–1.32 |
| CLC | | | 1.9–2.1 | | 0.1–0.2 | 0.1–0.6 |
| MAG | | | 1.9–2.1 | | 0.1–0.2 | 0.7–1.4 |

In the table the annotation [b] indicates that the drag coefficient has been corrected to allow for blockage, while [e] indicates a result which is affected by a known error, for instance in the boundary conditions. Some experimental results are given below the LES results: Lyn *et al.* (1995) (LYN) and Durao *et al.* (1988) (DUR) give the recirculation length and Strouhal number. Lee (1975) (LEE) gives a mean drag coefficient that agrees with a value given by Vickery (1966) (VIC), within the range found by Cheng *et al.* (1992) (CLC) and of McLean and Gartshore (1992) (MAG). These experiments were conducted under various conditions of free-stream turbulence and Reynolds numbers, but serve to indicate the general level of these quantities. There is broad agreement on the r.m.s. fluctuation of $C_D$, while the r.m.s. fluctuation of $C_L$ is less clear from the experiments; it is probably in the region of 1.3.

It is not possible to show detailed output from all twenty datasets; we therefore select a subset for more detailed study. As the baseline solution (and to provide a bridge with the Rottach-Egern workshop) results from UK1 are included, and also those from UK3 which represents a significant advance in resolution. The results from Groningen (GRO) are based on a short sampling time and second-order moments have not converged sufficiently to allow their inclusion; some results for the mean streamwise velocity only will be compared. Results are given for the finest resolution simulation of the Niigata-Tokyo group (NT7), for the simulation of the University of Illinois (IOU), and for the finest resolution simulation from the Institute of Industrial Science (IS3). The Tokyo Institute of Technology results, obtained on nested meshes, are included. Of the datasets submitted by the Science University of Tokyo group, the pure UTOPIA scheme appears to perform marginally better than the mixed scheme, Smagorinsky model, or the variable-alpha scheme, but is identical to a computation considered in the Rottach-Egern workshop; we therefore concentrate on the results from the variable-alpha scheme, ST5.

Thus in addition to UK1, we will present detailed results on one simulation from each group: UK3, GRO, NT7, UOI, IS3, TIT and ST3. These datasets are highlighted in bold in Table 3.

Many of the LES have difficulty in estimating the recirculation length. As we shall see, even quite credible streamwise profiles of $U$ may still cross the $U = 0$ axis at the wrong position. Given the difficulty of getting $l_r$ right, the performance of simulation NT7 seems impressive: this simulation predicts most of these integral parameters correctly, apart from the r.m.s. lift which may be low. However, we shall see that in other respects this simulation may be at fault. The group give some mean integral quantities as a diagnostic for their test runs NT1 to NT6; these results are summarised in Table 4. The lift coefficients, also estimated, were acceptably low. Several runs appear to be heading towards good mean values on the basis of these criteria, including NT6.

TABLE 4. Recirculation lengths and drag coefficients for runs NT1 to NT6

| run: | NT1 | NT2 | NT3 | NT4 | NT5 | NT6 |
|---|---|---|---|---|---|---|
| $l_r$ | 1.11 | 1.37 | 1.28 | 1.45 | 1.43 | 1.31 |
| $C_D$ | 2.10 | 2.08 | 2.06 | 2.12 | 2.09 | 2.03 |

The Strouhal number was found to be relatively easy to predict correctly both in this exercise and at Rottach-Egern. The rather high values found by the Science University group in runs ST2 and ST5 (also ST3 to some extent) are worrying, possibly indicating a deleterious effect of resolution or the special numerical schemes, including the variable-alpha scheme.

## 4. Discussion

Figure 2 shows the streamwise distributions of normalised mean streamwise velocity and turbulence intensities over selected ranges of $x$ downstream of the cylinder along the centreline. It is immediately apparent that the problem of predicting the recovery of the centreline velocity (2a, top), noted at Rottach-Egern, has not been solved conclusively. The experiment shows the velocity reaching about 0.6 of the upstream free-stream level and then apparently tending to level off (though data were not taken beyond $x/d = 8$). Some of the LES, in particular UK1, UK3 and NT7, show the velocity approaching the original level; others such as TIT and UOI show a distinct decline beyond about $x/d = 5$, which is difficult to understand physically although it is not necessarily incompatible with the experimental data shown, which appears to be levelling off at $x/d = 8$.

Figure 2b (top) is a view of $U$ in the recirculation zone and up to $x/d = 3$. We note that those simulations having the lowest velocity at their exit planes are those with the shortest recirculation lengths (TIT and UOI) while some simulations that recover towards $U/U_0 = 1$, such as NT7 and UK3, have longer recirculation lengths. The simulation GRO, without a subgrid model, has a long recirculation length but a reasonable recovery further downstream — a result that is confirmed by the comparisons made by Wang and Vanka (1996) with and without a subgrid model, though in their case the rather short recirculation was actually improved by removing the model.

It would appear that the simultaneous prediction of the recirculation bubble length and the final recovery of the centreline velocity is still diffi-

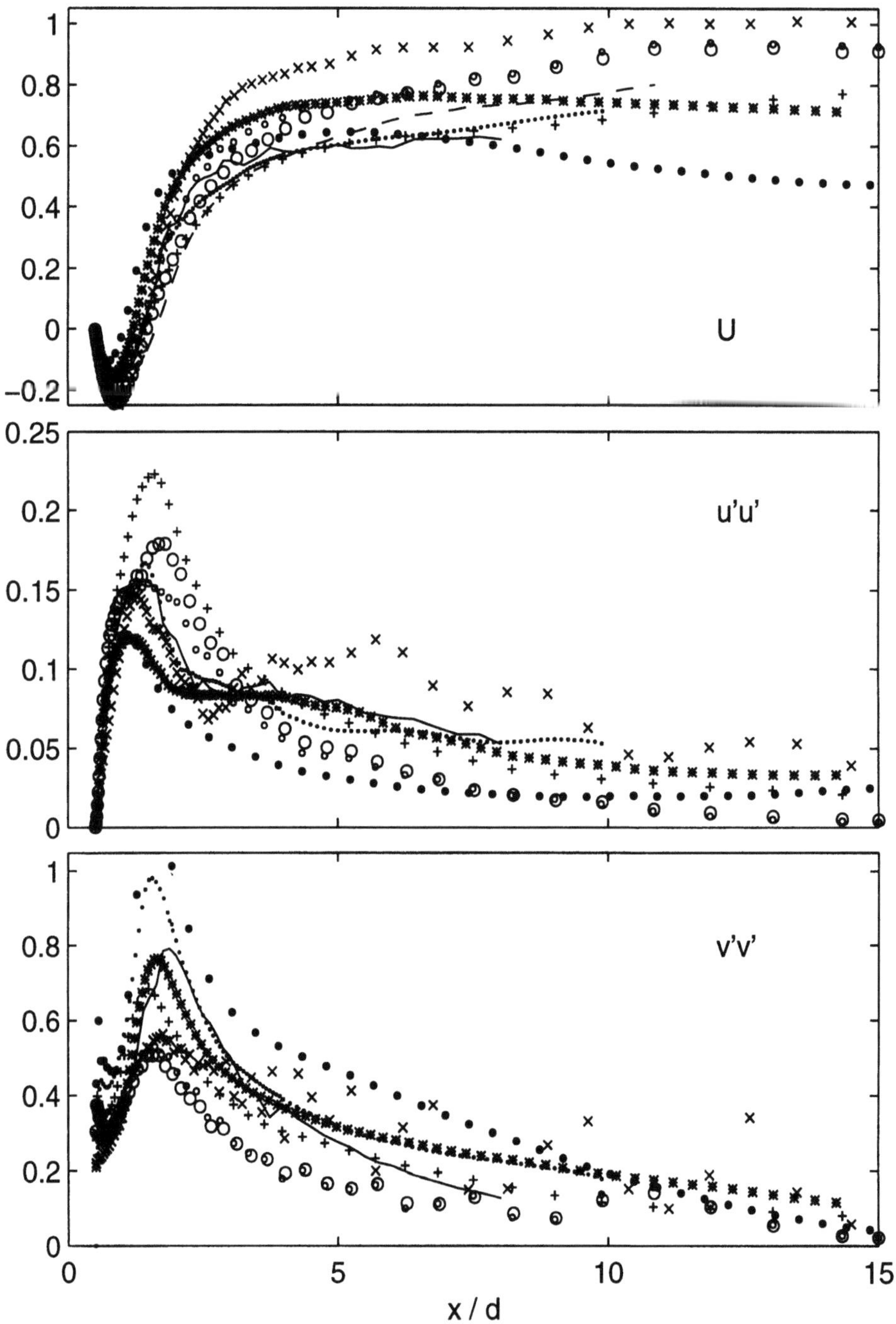


*Figure 2.* (a) Streamwise distributions of mean $U$ and stresses along the centreline $y = 0$ for $x < 15$; Solid line, experiment of Lyn *et al.*; symbols, simulations: GRO dashed line ($U$ only); UK1 ∘ ; UK3 ○ ; UOI $*$ ; NT7 $\times$ ; IS3 $+$ ; TIT • ; ST5 ● .

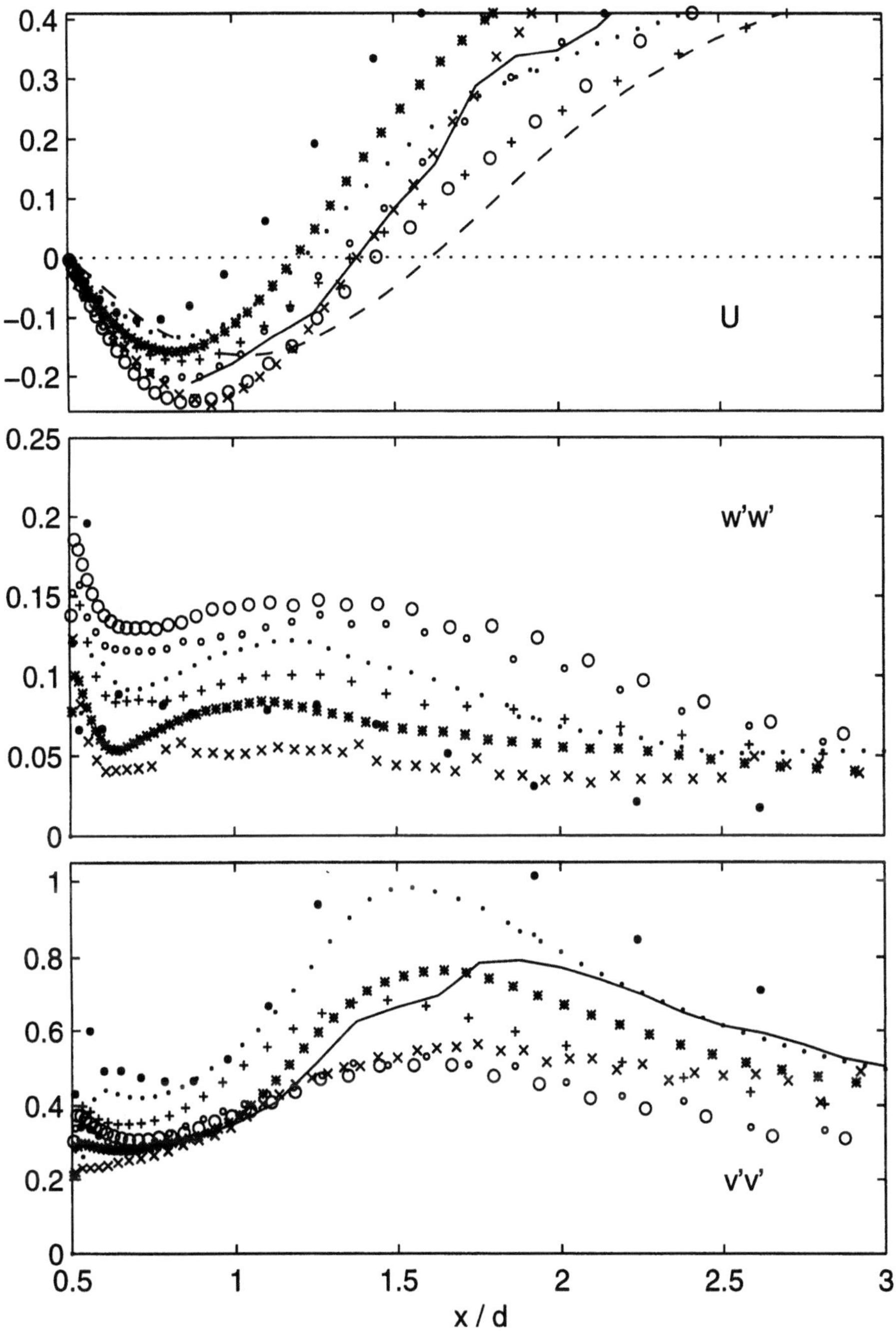


*Figure 2.* (b) Streamwise distributions of mean $U$ and stresses along the centreline $y = 0$ for $x < 3$. As Fig. 2a

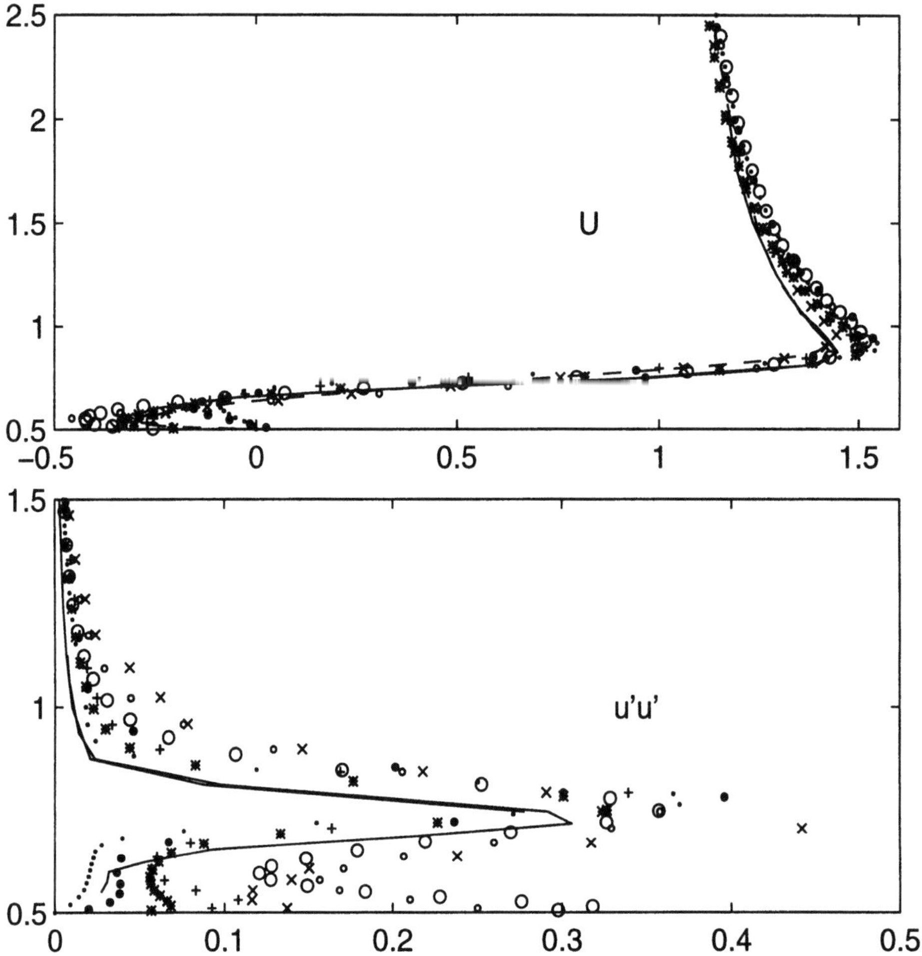


*Figure 3.* (a) Profiles of mean $U$ and stress $u'^2$ at $x = 0$, the cylinder centre. Solid line, experiment of Lyn *et al.*; symbols, simulations, as Figure 2. The vertical co-ordinate is $y$.

cult. The diffusion of streamwise momentum towards the centre is affected by the subgrid terms and presumably also by numerical effects which are not yet fully elucidated. It is possible that the stretching of the $x$ mesh and the resolution further downstream, which differ between the groups, is a major factor influencing the numerical diffusion, as suggested by Pourquié *et al.* (following paper).

Of the submissions shown here, IS3 has both a convincing centreline recovery and recirculation length; however a close inspection suggests there may be some good fortune in the latter prediction, and the high drag coefficient of this run is not reassuring. It is certainly premature to recommend

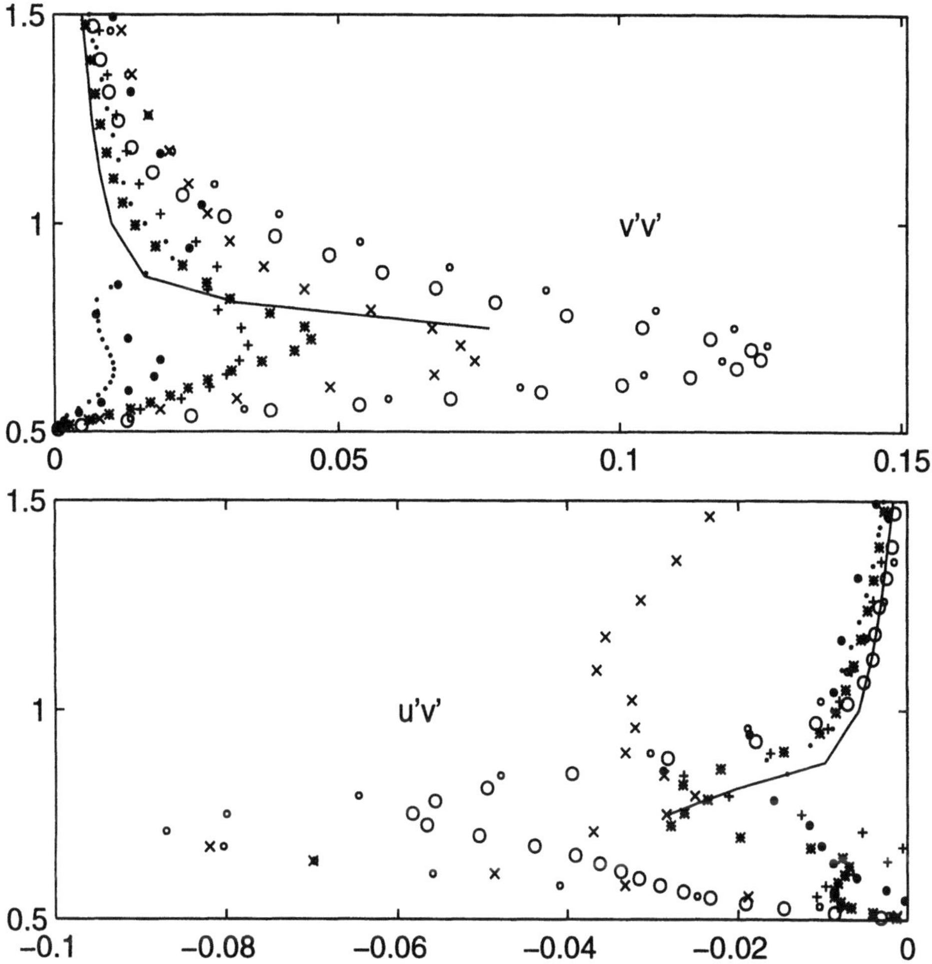


*Figure 3.* (b) Profiles of turbulence stresses at $x = 0$, the cylinder centre, as labelled. As Figure 3a.

a particular method from these results.

The predictions of stresses along the centreline also shown in Figure 2 confirm the view that no simulation is perfect. These are the full stresses comprising turbulent and coherent motions. NT7 suffers from some statistical scatter, and other simulations, while better converged statistically, fail to reveal systematic advantages or defects. There is no experimental data for $w'^2$, nor predictions of stresses from GRO. The expanded views of $v'^2$, Fig. 2b, reveal a tendency for all the LES to place the peak slightly closer to the rear face of the cylinder (at $x/d = 0.5$) than the experiment. The fact that all the simulations fail to resolve the inner peaks in $v'$ and, especially,

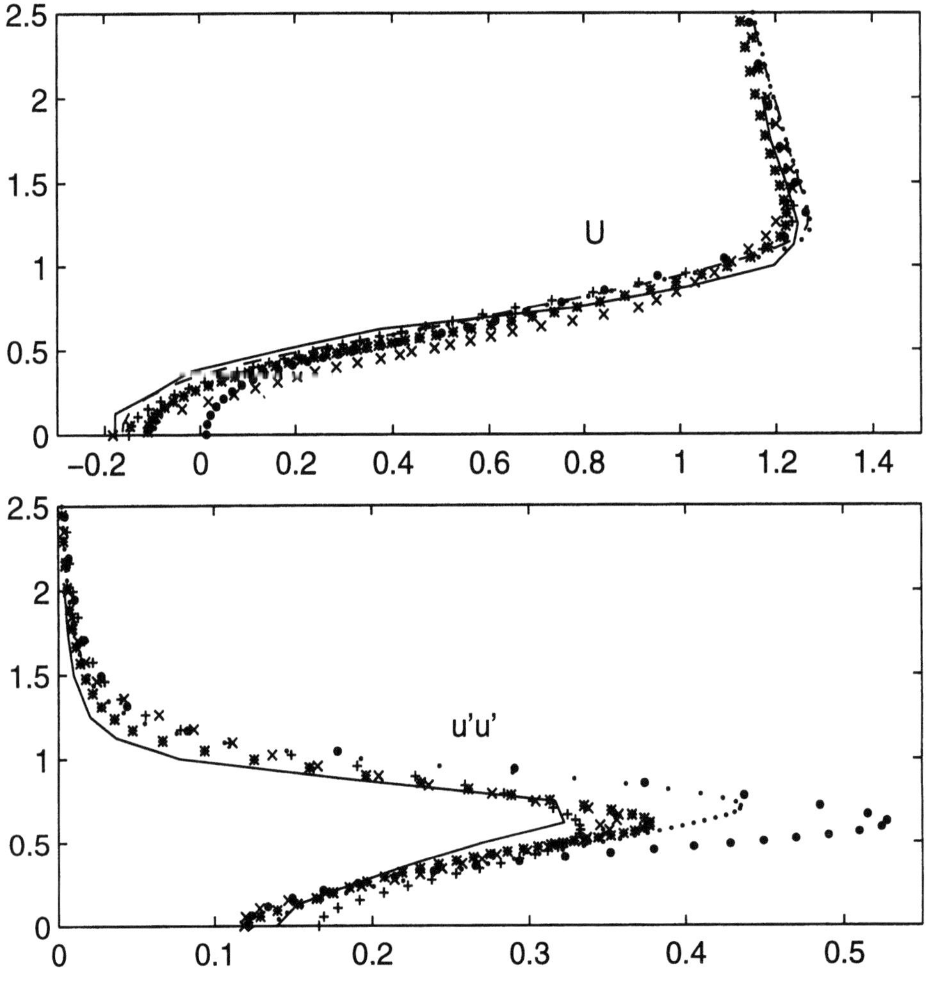


*Figure 4.* (a) Profiles of mean $U$ and stress $u'^2$ at $x = 1$, in the recirculation, as labelled. Solid line, experiment of Lyn *et al.*; symbols, simulations, as Figure 2. The vertical co-ordinate is $y$.

$w'$ also shows that we have inadequate resolution on the back face of the cylinder.

Figure 3 shows profiles in the centre of the top surface of the cylinder. Agreement for $U$ is adequate, but for the components of the stress the variable behaviour of the different numerical schemes and models is abundantly clear. Some methods predict a large spike of $u'^2$ very close to the wall, presumably the result of reverse shear at the wall inside the recirculation bubble, in contrast to the experiment. Again no one simulation shows a clear advantage in all the quantities plotted.

Figure 4, giving profiles at $x = 1$ which cuts the mean recirculation,

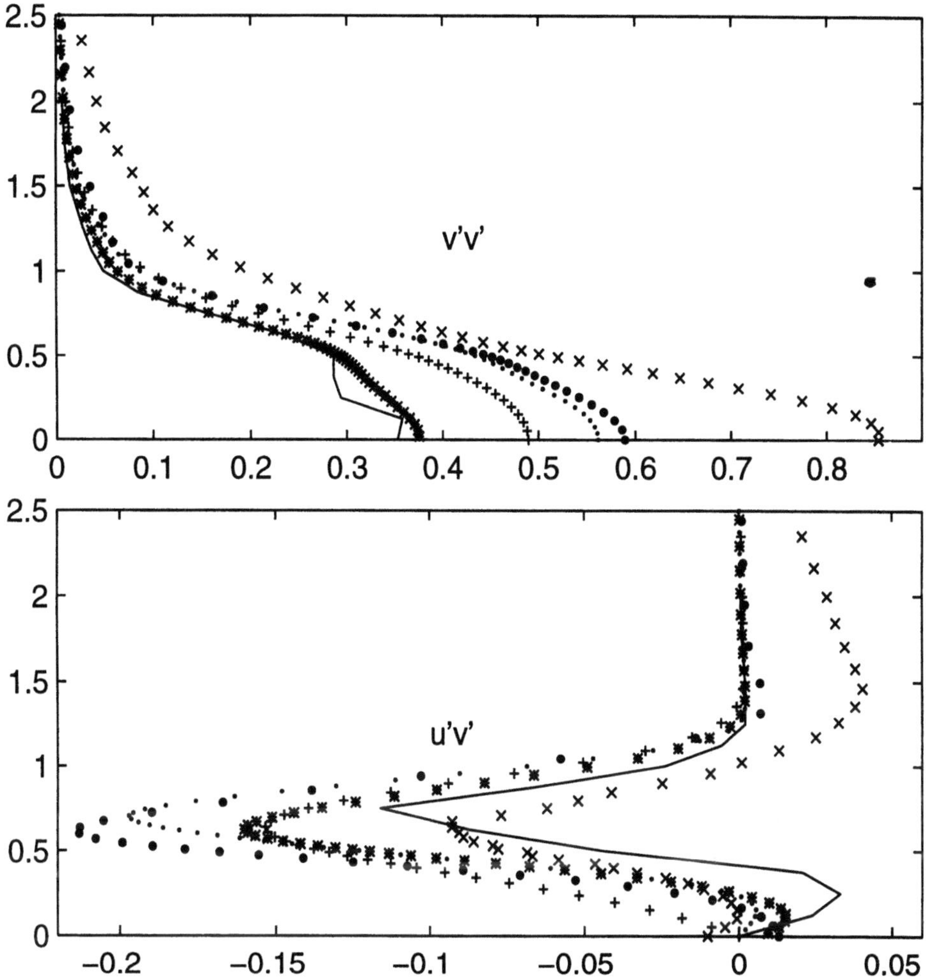


*Figure 4.* (b) Profiles of mean stresses at $x = 1$, as labelled.

reveals that at this point there is some agreement on the shape of the mean velocity profile, but already great variation in the predicted vertical fluctuation and the principal shear stress. Clearly, with such a variation in the magnitude (and also extent) of vertical mixing, the simulations will inevitably have widely divergent rates of filling of the wake. At this station, the results of the UOI simulation are qualitatively and quantitatively the closest to the experiment.

By the $x = 5$ station, the effects of the different rates of filling of the wake are readily apparent in the mean velocity profiles, Figure 5. Since even the mean shear rates are quite different between the simulations, we are not surprised to find qualitatively divergent pictures from the various groups.

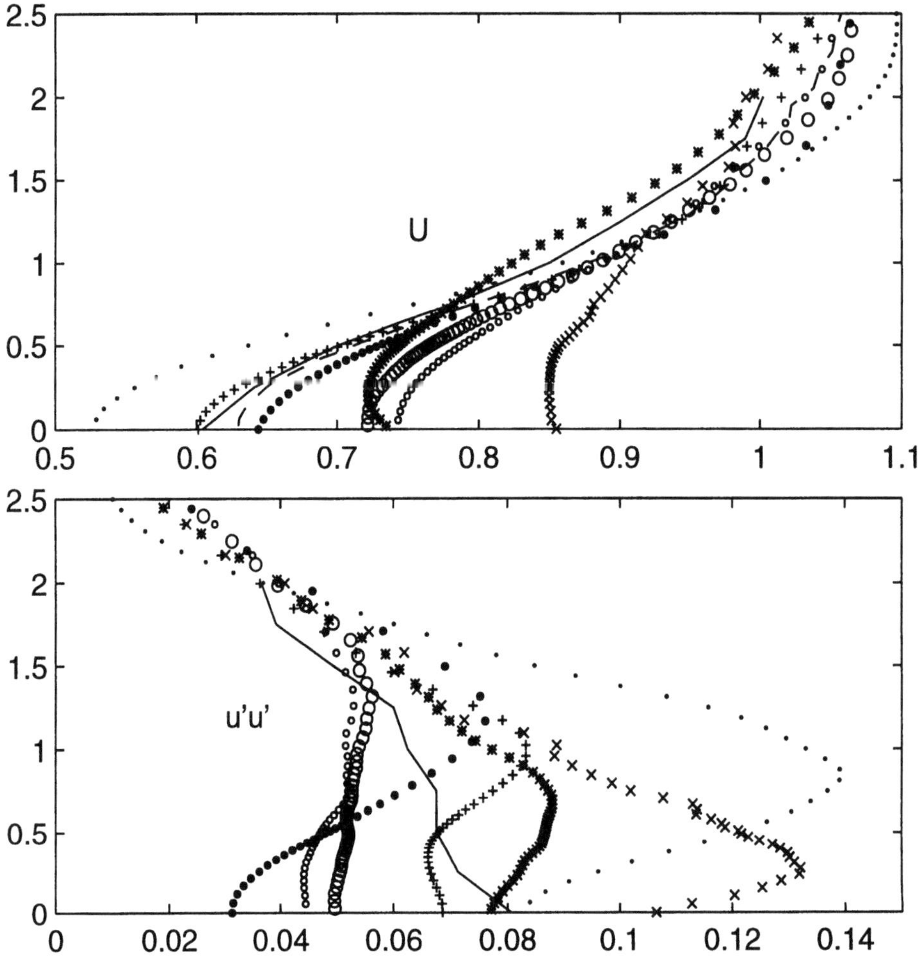


*Figure 5.* (a) Profiles of mean $U$ and stress $u'^2$ at $x = 5$, beyond the recirculation. Solid line, experiment of Lyn *et al.*; symbols, simulations, as Figure 2. The vertical co-ordinate is $y$.

Most of the simulations seem to suggest a central dip in $u'$ at this stage, unfortunately not supported by the experimental data used for comparison. The vertical fluctuations are still widely disparate, though the picture for $v'$ is not simply a reflection of the situation obtaining at $x = 1$; NT7, which overpredicts the $v'$ fluctuations at $x = 1$ has dropped back by $x = 5$, and in fact has a worryingly low level of Reynolds stress in the central portion of the wake.

Contours of these quantities in the $x$–$y$ plane were produced from submitted data and discussed at the workshop. The foregoing comments reflect insights gained from these two-dimensional distributions.

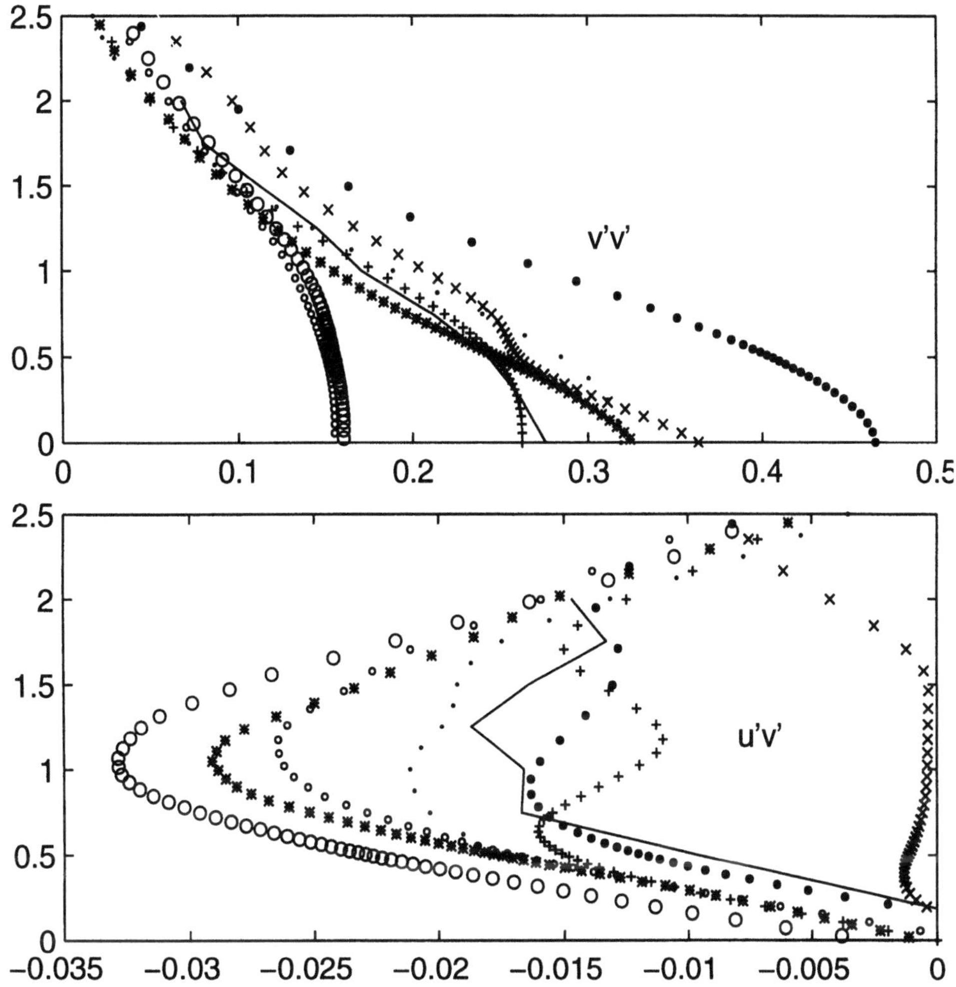


*Figure 5.* (b) Profiles of mean stresses at $x = 5$, as labelled.

## 5. Conclusions

In the light of the great variation in the predictions of the simulations around the cylinder, it is still not possible to make any recommendations apart from the obvious one that higher resolution is required at the walls. For the near wake region, IS3 and UOI, two simulations performed completely independently using higher order upwind schemes, dynamic models and somewhat similar meshing, produce reasonable quantitative predictions on many scores and an overall impression of qualitative agreement. We might be tempted to conclude that, at least for the wake region, this combination of methods has much to recommend it, though the data produced in this exercise are not extensive enough to indicate a clear advantage of particular methods or models at present.

These simulations, which give reasonable qualitative predictions in the wake while still having a number of clear faults, are among those using higher spanwise resolution. Only the rather special simulations GRO and TIT use higher spanwise resolution, and the latter only for its restricted, embedded meshes. In the light of the structural differences shown by Kogaki *et al.* (following paper) as the spanwise resolution is changed, it appears that the spanwise meshing used by many of the submissions is inadequate to deal with the breakdown of the flow to three dimensions behind the cylinder. It would be wise, in any future studies of this test case, to use a minimum of 32 meshes across the domain width of $4d$.

The other main conclusion is less firm, being based on the wide variation of predictions of the vertical fluctuation, resolved Reynolds stress, and rate of filling of the wake. The latter disagreement is presumably simply a natural consequence of the former. While different subgrid models or discretisation schemes may be partly to blame for the wide divergence of the predictions, another possibility is the different rates of stretching of the $x$ mesh in the region downstream of the cylinder. M. Pourquié pointed out at the workshop that his centreline filling can be brought into line with the experiment simply by retaining a fine $x$ mesh (with low stretching); his remarks were based on a further simulation, UK4, not included in this study and to be completed. It is important to verify if this is indeed the case independently of other variations such as the model or numerical scheme employed. Discussion at the workshop mentioned that rapid stretching of the mesh in the streamwise dimension can have radical and as yet poorly understood effects on simulated turbulence, as resolved eddies are advected onto a mesh that can no longer resolve them.

The flow over and following the square cylinder is therefore proving a major challenge to current LES techniques, a stimulus to further investigation and a problem to which, at present, we cannot claim to have a definitive solution.

## Acknowledgements

The author acknowledges his debt to the separate contributors to this exercise, who produced amd submitted a large amount of data in remarkable accord with the required formats. In particular, the contribution of Dr Pourquié is acknowledged in providing datasets UK1 and UK3, setting up the test case LES2 and co-operating in presenting the background of the exercise. Thanks also go to Professors Rodi and Ferziger for their initiative in first setting this test case.

## References

CHENG C.M., LU P.C. & CHENG R.H. (1992) Wind loads on square cylinders in homogeneous turbulent flows, *J. Wind Eng. Ind. Aero.* **41**, 729.

DURAO D.F.G., HEITOR M.V. & PEREIRA J.C.F. (1988) Measurements of turbulent and periodic flows around a square cross-section cylinder. *Exp. Fluids* **6**, 298.

KAWASHIMA N. & KAWAMURA H. (1997) Numerical analysis of LES of flow past a long square cylinder. Following in this volume.

KOGAKI T., KOBAYASHI T. & N. TANIGUCHI (1997) LES of flow around a square cylinder. Following in this volume.

LEE B.E. (1975) The effect of turbulence on the surface pressure field of a square prism, *J. Fluid Mech.* **69**, 263.

LYN D.A. & RODI W. (1994) The flapping shear layer formed by flow separation from the forward corner of a square cylinder, *J. Fluid Mech.* **267**, 353.

LYN D.A., EINAV S., RODI W. & PARK J.H. (1995) A laser-Doppler velocimetry study of ensemble averaged characteristics of the turbulent near wake of a square cylinder, *J. Fluid Mech* **304**, 285.

MCLEAN I. & GARTSHORE I. (1992) Spanwise correlations of pressure on a rigid square cylinder, *J. Wind Eng. Ind. Aero.* **41** 797.

MURAKAMI S., MOCHIDA A., TOMINAGA Y.& IIZUKA S. (1997) LES analysis of turbulent flow past a square cylinder using various SGS models. Following in this volume.

NOZAWA K. & TAMURA T. (1997) LES of flow past a square cylinder using embedded meshes. Following in this volume.

RODI W., FERZIGER J.H., BREUER M. & POURQUIÉ M. (1996) Status of Large Eddy Simulation: Results of a Workshop. *J. Fluids Eng.* To appear.

VERSTAPPEN R.W.C.P. & VELDMAN A.E.P. (1997) Fourth-order DNS of flow past a square cylinder: first results. Following in this volume.

VICKERY B.J. (1966) Fluctuating lift and drag on a long square cylinder of square cross-section in a smooth and in a turbulent stream, *J. Fluid Mech.* **25**, 481.

VREMAN B., GEURTS B. & KUERTEN H. (1994) On the formulation of the dynamic mixed subgrid scale model. *Phys. Fluids* **6** (12) 4057.

WANG G. & VANKA S.P. (1996) Large Eddy Simulations of High Reynolds Number Turbulent Flow Over a Square Cylinder. *Report CFD 96-02*, Computational Fluid Dynamics Laboratory, Dept. Mech & Ind. Engineering, University of Illinois at Urbana-Champaign, IL 61801.

WANG G. & VANKA S.P. (1997) LES of flow over a square cylinder. Following in this volume.

WERNER H. & WENGLE H. (1993) Large-eddy simulation of turbulent flow over and around a cube in a plane channel, *Turbulent Shear Flows 8* (ed. Durst *et al.*) Springer-Verlag, Berlin. pp 155–168.

ZANG Y., STREET R.L. & KOSEFF J.R. (1993) A dynamic mixed subgrid-scale model and its application to turbulent recirculating flows, *Phys. Fluids* **A5** (12) 3186.

# COMPUTED TEST CASE: SQUARE CYLINDER

M. POURQUIE, M. BREUER AND W. RODI
*Institute for Hydromechanics,*
*University of Karlsruhe, Germany*

## 1. Overview

Three calculations made with the LESOCC code have been completed. LESOCC is a second order accurate finite volume method on a collocated grid in general curvilinear coordinates, used for the calculations at the Rottach–Egern workshop. Two were presented at the Rottach-Egern workshop (UKAHY1 and UKAHY2, called UK1 and UK2 here). It was decided to continue runs with the UK2 grid, since in the case of the UK1 grid the near wall boundary resolution was very probably inadequate. The main aim is to check some of the conclusions, reached at the Rottach-Egern workshop. Among others, conclusions were made regarding the use of wall laws. UK3 is new; it is equivalent to UK2 except that no wall function was used but a simple no-slip BC. It allows conclusions regarding the use of a wall law.

Results for UK2 and UK3 (wall law versus no-slip) look qualitatively the same. Our conclusion so far is that the effect of the wall law on at least some of the bulk parameters is indeed small. Two examples are discussed, namely recirculation length and drag coefficient (see the table in the next section). One conclusion made at the Rottach–Egern workshop, namely that the recirculation length is shorter for simulations using no wall function, is not in agreement with what is found here: the recirculation lengths for UK2 and UK3 are only marginally different. A possibility is that one finds a shorter recirculation region when no wall law is used and one has at the same time a coarse grid near the cylinder, which was the case for the calculations presented at the Rottach–Egern workshop. Since the drag coefficient is related to the recirculation length (a shorter recirculation length corresponding with a higher drag coefficient and vice versa) an analogous conclusion for the drag coefficient, namely that using no wall law means obtaining a higher drag, also does not hold provided enough resolution is used.



*J.-P. Chollet et al. (eds.), Direct and Large-Eddy Simulation II,* 375-379.

All of the runs made so far show the same, quick recovery of the mean velocity component in the main stream direction. This is connected with the resolution far away from the cylinder.

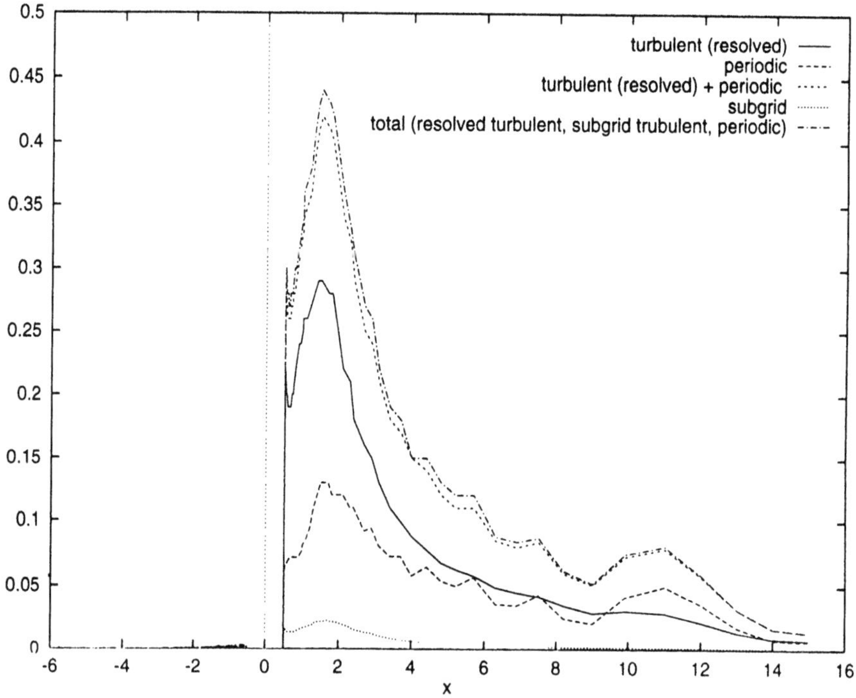


*Figure 1.* Values for the fluctuations (total, turbulent, periodic, subgrid) on the centerline: UK3

The influence of the SG terms is more fully investigated. Using Mason's (1983) estimate for the subgrid energy, we get the following picture, Figure 1, for the relative order of magnitude of SGE and resolved energy on the centerline. We distinguish between resolved (total), periodic, turbulent and subgrid component. One question here is what to compare the SG part with: with the total fluctuation or only with the turbulent. For this reason the turbulent part was included in the figure, although the phase averaging method used to obtain the turbulent part separately has some points which need further discussion. Moreover, other estimates than Mason's may give higher values for the SG part. However, we can assess whether the SGE is small compared to the resolved energy (on the centerline). Of course, for an LES one wants the SG energy to be much smaller than the resolved energy. If we use the criterion that the SG part of the turbulent kinetic energy should be less than 10% of the total fluctuation or even only the turbulent part, then we meet this criterion (on the centerline) if we use Mason's estimate.

Summarizing, some conclusions of the Tegernsee Workshop have been considered using calculations UK2 and UK3 using respectively Smagorinsky wall–law and Smagorinsky no–slip. Moreover, the SG energy content on the centerline has been found to be small enough compared to the resolved motion to meet the 10% criterion if we use Mason's estimate and compare SG energy with only the turbulent part of the resolved energy.

## 2. Details of calculations

### 2.1. GEOMETRY OF COMPUTATIONAL DOMAIN AND GRID

The calculation domain, viewed from the center of the cylinder is given by: $4D$ in periodic direction, $5D$ in front of the origin (= $4.5D$ in front of cylinder), $15D$ after the origin (= $14.5D$ after the back of the cylinder). The grids are equidistant in the direction of the cylinder axis. In the other two directions they are non–equidistant, generated with the aid of geometric series, separately for the region in front of the cylinder, above and below the cylinder, and behind the cylinder. A maximum distance from the wall is given for the grid cells nearest to the cylinder wall, and the stretching factor is not allowed to become larger than 1.1. As a result, the following three grids are applied:

| Key | Total no. of grid points | grid points left–right | grid points bottom–top | along axis | grid size near wall |
|---|---|---|---|---|---|
| UK1 | $109 \times 105 \times 20$ | $33, 29, 45$ | $37, 29, 37$ | 20 | 0.02 |
| UK2 | $146 \times 146 \times 20$ | $52, 40, 52$ | $52, 40, 52$ | 20 | 0.01 |
| UK3 | $146 \times 146 \times 20$ | $52, 40, 52$ | $52, 40, 52$ | 20 | 0.01 |

Resolution on the cylinder:

| Key | Grid points on cylinder |
|---|---|
| UK1 | $29 \times 29$ |
| UK2 | $40 \times 40$ |
| UK3 | $40 \times 40$ |

Stretching factors are always smaller than 1.1.

## BOUNDARY CONDITIONS

At the inflow plane constant velocity is imposed (no perturbations added). A convective boundary condition is utilized at the outflow boundary.

Werner & Wengle's wall function approach (1993) is applied on the cylinder walls for UK1 and UK2; for UK3 we used a simple no-slip BC. At the lateral walls free–slip conditions are used. Periodic boundary conditions are applied in the spanwise direction of the cylinder.

## 2.2. NUMERICAL METHOD

The code developed at the Institute for Hydromechanics, University of Karlsruhe, Germany, is based on a finite–volume method for solving the incompressible Navier–Stokes equations on general body–fitted, curvilinear grids (LESOCC = **L**arge **E**ddy **S**imulation **O**n **C**urvilinear **C**oordinates). A non–staggered, cell–centered grid arrangement is used. Both convective and viscous fluxes are approximated by central differences of second order accuracy. The temporal discretization consists of a predictor–corrector scheme, where the predictor step is an explicit Adams–Bashforth scheme for the momentum equations (second order in time) and the corrector step covers the implicit solution of the Poisson equation for the pressure correction (SIMPLE). The linear system is solved by the strongly implicit procedure of Stone which is accelerated by a FAS multigrid technique. In order to avoid decoupling of pressure and velocity on the non–staggered grid the momentum interpolation proposed by Rhie and Chow (1983) is applied.

## 2.3. SUBGRID–SCALE MODELS

Two different subgrid–scale models are implemented, the standard Smagorinsky model with Van Driest damping ($l = C_s\Delta(1 - \exp(-y^+/25)^3)^{0.5}$) and the dynamic model. In the calculations reported here, only the Smagorinsky model is used.

## 2.4. COMPUTER RESOURCES

All computations are performed on the mainframe Fujitsu VP S600/20 installed at the Computer Center of the University of Karlsruhe. It has a peak performance of 5 Gflops and a core memory of 2 Gbytes. The code written for curvilinear body–fitted coordinates is much more general applicable than for the test cases of the workshop which can be done by applying simple Cartesian grids. It should be noted that *no* tuning is done for the specific test cases which results in higher computing costs compared to a program written especially for Cartesian coordinates and applying special algorithms.

| Key | time step $\Delta t$ | time for sampling | simulation time | time steps | CPU time (hours) |
|---|---|---|---|---|---|
| UK1 | $10^{-3}$ | $104t*$ | 200 t* | 104000 | $\approx 33$ |
| UK2 | $4 \times 10^{-4}$ | $64t*$ | 128 t* | 160000 | $\approx 72$ |
| UK3 | $4 \times 10^{-4}$ | $96t*$ | 140 t* | 175000 | $\approx 80$ |

Sampling rate was every t*/20.

### 2.5. BULK QUANTITIES

| calculation | rec len. | St | Cd | Cl | rms Cd | rms Cl. |
|---|---|---|---|---|---|---|
| UK1 | 1.32 | .13 | 2.20 | -0.02 | .14 | 1.01 |
| UK2 | 1.46 | .13 | 2.30 | -0.04 | .14 | 1.15 |
| UK3 | 1.44 | .13 | 2.23 | -0.05 | .13 | 1.02 |

TABLE 1. Some bulk quantities for three cylinder calculations

## References

Rhie, C.M., Chow, W.L.: *A numerical study of the turbulent flow past an isolated airfoil with trailing edge separation,* AIAA-J., Vol. 21, pp. 1225–1532, (1983).

Werner, H., Wengle, H.: *Large-Eddy Simulation of Turbulent Flow Over and Around a Cube in a Plane Channel,* Turbulent Shear Flows 8, (ed. Durst *et al.*), Springer-Verlag, (1993).

# FOURTH-ORDER DNS OF FLOW PAST A SQUARE CYLINDER: FIRST RESULTS

R.W.C.P. VERSTAPPEN AND A.E.P. VELDMAN
*Department of Mathematics, University of Groningen, P.O.Box 800, 9700 AV Groningen, The Netherlands*

**Abstract.** In this paper we present some initial results of a fourth-order direct numerical simulation of flow past a square cylinder at $Re = 22,000$. The flow is identical to the second test case which is considered at this workshop. Mean velocities, the mean Strouhal number, the mean drag coefficient Cd, the mean lift coefficient Cl and the rms fluctuations of Cd and Cl are computed.

## 1. Introduction

In another paper in this volume (Verstappen and Veldman, 1996), we present a comparison between two DNS methods: a well-known second-order finite volume method and a fourth-order finite volume method which is constructed as the Richardson extrapolate of the second-order method. These two approaches are compared for a turbulent flow in a cubical lid-driven cavity at Re = 10,000. Experimental results are available for comparison. The fourth-order method turns out to be the best (see e.g. Figure 3 of that paper).

In this paper we consider a test case of this workshop, the flow past a square cylinder at $Re = 22,000$ at zero angle of attack. Also for this test case we find the fourth-order method performs better than the second-order method. On a $280 \times 210 \times 64$ grid all our second-order simulations failed due to insufficient spatial resolution, while the fourth-order simulation method performed well.

However, even using a fourth-order DNS method the flow past a square cylinder at $Re = 22,000$ is an expensive nut to crack: one shedding cycle takes about 40 hours on 16 nodes of a Cray J932. Up till now, the fourth-order DNS has taken several hundreds of CPU-hours. The simulation has



*J.-P. Chollet et al. (eds.), Direct and Large-Eddy Simulation II,* 381-384.

not finished completely: the start-up of the flow plus three shedding cycles have been computed, and limited results based on this part of the simulation are presented.

Unfortunately we cannot compute sensible average values of turbulence intensities or Reynolds stresses from the three shedding cycles that are available now: a period of three shedding cycles is simply too short. However, some first results — mean velocities, the mean Strouhal number, the mean drag coefficient Cd, the mean lift coefficient Cl and the root-mean-square fluctuations of Cd and of Cl — can be computed. These results are presented in Section 3. Before that, some details of the simulation are given.

## 2. Details of the simulation

We use for this test case the prescribed coordinate system, which has its origin at the centre of the cylinder, and normalise all quantities by the cylinder width and the inflow velocity. The streamwise direction is denoted by $x$, the lateral direction by $y$ and the spanwise (periodic) direction by $z$. The components of the velocity in these directions are $u$, $v$ and $w$, respectively.

The inflow boundary is located at $x = -7$, i.e., six and a half diameters upstream of the cylinder. The inflow condition is $u = 1$, $v = 0$, $w = 0$. The lateral boundaries are located at $y = \pm 7$. At these boundaries we have imposed $u_{yy} = w_{yy} = p_y = 0$, where $p$ is the pressure. The outflow boundary is located at $x = 20$. The outflow conditions are $v_{xx} = w_{xx} = 0$ and $p_x =$ constant, where the constant is determined such that the mass inflow equals the mass outflow at each time-step; this constant is approximately zero. In addition, in a buffer zone ($x = 15.0$–$20.0$) the Reynolds number is decreased from 22,000 to 1,000 to suppress (unphysical) waves which are reflected by the artificial outflow boundary. The spanwise boundaries are four diameters apart. No-slip boundaries are imposed at the cylinder surface.

We have used a $280 \times 210 \times 64$ (stretched and staggered) grid to cover the computational domain. The first meshpoint is spaced $5 \times 10^{-3} d$ from the cylinder surface. The grid is stretched away from the cylinder surface using a sinh function; the ratio of the largest to smallest gridsize is approximately 200 (in $x$) and 100 (in $y$). The time-step equals $10^{-3}$, and statistics are sampled at each time-step.

The numerical method is identical to the one that is described in Verstappen and Veldman (1996). It should be emphasized that in that paper, and also here, we have used constant weights (9/8 and -1/8) in the Richardson extrapolation. The reason for this is that the higher-order discretization thus obtained is conservative. If the weights are adapted to the stretching of the grid to obtain a formally more accurate scheme, the result is non-conservative.

The convective flux is integrated in time using a modified second-order Adams-Bashforth method. According to Adams-Bashforth we ought to take $\frac{3}{2}f(u^n) - \frac{1}{2}f(u^{n-1})$; instead we take $f(\frac{3}{2}u^n - \frac{1}{2}u^{n-1})$. The two approaches differ when the right-hand side $f$ is non-linear. For instance for $f(u) = \lambda(t)u$ with $\lambda(t) < 0$ and $u^n > 0$ the numerical solution $u^{n+1}$ of $u_t = f(u)$ is smaller than $u^n$ (for any time step) as it should be, while the solution obtained with the original Adams-Bashforth method does not satisfy the inequality $u^{n+1} < u^n$ unconditionally.

The discrete Poisson equation for the pressure is solved using a combination of a fast Fourier transform method in the spanwise direction and a modified incomplete Choleski conjugate gradient method in the resulting spectral space. After the Fourier transform (which can be computed in parallel over the $x$ and $y$ directions), the discrete Poisson equation falls apart into a set of mutually independent 2D equations with additions to their diagonals due to the Fourier transform. This set of mutually independent equations is distributed over the processors and solved. Thus a 100% parallel implementation is achieved.

## 3. Predicted data

In this section averages are shown which are computed over three shedding cycles and the spanwise direction. Velocities are also averaged over the top and bottom halves. Table 1 shows the mean Strouhal number, the mean drag coefficient Cd, the mean lift coefficient Cl and the root-mean-square fluctuations of Cd and Cl. The experiment is described in Lyn et al. (1995).

TABLE 1. Comparison of DNS with experiment

| | DNS | Experiment |
|---|---|---|
| Mean Strouhal number | 0.133 | 0.130 - 0.137 |
| Mean drag coefficient Cd | 2.09 | 1.9 - 2.1 |
| Mean lift coefficient Cl | 0.005 | - |
| Rms fluctuation of Cd | 0.178 | 0.1 - 0.2 |
| Rms fluctuation of Cl | 1.45 | 0.6 - 1.4 |

It may be noted that the mean lift coefficient has not been measured; it should be zero. So far, we have only computed three full shedding cycles, yet the computed values of all bulk quantities falls within the range set by the experiment except for the root-mean-square of the fluctuations of the lift coefficient Cl. The rms fluctuation of Cl seems to be slightly overestimated by the DNS.

Figure 1 shows the mean streamwise velocity at four locations past the cylinder.

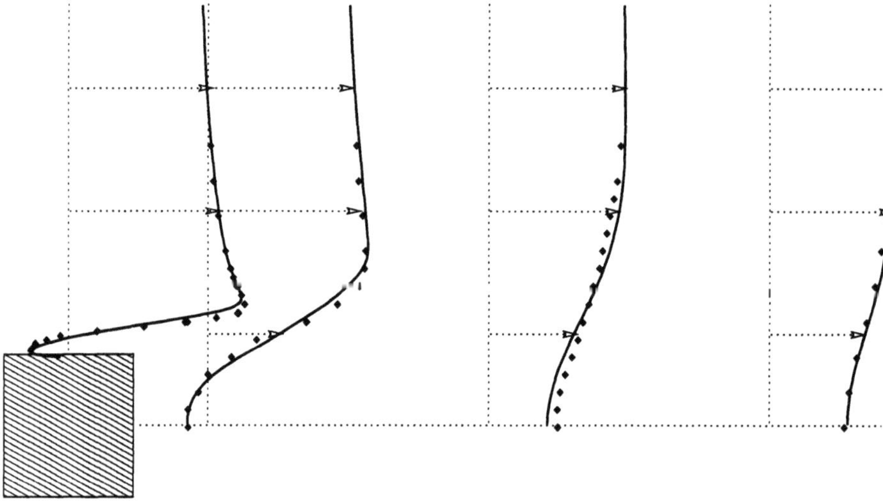

*Figure 1.* A comparison of mean velocities of the DNS with experimental results. The experimental data is taken from ERCOFTAC Database Case 43; see also Lyn *et al.* (1995). Shown is the mean streamwise velocity at x=0, x=1, x=3 and x=5. Continuous lines, DNS; dots, experimental data.

## Acknowledgements

The Stichting Nationale Computerfaciliteiten (National Computing Facilities Foundation, NCF) with financial support from the Nederlandse Organisatie voor Wetenschappelijk Onderzoek (Netherlands Organization for Scientific Research, NWO) is gratefully acknowledged for the use of supercomputer facilities.

## References

Verstappen, R.W.C.P. and Veldman, A.E.P. (1996), A comparison of low-order DNS, high-order DNS and LES. *Direct and Large-Eddy Simulation II*, (eds J.P. Chollet *et al.*) This Volume.

Lyn, D.A., Einav S., Rodi W. and Park J.H. (1995), A laser-Doppler velocimetry study of ensemble averaged characteristics of the turbulent near wake of a square cylinder, *Journal of Fluid Mechanics* **304**, 285.

# LES ANALYSIS OF TURBULENT FLOW PAST A SQUARE CYLINDER USING VARIOUS SGS MODELS

S. MURAKAMI AND S. IIZUKA
*I.I.S., University of Tokyo,*
*7-22-1, Minato-ku, Roppongi, Tokyo, 106, Japan*
AND
A. MOCHIDA AND Y. TOMINAGA
*Niigata Institute of Technology,*
*1719, Fujihashi, Kashiwazaki, Niigata, 945-11, Japan*

## 1. Outline of computations

The turbulent vortex shedding flow past a two-dimensional square cylinder at Re=2.2x10$^4$ (test case LES2) was analyzed by Large Eddy Simulation using various dynamic subgrid-scale (SGS) models [1-5]. The types of SGS model used are as follows: the static type of conventional Smagorinsky model (S model, eqns.(1) and (2) in the Appendix (case 1)); the Dynamic Smagorinsky model (DS model, eqn.(6) in the Appendix (case 2)); the Dynamic Mixed model (DM model, eqn.(10) in the Appendix (case 3)); and the Lagrangian Dynamic Mixed model (LDM model, eqn.(16) in the Appendix (case 4)).

Table 1 lists the 7 cases computed. Computations were carried out for three different computational grids (grids A-C), two different grid systems (colocated and staggered grids) and four different SGS models. Details of SGS models compared here are given in the Appendix. Preliminary computations were carried out on a relatively coarse grid (grid A) using four types of the SGS models in cases 1-4. The relative performance of these four SGS models was clarified by comparison with those given from the experiment by Lyn *et al.* [7]. As is described later, the LDM model (case 4) provided the best results.

A colocated grid was used in cases 1-4. In case 5, a computation using a staggered grid was carried out in order to assess the difference in prediction accuracy between the colocated grid and staggered grid. In case 6, we carried out the computation with the same conditions as those of the computation UKAHY1 by the University of Karlsruhe group for the 1995 Workshop of LES of Bluff Body Flows (Rottach-Egern workshop) [6]. The results of case 6 were compared with those of case 1. Finally, a computation using the LDM model was carried out with a grid spacing (grid C) which was finer than grid A used in



*J.-P. Chollet et al. (eds.), Direct and Large-Eddy Simulation II,* 385-395.

Table 1 Computed cases

| case | SGS model | grid | computational domain | number of grid point | $h_w$ | average time (Note 1) |
|---|---|---|---|---|---|---|
| 1 | S $(0.13, sf_\mu)^{*1}$ | colocated | 20 ($x_1$) × 14 ($x_2$) × 2 ($x_3$) | 104 ($x_1$) × 68 ($x_2$) × 10 ($x_2$) [grid A] | 0.022 | ≈ 61 |
| 2 | DS | colocated | | | 0.022 | ≈ 15 |
| 3 | DM | colocated | | | 0.022 | ≈ 15 |
| 4 | LDM | colocated | | | 0.022 | ≈ 15 |
| 5 | LDM | staggered | | | 0.022 | ≈ 15 |
| 6 | S $(0.1, mf_\mu)^{*2}$ | colocated | 20 ($x_1$) × 14 ($x_2$) × 4 ($x_3$) | 107 ($x_1$) × 103 ($x_2$) × 14 ($x_3$) [grid B] | 0.02 | ≈ 15 |
| 7 | LDM | staggered | | 140 ($x_1$) × 103 ($x_2$) × 32 ($x_3$) [grid C] | 0.02 | ≈ 61 |

$h_w$ : the grid interval adjacent to the cylinder wall
*1 S (Cs, damping function*2)
*2 $sf_\mu$ : standard Van Driest damping $(=1-\exp(-x_n^+/25))$,
$mf_\mu$ : modified Van Driest damping $(=(1-\exp(-x_n^+/25)^3)^{0.5})$
Values are made dimensionless by the cylinder width, D, and the inflow velocity, $U_0$.

cases 1-5 or grid B for case 6.

A second-order centered difference scheme was adopted for the spatial derivatives. For time advancement, the second-order Adams-Bashforth scheme was used for the convection terms and the Crank-Nicolson scheme for the diffusion terms. The interval for time advancement was $1.0\times10^{-3}$ (except $2.0\times10^{-4}$ for cases 2 and 3) in non-dimensional time units based on $U_0$ and D.

At the inflow boundary, the approach flow was set to be constant and uniform and no velocity fluctuations exist. A convective condition was used at the outflow boundary of the computational domain. Symmetry conditions were employed for the lateral boundaries, and periodicity conditions were imposed for the boundary planes perpendicular to the cylinder axis. For the boundary condition at the solid walls, Werner and Wengle's approach [8] was adopted.

## 2. Comparison of various SGS models with grid A (cases 1-4)

Firstly, the relative performance of various SGS models, i.e., S, DS, DM and LDM, are compared using grid A (cases 1-4, cf. Table 1).

Fig.1 shows the time-averaged velocity $\langle \bar{u}_1 \rangle$ along the centerline. In front of the cylinder, the results are not influenced by the difference of the SGS models used, but there are fairly large differences in the wake region. Case 1 (S model) underestimates the length of the reverse flow region considerably. Both cases 2 (DS model) and 3 (DM model) show better agreement with the experiment than does the S model, and there is little difference between the results obtained with these two types of dynamic SGS models in the region of $0.5<x_1<1.5$. Further downstream ($x_1>1.5$), case 3 (DM model) seems to give a more reasonable result than does case 2 (DS model).

Fig.2 compares the time-averaged velocity $\langle \bar{u}_1 \rangle$ along the centerline between case 3 (DM model) and case 4 (LDM model). In the LDM model, the model coefficient C ($C=Cs^2$) is calculated using the averaged quantities along the path line following the approach developed by Meneveau *et al.* [5] (eqns.(12)-(17) in the Appendix). Although both results show good agreement with the experiment, case 4 (LDM model) provides more accurate results than case 3 (DM model). The LDM model seems to provide the best result of these four SGS models. Furthermore, the use of the LDM model contributes to a remarkable improvement of calculation stability, as shown in Fig.3. Consequently, the CPU time was reduced by 33% in comparison with the DM model.

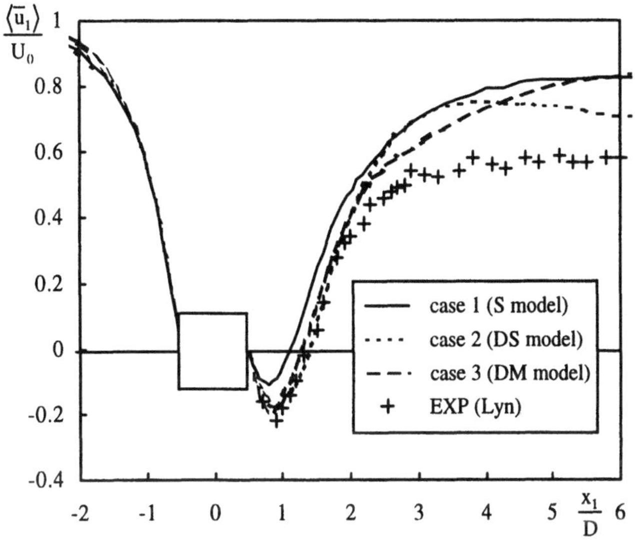


Fig.1 Comparison of the time-averaged velocity $\langle \bar{u}_1 \rangle$ along the center line for S, DS and DM models

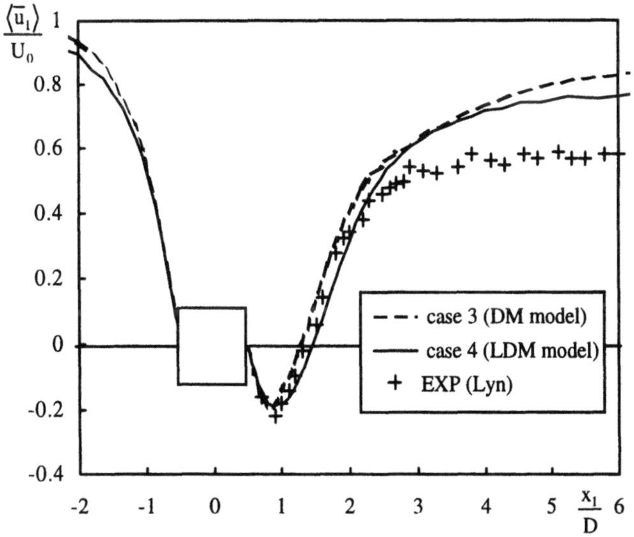


Fig.2 Comparison of the time-averaged velocity $\langle \bar{u}_1 \rangle$

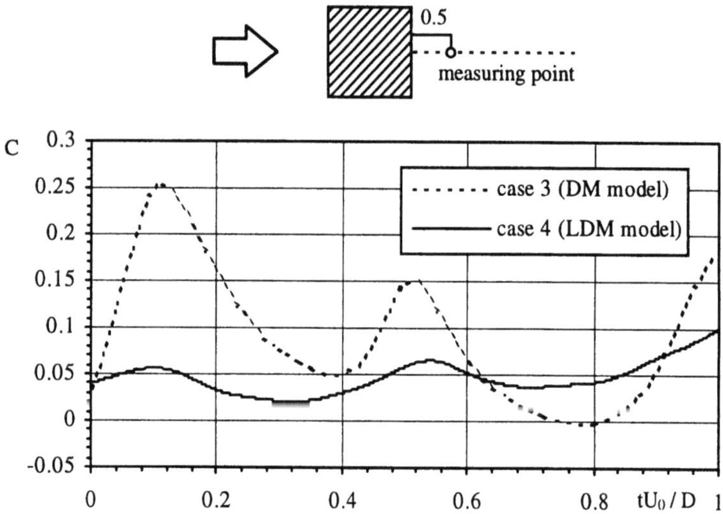


Fig.3 Comparison of the time-history of model coefficient C between DM and LDM models

## 3. Influence of difference in grid spacings (cases 1 and 6)

The conditions for computation (the grid spacing, SGS model, etc.) employed in case 6 are almost the same as those in the computation of UKAHY1 presented at the Rottach-Egern workshop [6].

The result of this reference case (case 6) is compared with case 1 in Fig.4. Spatial oscillation of the velocity is observed in the region of $x_1>2$ in case 6, while this oscillation does not appear in the result of case 1. As described in Note 3, this difference is mainly caused by the difference in the grid spacings for these cases. The spacing of grid B (case 6) in the streamwise ($x_1$) direction is coarser than that of grid A (case 1) in the region behind the cylinder ($x_1>1$). The spacing in the spanwise ($x_3$) direction of grid B (case 6) is also coarser than that of grid A (case 1) for the whole domain. With the coarse spacing of grid B, spatial oscillation of velocity was caused in case 6 in the region behind the cylinder, as indicated in Fig.4. On the other hand, the resolution of grid B in the $x_1$ and $x_2$ directions is finer than that of grid A in the vicinity of the cylinder wall ($-1<x_1<1$). Hence, the computation with grid B provided more accurate results in the vicinity of the cylinder wall in comparison with the results using grid A. Considering these points, a new grid (grid C) was designed. The resolution of grid C in the $x_1$ and $x_2$ directions is almost the same as that of grid B in the vicinity of the cylinder wall, but much finer than that of grid B in the region behind the cylinder. In this region, the resolution of grid C in the $x_1$ direction becomes identical to that of grid A. Furthermore, the grid spacing in the spanwise ($x_3$) direction is much finer in grid C than those in grids A or B.

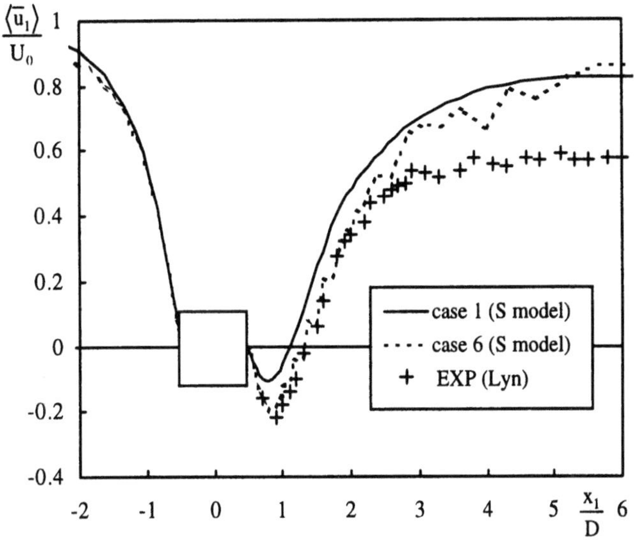


Fig.4 Comparison of the time-averaged velocity $\langle \bar{u}_1 \rangle$ along the center line between cases 1 and 6

## 4. Performance of LDM model with grid C (case 7)

Hereafter the results of the computation using the LDM model with grid C (case 7) are shown. In this case the staggered grid was used (Note 4). Integral parameters are compared in Table 2. The statistical quantities were determined by using the time-history of predicted flowfields during 8 vortex shedding periods for this case. The correspondence between the results of case 7 and experiments is satisfactory except for the value of $C_{Drms}$. Fig.5 compares the time-averaged velocity $\langle \bar{u}_1 \rangle$ along centerline. The result of case 7 reproduced the velocity distribution in the reverse flow region behind the cylinder accurately.

Table 2. Integral parameters

| | $L_1 \times L_2 \times L_3$ | $N_1 \times N_2 \times N_3$ | $h_w$ | St | $<C_D>$ | $C_{Drms}$ | $C_{Lrms}$ |
|---|---|---|---|---|---|---|---|
| case 7 (LDM model) | 20D × 14D × 4D | 140 × 103 × 32 | 0.02 | 0.131 | 2.05 | 0.12 | 1.39 |
| Lyn [7] | Experiment ($Re=2.2 \times 10^4$) | | | 0.132 | 2.1 | — | — |
| Vickey [12] | Experiment ($Re=1 \times 10^5$) | | | — | 2.05 | — | 1.32 |
| Lee [13] | Experiment ($Re=1.76 \times 10^5$) | | | — | 2.05 | 0.23 | 1.22 |

$L_i$ : computational domain in $x_i$ direction ( i=1,2,3)
$N_i$ : mesh resolution in $x_i$ direction
St : Strouhal number
$C_D$ : drag coefficient
$C_L$ : lift coefficient
$h_w$ : the grid interval in the normal direction adjacent to the cylinder wall

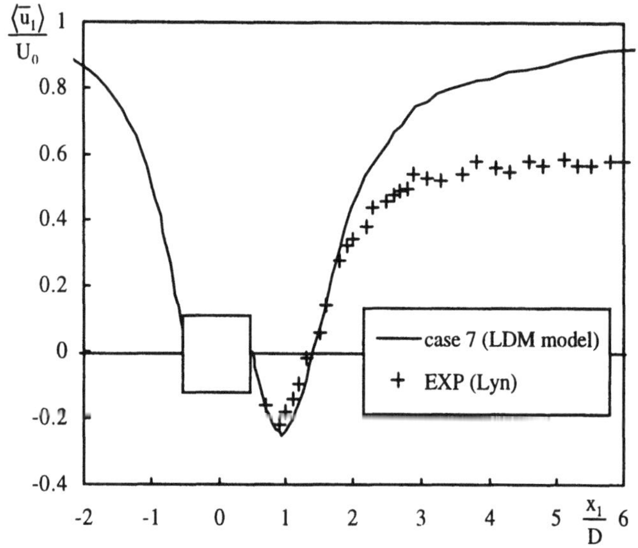


Fig.5 Comparison of the time-averaged velocity $\langle \bar{u}_1 \rangle$ along the centerline of LDM model (case 7)

## 5. Conclusions

(1) Dynamic SGS models provide much more accurate predictions of the sizes of the reverse flow region behind the cylinder than does the static type of conventional Smagorinsky model (S model).

(2) The method for stabilization by averaging over particle trajectories employed in the LDM model can contribute a remarkable improvement of calculation stability. In the experience of our group, the LDM model seems to be most suited for the analysis of the flowfield around a cylinder, since it provides good calculation stability and also good prediction accuracy.

## Note 1

The time-averaged values were determined by the time-averaging over 8 vortex shedding periods (61 in the non-dimensional time scale) in cases 1 and 8 as well as by averaging over the spanwise direction.

Fig.6 compares the time-averaged velocity determined by time-averaging over 2 shedding periods with that averaged over 8 shedding periods for case 1 (S model). The difference between values based on these two different averaging times is very small as far as time-averaged velocity is concerned. Thus, the averaging time for time-averaged values was reduced to 2 vortex shedding periods (15 in the non-dimensional time scale) in cases 2-6.

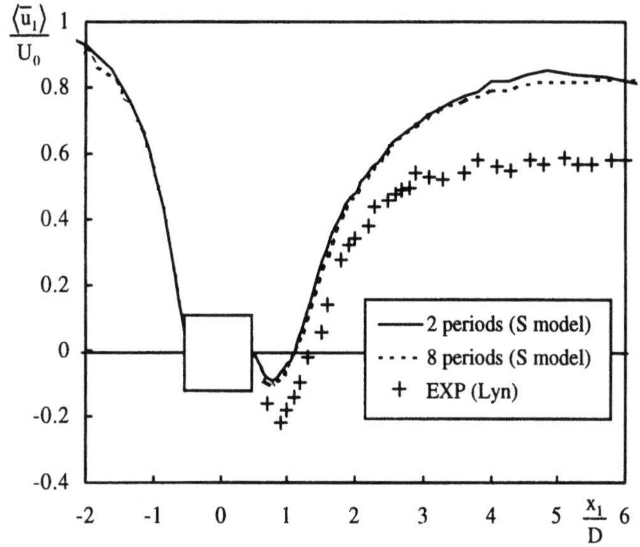


Fig.6 Comparison of the time-averaged velocity $\langle \bar{u}_1 \rangle$ along the centerline determined by time averaging over 2 shedding periods and 8 shedding periods

## Note 2

In Werner and Wengle's approach [8], a linear or 1/7 power-law distribution of the instantaneous velocity is assumed :

$$\frac{\bar{u}}{u*} = x_n^+ \qquad (x_n^+ \le 11.81),$$

$$\frac{\bar{u}}{u*} = 8.3 x_n^{+1/7} \qquad (x_n^+ > 11.81).$$

This wall function becomes identical to the no-slip boundary condition when $x_n^+ \le 11.81$. Fig.7 illustrates the distribution of time-averaged values of $x_n^+$ of the grid points adjacent to the cylinder walls. In most areas, the value of $x_n^+$ is lower than 11.81. Thus the linear-law, i.e., the no-slip boundary condition, is applied in these areas. At corner areas, the value of $x_n^+$ exceeds 11.81 a little.

## Note 3

As shown in Table 1, four factors, (grid spacings, values of Cs, wall damping functions and sizes of computational domain in the spanwise direction) are different between cases 1 and 6. In the earlier research by our group (Murakami *et al.* [9]), the influence of the Cs value and the size of computational domain was examined. In that study, computations were carried out for two different values of Cs (0.13 and 0.1 adopted in cases 1 and 6 respectively) with grid A. As shown in [9], the difference between the results of these two cases was very

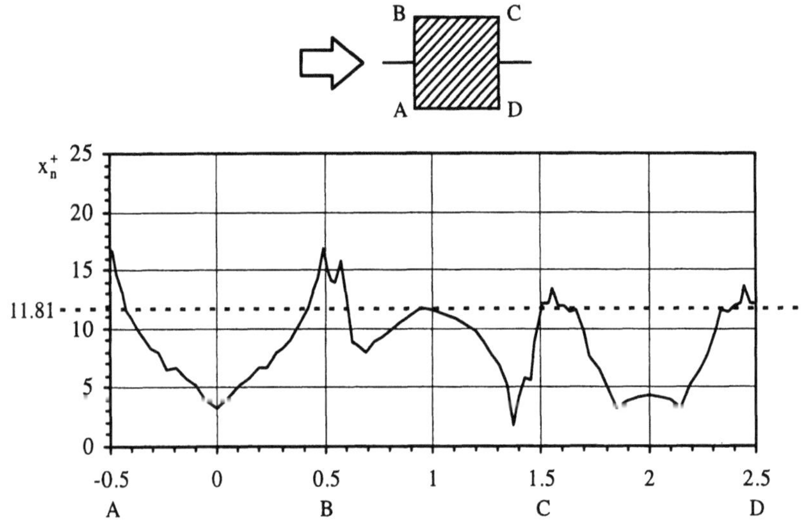


Fig.7 Distribution of time-averaged values of $x_n^+$ (case 7)

small. Furthermore, two computations were carried out and compared for two sizes of different computational domain in the spanwise direction, 2D and 4D, in [9]. Regarding the time-averaged velocity $\langle \bar{u}_1 \rangle$, the result of the case with a larger computational domain (4D) provided a slightly smaller reverse flow region behind the cylinder. However, we noted that there were no outstanding differences for each computational domain size.

We have also carried out another computation (case 6') in which only the damping function was changed from $mf_\mu$ $(=(1\text{-exp}(-x_n^+/25)^3)^{0.5})$ adopted in case 6 to the function used in case 1 ($f_\mu$=1-exp($-x_n^+$/25)). Through the comparison between the results of cases 6 and 6', it was confirmed that the results were not much affected by the difference in the forms of wall damping functions.

The remaining factor that differs between cases 1 and 6 is grid spacing. We consider that the difference in the grid spacing is the main factor contributing to the differences between the results for these cases.

## Note 4

A colocated grid (Rhie and Chow [10]) was used for cases 1-4 and 6. However, recently it was reported that the result based on a staggered grid provides more accurate prediction than that based on a colocated grid in a LES computation of channel flow, because the result based on colocated grid includes some numerical errors (Ooka *et al.* [11]).

In this study, we compared the result based on colocated grid (case 4) with that based on staggered grid (case 5). As for the time-averaged velocity $\langle \bar{u}_1 \rangle$, both results are almost the same, but case 5 is slightly better (figures are omitted here). So we used the staggered grid for case 7.

## Nomenclature

$x_i$ : spatial coordinates (i=1,2,3 :streamwise, lateral, spanwise)
$u_i$ : components of the velocity vector
$f$ : instantaneous value of a quantity, $f$ : filtered value of $f$,
$\langle f \rangle$ : time averaged value of $f$
$x_n$ : distance from the wall, $x_n^+ = x_n u^*/\nu$, $u^*$ : friction velocity
D : width of the square cylinder,
$U_0$ : $\langle u_1 \rangle$ value at inflow of computational domain

Values are made dimensionless using D, $U_0$ and air density $\rho$.

## References


[1] Germano,M., U.Piomelli, P.Moin and W.H.Cabot, A dynamic subgrid-scale eddy viscosity model, Phys. Fluids, A3, 1760 (1991).

[2] Lilly,D.K., A proposed modification of the Germano subgrid-scale closure method, Phys. Fluids A4, 633 (1992).

[3] Zang,Y., R.L.Street and J.R.Koseff, A dynamic mixed subgrid-scale model and its application to turbulent recirculating flows, Phys. Fluids A5 (12) pp 3186-3196 (1993).

[4] Vreman,B., B.Geurts and H.Kuerten, On the formulation of the dynamic mixed subgrid-scale model, Phys. Fluids 6 (12), pp.4057-4059 (1994).

[5] Meneveau,C., T.S.Lund and W.Cabot, A lagrangian dynamic subgrid-scale model of turbulence, Proceedings of the Summer Program 1994, Center for Turbulence Research, pp.2746-2757 (1991).

[6] LES workshop of Flows past Bluff Bodies, June 26-28 Rottach-Egern, Tegernsee, Germany, organized by W.Rodi and J.H.Ferziger (1995).

[7] Lyn, D.A., Einav S., Rodi W. and Park J.H. (1995), A laser-Doppler velocimetry study of ensemble averaged characteristics of the turbulent near wake of a square cylinder, J. Fluid Mech. vol. 304, p.285 (1995).

[8] Werner,H. and H.Wengle, Large-eddy simulation of turbulent flow over and around a cube in a plate channel, 8th Symp. on Turbulent Shear Flows 19-4 (1991).

[9] Murakami,S., W.Rodi, A.Mochida and S.Sakamoto, Large eddy simulation of turbulent vortex shedding flow past 2D square cylinders, in: Proc. Symp. on Engineering Applications of large Eddy Simulations (FED-Vol.162). ASME p.113 (1991).

[10] Rhie,C.M. and W.L.Chow, Numerical study of the turbulent flow past an airfoil with trailing edge separation, AIAA Journal, vol. 21, No.11 (1983).

[11] Ooka,R., S.Murakami and A.Mochida, Study on conservation property of kinetic energy of LES with colocated grid, submitted to International Journal of Numerical Method in Fluids.

[12] Vickery,B.J., Fluctuating lift and drag on a long cylinder of square cross section in a smooth and a turbulent stream, J. Fluid Mech. 25, p.481 (1975).

[13] Lee,B.E., The effect of turbulence on the surface pressure field of square prisms, J. Fluid Mech. 69, p.263 (1975).

[14] Bardina,J., J.H.Ferziger and W.C.Reynolds, Improved Subgrid-Scale Models for Large-Eddy Simulation, AIAA paper-80 (1981).

[15] Akselvoll,K., and P.Moin, Large eddy simulation of a backward facing step flow, Eng. Turbulence Modelling and Experiments 2 pp.303-313 (1993).

## Appendix: SGS models

In the Smagorinsky model, the anisotropic part of the SGS stress $\tau_{ij}$ is modeled as follows:

$$\tau_{ij} - \frac{1}{3}\delta_{ij}\tau_{kk} = -2C\overline{\Delta}^2|\overline{S}|\overline{S}_{ij} = -2\nu_{SGS}\overline{S}_{ij} \quad (1)$$

where the SGS viscosity $\nu_{SGS}$ is

$$\nu_{SGS} = C\overline{\Delta}^2|\overline{S}| \quad (C = Cs^2) \quad (2)$$

Here, $\overline{\phantom{x}}$ (overbar) denotes the grid-filtered values and $\overline{\Delta}$ is the width of the grid-filter, $S_{ij}$ is the resolved-scale strain rate tensor,

$$\overline{S}_{ij} = \frac{1}{2}\left(\frac{\partial \overline{u}_i}{\partial x_j} + \frac{\partial \overline{u}_j}{\partial x_i}\right), \quad |\overline{S}| = (2\overline{S}_{ij}\overline{S}_{ij})^{1/2} \quad (3)$$

In the S model (cases 1 and 6), C is treated as a constant. The value of 0.0169 and 0.01 were selected for cases 1 and 6, respectively. These values correspond to 0.13 and 0.1 of the so-called Smagorinsky constant Cs (cf. Table 1). $\overline{\Delta}$ is multiplied by the Van Driest wall damping function $f_\mu$, $1\text{-}\exp(-x_n^+/25)$ in case 1 and $mf_\mu$, $\{1\text{-}\exp(-x_n^+/25)^3\}^{0.5}$ in case 6 in order to account for the near wall effect in S model, while $f_\mu$ is not necessary in dynamic SGS models.

In dynamic SGS models, the model coefficient C is determined dynamically. Following Germano [1], a test filter (denoted as $\hat{\phantom{x}}$ ) is introduced to derive an expression for C. The width of the test-filter is taken to be twice the width of the grid-filter. Germano *et al.* defined the resolved turbulent stress as follows :

$$£_{ij} = \widehat{\overline{u}_i\overline{u}_j} - \hat{\overline{u}}_i\hat{\overline{u}}_j \quad (4)$$

$£_{ij}$ can be related to the SGS stress $\tau_{ij}$ and the subtest stress $T_{ij} = \widehat{u_iu_j} - \hat{\overline{u}}_i\hat{\overline{u}}_j$ by

$$£_{ij} = T_{ij} - \hat{\tau}_{ij} \quad (5)$$

The DS model (case 2) employs the dynamic procedure using a least square method proposed by Lilly [2] to determine the coefficient C :

$$C(\vec{x},t) = -\frac{1}{2}\frac{£_{ij}M_{ij}}{M_{kl}^2} \quad (6)$$

where $M_{ij} = \hat{\overline{\Delta}}^2|\hat{\overline{S}}|\hat{\overline{S}}_{ij} - \overline{\Delta}^2\widehat{|\overline{S}|\overline{S}_{ij}}$ (7)

In DM model (case 3) a linear combination of the dynamic Smagorinsky model and the scale similarity model is adopted [4, 14]. In the DM model, the anisotropic part of $\tau_{ij}$ is expressed as [4, 14] :

$$\tau_{ij} - \frac{1}{3}\delta_{ij}\tau_{kk} = -2\nu_{SGS}\overline{S}_{ij} + B_{ij} - \frac{1}{3}\delta_{ij}B_{kk} \quad (8)$$

where $B_{ij} = \overline{\overline{u}_i\overline{u}_j} - \overline{\overline{u}}_i\overline{\overline{u}}_j$ (9)

The first and the second terms on the right-hand of eqn.(8) derive from the

Smagorinsky and the scale similarity models respectively. C is determined by

$$C(\vec{x},t) = -\frac{1}{2}\frac{M_{ij}(\pounds_{ij} - H_{ij})}{M_{kl}^2} \tag{10}$$

where $H_{ij}$ is the term originating from the scale similarity model, written as [4] ;

$$H_{ij} = \widehat{\hat{\bar{u}}_i\hat{\bar{u}}_j} - \hat{\hat{\bar{u}}}_i\hat{\hat{\bar{u}}}_j - (\widehat{\bar{\bar{u}}_i\bar{\bar{u}}_j} - \widehat{\bar{\bar{u}}_i\bar{\bar{u}}_j}) \tag{11}$$

The value of C obtained from eqns.(6) and (10) can be either positive or negative. A negative value of C implies a locally negative value of eddy-viscosity which causes the numerical instability. In cases 2 and 3, we set the coefficient C equal to zero wherever C was estimated to be negative (clipping procedure).

Previous authors have used averaging over homogeneous directions to avoid the numerical instability (Germano *et al.* [1], Akselvoll *et al.* [15], Zang *et al.* [3] ). The disadvantages of this treatment is that the plane-averaging can only be applied to flowfields that have a homogeneous direction. This treatment cannot be applied to three-dimensional flowfields such as a flow around a cube; thus this treatment excludes the application of LES to more challenging flows of engineering interest. In the Lagrangian Dynamic Smagorinsky model (LDS model), the model coefficient C is calculated following the approach developed by Meneveau *et al.* [5] in which the residual in eqn.(5) is minimized along fluid trajectories rather than flow homogeneous directions, resulting in an expression for the model coefficient :

$$C(\vec{x},t) = -\frac{1}{2}\frac{I_{LM}}{I_{MM}} \tag{12}$$

The numerator and denominator of eqn.(12) are obtained using a simple time discretization, resulting in

$$I_{LM}^{n+1}(\vec{x}) = H\left\{\varepsilon \pounds_{ij}^{n+1} M_{ij}^{n+1}(\vec{x}) + (1-\varepsilon) I_{LM}^{n}(\vec{x} - \bar{u}_i^{\,n}\Delta t)\right\} \tag{13}$$

$$I_{MM}^{n+1}(\vec{x}) = \varepsilon M_{ij}^{n+1} M_{ij}^{n+1}(\vec{x}) + (1-\varepsilon) I_{MM}^{n}(\vec{x} - \bar{u}_i^{\,n}\Delta t) \tag{14}$$

where H{x} is the ramp function (H{x}=x if x>0 and zero otherwise), and

$$\varepsilon = \frac{\Delta t / T^n}{(1 + \Delta t / T^n)} \tag{15}$$

The technique of Lagrangian averaging can be easily added to the Dynamic Mixed model. We call it the LDM model.

$$C(\vec{x},t) = -\frac{1}{2}\frac{I_{LM} - I_{HM}}{I_{MM}} \tag{16}$$

$$\text{where } I_{HM}^{n+1}(\vec{x}) = \varepsilon H_{ij}^{n+1} M_{ij}^{n+1}(\vec{x}) + (1-\varepsilon) I_{HM}^{n}(\vec{x} - \bar{u}_i^{\,n}\Delta t) \tag{17}$$

The LDM model was used in cases 4, 5 and 7 in Table 1.

The time-scale T in eqn.(15) is defined as T=αn ($n=\bar{\Delta}\, I_{LM}^{-1/4}$). Meneveau *et al.* [5] recommend a value of 2 for α based on the filtered DNS data of isotropic turbulence. In the experience of our group, this value of 2 was not optimum for α for flow around a bluff body. Here, the value of 0.2 was selected for α (T=0.2n) in cases 4, 5 and 7 as a result of numerical experiments.

# LES OF FLOW OVER A SQUARE CYLINDER

G. WANG AND S.P. VANKA
*Department of Mechanical and Industrial Engineering*
*University of Illinois at Urbana-Champaign, IL 61801, USA*

**Abstract.** In the present study we have performed LES of flow over a square cylinder at a Reynolds number of 21,400. Several calculations with progressive improvements in numerical accuracy and grid resolution have been conducted, and the results from one of these are included in the test case exercise.

## 1. Introduction

The problem considered is the flow over a square cylinder that is placed normal to a uniform free stream. We present the results of one of several Large-Eddy Simulations of high Reynolds number turbulent flow over the square cylinder.

In the following sections, we provide details of the numerical schemes use in our study and of the implementation of the dynamic subgrid-scale model. A full report of the various simulations performed is given by Wang and Vanka (1996).

## 2. Governing Equations

In large eddy simulations, the resolved flow field is separated from the sub-grid field by applying a spatial filter to a continuous function in space and time. After applying a suitable filter to the continuity and momentum equations, the filtered equations of motion can be written as :

$$\frac{\partial \overline{u}_i}{\partial x_i} = 0 \tag{1}$$

$$\frac{\partial \overline{u}_i}{\partial t} + \frac{\partial}{\partial x_j}(\overline{u}_i \overline{u}_j) = -\frac{\partial \overline{p}}{\partial x_i} - \frac{\partial \tau_{ij}}{\partial x_j} + \frac{1}{Re}\frac{\partial^2 \overline{u}_i}{\partial x_j \partial x_j} \tag{2}$$



*J.-P. Chollet et al. (eds.), Direct and Large-Eddy Simulation II, 397-400.*

where the index $i = 1, 2, 3$ refers to the streamwise (x-direction), crosswise (y-direction) and spanwise (z-direction) directions, respectively. In the above equations, the velocities are non-dimensionalized by the freestream velocity, $U_0$, and the pressure is non-dimensionalized by the inlet dynamic pressure, $\rho {U_0}^2$.

The subgrid-scale stress $\tau_{ij}$ in equation (2) represents the effects of small scales, and must be modeled. The key to success in large eddy simulations is the accurate representation of the unresolved subgrid-scale motions. In the Smagorinsky model the subgrid stresses are related to the resolved strain rates through an isotropic viscosity, $\nu_T$. Thus,

$$\tau_{ij} - \frac{\delta_{ij}}{3}\tau_{kk} = -2\nu_T \overline{S}_{ij} \tag{3}$$

where $\delta_{ij}$ is the Kronecker delta. The eddy viscosity, $\nu_T$, and the large scale strain rate tensor, $\overline{S}_{ij}$, are defined as

$$\nu_T = C_s \overline{\Delta}^2 \sqrt{2\overline{S}_{ij}\overline{S}_{ij}} = C_s \overline{\Delta}^2 |\overline{S}| \tag{4}$$

$$\overline{S}_{ij} = \frac{1}{2}\left(\frac{\partial \overline{u}_i}{\partial x_j} + \frac{\partial \overline{u}_j}{\partial x_i}\right) \tag{5}$$

$C_s$ is the dimensionless model coefficient, $\overline{\Delta} = (\Delta x \Delta y \Delta z)^{1/3}$ is the grid filter width. In the dynamic model, the value of the $C_s$ coefficient is calculated at every time step, as the flow evolves, by examining the instantaneous energy transfers. A second filter, called the test filter (denoted by $\hat{}$ ), is applied to the resolved velocity field and the resolved turbulent stresses, $\mathcal{L}_{ij}$, the subgrid scale stresses, $\tau_{ij}$, and the subtest-scale stresses are related to determine the coefficient $C_s$. We currently use the modification proposed by Lilly (1992) to the original Germano model (1991) which provides the following expression for $C_s$.

$$C_s = -\frac{1}{2}\frac{< \mathcal{L}_{ij}\mathcal{M}_{ij} >}{< \mathcal{M}_{ij}\mathcal{M}_{ij} >} \tag{6}$$

where $<>$ denotes averaging in the spanwise direction. $\mathcal{L}_{ij}$ and $\mathcal{M}_{ij}$ are defined as :

$$\mathcal{L}_{ij} = \widehat{\overline{u}_i \overline{u}_j} - \hat{\overline{u}}_i \hat{\overline{u}}_j \tag{7}$$

$$\mathcal{M}_{ij} = \hat{\overline{\Delta}}^2 |\hat{\overline{S}}| \hat{\overline{S}}_{ij} - \overline{\Delta}^2 \widehat{|\overline{S}|\overline{S}_{ij}} \tag{8}$$

The test filter has been applied only in the spanwise (homogeneous) direction. Thus, $\hat{\overline{\Delta}}/\overline{\Delta}$, which is the ratio of the filter widths, is $2^{1/3}$ according to the definition of $\Delta = (\Delta x \Delta y \Delta z)^{1/3}$. Further, negative $C_s$ is truncated to ensure that the total viscosity $(\nu + \nu_T)$ remained positive.

## 3. Numerical Procedure

The above equations are numerically integrated in time by a two-stage fractional step procedure with Adams-Bashforth explicit differencing for both convection and diffusion terms.

$$\frac{\tilde{u}_i - u_i^n}{\Delta t} = \frac{3}{2} H_i^n - \frac{1}{2} H_i^{n-1} \tag{9}$$

$$\nabla \cdot (\nabla p) = \frac{1}{\Delta t} \frac{\partial \tilde{u}_i}{\partial x_i} \tag{10}$$

$$u_i^{n+1} = \tilde{u}_i - \frac{\partial p}{\partial x_i} \Delta t \tag{11}$$

where $\tilde{u}_i$ is the intermediate velocity field, and $H_i$ is given by

$$H_i = -\frac{\partial}{\partial x_j}(u_i u_j) - \frac{\partial \tau_{ij}}{\partial x_j} + \frac{1}{Re} \frac{\partial^2 u_i}{\partial x_j \partial x_j} \tag{12}$$

A collocated arrangement has been used for the velocities and pressure and the spatial derivatives are integrated with a finite-volume methodology. The values at the cell faces are evaluated with either third-, fifth-, or seventh-order accuracy for the convection terms and second-, fourth- or sixth-order accuracy for the diffusion terms.

The above equations are solved on a grid which may in general be curvilinear in the $x - y$ plane. Volume fluxes at the cell faces are given by the relation :

$$U = J(\xi_x u + \xi_y v) \tag{13}$$

$$V = J(\eta_x u + \eta_y v) \tag{14}$$

$$W = J(\gamma_z w) \tag{15}$$

where $\xi_x$, $\xi_y$, $\eta_x$, $\eta_y$, and $\gamma_z$ are the metrics and $J$ is the Jacobian of transformation. The volume fluxes at the cell faces are required to satisfy the continuity equation.

The pressure is computed from the pressure Poisson equation. The solution of the pressure equation on a general curvilinear grid requires an iterative procedure, and the same conjugate gradient method was used in the present computations on a Cartesian mesh. In order to enforce mass conservation, the Poisson equation for pressure was discretized using the finite-volume approach. In the computational space, the equation is integrated over a cell volume centered at the collocation point. Computation of the pressure is the most computationally intensive operation of the entire algorithm. The convergence rate of the iterative solver can be accelerated by the utilization of a preconditioner with the conjugate gradient method. In the present algorithm, provision is made to perform several Jacobi iterations alternating with every CG iteration to accelerate the convergence.

## 4. Simulation Results

The results from a single LES (Run 4 of Wang and Vanka (1996)) are included in this test case exercise. This run utilised a total of $192 \times 160 \times 48$ grid points on a domain extending from $-8d$ to $16d$ in the streamwise direction, from $-10d$ to $10d$ in the crosswise direction, and of dimension $\pi d$ in the spanwise direction. This simulation used fifth-order differencing for the convection terms and fourth-order differencing for the diffusion terms. The time step was $2 \times 10^{-3}$ and statistics were gathered over 11 shedding cycles (60000 time steps).

Selected results are given in the preceding review paper in this volume by Voke. For detailed comparisons of all the simulations in the study, see Wang and Vanka (1996) and Wang (1996).

## References

GERMANO, M. PIOMELLI, U., MOIN, P. and CABOT, W. (1991) A dynamic subgrid-scale eddy viscosity model, *Phys Fluids A*, **3**, p. 1760.

LILLY, D.K. (1992) A proposed modification of the Germano subgrid-scale closure method, *Phys Fluids A*, **4**, p. 633.

WANG G. and VANKA, S.P. (1996) Large eddy simulations of high Reynolds number flow over a square cylinder, Report CFD 96-02, Computational Fluid Dynamics Laboratory, Dept. Mech. and Ind. Eng., University of Illinios at Urbana-Champaign, IL 61801.

WANG, G. (1996) Large eddy simulations of turbulent flow over a square cylinder on a massively parallel computer, *Ph.D. Dissertation* (in preparation) Dept. Mech. and Ind. Eng., University of Illinios at Urbana-Champaign.

# LES OF FLOW AROUND A SQUARE CYLINDER

TETSUYA KOGAKI
*Graduate School, University of Tokyo*
*7-3-1 Hongou, Bunkyo-ku, 113 Tokyo, Japan*

AND

TOSHIO KOBAYASHI AND NOBUYUKI TANIGUCHI
*Institute of Industrial Science, University of Tokyo*
*7-22-1 Roppongi, Minato-ku, 106 Tokyo, Japan*

**Abstract.**
Large-eddy simulations (LES) of turbulent flow around a square cylinder at Reynolds number of $2.2\times10^4$ are conducted. The subgrid-scale (SGS) models used here are the standard Smagorinsky model and a dynamic mixed SGS model. Simulation results indicate that the spanwise flow structure behind the cylinder is highly influenced by the spanwise mesh resolution and that the artificial dissipative effects of upwind schemes cannot be ignored even in the case of higher order upwind schemes.

## 1. SGS model

The standard Smagorinsky model (Smagorinsky, 1963; Deardorff, 1970) is expressed as follows:

$$\tau_{ij} - \frac{1}{3}\delta_{ij}\tau_{kk} = -2\left(C_s\overline{\Delta}\right)^2 \left|\overline{S}\right| \overline{S}_{ij}, \tag{1}$$

$$\overline{S}_{ij} = \frac{1}{2}\left(\frac{\partial \overline{u}_i}{\partial x_j} + \frac{\partial \overline{u}_j}{\partial x_i}\right), \quad |S| = \sqrt{2\overline{S}_{ij}\overline{S}_{ij}}\,, \tag{2}$$

where the value of the model coefficient is $C_S$ set to be 0.13, which is conjectured by Mochida *et al.* (1993).

On the other hand, the model coefficient is computed dynamically in a dynamic SGS model (DSM) (Germano *et al.*, 1991). Zang *et al.* (1993) modified the DSM by incorporation of the mixed model of Bardina *et al.*



*J.-P. Chollet et al. (eds.), Direct and Large-Eddy Simulation II,* 401-408.

(1983). Moreover, Vreman *et al.* (1994) rewrote the dynamic mixed SGS model in order to remove a mathematical inconsistency (hereafter the dynamic version of mixed model developed by Vreman *et al.* is referred to as DMSM). SGS stresses of the DMSM are given by

$$\tau_{ij} - \frac{1}{3}\delta_{ij}\tau_{kk} = -2C\overline{\Delta}^2 \left|\overline{S}\right| \overline{S}_{ij} + L_{ij}^m - \frac{1}{3}\delta_{ij}L_{kk}^m , \tag{3}$$

where $L_{ij}^m$ is the modified Leonard term represented by

$$L_{ij}^m = \overline{\overline{u}_i\overline{u}_j} - \overline{\overline{u}}_i\overline{\overline{u}}_j . \tag{4}$$

Using the least-square approach suggested by Lilly (1992), the model coefficient $C$ of DMSM is computed as

$$C = -\frac{1}{2}\frac{\langle (L_{ij} - H_{ij})\, M_{ij} \rangle}{\langle M_{ij}M_{ij} \rangle} , \tag{5}$$

$$L_{ij} = \widetilde{\overline{u}_i\overline{u}_j} - \tilde{\overline{u}}_i\tilde{\overline{u}}_j , \tag{6}$$

$$H_{ij} = \widetilde{\overline{\tilde{\overline{u}}_i\tilde{\overline{u}}_j}} - \overline{\tilde{\overline{u}}}_i\overline{\tilde{\overline{u}}}_j - \left(\widetilde{\overline{\overline{u}_i\overline{u}_j}} - \widetilde{\overline{\overline{u}}_i\overline{\overline{u}}_j}\right) , \tag{7}$$

$$M_{ij} = \overline{\Delta}^2 \left(\alpha^2 \left|\tilde{\overline{S}}\right| \tilde{\overline{S}}_{ij} - \widetilde{\left|\overline{S}\right| \overline{S}_{ij}}\right) , \quad \alpha = \tilde{\overline{\Delta}}/\overline{\Delta} , \tag{8}$$

where brackets '⟨⟩' in eq. (5) denote averaging operations in homogeneous directions.

The dynamic versions of the SGS model require an explicit filter operation. The test-scale filtering and the grid-scale filtering with the top-hat filter or the Gaussian filter are evaluated using Simpson's rule as follows:

$$\tilde{\overline{f}} = \overline{f} + \frac{\tilde{\Delta}^2}{24} \cdot \nabla^2\overline{f} + \mathrm{O}\left(\tilde{\Delta}^4\right) , \tag{9}$$

$$\overline{\overline{f}} = \overline{f} + \frac{\overline{\Delta}^2}{24} \cdot \nabla^2\overline{f} + O\left(\overline{\Delta}^4\right) . \tag{10}$$

## 2. Methods

### 2.1. RESOLUTION

The computational domain, total numbers of node points and SGS models used in our computation are given in Table 1. The computational domain is the same one prescribed by the organizer except for the length from the

TABLE 1. Mesh Configurations

| Case | Domain | Resolution | SGS model |
|---|---|---|---|
| IS1 | 21.1D×14.0D×4.0D | 76,880 (82×63×16) | Smagorinsky |
| IS2 | 21.1D×14.0D×4.0D | 153,760 (82×63×32) | Smagorinsky |
| IS3 | 20.4D×14.0D×4.0D | 347,648 (112×104×32) | DMSM |

center of the cylinder to the outlet, which is extended up to more than $20D$ ($D$ = side of the cylinder). The mesh used in case IS3 has the same resolution in the $(x,y)$-plane as the mesh used by the University of Karlsruhe group (UK1).

## 2.2. NUMERICAL METHODS

The finite volume method (FVM) was employed and the coupling scheme between the continuity equation and the pressure was based on the HSMAC method proposed by Hirt *et al.* (1972). The second order upwind scheme QUICK for the convection terms and second order central differencing for the diffusion terms were used for case IS1 and IS2. A combination of the fifth order upwind scheme (Rai and Moin, 1991) and fourth order central differencing was used for case IS3. The time marching method was a semi-implicit method with the second order Adams-Bashforth method for the convection terms and the Crank-Nicolson scheme for the diffusion terms. To stabilize the dynamic mixed model, the model coefficient $C$ was averaged in the spanwise direction and the total viscosity was clipped to be non-negative. Moreover, in laminar flow regions where $M_{ij}$ becomes almost zero, the model coefficient became excessively large. Therefore, we set $C = 0$ wherever $M_{ij} \times M_{ij}$ is less than $10^{-6}$ (Mochida *et al.*, 1995).

## 2.3. BOUNDARY CONDITIONS

A no-slip condition was adopted on the surface of the cylinder, while free slip conditions were used on the top and bottom domain boundaries. At the entrance to the flow domain, a constant velocity with no perturbations was imposed. At the exit of the flow domain, a convective boundary condition was used (Pauley *et al.*, 1990; Dai *et al.*, 1992):

$$\frac{\partial \overline{u}_i}{\partial t} + U_C \frac{\partial \overline{u}_i}{\partial x_1} = 0 \,, \tag{11}$$

where $U_C$ is the convection velocity which we set equal to the mean velocity at the entrance. The spanwise boundary condition was periodic.

### 3. Results

Contours of the instantaneous streamwise component of the vorticity vector $q_x$ in the $(x,z)$-plane at the centerline are compared between the three cases in figure 1. Comparing IS1 and IS2, it is found that the size of the streamwise vortices behind the cylinder is highly affected by the number of node points in the spanwise direction. In the case of the flow around a circular cylinder, the mean spacing of the streamwise vortex pairs behind the cylinder observed experimentally was approximately one cylinder diameter over the wide range of Reynolds number 320 to 21,000 (Bays-Muchmore *et al.*, 1993). It is expected that the spanwise mesh resolution for IS1 was not sufficient to resolve the appropriate spanwise structure. The spanwise flow structures for case IS3 have the same tendency as the case IS2 near the cylinder up to $x \cong 2$; however the arrangement of the streamwise vortex pairs becomes unstable in the far wake region.

To evaluate the effects of the upwind schemes at the scalar level (Clark *et al.*, 1979), the GS energy equation with additional artificial terms in the upwind schemes is considered:

$$\frac{\partial}{\partial t}\left(\frac{1}{2}\overline{u}_i\overline{u}_i\right)+\overline{u}_j\frac{\partial}{\partial x_j}\left(\frac{1}{2}\overline{u}_i\overline{u}_i\right)=\frac{\partial}{\partial x_j}\left(T_{ij}\overline{u}_i\right)-\frac{2}{Re}\overline{S}_{ij}\overline{S}_{ij}+\tau_{ij}\overline{S}_{ij}+E\ , \tag{12}$$

where

$$T_{ij}=-\overline{p}\delta_{ij}+\frac{2}{Re}\overline{S}_{ij}-\tau_{ij}\ . \tag{13}$$

The second and third terms on the right-hand side of eq. (12) are the molecular viscous dissipation and the turbulent viscous dissipation evaluated by the SGS models, respectively. The artificial additional terms of QUICK and of the fifth order upwind scheme are given by

$$E=-\overline{u}_i\overline{u}_j\frac{\Delta x_j^2}{24}\frac{\partial^3\overline{u}_i}{\partial x_j^3}-\overline{u}_i\left|\overline{u}_j\right|\frac{\Delta x_j^3}{16}\frac{\partial^4\overline{u}_i}{\partial x_j^4}\qquad\text{(QUICK)}\ , \tag{14}$$

$$E=\overline{u}_i\left|\overline{u}_j\right|\frac{\Delta x_j^5}{60}\frac{\partial^6\overline{u}_i}{\partial x_j^6}\qquad\text{(5th order upwind scheme)}\ . \tag{15}$$

The additional artificial terms of QUICK have dispersive and dissipative effects on the numerical solution while the fifth order upwind scheme has only dissipative effects.

Figure 2 shows distributions of the instantaneous values of the additional artificial terms of QUICK compared with the total viscous dissipation in the GS energy eq. (12) for IS2. In the vicinity of the cylinder and the near wake region, the dissipative effect of QUICK is larger than the

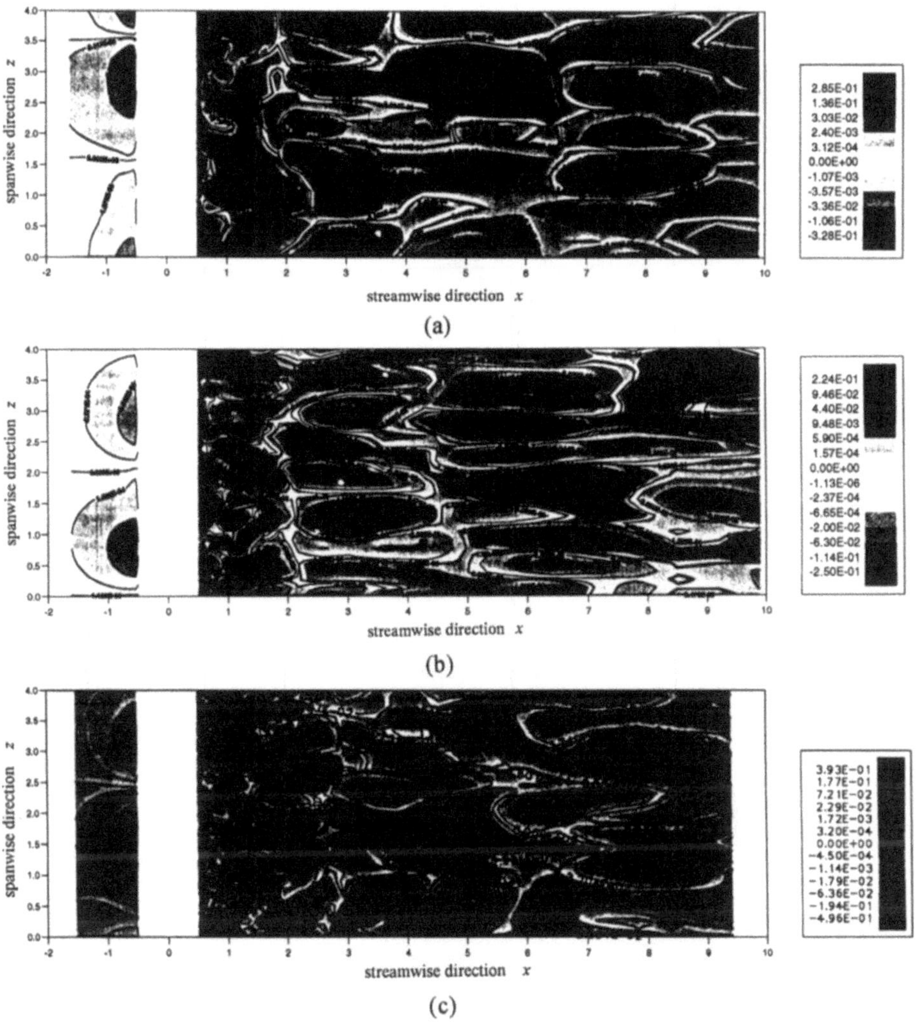


*Figure 1.* Contours of the streamwise component of the vorticity vector $q_x$ in the $(x,z)$-plane at the centerline. (a) IS1; (b) IS2; (c) IS3.

total viscous dissipation. In the case of the fifth order upwind scheme (figure 3), the region that exhibits high values of the additional artificial terms becomes small but the maximum values was comparable to that of QUICK ($E_{max} \cong 4 \times 10^{-1}$ for both cases of QUICK and the 5th order upwind scheme in the wake region).

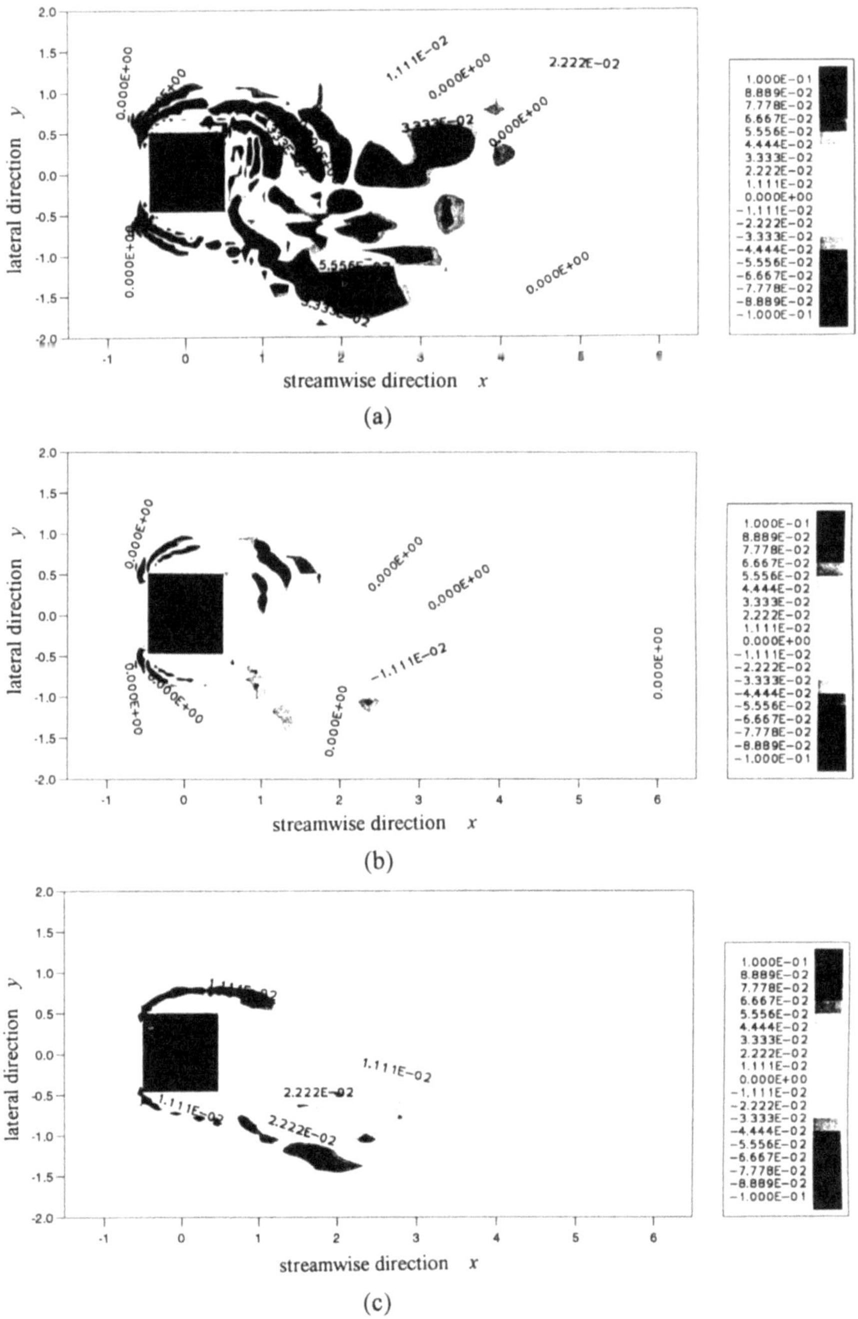


*Figure 2.* The distribution of the instantaneous values of the additional artificial terms of QUICK and the total viscous dissipation in the GS energy equation for IS2: (a) dissipative term; (b) dispersive term ; (c) total viscous dissipation.

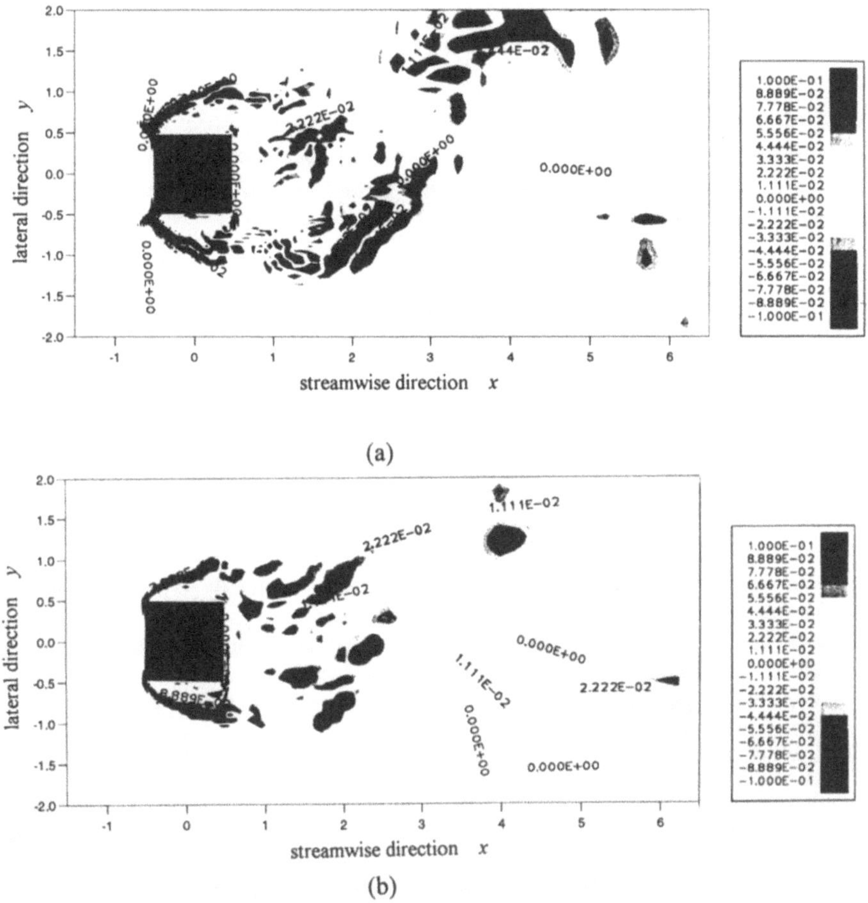


*Figure 3.* The distribution of the instantaneous values of the artificial additional term of the fifth order upwind scheme and the total viscous dissipation in the GS energy equation for IS3: (a) dissipative term; (b) total viscous dissipation.

## 4. Discussions and Conclusions

The effect of the spanwise mesh resolution was investigated using computational grids that have the same mesh resolution in the $(x, y)$-plane. The spanwise flow structure is highly affected by the number of spanwise node points; therefore more than 10 node points in the spanwise direction per cylinder length is appropriate to capture the proper spanwise structure.

Moreover, the qualitative effect of introducing upwind schemes for the convection terms was investigated in the GS energy equation. In the case of the second order upwind scheme QUICK (IS1 and IS2), the artificial dissipative effect due to the upwinding was larger than that of the Smagorinsky model. Therefore, the SGS model was covered by the upwinding. To reduce the effect of upwinding, the fifth order upwind scheme of Rai and Moin and a finer mesh resolution in the $(x, y)$-plane was adopted in case IS3.

The effect of the upwind scheme could not be ignored even in this case in the instantaneous GS energy equation. However, the profile of the time-averaged velocity and turbulence intensities were in fairly good agreement with the experimental data (Lyn, 1994) in all the cases. Therefore, in engineering terms, it is too early to conclude that the use of the upwind scheme is purposeless.

## References


Bardina, J., Ferziger, J. H. and Reynolds, W. C., Improved turbulence models based on large eddy simulation of homogeneous incompressible turbulent flows, *Ph.D. dissertation*, Dept. Mech. Eng. (1983), Stanford University

Bays-Muchmore, B and Ahmed, A., On streamwise vortices in turbulent wakes of cylinders, *Phys. Fluids* A5 (1993) 387–392.

Clark, R. A., Ferziger, J. H. and Reynolds, W. C., Evaluation of subgrid-scale models using an accurately simulated turbulent flow,*J. Fluid Mech.* **91** (1979) 1–16.

Dai, Y. and Kobayashi, T., Numerical analysis on outflow boundary condition of vortex convection with uniform mean flow, *Trans. of JSME* **58-546** B(1992, in Japanese), 313–320.

Deardorff, J. W., A numerical study of three-dimensional turbulent channel flow at large Reynolds numbers, *J. Fluid Mech.* **41** (1970), 453–480.

Germano, M., Piomelli, U., Moin, P., and Cabot W. H., A dynamic subgrid-scale eddy viscosity model, *Phys. Fluids* **A3** (1993), 1760–1765

Hirt, C. W. and Cook, J. L., Calculating three-dimensional flows around structures and over rough terrain, *J. Comput. Phys.* **10** (1972), 324–340.

Kogaki, T., Kobayashi, T. and Taniguchi, N., Large eddy simulation of flow around a rectangular cylinder, submitted to *Special Issue on Mathematical Modeling of Turbulent Flows* (ed. Daiguji *et al.*, Elsevier).

Lilly, D. K., A proposed modification of the Germano subgrid-scale closure method, *Phys. Fluids* **A4** (1992), 663–635

Lyn, D. A., Einav, S., Rodi, W. and Park, J. H. A laser-Doppler velocimetry study of ensemble-averaged characteristics of the turbulent near wake of a square cylinder, *Rept. SFB* **210/E/100** (1994).

Mochida, A., Murakami, S., Rodi, W. and Sakamoto, S., Large eddy simulation of vortex shedding flow past 2D square cylinder, *J. Wind Eng.* **55** (1993, in Japanese), 79–80.

Mochida, A., Murakami, S. and Tominaga, Y., Large eddy simulation of flow past 2D square cylinder using dynamic mixed SGS model, *Seisan-Kenkyu (Mon. J. Institute of Industrial Science, University of Tokyo)* **47** (1995, in Japanese), 79–84

Pauley, L. L., Moin, P. and Reynolds, W. C., The structure of two-dimensional separation, *J. Fluid Mech.* **220** (1990), 397–411.

Rai, M. M. and Moin, P., Direct Simulations of Turbulent Flow Using Finite-Difference Schemes, *J. Comput. Phys.* **96** (1991), 15–53

Smagorinsky, J., General circulation experiments with the primitive equations. I. The basic experiment, *Mon. Weath. Rev.* **91** (1963), 99–164.

Taniguchi, N., Dynamic SGS model by finite difference method, *Seisan-Kenkyu (Mon. J. Institute of Industrial Science, University of Tokyo)* **47** (1995, in Japanese), 120–123

Vreman, B, Geurts, B. and Kuerten, H., On the formulation of the dynamic mixed subgrid-scale model, *Phys. Fluids* **6** (1994), 4057–4059

Zang, Y., Street, R. L. and Koseff, J. R. A dynamic mixed subgrid-scale model and its application to turbulent recirculating flows, *Phys. Fluids* A5 (1993), 3186–3196.

# LES OF FLOW PAST A SQUARE CYLINDER USING EMBEDDED MESHES

KOJIRO NOZAWA
*Izumi Research Institute, Shimizu Corporation*
*2-2-2, Uchisaiwai-cho, Chiyoda-ku, Tokyo 100, Japan*

AND

TETSURO TAMURA
*Tokyo Institute of Technology,*
*4259, Nagatsuta-cho, Midori-ku, Yokohama-shi,*
*Kanagawa 226, Japan*

## 1. Introduction

This paper presents the result of a large-eddy simulation of the complicated turbulent flow past a square cylinder. The technique of embedded meshes is employed for the computation of turbulent flows with various scales. We also apply the 3rd-order upwind scheme (Kawamura *et al.*, 1984) to the convective terms to avoid numerical instability.

In the flow past a bluff cylinder, there is a large area of high shear rate which it must be very expensive to cover with a fine mesh. Without using any stabilizing technique, we can hardly avoid numerical instability. However, using a high-order upwind scheme to avoid numerical instability, there will be excess damping in the area where the flow is turbulent and the subgrid viscosity of LES must work, if the mesh is not sufficiently fine.

Figure 1 shows the results of LES for the $Re_\tau$=180 turbulent channel flow in which a 3rd-order upwind scheme is applied and the subgrid-scale eddy viscosity is generated by a dynamic procedure. In the simulation, by using a larger width filter of $2h$ or $4h$ ($h$ is grid size) for LES instead of $1h$ and reducing the coefficient of the numerical dissipation terms of the 3rd-order upwind scheme, it becomes closer to the DNS results. This means it is possible to suppress the numerical effects through the choice of the filter width.



*J.-P. Chollet et al. (eds.), Direct and Large-Eddy Simulation II*, 409-412.

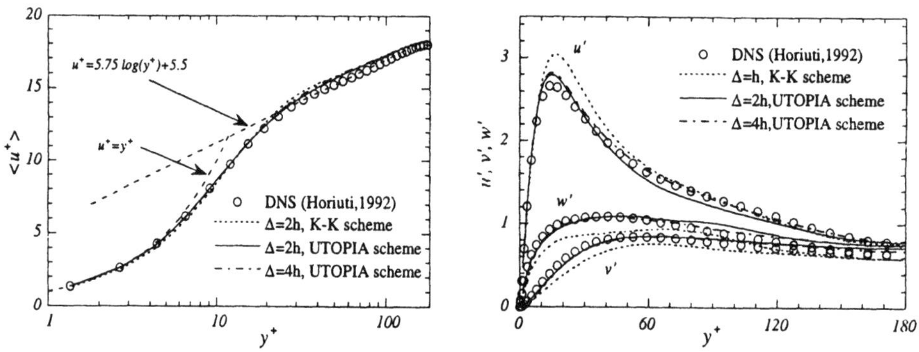


*Figure 1* $Re_\tau$=180 channel flows; left, mean-velocity profiles; right, root-mean-square velocity fluctuations normalized by the wall shear velocity.

Accordingly in the present calculation we refine the mesh by using embedded meshes and make the filter width to be double the grid size to reduce the influence of the numerical dissipation terms in the turbulent area. The flow structures in the wake of a square cylinder are investigated, especially focusing on the behavior of the separated shear layers.

## 2. Embedded mesh

The technique of embedded meshes is used over the computational domain $(15d \times 14d \times 4d)$. Figure 2 shows the embedded meshes sliced transversely in the spanwise direction, and the size and the location of each embedded mesh are shown in Table 1.

TABLE 1. Size and Location of Embedded Meshes

| | Size | Location of subdomain |
|---|---|---|
| inner mesh | 91 × 91 × 127 | $(-1d, 2d) \times (-1.5d, 1.5d) \times (0, 4d)$ |
| middle mesh | 113 × 97 × 57 | $(-3d, 4d) \times (-3d, 3d) \times (0, 4d)$ |
| outermost mesh | 121 × 114 × 27 | $(-5d, 10d) \times (-7d, 7d) \times (0, 4d)$ |

The inner mesh which surrounds the cylinder is stretched to refine the resolution of the near-wall region of the cylinder. The finest resolution at the cylinder surface is $d/200$ at the front, $d/50$ at the sides and $d/30$ at the rear face, and no-slip velocity boundary conditions are imposed on the surface of the cylinder. The other meshes have uniform grids. The advantage of using embedded meshes is to get high resolution while avoiding the disadvantage of using strongly flattened (anisotropic) grids in LES.

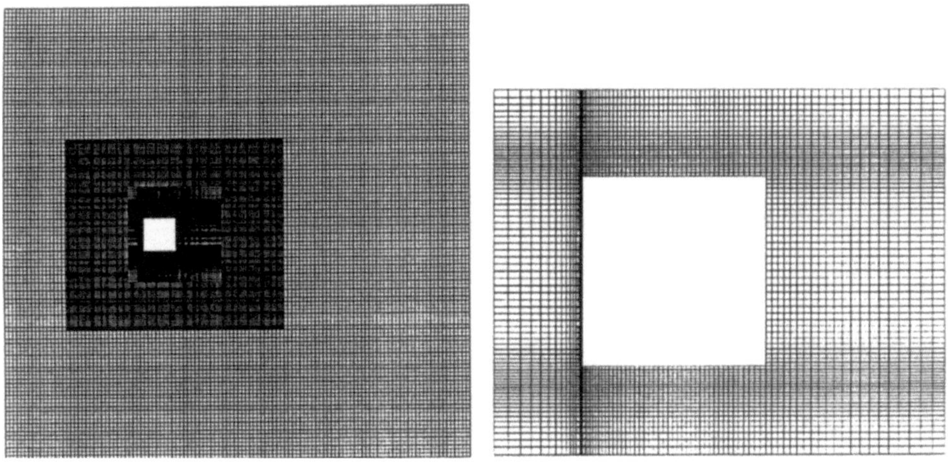

*Figure 2.* Embedded mesh: left, whole domain; right, vicinity of the square cylinder.

Using embedded meshes, the computed quantities must be exchanged on the interior boundaries. Since we use symmetric Dirichlet condition on the interior boundaries, we have to overlap the computational grids. The Dirichlet boundary condition is calculated using linear interpolation in the overlapped zone. All physical quantities except pressure are exchanged between meshes at every time step. The Poisson equation for pressure is solved by SOR, so we have to exchange pressure data at every iteration.

## 3. Numerical method

The numerical procedures are based on the MAC method. For the time marching, the first order explicit scheme is used. For discretization, the second-order central spatial differencing is employed except for the convective terms. A non-staggered mesh system is employed, where all physical quantities are defined at the same location for a grid.

In this simulation, the dynamic subgrid-scale model (Germano *et al.* 1991) based on the Smagorinsky model is applied. The coefficient $C$ is evaluated using the least square approach (Lilly, 1992). To stabilize the dynamic model, the coefficient $C$ is bounded to be non-negative but it is not averaged over the spanwise direction.

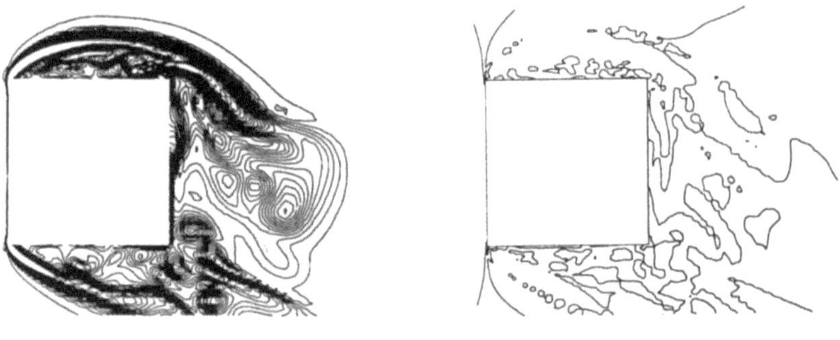

*Figure 3.* Instantaneous flow pattern around a square cylinder; left, vorticity($\omega_z$) contours; right, contours of $C$ (contour lines: $C$=0.01).

## 4. Computational conditions

The approaching flow is assumed to be uniform($U_0$) on the inflow surface boundary and the convective condition is employed on the outflow boundary. The convection velocity is set to be equal to the uniform inflow velocity. Periodic boundary conditions are used in both spanwise and transverse directions. The Reynolds number based on the incoming velocity $U_0$ and the depth of the square cylinder $d$ is 22,000.

The time step size of this calculation is $0.001(tU_0/B)$ and it cost about 19 hours per shedding cycle on one processor of a Fujitsu VX. Total sampling time is about $39(tU_0/d)$, which is equal to 5 shedding cycles, and the sampling rate of data is about $0.1(tU_0/d)$.

## 5. Predicted data

Time- and span-averaged values of the reattachment length, Strouhal number, drag coefficient $C_d$, lift coefficient $C_l$, rms fluctuation of $C_d$ and rms fluctuation of $C_l$ are give in the preceding paper in this volume by Voke.

Figure 3 shows the instantaneous pattern of vorticity and $C$ around the square cylinder (innermost mesh). The shear flow separating from the front corner of the square cylinder is in the early stage of transition to turbulence and $C$ is estimated below 0.01 in these areas.

## References

Germano, M., Piomelli, U., Moin, P. and Cabot, W. H., Phys. Fluids, A3 (1991), 1760.
Kawamura, T. and Kuwahara, K., AIAA-84-0340, (1984).
Lilly, D. K., Phys. Fluids, A4 (1992), 633.

# NUMERICAL ANALYSIS OF LES OF FLOW PAST A LONG SQUARE CYLINDER

NORIFUSA KAWASHIMA AND HIROSHI KAWAMURA
*Department of Mechanical Engineering,*
*Science University of Tokyo*
*Noda-shi, Chiba-ken, 278 Japan*

**Abstract.** The flow past a bluff body is frequently associated with periodic vortex shedding. In this paper, the turbulent flow past a square cylinder is simulated by larg-eddy simulation. The Smagorinsky model (1963) is adopted as the SGS model for the subgrid Reynolds stress term. In addition, an upwind-biased scheme is used for discretizing the convection term to stabilize the numerical calculation. The effect of the upwinding is found to be rather significant. The magnitude of the numerical viscosity term of the upwind scheme is controlled to minimize its effect.

## 1. Introduction

Most practical engineering flows are turbulent and complex in geometry. Large eddy simulation (LES) provides an effective tool for complex turbulent flows. Subgrid scale (SGS) models are introduced to represent the unresolved subgrid scale motions. The SGS model used most widely is the Smagorinsky model (1963). On the other hand, the dynamic SGS model proposed by Germano *et al.* (1991) has been applied for determining the unknown variable model coefficient from the information of the resolved velocity field.

For the flow past a square cylinder, Mochida *et al.* (1992) simulated the flow using the LES method. A workshop on LES was held coordinated by Rodi and Ferziger in 1995, in which the flow past a square cylinder was one of the cases in the workshop. Several groups submitted the time-averaged and the phase-averaged velocities and turbulent intensities, and compared their results with the experimental data (Lyn *et al.* 1994).



*J.-P. Chollet et al. (eds.), Direct and Large-Eddy Simulation II,* 413-422.

The Smagorinsky model was applied by Breuer *et al.* ($C_s = 0.1$) and by Kobayashi *et al.* ($C_s = 0.13$), using respectively the 2nd order central and QUICK schemes for the convection term. Wang and Vanka, Archambeau *et al.* and Tamura *et al.* employed the dynamic model, and their calculations were performed with QUICK, the 2nd order central and the 3rd order upwind schemes respectively. The present authors' group calculated the flow using the 3rd order upwind scheme (UTOPIA) for the convection term without the SGS model to investigate the influence of the numerical viscosity on the result.

The main incentive of LES is to perform turbulent simulation with the use of a rather coarse mesh, because if a fine enough mesh could be adopted one could perform a direct numerical simulation. Since it is inevitable to utilize a rather coarse mesh, LES is often associated with problems of numerical instability. A common way to resolve this problem is to introduce an upwind-biased scheme such as QUICK or UTOPIA in the convection term of the momentum equation. The upwind-biased scheme, however, introduces numerical viscosity and contaminates the effect of the SGS model. This is a dilemma in LES work.

In this paper, the turbulent flow past a square cylinder is analyzed by LES using the Smagorinsky model (1963) with the non-slip condition over the solid surface. An upwind-biased scheme is introduced to stabilize the numerical calculation, and an attempt is made to minimize its effect on the numerical results.

## 2. Fundamental equations

Unsteady incompressible flow is considered. The continuity and grid-filtered Navier-Stokes equations are

$$\frac{\partial \overline{u_i}}{\partial x_i} = 0, \tag{1}$$

$$\frac{\partial \overline{u_i}}{\partial t} + \overline{u_j}\frac{\partial \overline{u_i}}{\partial x_j} = -\frac{1}{\rho}\frac{\partial \overline{P}}{\partial x_i} + \nu \frac{\partial^2 \overline{u_i}}{\partial x_j \partial x_j} - \frac{\partial \tau_{ij}}{\partial x_j}, \tag{2}$$

where

$$\begin{aligned} \tau_{ij} &= \overline{u_i u_j} - \overline{u_i}\,\overline{u_j} \\ &= \underbrace{\overline{\overline{u_i}\,\overline{u_j}} - \overline{u_i}\,\overline{u_j}}_{L_{ij}} + \underbrace{\overline{\overline{u_i} u_j'} + \overline{u_i' \overline{u_j}}}_{C_{ij}} + \underbrace{\overline{u_i' u_j'}}_{R_{ij}}. \end{aligned} \tag{3}$$

We assume that the summation of the Leonard term $L_{ij}$ and the cross term $C_{ij}$ is zero. The standard Smagorinsky model (1963) is adopted as the SGS model for the Reynolds stress term $R_{ij}$,

$$R_{ij} = \overline{u'_i u'_j} = -2\nu_t \overline{S_{ij}} + \frac{2}{3} K_{SGS} \delta_{ij}, \tag{4}$$

$$\nu_t = (C_s \Delta)^2 (2\overline{S_{ij}}\,\overline{S_{ij}})^{\frac{1}{2}}, \tag{5}$$

$$K_{SGS} = \frac{\nu_t^2}{(C_K \Delta)^2}, \tag{6}$$

where $K_{SGS}$ is the subgrid scale turbulent kinetic energy. $\overline{S_{ij}}$ and $\Delta$ are the grid scale (GS) strain-rate tensor and the grid-filter width, respectively,

$$\overline{S_{ij}} = \frac{1}{2}\left(\frac{\partial \overline{u_i}}{\partial x_j} + \frac{\partial \overline{u_j}}{\partial x_i}\right) \tag{7}$$

$$\Delta = (\Delta x_1 \Delta x_2 \Delta x_3)^{\frac{1}{3}} . \tag{8}$$

Here, $\Delta x_i (i = 1, 2, 3)$ is the grid spacing in the $i$ direction. In eqs. (5) and (6), the coefficients $C_s$ and $C_K$ are given as

$$C_s = 0.10, \qquad C_K = 0.094. \tag{9}$$

To take account of near wall effects, $\Delta$ is multiplied by the van Driest wall damping function:

$$f_w = 1 - \exp(-y^+/A^+). \tag{10}$$

The common value adopted for the empirical coefficient $A^+$ is 26.

## 3. Calculation methods and conditions

The governing equations are discretized on a staggered grid using the finite difference method. Second order central differencing is adopted for both the pressure and diffusion terms.

The coupling scheme between the continuity and momentum equations is based on the fractional step method. Time marching is performed with use of the 2nd order explicit Adams-Bashforth scheme for both the convection and viscous terms.

The calculation domain is shown in Fig. 1. The number of grids is $109 \times 78 \times 20$ for cases 1 and 2, $125 \times 78 \times 20$ for case 3, and $107 \times 103 \times 20$ for cases 4 and 5. The Reynolds number based on the inflow mean velocity and the square cylinder side length is $22,000$. The time increment $\Delta t$ is 0.0005 for all cases.

The non-slip condition is applied on the surface of the cylinder, while free-slip conditions are adopted on the top and bottom domain boundaries.

TABLE 1. Parameter $\alpha$

| | SGS | Parameter $\alpha$ |
|---|---|---|
| case 1 | yes | no ( 2nd order central scheme ) |
| case 2 | yes | $\alpha = 1.0$ : ( whole domain ) |
| case 3 | no | $\alpha = 1.0$ : ( whole domain ) |
| case 4 | yes | $\alpha = 1.0$ : ( $-3.5 \le y/D \le 3.5$ ) |
| | | $\alpha = 0.0$ : ( otherwise ) |
| case 5 | yes | $\alpha = \alpha_0 \exp(-\nu_t/100\nu)$ ( $\alpha_0 = 1.0$ ) |

The periodic boundary condition is used in the spanwise direction. At the inflow boundary, constant velocity without perturbation is imposed. For the exit boundary condition, we assume a zero gradient condition for velocity and a constant value for pressure.

Firstly, a calculation has been made using the Smagorinsky model (1963) and with no use of the upwind scheme (case 1). The results, however, suffered from numerical instability. To avoid the numerical instability, the upwind-biased scheme is introduced into the convection term of the Navier-Stokes equation. It is represented by

$$f\left.\frac{\partial\phi}{\partial x}\right|_{x=i} = f_i\frac{-\phi_{i+2}+8\phi_{i+1}-8\phi_{i-1}+\phi_{i-2}}{12\Delta x} + \alpha\left|f_i\right|\frac{\phi_{i+2}-4\phi_{i+1}+6\phi_i-4\phi_{i-1}+\phi_{i-2}}{12\Delta x}, \quad (11)$$

$\alpha$ is a parameter to control the magnitude of the upwinding. When $\alpha = 0$, it tends to the 4th order central scheme; when $\alpha = 1$, to UTOPIA. The upwind scheme is introduced in the whole region to stabilize the calculation (case 2). On the other hand, the calculation can proceed even without introducing the SGS model (case 3). Thus, the effect of the upwind scheme is quite significant. Then an attempt is made to minimize its effect through control of the magnitude of the parameter $\alpha$ (cases 4 and 5). The model and calculation conditions are summarized in Table 1.

## 4. Results

An instantaneous velocity field from case 1 is shown in figure 2. Numerical oscillation is observed in the region at the front of the square cylinder. The upwind scheme was introduced for the convection term to prevent occurrence of this numerical instability, as given in Table 1.

The Strouhal number $St$ and the drag coefficient $C_D$ are given in Table 2. The LES calculations (cases 2 to 5) give fairly good agreement with the

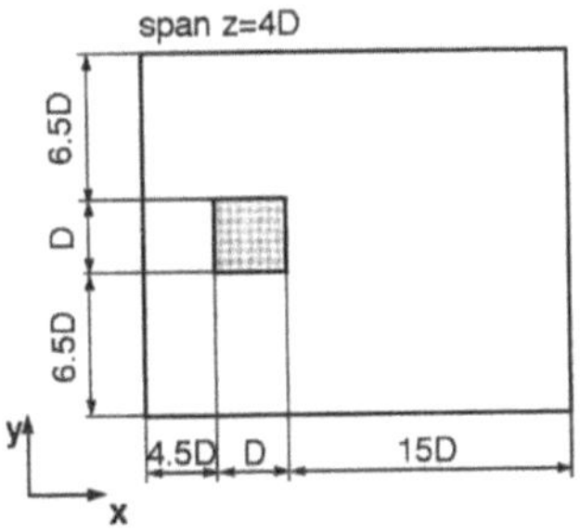


*Figure 1.* Geometry of the domain.

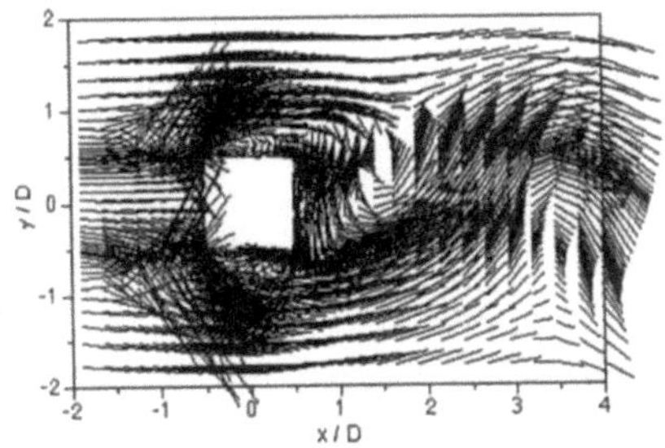


*Figure 2.* Velocity vectors for case 1.

experimental data (Lyn *et al.* 1994) but are somewhat higher than the experiment.

Figure 3 gives the time averaged streamwise velocity $\overline{u}$ along the vertical line at the center ($x = 0$) of the cylinder top surface. In the following, the overbar and the prime mean the time averaged value on the grid scale and the difference from it, respectively. (Note the different definitions from Eqs.(1) to (4).) In the vicinity of the wall, the calculations underpredict $\overline{u}$ in comparison with the experimental data (Lyn *et al.* 1994). The numerical viscosity in cases 2, 4 and 5 are different, but they do not affect the results in this region much. The cases with the SGS model give a slightly better result. Closer inspection indicates that the reattachment does not take place on the top (or bottom) surface of the cylinder in the experiment while it does in the LES calculations. Moreover a small counterclockwise 'sub-recirculation' is observed in the calculations with the SGS model (cases 2, 4 and 5). These are the reasons for the discrepancies in the present calculations.

Figure 4 shows the distribution of the time averaged streamwise velocity $\overline{u}$ along the centerline. All the calculations underpredict the length of the time averaged separation behind the square cylinder. In front of the square cylinder, the results are not much influenced by the calculation methods, but fairly large differences are found in the wake region. The centerline

TABLE 2. Strouhal number and drag coefficient

| | $St$ | $C_D$ | | $St$ | $C_D$ |
|---|---|---|---|---|---|
| Exp. | 0.132 | 2.1 | case 4 | 0.139 | 2.73 |
| case 2 | 0.160 | 2.72 | case 5 | 0.161 | 2.78 |
| case 3 | 0.150 | 2.65 | | | |

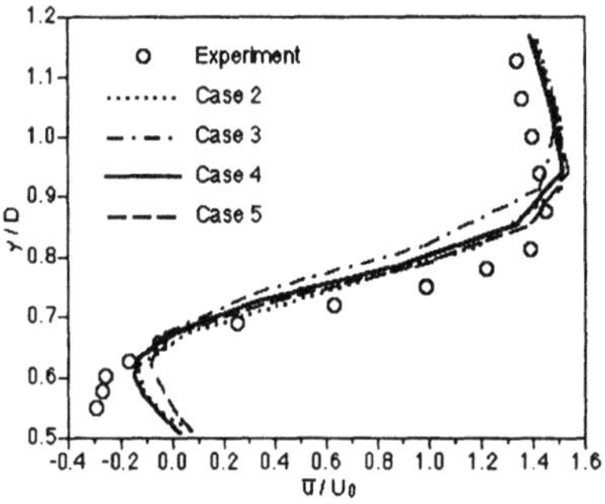


*Figure 3.* Time-averaged velocity $\overline{u}$ profile at the cylinder center $x = 0$.

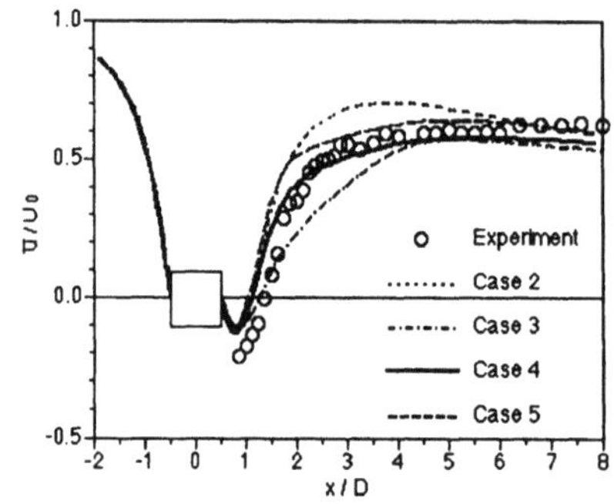


*Figure 4.* Time-averaged velocity $u$ along the centerline $y = 0$.

velocities calculated with the SGS model (cases 2, 4 and 5) approach the free-stream velocity faster than is measured by the experiment (Lyn *et al.* 1994). The centerline velocity of the case 2 (SGS+UTOPIA) first overshoots and then asymptotes to the experimental level. The result of case 3 (UTOPIA only) is lower than the experiment. Among the present calculations, the results with the controlled magnitude of the upwind scheme (cases 4 and 5) give the best agreement with the experiment.

The contours of the mean streamwise velocity $\overline{u}$ are shown in figure 7. The behavior of the calculated results is similar to that of the experimental data. Note that in the region behind the square cylinder, the width of the interval in each contour is different.

Contours of the mean streamwise normal stress $\overline{u'u'}$ are shown in figure 8. Two peaks are observed close to the top and bottom surfaces of the square cylinder, the position and the magnitude of the peak agreeing fairly well with the experimental data. All the calculations give somewhat higher peaks than the experiment.

Contours of the mean crossflow normal stress $\overline{v'v'}$ are shown in figure 9. The position of the maximum point is very close to that of the experiment; however, its peak value is significantly higher than the experimental data. In addition, the fluctuation decays more slowly than the experiment with increasing $x$. This means that the vertical oscillation of the wake behind the square cylinder is more prominent than that of the experiment.

Figure 10 shows the averaged Reynolds shear stress $\overline{u'v'}$. The peak values of the calculations with the SGS model are higher than those of the experiment. In this respect, the calculation with UTOPIA only (case 3) gives the best agreement with the experiment.

Figure 5 shows contours of the instantaneous $\nu_t/\nu$ for case 2. The value is low in the vicinity of the cylinder surface. In the region with the periodic

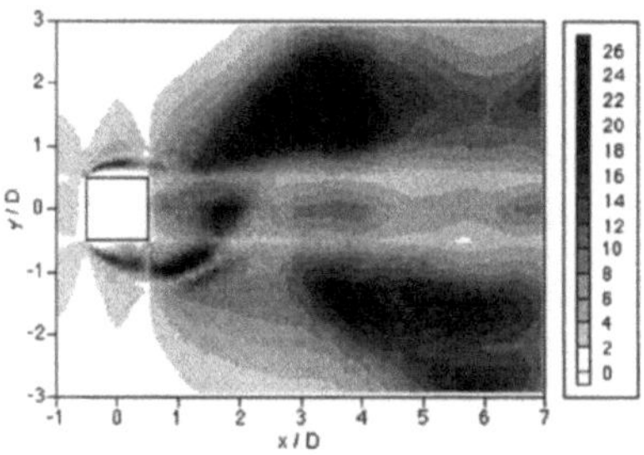

*Figure 5.* Contours of instantaneous $\nu_t/\nu$ for case 2

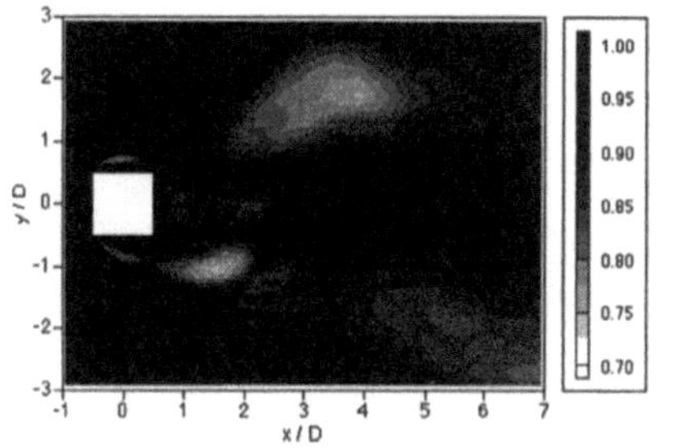

*Figure 6.* Contours of instantaneous $\alpha$ for case 5

vortex, the magnitude of $\nu_t/\nu$ is high; in other words, the effect of the SGS model is larger in this region than in the close vicinity of the cylinder.

An attempt has been made to reduce the effect of the upwinding. In case 4, the upwind scheme is introduced only in the central region of the domain, i.e. $-3.5 \leq y/D \leq 3.5$; if the region of nonzero $\alpha$ is restricted further, the numerical calculation becomes unstable. In case 5, the magnitude of parameter $\alpha$ is decreased as a function of $\nu_t/\nu$ as given in Table 1. Though the parameter $\alpha$ is not uniform in cases 4 and 5 (see Figure 6), the calculations proceed stably and their results are in better agreement with the experiment. A refinement of the method to control $\alpha$ is left for further investigation.

## 5. Conclusions

The turbulent flow past a square cylinder has been calculated by LES using the Smagorinsky model. Because of a rather coarse mesh, which is unavoidable in LES, numerical instability was observed in the front region of the cylinder. To avoid the instability, an upwind-biased scheme (UTOPIA) was introduced in the whole calculation region; however, its effect on the calculated flow field was rather significant. Thus an effort was made to reduce the region in which the upwind scheme is introduced. The calculation was numerically stable and an improvement was obtained compared with LES using the upwind scheme in the whole region.

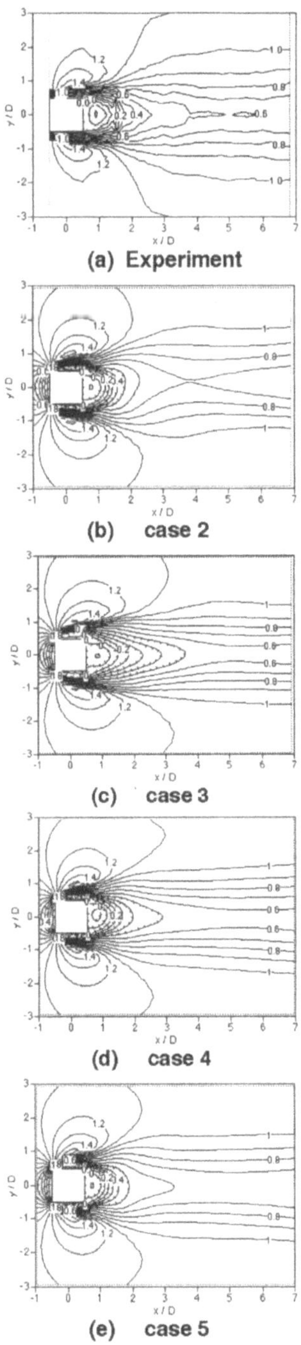


*Figure 7.* Time-averaged velocity $\overline{u}$.

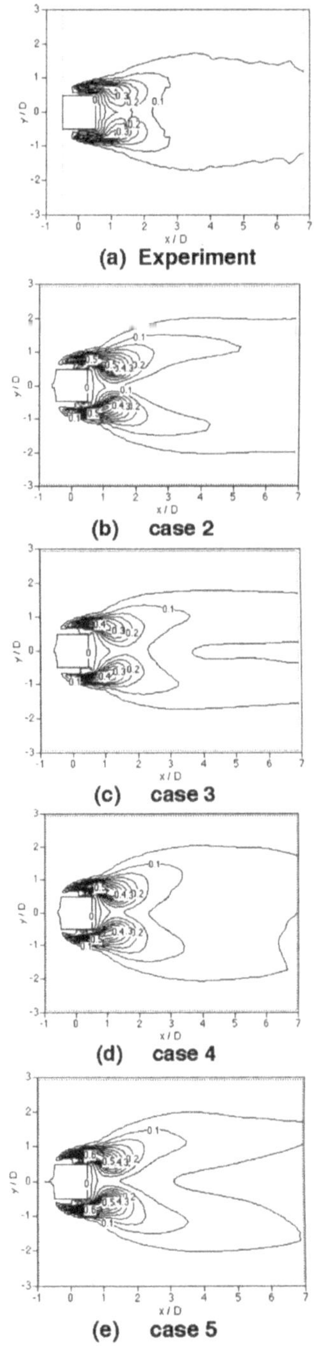


*Figure 8.* Time-averaged stress $\overline{u'u'}$.

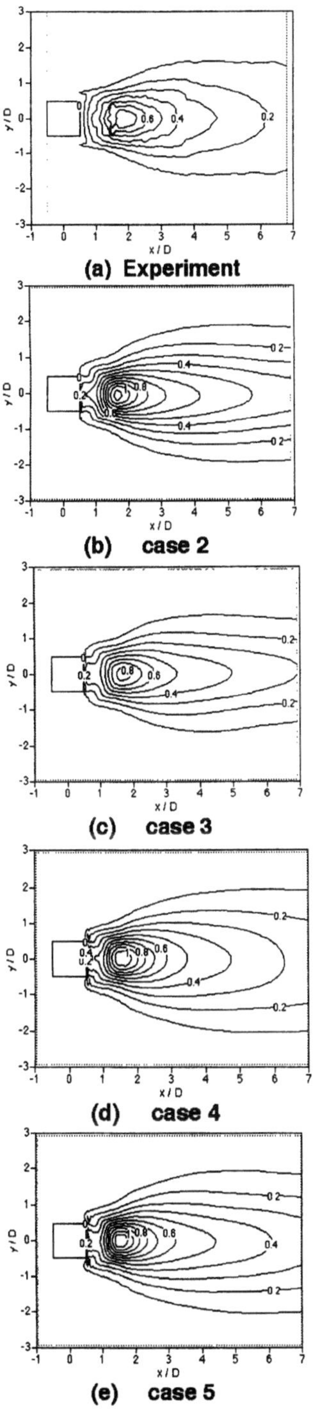


*Figure 9.* Time-averaged stress $\overline{v'v'}$

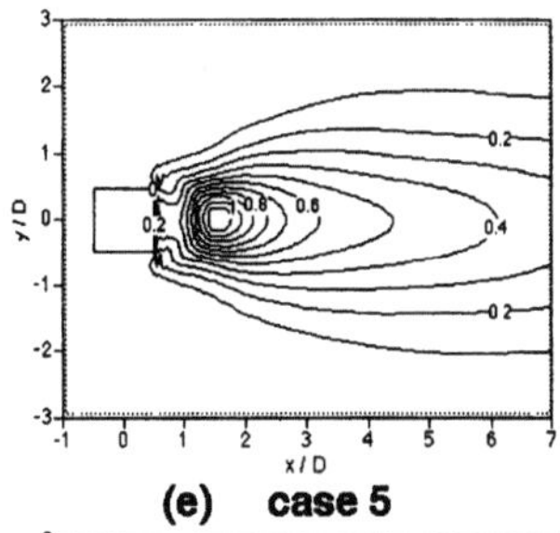


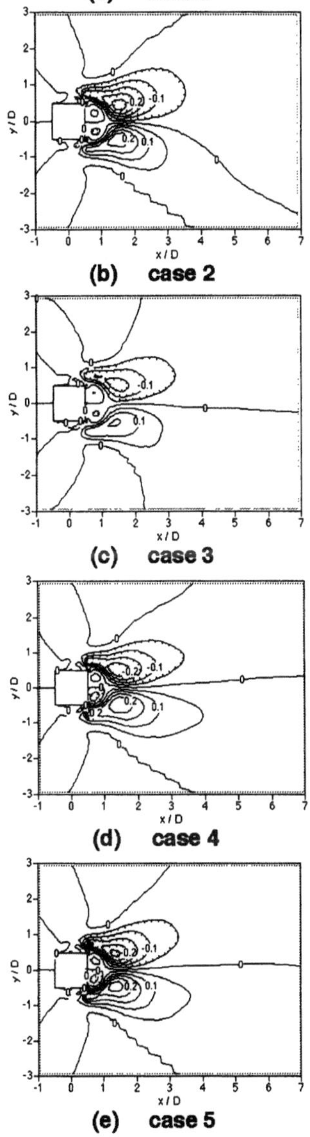


*Figure 10.* Time-averaged stress $\overline{u'v'}$

## References


Archambeau, F., Laurence, D. and Leschziner M. A., (1995) Test case A: Vortex Shedding behind a Square Prism at $Re = 21,400$, *Workshop on Large Eddy Simulation of Flows past Bluff Bodies*, Rottach-Egern, Germany.

Breuer, M., Pourquie, M. and Rodi, W., (1995) Description of the LES Method for the Workshop on Large Eddy Simulation of Flows past Bluff Bodies, *Workshop (ibid.)*.

Germano, M., Piomelli, U., Moin, P. and Cabot, W. H. (1991) A Dynamic Subgrid-scale Eddy Viscosity Model, *Phys. Fluids*, A**3**, pp 1760–1765.

Kawamura, H. and Kawashima, N., (1995) Vortex Shedding past a Square Cylinder, *Workshop (ibid.)*.

Kobayashi, T., Taniguchi, N. and Kogaki, T., (1995) Description of Numerical Methods, Boundary Conditions and SGS Models for Workshop on Large Eddy Simulation of Flow past Bluff Bodies *Workshop (ibid.)*.

Lyn, D. A., Einav, S., Rodi, W. and Park, J. H. (1994) A Laser Doppler Velocimetry Study of Ensemble-Averaged Characteristics of the Turbulent near Wake of a Square Cylinder, Rept. SFB 210/E/100.

Mochida, A., Murakami, S., Rodi, W. and Sakamoto, S., (1992) Comparison of LES with DSM, $k - \varepsilon$ and Experiments for Turbulent Vortex Shedding Flow past 2D Square Cylinder, *Proc. 6th Symp. on Computational Fluid Dynamics (in Japanese)*, pp 277–280.

Smagorinsky, J. (1963) General Circulation Experiments with the Primitive Equations. Part I. the Basic Experiment, *Monthly Weather Review*, **91**, pp 99–164.

Tamura, T., Takakuwa, A. and Nozawa, K., (1995) Test case A: Vortex Shedding past a Square Cylinder, *Workshop (ibid.)*.

Wang, G. and Vanka S. P., (1995) Description of Large Eddy Simulation of Flow past a Square Cylinder, *Workshop (ibid.)*.

## *ERCOFTAC* SERIES

1. A. Gyr and F.-S. Rys (eds.): *Diffusion and Transport of Pollutants in Atmospheric Mesoscale Flow Fields.* 1995 ISBN 0-7923-3260-1
2. M. Hallbäck, D.S. Henningson, A.V. Johansson and P.H. Alfredsson (eds.): *Turbulence and Transition Modelling.* Lecture Notes from the ERCOFTAC/IUTAM Summerschool held in Stockholm. 1996 ISBN 0-7923-4060-4
3. P. Wesseling (ed.): *High Performance Computing in Fluid Dynamics.* Proceedings of the Summerschool held in Delft, The Netherlands. 1996 ISBN 0-7923-4063-9
4. Th. Dracos (ed.): *Three-Dimensional Velocity and Vorticity Measuring and Image Analysis Techniques.* Lecture Notes from the Short Course held in Zürich, Switzerland. 1996 ISBN 0-7923-4256-9
5. J.-P. Chollet, P.R. Voke and L. Kleiser (eds.): *Direct and Large-Eddy Simulation II.* Proceedings of the ERCOFTAC Workshop held in Grenoble, France. 1997 ISBN 0-7923-4687-4

KLUWER ACADEMIC PUBLISHERS – DORDRECHT / BOSTON / LONDON

# Mechanics

## *FLUID* MECHANICS AND ITS APPLICATIONS

*Series Editor:* R. Moreau

*Aims and Scope of the Series*

The purpose of this series is to focus on subjects in which fluid mechanics plays a fundamental role. As well as the more traditional applications of aeronautics, hydraulics, heat and mass transfer etc., books will be published dealing with topics which are currently in a state of rapid development, such as turbulence, suspensions and multiphase fluids, super and hypersonic flows and numerical modelling techniques. It is a widely held view that it is the interdisciplinary subjects that will receive intense scientific attention, bringing them to the forefront of technological advancement. Fluids have the ability to transport matter and its properties as well as transmit force, therefore fluid mechanics is a subject that is particularly open to cross fertilisation with other sciences and disciplines of engineering. The subject of fluid mechanics will be highly relevant in domains such as chemical, metallurgical, biological and ecological engineering. This series is particularly open to such new multidisciplinary domains.

1. M. Lesieur: *Turbulence in Fluids*. 2nd rev. ed., 1990 ISBN 0-7923-0645-7
2. O. Métais and M. Lesieur (eds.): *Turbulence and Coherent Structures*. 1991 ISBN 0-7923-0646-5
3. R. Moreau: *Magnetohydrodynamics*. 1990 ISBN 0-7923-0937-5
4. E. Coustols (ed.): *Turbulence Control by Passive Means*. 1990 ISBN 0-7923-1020-9
5. A.A. Borissov (ed.): *Dynamic Structure of Detonation in Gaseous and Dispersed Media*. 1991 ISBN 0-7923-1340-2
6. K.-S. Choi (ed.): *Recent Developments in Turbulence Management*. 1991 ISBN 0-7923-1477-8
7. E.P. Evans and B. Coulbeck (eds.): *Pipeline Systems*. 1992 ISBN 0-7923-1668-1
8. B. Nau (ed.): *Fluid Sealing*. 1992 ISBN 0-7923-1669-X
9. T.K.S. Murthy (ed.): *Computational Methods in Hypersonic Aerodynamics*. 1992 ISBN 0-7923-1673-8
10. R. King (ed.): *Fluid Mechanics of Mixing*. Modelling, Operations and Experimental Techniques. 1992 ISBN 0-7923-1720-3
11. Z. Han and X. Yin: *Shock Dynamics*. 1993 ISBN 0-7923-1746-7
12. L. Svarovsky and M.T. Thew (eds.): *Hydroclones*. Analysis and Applications. 1992 ISBN 0-7923-1876-5
13. A. Lichtarowicz (ed.): *Jet Cutting Technology*. 1992 ISBN 0-7923-1979-6
14. F.T.M. Nieuwstadt (ed.): *Flow Visualization and Image Analysis*. 1993 ISBN 0-7923-1994-X
15. A.J. Saul (ed.): *Floods and Flood Management*. 1992 ISBN 0-7923-2078-6
16. D.E. Ashpis, T.B. Gatski and R. Hirsh (eds.): *Instabilities and Turbulence in Engineering Flows*. 1993 ISBN 0-7923-2161-8
17. R.S. Azad: *The Atmospheric Boundary Layer for Engineers*. 1993 ISBN 0-7923-2187-1
18. F.T.M. Nieuwstadt (ed.): *Advances in Turbulence IV*. 1993 ISBN 0-7923-2282-7
19. K.K. Prasad (ed.): *Further Developments in Turbulence Management*. 1993 ISBN 0-7923-2291-6
20. Y.A. Tatarchenko: *Shaped Crystal Growth*. 1993 ISBN 0-7923-2419-6

Kluwer Academic Publishers – Dordrecht / Boston / London

# Mechanics

## *FLUID* MECHANICS AND ITS APPLICATIONS

*Series Editor:* R. Moreau

21. J.P. Bonnet and M.N. Glauser (eds.): *Eddy Structure Identification in Free Turbulent Shear Flows.* 1993 ISBN 0-7923-2449-8
22. R.S. Srivastava: *Interaction of Shock Waves.* 1994 ISBN 0-7923-2920-1
23. J.R. Blake, J.M. Boulton-Stone and N.H. Thomas (eds.): *Bubble Dynamics and Interface Phenomena.* 1994 ISBN 0-7923-3008-0
24. R. Benzi (ed.): *Advances in Turbulence V.* 1995 ISBN 0-7923-3032-3
25. B.I. Rabinovich, V.G. Lebedev and A.I. Mytarev: *Vortex Processes and Solid Body Dynamics.* The Dynamic Problems of Spacecrafts and Magnetic Levitation Systems. 1994 ISBN 0-7923-3092-7
26. P.R. Voke, L. Kleiser and J.-P. Chollet (eds.): *Direct and Large-Eddy Simulation I.* Selected papers from the First ERCOFTAC Workshop on Direct and Large-Eddy Simulation. 1994 ISBN 0-7923-3106-0
27. J.A. Sparenberg: *Hydrodynamic Propulsion and its Optimization.* Analytic Theory. 1995 ISBN 0-7923-3201-6
28. J.F. Dijksman and G.D.C. Kuiken (eds.): *IUTAM Symposium on Numerical Simulation of Non-Isothermal Flow of Viscoelastic Liquids.* Proceedings of an IUTAM Symposium held in Kerkrade, The Netherlands. 1995 ISBN 0-7923-3262-8
29. B.M. Boubnov and G.S. Golitsyn: *Convection in Rotating Fluids.* 1995 ISBN 0-7923-3371-3
30. S.I. Green (ed.): *Fluid Vortices.* 1995 ISBN 0-7923-3376-4
31. S. Morioka and L. van Wijngaarden (eds.): *IUTAM Symposium on Waves in Liquid/Gas and Liquid/Vapour Two-Phase Systems.* 1995 ISBN 0-7923-3424-8
32. A. Gyr and H.-W. Bewersdorff: *Drag Reduction of Turbulent Flows by Additives.* 1995 ISBN 0-7923-3485-X
33. Y.P. Golovachov: *Numerical Simulation of Viscous Shock Layer Flows.* 1995 ISBN 0-7923-3626-7
34. J. Grue, B. Gjevik and J.E. Weber (eds.): *Waves and Nonlinear Processes in Hydrodynamics.* 1996 ISBN 0-7923-4031-0
35. P.W. Duck and P. Hall (eds.): *IUTAM Symposium on Nonlinear Instability and Transition in Three-Dimensional Boundary Layers.* 1996 ISBN 0-7923-4079-5
36. S. Gavrilakis, L. Machiels and P.A. Monkewitz (eds.): *Advances in Turbulence VI.* Proceedings of the 6th European Turbulence Conference. 1996 ISBN 0-7923-4132-5
37. K. Gersten (ed.): *IUTAM Symposium on Asymptotic Methods for Turbulent Shear Flows at High Reynolds Numbers.* Proceedings of the IUTAM Symposium held in Bochum, Germany. 1996 ISBN 0-7923-4138-4
38. J. Verhás: *Thermodynamics and Rheology.* 1997 ISBN 0-7923-4251-8
39. M. Champion and B. Deshaies (eds.): *IUTAM Symposium on Combustion in Supersonic Flows.* Proceedings of the IUTAM Symposium held in Poitiers, France. 1997 ISBN 0-7923-4313-1
40. M. Lesieur: *Turbulence in Fluids.* Third Revised and Enlarged Edition. 1997 ISBN 0-7923-4415-4; Pb: 0-7923-4416-2

# Mechanics

***FLUID* MECHANICS AND ITS APPLICATIONS**

*Series Editor:* R. Moreau

41. L. Fulachier, J.L. Lumley and F. Anselmet (eds.): *IUTAM Symposium on Variable Density Low-Speed Turbulent Flows*. Proceedings of the IUTAM Symposium held in Marseille, France. 1997 ISBN 0-7923-4602-5

Kluwer Academic Publishers – Dordrecht / Boston / London

MIX
Papier aus verantwortungsvollen Quellen
Paper from responsible sources
FSC® C105338

If you have any concerns about our products,
you can contact us on
**ProductSafety@springernature.com**

In case Publisher is established outside the EU,
the EU authorized representative is:
**Springer Nature Customer Service Center GmbH**
**Europaplatz 3, 69115 Heidelberg, Germany**

Printed by Libri Plureos GmbH
in Hamburg, Germany